Graduate Texts in Mathematics 207

Graduate Texts in Mathematics

1 TAKEUTI/ZARING. Introduction to Axiomatic Set Theory. 2nd ed.
2 OXTOBY. Measure and Category. 2nd ed.
3 SCHAEFER. Topological Vector Spaces. 2nd ed.
4 HILTON/STAMMBACH. A Course in Homological Algebra. 2nd ed.
5 MAC LANE. Categories for the Working Mathematician. 2nd ed.
6 HUGHES/PIPER. Projective Planes.
7 SERRE. A Course in Arithmetic.
8 TAKEUTI/ZARING. Axiomatic Set Theory.
9 HUMPHREYS. Introduction to Lie Algebras and Representation Theory.
10 COHEN. A Course in Simple Homotopy Theory.
11 CONWAY. Functions of One Complex Variable I. 2nd ed.
12 BEALS. Advanced Mathematical Analysis.
13 ANDERSON/FULLER. Rings and Categories of Modules. 2nd ed.
14 GOLUBITSKY/GUILLEMIN. Stable Mappings and Their Singularities.
15 BERBERIAN. Lectures in Functional Analysis and Operator Theory.
16 WINTER. The Structure of Fields.
17 ROSENBLATT. Random Processes. 2nd ed.
18 HALMOS. Measure Theory.
19 HALMOS. A Hilbert Space Problem Book. 2nd ed.
20 HUSEMOLLER. Fibre Bundles. 3rd ed.
21 HUMPHREYS. Linear Algebraic Groups.
22 BARNES/MACK. An Algebraic Introduction to Mathematical Logic.
23 GREUB. Linear Algebra. 4th ed.
24 HOLMES. Geometric Functional Analysis and Its Applications.
25 HEWITT/STROMBERG. Real and Abstract Analysis.
26 MANES. Algebraic Theories.
27 KELLEY. General Topology.
28 ZARISKI/SAMUEL. Commutative Algebra. Vol.I.
29 ZARISKI/SAMUEL. Commutative Algebra. Vol.II.
30 JACOBSON. Lectures in Abstract Algebra I. Basic Concepts.
31 JACOBSON. Lectures in Abstract Algebra II. Linear Algebra.
32 JACOBSON. Lectures in Abstract Algebra III. Theory of Fields and Galois Theory.
33 HIRSCH. Differential Topology.
34 SPITZER. Principles of Random Walk. 2nd ed.
35 ALEXANDER/WERMER. Several Complex Variables and Banach Algebras. 3rd ed.
36 KELLEY/NAMIOKA et al. Linear Topological Spaces.
37 MONK. Mathematical Logic.
38 GRAUERT/FRITZSCHE. Several Complex Variables.
39 ARVESON. An Invitation to C^*-Algebras.
40 KEMENY/SNELL/KNAPP. Denumerable Markov Chains. 2nd ed.
41 APOSTOL. Modular Functions and Dirichlet Series in Number Theory. 2nd ed.
42 SERRE. Linear Representations of Finite Groups.
43 GILLMAN/JERISON. Rings of Continuous Functions.
44 KENDIG. Elementary Algebraic Geometry.
45 LOÈVE. Probability Theory I. 4th ed.
46 LOÈVE. Probability Theory II. 4th ed.
47 MOISE. Geometric Topology in Dimensions 2 and 3.
48 SACHS/WU. General Relativity for Mathematicians.
49 GRUENBERG/WEIR. Linear Geometry. 2nd ed.
50 EDWARDS. Fermat's Last Theorem.
51 KLINGENBERG. A Course in Differential Geometry.
52 HARTSHORNE. Algebraic Geometry.
53 MANIN. A Course in Mathematical Logic.
54 GRAVER/WATKINS. Combinatorics with Emphasis on the Theory of Graphs.
55 BROWN/PEARCY. Introduction to Operator Theory I: Elements of Functional Analysis.
56 MASSEY. Algebraic Topology: An Introduction.
57 CROWELL/FOX. Introduction to Knot Theory.
58 KOBLITZ. p-adic Numbers, p-adic Analysis, and Zeta-Functions. 2nd ed.
59 LANG. Cyclotomic Fields.
60 ARNOLD. Mathematical Methods in Classical Mechanics. 2nd ed.
61 WHITEHEAD. Elements of Homotopy Theory.
62 KARGAPOLOV/MERLZJAKOV. Fundamentals of the Theory of Groups.
63 BOLLOBAS. Graph Theory.
64 EDWARDS. Fourier Series. Vol. I. 2nd ed.
65 WELLS. Differential Analysis on Complex Manifolds. 2nd ed.

(continued after index)

Chris Godsil
Gordon Royle

Algebraic Graph Theory

With 120 Illustrations

Springer

Chris Godsil
Department of Combinatorics
and Optimization
University of Waterloo
Waterloo, Ontario N2L 3G1
Canada
cgodsil@math.uwaterloo.ca

Gordon Royle
Department of Computer Science
University of Western Australia
Nedlands, Western Australia 6907
Australia
gordon@cs.uwa.edu.au

Mathematics Subject Classification (2000): 05Cxx, 05Exx

Library of Congress Cataloging-in-Publication Data
Godsil, C.D. (Christopher David), 1949-
Algebraic graph theory / Chris Godsil, Gordon Royle.
cm. – (Graduate texts in mathematics; 207)
Includes bibliographical references and index.
ISBN 0-387-95241-1 (hc. : alk. paper)
ISBN 0-387-95220-9 (pbk.: alk. paper)
Graph theory I. Royle, Gordon. II. Title. III. Series.
QA166 .G63 2001
511'.5—dc21 00-053776

Godsil and Royle: Algebraic Graph Theory

First Indian Reprint 2006
Third Indian Reprint 2013

ISBN 978-81-8128-262-0

This edition is manufactured in India for sale only in India, Pakistan, Bangladesh, Nepal and Sri Lanka and any other country as authorized by the publisher.

This edition is published by Springer (India) Private Limited,
A part of Springer Science+Business Media, Registered Office: 7th Floor,
Vijaya Building, 17 Barakhamba Road, New Delhi – 110 001, India.

Printed in India by Rashtriya Printers, Delhi.

To Gillian and Jane

Preface

Many authors begin their preface by confidently describing how their book arose. We started this project so long ago, and our memories are so weak, that we could not do this truthfully. Others begin by stating why they decided to write. Thanks to Freud, we know that unconscious reasons can be as important as conscious ones, and so this seems impossible, too. Moreover, the real question that should be addressed is why the reader should struggle with this text.

Even that question we cannot fully answer, so instead we offer an explanation for our own fascination with this subject. It offers the pleasure of seeing many unexpected and useful connections between two beautiful, and apparently unrelated, parts of mathematics: algebra and graph theory. At its lowest level, this is just the feeling of getting something for nothing. After devoting much thought to a graph-theoretical problem, one suddenly realizes that the question is already answered by some lonely algebraic fact. The canonical example is the use of eigenvalue techniques to prove that certain extremal graphs cannot exist, and to constrain the parameters of those that do. Equally unexpected, and equally welcome, is the realization that some complicated algebraic task reduces to a question in graph theory, for example, the classification of groups with BN pairs becomes the study of generalized polygons.

Although the subject goes back much further, Tutte's work was fundamental. His famous characterization of graphs with no perfect matchings was proved using Pfaffians; eventually, proofs were found that avoided any reference to algebra, but nonetheless, his original approach has proved fruitful in modern work developing parallelizable algorithms for determining the maximum size of a matching in a graph. He showed that the order of the vertex stabilizer of an arc-transitive cubic graph was at most 48. This is still the most surprising result on the autmomorphism groups of graphs, and it has stimulated a vast amount of work by group theorists interested in deriving analogous bounds for arc-transitive graphs with valency greater than three. Tutte took the chromatic polynomial and gave us back the Tutte polynomial, an important generalization that we now find is related to the surprising developments in knot theory connected to the Jones polynomial.

But Tutte's work is not the only significant source. Hoffman and Singleton's study of the maximal graphs with given valency and diameter led them to what they called Moore graphs. Although they were disappointed in that, despite the name, Moore graphs turned out to be very rare, this was nonetheless the occasion for introducing eigenvalue techniques into the study of graph theory.

Moore graphs and generalized polygons led to the theory of distance-regular graphs, first thoroughly explored by Biggs and his collaborators. Generalized polygons were introduced by Tits in the course of his fundamental work on finite simple groups. The parameters of finite generalized polygons were determined in a famous paper by Feit and Higman; this can still be viewed as one of the key results in algebraic graph theory. Seidel also played a major role. The details of this story are surprising: His work was actually motivated by the study of geometric problems in general metric spaces. This led him to the study of equidistant sets of points in projective space or, equivalently, the subject of equiangular lines. Extremal sets of equiangular lines led in turn to regular two-graphs and strongly regular graphs. Interest in strongly regular graphs was further stimulated when group theorists used them to construct new finite simple groups.

We make some explanation of the philosophy that has governed our choice of material. Our main aim has been to present and illustrate the main tools and ideas of algebraic graph theory, with an emphasis on current rather than classical topics. We place a strong emphasis on concrete examples, agreeing entirely with H. Lüneburg's admonition that "...the goal of theory is the mastering of examples." We have made a considerable effort to keep our treatment self-contained.

Our view of algebraic graph theory is inclusive; perhaps some readers will be surprised by the range of topics we have treated—fractional chromatic number, Voronoi polyhedra, a reasonably complete introduction to matroids, graph drawing—to mention the most unlikely. We also find occasion to discuss a large fraction of the topics discussed in standard graph theory texts (vertex and edge connectivity, Hamilton cycles, matchings, and colouring problems, to mention some examples).

We turn to the more concrete task of discussing the contents of this book. To begin, a brief summary: automorphisms and homomorphisms, the adjacency and Laplacian matrix, and the rank polynomial.

In the first part of the book we study the automorphisms and homomorphisms of graphs, particularly vertex-transitive graphs. We introduce the necessary results on graphs and permutation groups, and take care to describe a number of interesting classes of graphs; it seems silly, for example, to take the trouble to prove that a vertex-transitive graph with valency k has vertex connectivity at least $2(k+1)/3$ if the reader is not already in position to write down some classes of vertex-transitive graphs. In addition to results on the connectivity of vertex-transitive graphs, we also present material on matchings and Hamilton cycles.

There are a number of well-known graphs with comparatively large automorphism groups that arise in a wide range of different settings—in particular, the Petersen graph, the Coxeter graph, Tutte's 8-cage, and the Hoffman–Singleton graph. We treat these famous graphs in some detail. We also study graphs arising from projective planes and symplectic forms over 4-dimensional vector spaces. These are examples of generalized polygons, which can be characterized as bipartite graphs with diameter d and girth $2d$. Moore graphs can be defined to be graphs with diameter d and girth $2d+1$. It is natural to consider these two classes in the same place, and we do so.

We complete the first part of the book with a treatment of graph homomorphisms. We discuss Hedetniemi's conjecture in some detail, and provide an extensive treatment of cores (graphs whose endomorphisms are all automorphisms). We prove that the complement of a perfect graph is perfect, offering a short algebraic argument due to Gasparian. We pay particular attention to the Kneser graphs, which enables us to treat fractional chromatic number and the Erdős–Ko–Rado theorem. We determine the chromatic number of the Kneser graphs (using Borsuk's theorem).

The second part of our book is concerned with matrix theory. Chapter 8 provides a course in linear algebra for graph theorists. This includes an extensive, and perhaps nonstandard, treatment of the rank of a matrix. Following this we give a thorough treatment of interlacing, which provides one of the most powerful ways of using eigenvalues to obtain graph-theoretic information. We derive the standard bounds on the size of independent sets, but also give bounds on the maximum number of vertices in a bipartite induced subgraph. We apply interlacing to establish that certain carbon molecules, known as fullerenes, satisfy a stability criterion. We treat strongly regular graphs and two-graphs. The main novelty here is a careful discussion of the relation between the eigenvalues of the subconstituents of a strongly regular graph and those of the graph itself. We use this to study the strongly regular graphs arising as the point graphs of generalized quadrangles, and characterize the generalized quadrangles with lines of size three.

The least eigenvalue of the adjacency matrix of a line graph is at least -2. We present the beautiful work of Cameron, Goethals, Shult, and Seidel, characterizing the graphs with least eigenvalue at least -2. We follow the original proof, which reduces the problem to determining the generalized quadrangles with lines of size three and also reveals a surprising and close connection with the theory of root systems.

Finally we study the Laplacian matrix of a graph. We consider the relation between the second-largest eigenvalue of the Laplacian and various interesting graph parameters, such as edge-connectivity. We offer several viewpoints on the relation between the eigenvectors of a graph and various natural graph embeddings. We give a reasonably complete treatment of the cut and flow spaces of a graph, using chip-firing games to provide a novel approach to some aspects of this subject.

The last three chapters are devoted to the connection between graph theory and knot theory. The most startling aspect of this is the connection between the rank polynomial and the Jones polynomial.

For a graph theorist, the Jones polynomial is a specialization of a straightforward generalization of the rank polynomial of a graph. The rank polynomial is best understood in the context of matroid theory, and consequently our treatment of it covers a significant part of matroid theory. We make a determined attempt to establish the importance of this polynomial, offering a fairly complete list of its remarkable applications in graph theory (and coding theory). We present a version of Tutte's theory of rotors, which allows us to construct nonisomorphic 3-connected graphs with the same rank polynomial.

After this work on the rank polynomial, it is not difficult to derive the Jones polynomial and show that it is a useful knot invariant. In the last chapter we treat more of the graph theory related to knot diagrams. We characterize Gauss codes and show that certain knot theory operations are just topological manifestations of standard results from graph theory, in particular, the theory of circle graphs.

As already noted, our treatment is generally self-contained. We assume familiarity with permutations, subgroups, and homomorphisms of groups. We use the basics of the theory of symmetric matrices, but in this case we do offer a concise treatment of the machinery. We feel that much of the text is accessible to strong undergraduates. Our own experience is that we can cover about three pages of material per lecture. Thus there is enough here for a number of courses, and we feel this book could even be used for a first course in graph theory.

The exercises range widely in difficulty. Occasionally, the notes to a chapter provide a reference to a paper for a solution to an exercise; it is then usually fair to assume that the exercise is at the difficult end of the spectrum. The references at the end of each chapter are intended to provide contact with the relevant literature, but they are not intended to be complete.

It is more than likely that any readers familiar with algebraic graph theory will find their favourite topics slighted; our consolation is the hope that no two such readers will be able to agree on where we have sinned the most.

Both authors are human, and therefore strongly driven by the desire to edit, emend, and reorganize anyone else's work. One effect of this is that there are very few places in the text where either of us could, with any real confidence or plausibility, blame the other for the unfortunate and inevitable mistakes that remain. In this matter, as in others, our wives, our friends, and our students have made strenuous attempts to point out, and to eradicate, our deficiencies. Nonetheless, some will still show through, and so we must now throw ourselves on our readers' mercy. We do intend, as an exercise in public self-flagellation, to maintain a webpage listing corrections at `http://quoll.uwaterloo.ca/agt/`.

A number of people have read parts of various versions of this book and offered useful comments and advice as a result. In particular, it is a pleasure to acknowledge the help of the following: Rob Beezer, Anthony Bonato, Dom de Caen, Reinhard Diestel, Michael Doob, Jim Geelen, Tommy Jensen, Bruce Richter.

We finish with a special offer of thanks to Norman Biggs, whose own *Algebraic Graph Theory* is largely responsible for our interest in this subject.

Chris Godsil
Gordon Royle

Waterloo
Perth

Contents

1
Graphs

In this chapter we undertake the necessary task of introducing some of the basic notation for graphs. We discuss mappings between graphs—isomorphisms, automorphisms, and homomorphisms—and introduce a number of families of graphs. Some of these families will play a significant role in later chapters; others will be used to illustrate definitions and results.

1.1 Graphs

A *graph* X consists of a *vertex* set $V(X)$ and an *edge* set $E(X)$, where an edge is an unordered pair of distinct vertices of X. We will usually use xy rather than $\{x, y\}$ to denote an edge. If xy is an edge, then we say that x and y are *adjacent* or that y is a *neighbour* of x, and denote this by writing $x \sim y$. A vertex is *incident* with an edge if it is one of the two vertices of the edge. Graphs are frequently used to model a binary relationship between the objects in some domain, for example, the vertex set may represent computers in a network, with adjacent vertices representing pairs of computers that are physically linked.

Two graphs X and Y are equal if and only if they have the same vertex set and the same edge set. Although this is a perfectly reasonable definition, for most purposes the model of a relationship is not essentially changed if Y is obtained from X just by renaming the vertex set. This motivates the following definition: Two graphs X and Y are *isomorphic* if there is a bijection, φ say, from $V(X)$ to $V(Y)$ such that $x \sim y$ in X if and only if $\varphi(x) \sim \varphi(y)$ in Y. We say that φ is an *isomorphism* from X to Y. Since φ is a bijection, it has an inverse, which is an isomorphism from Y to X. If X and Y are isomorphic, then we write $X \cong Y$. It is normally appropriate to treat isomorphic graphs as if they were equal.

It is often convenient, interesting, or attractive to represent a graph by a picture, with points for the vertices and lines for the edges, as in Figure 1.1. Strictly speaking, these pictures do not define graphs, since the vertex set is not specified. However, we may assign distinct integers arbitrarily to the points, and the edges can then be written down as ordered pairs. Thus the diagram determines the graph up to isomorphism, which is usually all that matters. We emphasize that in a picture of a graph, the positions of the points and lines do not matter—the only information it conveys is which pairs of vertices are joined by an edge. You should convince yourself that the two graphs in Figure 1.1 are isomorphic.

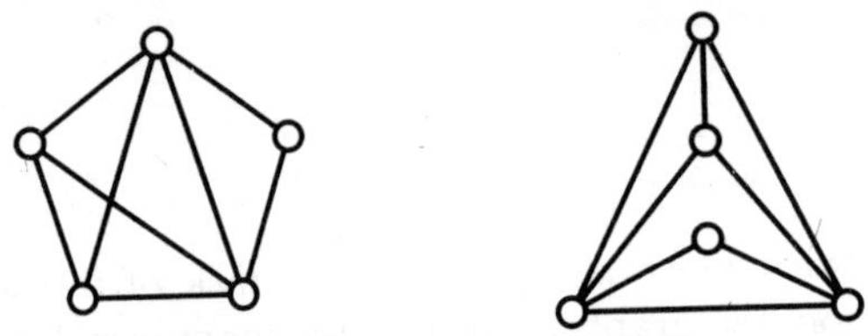

Figure 1.1. Two graphs on five vertices

A graph is called *complete* if every pair of vertices are adjacent, and the complete graph on n vertices is denoted by K_n. A graph with no edges (but at least one vertex) is called *empty*. The graph with no vertices and no edges is the *null graph*, regarded by some authors as a pointless concept. Graphs as we have defined them above are sometimes referred to as *simple graphs*, because there are some useful generalizations of this definition. For example, there are many occasions when we wish to use a graph to model an asymmetric relation. In this situation we define a *directed graph* X to consist of a vertex set $V(X)$ and an arc set $A(X)$, where an *arc*, or *directed edge*, is an ordered pair of distinct vertices. In a drawing of a directed graph, the direction of an arc is indicated with an arrow, as in Figure 1.2. Most graph-theoretical concepts have intuitive analogues for directed graphs. Indeed, for many applications a simple graph can equally well be viewed as a directed graph where (y, x) is an arc whenever (x, y) is an arc.

Throughout this book we will explicitly mention when we are considering directed graphs, and otherwise "graph" will refer to a simple graph. Although the definition of graph allows the vertex set to be infinite, we do not consider this case, and so all our graphs may be assumed to be *finite*—an assumption that is used implicitly in a few of our results.

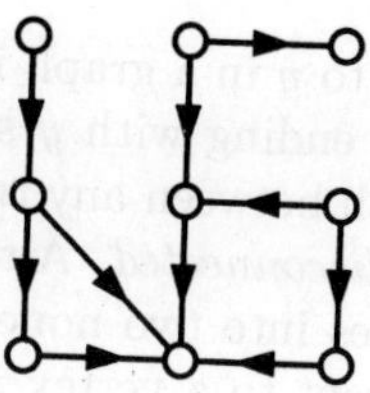

Figure 1.2. A directed graph

1.2 Subgraphs

A *subgraph* of a graph X is a graph Y such that

$$V(Y) \subseteq V(X), \qquad E(Y) \subseteq E(X).$$

If $V(Y) = V(X)$, we call Y a *spanning subgraph* of X. Any spanning subgraph of X can be obtained by deleting some of the edges from X. The first drawing in Figure 1.3 shows a spanning subgraph of a graph. The number of spanning subgraphs of X is equal to the number of subsets of $E(X)$.

A subgraph Y of X is an *induced subgraph* if two vertices of $V(Y)$ are adjacent in Y if and only if they are adjacent in X. Any induced subgraph of X can be obtained by deleting some of the vertices from X, along with any edges that contain a deleted vertex. Thus an induced subgraph is determined by its vertex set: We refer to it as the subgraph of X induced by its vertex set. The second drawing in Figure 1.3 shows an induced subgraph of a graph. The number of induced subgraphs of X is equal to the number of subsets of $V(X)$.

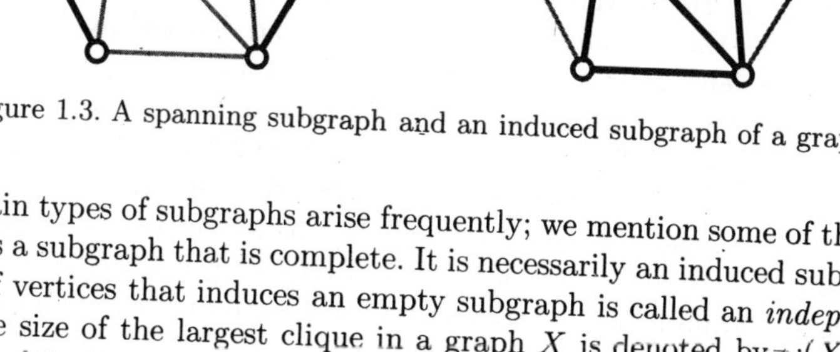

Figure 1.3. A spanning subgraph and an induced subgraph of a graph

Certain types of subgraphs arise frequently; we mention some of these. A *clique* is a subgraph that is complete. It is necessarily an induced subgraph. A set of vertices that induces an empty subgraph is called an *independent set*. The size of the largest clique in a graph X is denoted by $\omega(X)$, and the size of the largest independent set by $\alpha(X)$. As we shall see later, $\alpha(X)$ and $\omega(X)$ are important parameters of a graph.

A *path* of length r from x to y in a graph is a sequence of $r+1$ distinct vertices starting with x and ending with y such that consecutive vertices are adjacent. If there is a path between any two vertices of a graph X, then X is *connected*, otherwise *disconnected*. Alternatively, X is disconnected if we can partition its vertices into two nonempty sets, R and S say, such that no vertex in R is adjacent to a vertex in S. In this case we say that X is the *disjoint union* of the two subgraphs induced by R and S. An induced subgraph of X that is maximal, subject to being connected, is called a *connected component* of X. (This is almost always abbreviated to "component.")

A *cycle* is a connected graph where every vertex has exactly two neighbours; the smallest cycle is the complete graph K_3. The phrase "a cycle in a graph" refers to a subgraph of X that is a cycle. A graph where each vertex has at least two neighbours must contain a cycle, and proving this fact is a traditional early exercise in graph theory. An *acyclic graph* is a graph with no cycles, but these are usually referred to by more picturesque terms: A connected acyclic graph is called a *tree*, and an acyclic graph is called a *forest*, since each component is a tree. A spanning subgraph with no cycles is called a *spanning tree*. We see (or you are invited to prove) that a graph has a spanning tree if and only if it is connected. A *maximal spanning forest* in X is a spanning subgraph consisting of a spanning tree from each component.

1.3 Automorphisms

An isomorphism from a graph X to itself is called an *automorphism* of X. An automorphism is therefore a permutation of the vertices of X that maps edges to edges and nonedges to nonedges. Consider the set of all automorphisms of a graph X. Clearly the identity permutation is an automorphism, which we denote by e. If g is an automorphism of X, then so is its inverse g^{-1}, and if h is a second automorphism of X, then the product gh is an automorphism. Hence the set of all automorphisms of X forms a group, which is called the *automorphism group* of X and denoted by $\mathrm{Aut}(X)$. The *symmetric group* $\mathrm{Sym}(V)$ is the group of all permutations of a set V, and so the automorphism group of X is a subgroup of $\mathrm{Sym}(V(X))$. If X has n vertices, then we will freely use $\mathrm{Sym}(n)$ for $\mathrm{Sym}(V(X))$.

In general, it is a nontrivial task to decide whether two graphs are isomorphic, or whether a given graph has a nonidentity automorphism. Nonetheless there are some cases where everything is obvious. For example, every permutation of the vertices of the complete graph K_n is an automorphism, and so $\mathrm{Aut}(K_n) \cong \mathrm{Sym}(n)$.

The image of an element $v \in V$ under a permutation $g \in \mathrm{Sym}(V)$ will be denoted by v^g. If $g \in \mathrm{Aut}(X)$ and Y is a subgraph of X, then we define

Y^g to be the graph with

$$V(Y^g) = \{x^g : x \in V(Y)\}$$

and

$$E(Y^g) = \{\{x^g, y^g\} : \{x, y\} \in E(Y)\}.$$

It is straightforward to see that Y^g is isomorphic to Y and is also a subgraph of X.

The *valency* of a vertex x is the number of neighbours of x, and the maximum and minimum valency of a graph X are the maximum and minimum values of the valencies of any vertex of X.

Lemma 1.3.1 *If x is a vertex of the graph X and g is an automorphism of X, then the vertex $y = x^g$ has the same valency as x.*

Proof. Let $N(x)$ denote the subgraph of X induced by the neighbours of x in X. Then

$$N(x)^g = N(x^g) = N(y),$$

and therefore $N(x)$ and $N(y)$ are isomorphic subgraphs of X. Consequently they have the same number of vertices, and so x and y have the same valency.

□

This shows that the automorphism group of a graph permutes the vertices of equal valency among themselves. A graph in which every vertex has equal valency k is called *regular* of valency k or k-regular. A 3-regular graph is called *cubic*, and a 4-regular graph is sometimes called *quartic*. In Chapter 3 and Chapter 4 we will be studying graphs with the very special property that for any two vertices x and y, there is an automorphism g such that $x^g = y$; such graphs are necessarily regular.

The *distance* $d_X(x, y)$ between two vertices x and y in a graph X is the length of the shortest path from x to y. If the graph X is clear from the context, then we will simply use $d(x, y)$.

Lemma 1.3.2 *If x and y are vertices of X and $g \in \mathrm{Aut}(X)$, then $d(x, y) = d(x^g, y^g)$.*

□

The *complement* $\overline{X}$ of a graph X has the same vertex set as X, where vertices x and y are adjacent in $\overline{X}$ if and only if they are not adjacent in X (see Figure 1.5).

Lemma 1.3.3 *The automorphism group of a graph is equal to the automorphism group of its complement.*

□

If X is a directed graph, then an automorphism is a permutation of the vertices that maps arcs onto arcs, that is, it preserves the directions of the edges.

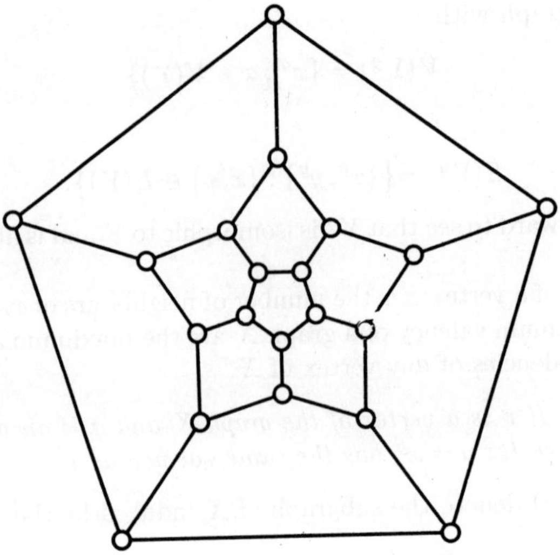

Figure 1.4. The dodecahedron is a cubic graph

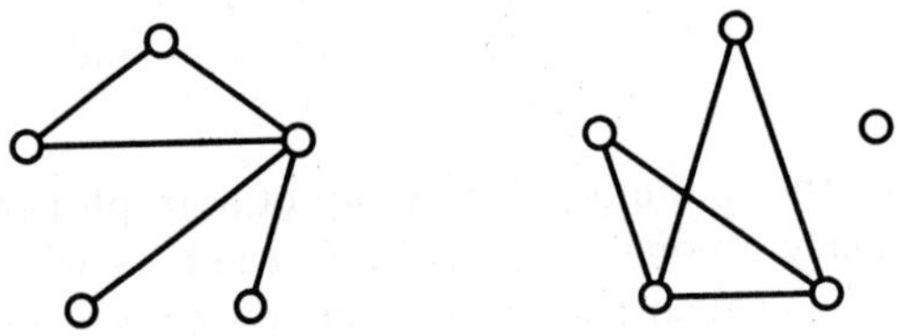

Figure 1.5. A graph and its complement

1.4 Homomorphisms

Let X and Y be graphs. A mapping f from $V(X)$ to $V(Y)$ is a *homomorphism* if $f(x)$ and $f(y)$ are adjacent in Y whenever x and y are adjacent in X. (When X and Y have no loops, which is our usual case, this definition implies that if $x \sim y$, then $f(x) \neq f(y)$.)

Any isomorphism between graphs is a homomorphism, and in particular any automorphism is a homomorphism from a graph to itself. However there are many homomorphisms that are not isomorphisms, as the following example illustrates. A graph X is called *bipartite* if its vertex set can be partitioned into two parts V_1 and V_2 such that every edge has one end in V_1 and one in V_2. If X is bipartite, then the mapping from $V(X)$ to $V(K_2)$ that sends all the vertices in V_i to the vertex i is a homomorphism from X to K_2.

This example belongs to the best known class of homomorphisms: proper colourings of graphs. A *proper colouring* of a graph X is a map from $V(X)$ into some finite set of colours such that no two adjacent vertices are assigned the same colour. If X can be properly coloured with a set of k colours, then we say that X can be properly k-coloured. The least value of k for which X can be properly k-coloured is the *chromatic number* of X, and is denoted by $\chi(X)$. The set of vertices with a particular colour is called a *colour class* of the colouring, and is an independent set. If X is a bipartite graph with at least one edge, then $\chi(X) = 2$.

Lemma 1.4.1 *The chromatic number of a graph X is the least integer r such that there is a homomorphism from X to K_r.*

Proof. Suppose f is a homomorphism from the graph X to the graph Y. If $y \in V(Y)$, define $f^{-1}(y)$ by

$$f^{-1}(y) := \{x \in V(X) : f(x) = y\}.$$

Because y is not adjacent to itself, the set $f^{-1}(y)$ is an independent set. Hence if there is a homomorphism from X to a graph with r vertices, the r sets $f^{-1}(y)$ form the colour classes of a proper r-colouring of X, and so $\chi(X) \leq r$. Conversely, suppose that X can be properly coloured with the r colours $\{1, \ldots, r\}$. Then the mapping that sends each vertex to its colour is a homomorphism from X to the complete graph K_r. □

A *retraction* is a homomorphism f from a graph X to a subgraph Y of itself such that the restriction $f \restriction Y$ of f to $V(Y)$ is the identity map. If there is a retraction from X to a subgraph Y, then we say that Y is a *retract* of X. If the graph X has a clique of size $k = \chi(X)$, then any k-colouring of X determines a retraction onto the clique.

Figure 1.6 shows the 5-*prism* as it is normally drawn, and then drawn to display a retraction (each vertex of the outer cycle is fixed, and each vertex of the inner cycle is mapped radially outward to the nearest vertex on the outer cycle).

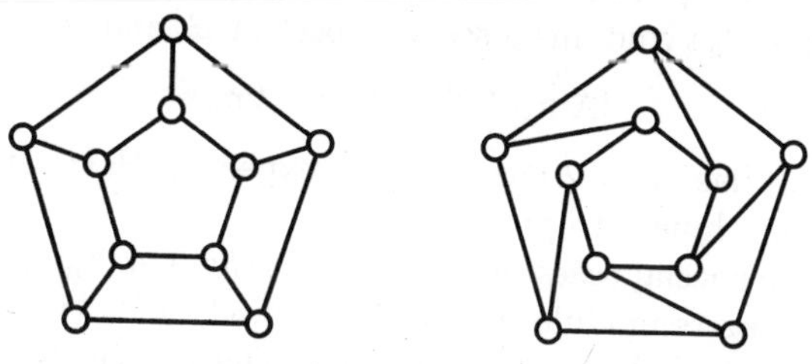

Figure 1.6. A graph with a retraction onto a 5-cycle

In Chapter 3 we will need to consider homomorphisms between directed graphs. If X and Y are directed graphs, then a map f from $V(X)$ to $V(Y)$ is a homomorphism if $(f(x), f(y))$ is an arc of Y whenever (x, y) is an arc of X. In other words, a homomorphism must preserve the sense of the directed edges.

In Chapter 6 we will relax the definition of a graph still further, so that the two ends of an edge can be the same vertex, rather than two distinct vertices. Such edges are called *loops*, and if loops are permitted, then the properties of homomorphisms are quite different. For example, a property of homomorphisms of simple graphs used in Lemma 1.4.1 is that the preimage of a vertex is an independent set. If loops are present, this is no longer true: A homomorphism can map any set of vertices onto a vertex with a loop.

A homomorphism from a graph X to itself is called an *endomorphism*, and the set of all endomorphisms of X is the *endomorphism monoid* of X. (A *monoid* is a set that has an associative binary multiplication defined on it and an identity element.) The endomorphism monoid of X contains its automorphism group, since an automorphism is an endomorphism.

1.5 Circulant Graphs

We now introduce an important class of graphs that will provide useful examples in later sections.

First we give a more elaborate definition of a cycle. The *cycle* on n vertices is the graph C_n with vertex set $\{0, \ldots, n-1\}$ and with i adjacent to j if and only if $j - i \equiv \pm 1 \bmod n$.

We determine some automorphisms of the cycle. If g is the element of $\mathrm{Sym}(n)$ that maps i to $(i+1) \bmod n$, then $g \in \mathrm{Aut}(C_n)$. Therefore $\mathrm{Aut}(C_n)$ contains the cyclic subgroup

$$R = \{g^m : 0 \le m \le n-1\}.$$

It is also easy to verify that the permutation h that maps i to $-i \bmod n$ is an automorphism of C_n. Notice that $h(0) = 0$, so h fixes a vertex of C_n. On the other hand, the nonidentity elements of R are fixed-point-free automorphisms of C_n. Therefore, h is not a power of g, and so $h \notin R$. It follows that $\mathrm{Aut}(C_n)$ contains a second coset of R, and therefore

$$|\mathrm{Aut}(C_n)| \ge 2|R| = 2n.$$

In fact, $\mathrm{Aut}(C_n)$ has order $2n$ as might be expected. However, we have not yet set up the machinery to prove this.

The cycles are special cases of *circulant graphs*. Let $\mathbb{Z}_n$ denote the additive group of integers modulo n. If C is a subset of $\mathbb{Z}_n \setminus 0$, then construct a directed graph $X = X(\mathbb{Z}_n, C)$ as follows. The vertices of X are the elements of $\mathbb{Z}_n$ and (i, j) is an arc of X if and only if $j - i \in C$. The graph $X(\mathbb{Z}_n, C)$ is called a circulant of order n, and C is called its *connection set*.

Suppose that C has the additional property that it is closed under additive inverses, that is, $-c \in C$ if and only if $c \in C$. Then (i, j) is an arc if and only if (j, i) is an arc, and so we can view X as an undirected graph.

It is easy to see that the permutation that maps each vertex i to $i + 1$ is an automorphism of X. If C is inverse-closed, then the mapping that

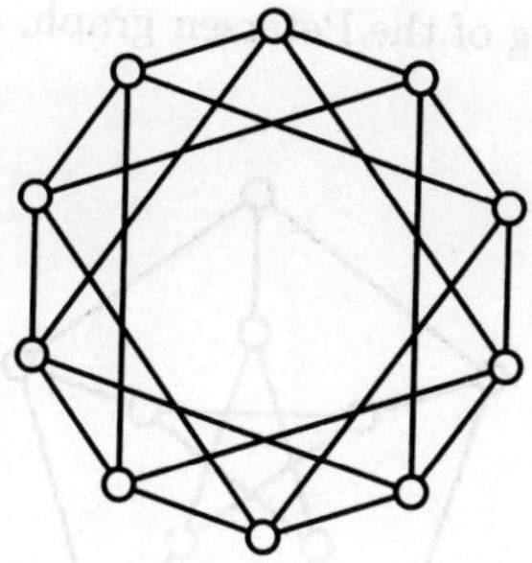

Figure 1.7. The circulant $X(\mathbb{Z}_{10}, \{-1, 1, -3, 3\})$

sends i to $-i$ is also an automorphism. Therefore, if X is undirected, its automorphism group has order at least $2n$.

The cycle C_n is a circulant of order n, with connection set $\{1, -1\}$. The complete and empty graphs are also circulants, with $C = \mathbb{Z}_n$ and $C = \emptyset$ respectively, and so the automorphism group of a circulant of order n can have order much greater than $2n$.

1.6 Johnson Graphs

Next we consider another family of graphs $J(v, k, i)$ that will recur throughout this book. These graphs are important because they enable us to translate many combinatorial problems about sets into graph theory.

Let v, k, and i be fixed positive integers, with $v \geq k \geq i$; let Ω be a fixed set of size v; and define $J(v, k, i)$ as follows. The vertices of $J(v, k, i)$ are the subsets of Ω with size k, where two subsets are adjacent if their intersection has size i. Therefore, $J(v, k, i)$ has $\binom{v}{k}$ vertices, and it is a regular graph with valency

$$\binom{k}{i}\binom{v-k}{k-i}.$$

As the next result shows, we can assume that $v \geq 2k$.

Lemma 1.6.1 *If $v \geq k \geq i$, then $J(v, k, i) \cong J(v, v-k, v-2k+i)$.*

Proof. The function that maps a k-set to its complement in Ω is an isomorphism from $J(v, k, i)$ to $J(v, v-k, v-2k+i)$; you are invited to check the details. □

For $v \geq 2k$, the graphs $J(v, k, k-1)$ are known as the *Johnson graphs*, and the graphs $J(v, k, 0)$ are known as the *Kneser graphs*, which we will study in some depth in Chapter 7. The Kneser graph $J(5, 2, 0)$ is one of the most famous and important graphs and is known as the *Petersen graph*.

Figure 1.8 gives a drawing of the Petersen graph, and Section 4.4 examines it in detail.

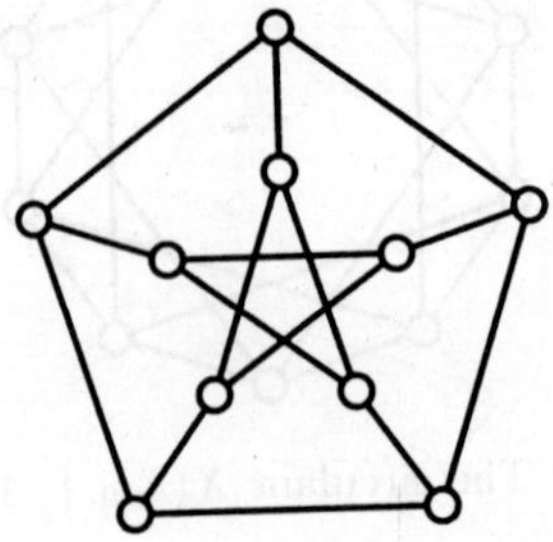

Figure 1.8. The Petersen graph $J(5,2,0)$

If g is a permutation of Ω and $S \subseteq \Omega$, then we define S^g to be the subset

$$S^g := \{s^g : s \in S\}$$

It follows that each permutation of Ω determines a permutation of the subsets of Ω, and in particular a permutation of the subsets of size k. If S and T are subsets of Ω, then

$$|S \cap T| = |S^g \cap T^g|,$$

and so g is an automorphism of $J(v,k,i)$. Thus we obtain the following.

Lemma 1.6.2 *If* $v \geq k \geq i$, *then* $\mathrm{Aut}(J(v,k,i))$ *contains a subgroup isomorphic to* $\mathrm{Sym}(v)$. □

Note that $\mathrm{Aut}(J(v,k,i))$ is a permutation group acting on a set of size $\binom{v}{k}$, and so when $k \neq 1$ or $v-1$, it is not actually equal to $\mathrm{Sym}(v)$. Nevertheless, it is true that $\mathrm{Aut}(J(v,k,i))$ is usually isomorphic to $\mathrm{Sym}(v)$, although this is not always easy to prove.

1.7 Line Graphs

The *line graph* of a graph X is the graph $L(X)$ with the edges of X as its vertices, and where two edges of X are adjacent in $L(X)$ if and only if they are incident in X. An example is given in Figure 1.9 with the graph in grey and the line graph below it in black.

The *star* $K_{1,n}$, which consists of a single vertex with n neighbours, has the complete graph K_n as its line graph. The path P_n is the graph with vertex set $\{1,\ldots,n\}$ where i is adjacent to $i+1$ for $1 \leq i \leq n-1$. It has line graph equal to the shorter path P_{n-1}. The cycle C_n is isomorphic to its own line graph.

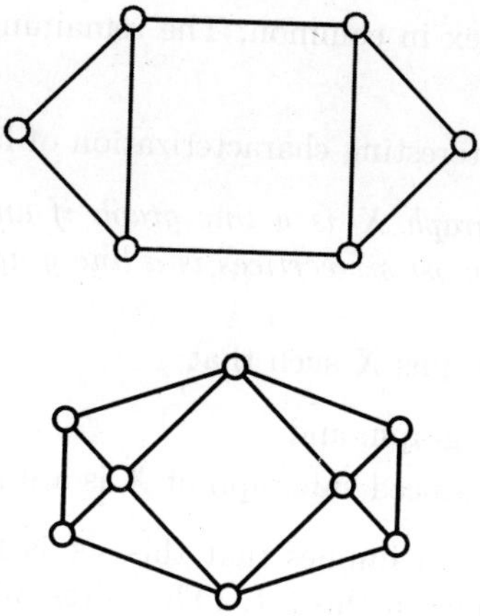

Figure 1.9. A graph and its line graph

Lemma 1.7.1 *If X is regular with valency k, then $L(X)$ is regular with valency $2k-2$.* □

Each vertex in X determines a clique in $L(X)$: If x is a vertex in X with valency k, then the k edges containing x form a k-clique in $L(X)$. Thus if X has n vertices, there is a set of n cliques in $L(X)$ with each vertex of $L(X)$ contained in at most two of these cliques. Each edge of $L(X)$ lies in exactly one of these cliques. The following result provides a useful converse:

Theorem 1.7.2 *A nonempty graph is a line graph if and only if its edge set can be partitioned into a set of cliques with the property that any vertex lies in at most two cliques.* □

If X has no *triangles* (that is, cliques of size three), then any vertex of $L(X)$ with at least two neighbours in one of these cliques must be contained in that clique. Hence the cliques determined by the vertices of X are all maximal.

It is both obvious and easy to prove that if $X \cong Y$, then $L(X) \cong L(Y)$. However, the converse is false: K_3 and $K_{1,3}$ have the same line graph, namely K_3. Whitney proved that this is the only pair of connected counterexamples. We content ourselves with proving the following weaker result.

Lemma 1.7.3 *Suppose that X and Y are graphs with minimum valency four. Then $X \cong Y$ if and only if $L(X) \cong L(Y)$.*

Proof. Let C be a clique in $L(X)$ containing exactly c vertices. If $c > 3$, then the vertices of C correspond to a set of c edges in X, meeting at a common vertex. Consequently, there is a bijection between the vertices of X and the maximal cliques of $L(X)$ that takes adjacent vertices to pairs of cliques with a vertex in common. The remaining details are left as an exercise. □

There is another interesting characterization of line graphs:

Theorem 1.7.4 *A graph X is a line graph if and only if each induced subgraph of X on at most six vertices is a line graph.* □

Consider the set of graphs X such that

(a) X is not a line graph, and

(b) every proper induced subgraph of X is a line graph.

The previous theorem implies that this set is finite, and in fact there are exactly nine graphs in this set. (The notes at the end of the chapter indicate where you can find the graphs themselves.)

We call a bipartite graph *semiregular* if it has a proper 2-colouring such that all vertices with the same colour have the same valency. The cheapest examples are the *complete bipartite graphs* $K_{m,n}$ which consist of an independent set of m vertices completely joined to an independent set of n vertices.

Lemma 1.7.5 *If the line graph of a connected graph X is regular, then X is regular or bipartite and semiregular.*

Proof. Suppose that $L(X)$ is regular with valency k. If u and v are adjacent vertices in X, then their valencies sum to $k+2$. Consequently, all neighbours of a vertex u have the same valency, and so if two vertices of X share a common neighbour, then they have the same valency. Since X is connected, this implies that there are at most two different valencies.

If two adjacent vertices have the same valency, then an easy induction argument shows that X is regular. If X contains a cycle of odd length, then it must have two adjacent vertices of the same valency, and so if it is not regular, then it has no cycles of odd length. We leave it as an exercise to show that a graph is bipartite if and only if it contains no cycles of odd length. □

1.8 Planar Graphs

We have already seen that graphs can conveniently be given by drawings where each vertex is represented by a point and each edge uv by a line connecting u and v. A graph is called *planar* if it can be drawn without crossing edges.

Although this definition is intuitively clear, it is topologically imprecise. To make it precise, consider a function that maps each vertex of a graph X to a distinct point of the plane, and each edge of X to a continuous non

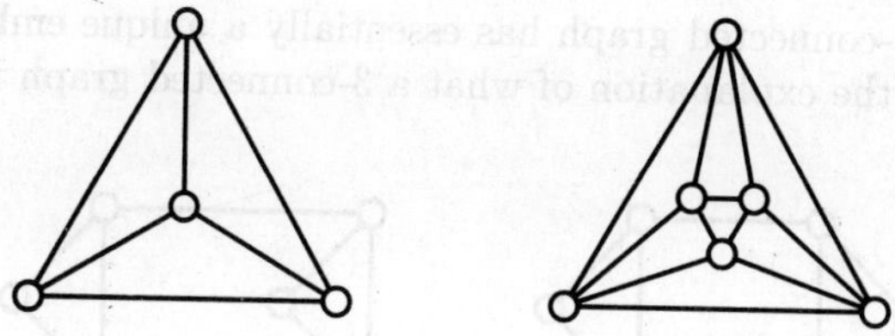

Figure 1.10. Planar graphs K_4 and the octahedron

self-intersecting curve in the plane joining its endpoints. Such a function is called a *planar embedding* if the curves corresponding to nonincident edges do not meet, and the curves corresponding to incident edges meet only at the point representing their common vertex. A graph is planar if and only if it has a planar embedding. Figure 1.10 shows two planar graphs: the complete graph K_4 and the *octahedron*.

A *plane graph* is a planar graph together with a fixed embedding. The edges of the graph divide the plane into regions called the *faces* of the plane graph. All but one of these regions is bounded, with the unbounded region called the *infinite* or *external face*. The length of a face is the number of edges bounding it.

Euler's famous formula gives the relationship between the number of vertices, edges, and faces of a connected plane graph.

Theorem 1.8.1 (Euler) *If a connected plane graph has n vertices, e edges and f faces, then*

$$n - e + f = 2. \qquad \square$$

A *maximal planar graph* is a planar graph X such that the graph formed by adding an edge between any two nonadjacent vertices of X is not planar. If an embedding of a planar graph has a face of length greater than three, then an edge can be added between two vertices of that face. Therefore, in any embedding of a maximal planar graph, every face is a triangle. Since each edge lies in two faces, we have

$$2e = 3f,$$

and so by Euler's formula,

$$e = 3n - 6.$$

A planar graph on n vertices with $3n-6$ edges is necessarily maximal; such graphs are called *planar triangulations* . Both the graphs of Figure 1.10 are planar triangulations.

A planar graph can be embedded into the plane in infinitely many ways. The two embeddings of Figure 1.11 are easily seen to be combinatorially different: the first has faces of length 3, 3, 4, and 6 while the second has faces of lengths 3, 3, 5, and 5. It is an important result of topological graph theory that a 3-connected graph has essentially a unique embedding. (See Section 3.4 for the explanation of what a 3-connected graph is.)

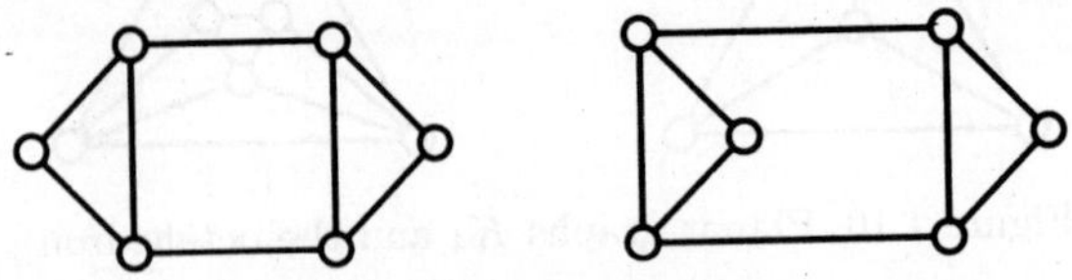

Figure 1.11. Two plane graphs

Given a plane graph X, we can form another plane graph called the *dual graph* X^*. The vertices of X^* correspond to the faces of X, with each vertex being placed in the corresponding face. Every edge e of X gives rise to an edge of X^* joining the two faces of X that contain e (see Figure 1.12).

Notice that two faces of X may share more than one common edge, in which case the graph X^* may contain *multiple edges*, meaning that two vertices are joined by more than one edge. This requires the obvious generalization to our definition of a graph, but otherwise causes no difficulties. Once again, explicit warning will be given when it is necessary to consider graphs with multiple edges.

Since each face in a planar triangulation is a triangle, its dual is a cubic graph. Considering the graphs of Figure 1.10, it is easy to check that K_4 is isomorphic to its dual; such graphs are called *self-dual*. The dual of the octahedron is a bipartite cubic graph on eight vertices known as the *cube*, which we will discuss further in Section 3.1.

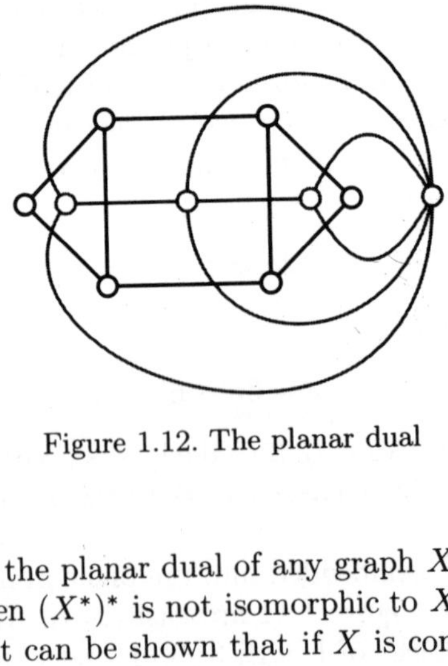

Figure 1.12. The planar dual

As defined above, the planar dual of any graph X is connected, so if X is not connected, then $(X^*)^*$ is not isomorphic to X. However, this is the only difficulty, and it can be shown that if X is connected, then $(X^*)^*$ is isomorphic to X.

The notion of embedding a graph in the plane can be generalized directly to embedding a graph in any surface. The dual of a graph embedded in any surface is defined analogously to the planar dual.

The *real projective plane* is a nonorientable surface, which can be represented on paper by a circle with diametrically opposed points identified. The complete graph K_6 is not planar, but it can be embedded in the projective plane, as shown in Figure 1.13. This embedding of K_6 is a triangulation in the projective plane, so its dual is a cubic graph, which turns out to be the Petersen graph.

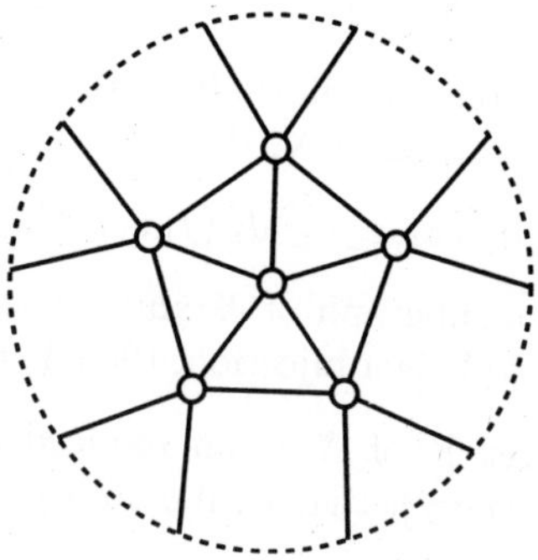

Figure 1.13. An embedding of K_6 in the projective plane

The *torus* is an orientable surface, which can be represented physically in Euclidean 3-space by the surface of a torus, or doughnut. It can be represented on paper by a rectangle where the points on the bottom side are identified with the points directly above them on the top side, and the points of the left side are identified with the points directly to the right of them on the right side. The complete graph K_7 is not planar, nor can it be embedded in the projective plane, but it can be embedded in the torus as shown in Figure 1.14 (note that due to the identification the four "corners" are actually the same point). This is another triangulation; its dual is a cubic graph known as the Heawood graph, which is discussed in Section 5.10.

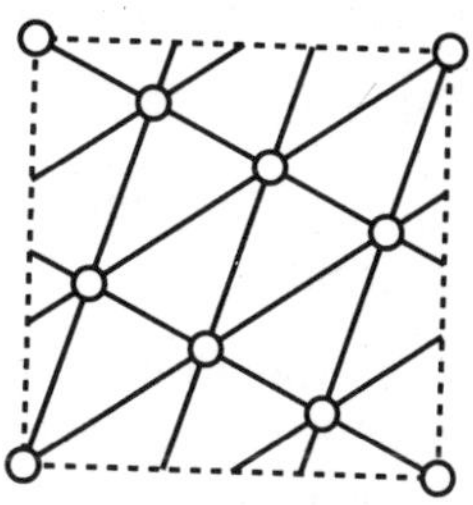

Figure 1.14. An embedding of K_7 in the torus

Exercises

1. Let X be a graph with n vertices. Show that X is complete or empty if and only if every transposition of $\{1, \dots, n\}$ belongs to $\mathrm{Aut}(X)$.

2. Show that X and $\overline{X}$ have the same automorphism group, for any graph X.

3. Show that if x and y are vertices in the graph X and $g \in \mathrm{Aut}(X)$, then the distance between x and y in X is equal to the distance between x^g and y^g in X.

4. Show that if f is a homomorphism from the graph X to the graph Y and x_1 and x_2 are vertices in X, then

$$d_X(x_1, x_2) \geq d_Y(f(x_1), f(x_2)).$$

5. Show that if Y is a subgraph of X and f is a homomorphism from X to Y such that $f \restriction Y$ is a bijection, then Y is a retract.

6. Show that a retract Y of X is an induced subgraph of X. Then show that it is *isometric*, that is, if x and y are vertices of Y, then $d_X(x, y) = d_Y(x, y)$.

7. Show that any edge in a bipartite graph X is a retract of X.

8. The *diameter* of a graph is the maximum distance between two distinct vertices. (It is usually taken to be infinite if the graph is not connected.) Determine the diameter of $J(v, k, k-1)$ when $v > 2k$.

9. Show that $\mathrm{Aut}(K_n)$ is not isomorphic to $\mathrm{Aut}(L(K_n))$ if and only if $n = 2$ or 4.

10. Show that the graph $K_5 \setminus e$ (obtained by deleting any edge e from K_5) is not a line graph.

11. Show that $K_{1,3}$ is not an induced subgraph of a line graph.

12. Prove that any induced subgraph of a line graph is a line graph.

13. Prove Krausz's characterization of line graphs (Theorem 1.7.2).

14. Find all graphs G such that $L(G) \cong G$.

15. Show that if X is a graph with minimum valency at least four, $\mathrm{Aut}(X)$ and $\mathrm{Aut}(L(X))$ are isomorphic.

16. Let S be a set of nonzero vectors from an m-dimensional vector space. Let $X(S)$ be the graph with the elements of S as its vertices, with two vectors x and y adjacent if and only if $x^T y \neq 0$. (Call $X(S)$ the "nonorthogonality" graph of S.) Show that any independent set in $X(S)$ has cardinality at most m.

17. Let X be a graph with n vertices. Show that the line graph of X is the nonorthogonality graph of a set of vectors in $\mathbb{R}^n$.
18. Show that a graph is bipartite if and only if it contains no odd cycles.
19. Show that a tree on n vertices has $n-1$ edges.
20. Let X be a connected graph. Let $T(X)$ be the graph with the spanning trees of X as its vertices, where two spanning trees are adjacent if the symmetric difference of their edge sets has size two. Show that $T(X)$ is connected.
21. Show that if two trees have isomorphic line graphs, they are isomorphic.
22. Use Euler's identity to show that K_5 is not planar.
23. Construct an infinite family of self-dual planar graphs.
24. A graph is *self-complementary* if it is isomorphic to its complement. Show that $L(K_{3,3})$ is self-complementary.
25. Show that if there is a self-complementary graph X on n vertices, then $n \equiv 0, 1 \mod 4$. If X is regular, show that $n \equiv 1 \mod 4$.
26. The *lexicographic product* $X[Y]$ of two graphs X and Y has vertex set $V(X) \times V(Y)$ where $(x, y) \sim (x', y')$ if and only if
    (a) x is adjacent to x' in X, or
    (b) $x = x'$ and y is adjacent to y' in Y.

    Show that the complement of the lexicographic product of X and Y is the lexicographic product of $\overline{X}$ and $\overline{Y}$.

Notes

For those readers interested in a more comprehensive view of graph theory itself, we recommend the books by West [6] and Diestel [2].

The problem of determining whether two graphs are isomorphic has a long history, as it has many applications—for example, among chemists who wish to tabulate all molecules in a certain class. All attempts to find a collection of easily computable graph parameters that are sufficient to distinguish any pair of nonisomorphic graphs have failed. Nevertheless the problem of determining graph isomorphism has not been shown to be NP-complete. It is considered a prime candidate for membership in the class of problems in NP that are neither NP-complete nor in P (if indeed NP $\neq$ P).

In practice, computer programs such as Brendan McKay's `nauty` [5] can determine isomorphisms between most graphs up to about 20000 vertices, though there are significant "pathological" cases where certain very highly structured graphs on only a few hundred vertices cannot be dealt with. Determining the automorphism group of a graph is closely related to determining whether two graphs are isomorphic. As we have already seen, it is often easy to find some automorphisms of a graph, but quite difficult to show that one has identified the full automorphism group of the graph. Once again, for moderately sized graphs with explicit descriptions, use of a computer is recommended.

Many graph parameters are known to be NP-hard to compute. For example, determining the chromatic number of a graph or finding the size of the maximum clique are both NP-hard.

Krausz's theorem (Theorem 1.7.2) comes from [4] and is surprisingly useful. A proof in English appears in [6], but you are better advised to construct your own. Beineke's result (Theorem 1.7.4) is proved in [1].

Most introductory texts on graph theory discuss planar graphs. For more complete information about embeddings of graphs, we recommend Gross and Tucker [3].

Part of the charm of graph theory is that it is easy to find interesting and worthwhile problems that can be attacked by elementary methods, and with some real prospect of success. We offer the following by way of example. Define the iterated line graph $L^n(X)$ of a graph X by setting $L^1(X)$ equal to $L(X)$ and, if $n > 1$, defining $L^n(X)$ to be $L(L^{n-1}(X))$. It is an open question, due to Ron Graham, whether a tree T is determined by the integer sequence

$$|V(L^n(T))|, \qquad n \geq 1.$$

References

[1] L. W. Beineke, *Derived graphs and digraphs*, Beiträge zur Graphentheorie, (1968), 17–33.

[2] R. Diestel, *Graph Theory*, Springer-Verlag, New York, 1997.

[3] J. L. Gross and T. W. Tucker, *Topological Graph Theory*, John Wiley & Sons Inc., New York, 1987.

[4] J. Krausz, *Démonstration nouvelle d'une théorème de Whitney sur les réseaux*, Mat. Fiz. Lapok, 50 (1943), 75–85.

[5] B. McKay, *nauty user's guide (version 1.5)*, tech. rep., Department of Computer Science, Australian National University, 1990.

[6] D. B. West, *Introduction to Graph Theory*, Prentice Hall Inc., Upper Saddle River, NJ, 1996.

2
Groups

The automorphism group of a graph is very naturally viewed as a group of permutations of its vertices, and so we now present some basic information about permutation groups. This includes some simple but very useful counting results, which we will use to show that the proportion of graphs on n vertices that have nontrivial automorphism group tends to zero as n tends to infinity. (This is often expressed by the expression "almost all graphs are asymmetric.") For a group theorist this result might be a disappointment, but we take its lesson to be that interesting interactions between groups and graphs should be looked for where the automorphism groups are large. Consequently, we also take the time here to develop some of the basic properties of transitive groups.

2.1 Permutation Groups

The set of all permutations of a set V is denoted by $\mathrm{Sym}(V)$, or just $\mathrm{Sym}(n)$ when $|V| = n$. A *permutation group* on V is a subgroup of $\mathrm{Sym}(V)$. If X is a graph with vertex set V, then we can view each automorphism as a permutation of V, and so $\mathrm{Aut}(X)$ is a permutation group.

A *permutation representation* of a group G is a homomorphism from G into $\mathrm{Sym}(V)$ for some set V. A permutation representation is also referred to as an *action* of G on the set V, in which case we say that G acts on V. A representation is *faithful* if its kernel is the identity group.

A group G acting on a set V induces a number of other actions. If S is a subset of V, then for any element $g \in G$, the translate S^g is again a subset of V. Thus each element of G determines a permutation of the subsets of V, and so we have an action of G on the power set 2^V. We can be more precise than this by noting that $|S^g| = |S|$. Thus for any fixed k, the action of G on V induces an action of G on the k-subsets of V. Similarly, the action of G on V induces an action of G on the ordered k-tuples of elements of V.

Suppose G is a permutation group on the set V. A subset S of V is G-*invariant* if $s^g \in S$ for all points s of S and elements g of G. If S is invariant under G, then each element $g \in G$ permutes the elements of S. Let $g \restriction S$ denote the *restriction* of the permutation g to S. Then the mapping

$$g \mapsto g \restriction S$$

is a homomorphism from G into $\mathrm{Sym}(S)$, and the image of G under this homomorphism is a permutation group on S, which we denote by $G \restriction S$. (It would be more usual to use G^S.)

A permutation group G on V is *transitive* if given any two points x and y from V there is an element $g \in G$ such that $x^g = y$. A G-invariant subset S of V is an *orbit* of G if $G \restriction S$ is transitive on S. For any $x \in V$, it is straightforward to check that the set

$$x^G := \{x^g : g \in G\}$$

is an orbit of G. Now, if $y \in x^G$, then $y^G = x^G$, and if $y \notin x^G$, then $y^G \cap x^G = \emptyset$, so each point lies in a unique orbit of G, and the orbits of G partition V. Any G-invariant subset of V is a union of orbits of G (and in fact, we could define an orbit to be a minimal G-invariant subset of V).

2.2 Counting

Let G be a permutation group on V. For any $x \in V$ the *stabilizer* G_x of x is the set of all permutations $g \in G$ such that $x^g = x$. It is easy to see that G_x is a subgroup of G. If $x_1, \dots, x_r$ are distinct elements of V, then

$$G_{x_1,\dots,x_r} := \bigcap_{i=1}^{r} G_{x_i}.$$

Thus this intersection is the subgroup of G formed by the elements that fix x_i for all i; to emphasize this it is called the *pointwise stabilizer* of $\{x_1, \dots, x_r\}$. If S is a subset of V, then the stabilizer G_S of S is the set of all permutations g such that $S^g = S$. Because here we are not insisting that every element of S be fixed this is sometimes called the *setwise stabilizer* of S. If $S = \{x_1, \dots, x_r\}$, then $G_{x_1,\dots,x_r}$ is a subgroup of G_S.

Lemma 2.2.1 *Let G be a permutation group acting on V and let S be an orbit of G. If x and y are elements of S, the set of permutations in G that* *map x to y is a right coset of G_x. Conversely, all elements in a right coset of G_x map x to the same point in S.*

Proof. Since G is transitive on S, it contains an element, g say, such that $x^g = y$. Now suppose that $h \in G$ and $x^h = y$. Then $x^g = x^h$, whence $x^{hg^{-1}} = x$. Therefore, $hg^{-1} \in G_x$ and $h \in G_x g$. Consequently, all elements mapping x to y belong to the coset $G_x g$.

For the converse we must show that every element of $G_x g$ maps x to the same point. Every element of $G_x g$ has the form hg for some element $h \in G_x$. Since $x^{hg} = (x^h)^g = x^g$, it follows that all the elements of $G_x g$ map x to x^g. □

There is a simple but very useful consequence of this, known as the orbit-stabilizer lemma.

Lemma 2.2.2 (Orbit-stabilizer) *Let G be a permutation group acting on V and let x be a point in V. Then*

$$|G_x|\,|x^G| = |G|$$

Proof. By the previous lemma, the points of the orbit x^G correspond bijectively with the right cosets of G_x. Hence the elements of G can be partitioned into $|x^G|$ cosets, each containing $|G_x|$ elements of G. □

In view of the above it is natural to wonder how G_x and G_y are related if x and y are distinct points in an orbit of G. To answer this we first need some more terminology. An element of the group G that can be written in the form $g^{-1}hg$ is said to be *conjugate* to h, and the set of all elements of G conjugate to h is the *conjugacy class* of h. Given any element $g \in G$, the mapping $\tau_g : h \mapsto g^{-1}hg$ is a permutation of the elements of G. The set of all such mappings forms a group isomorphic to G with the conjugacy classes of G as its orbits. If $H \subseteq G$ and $g \in G$, then $g^{-1}Hg$ is defined to be the subset

$$\{g^{-1}hg : h \in H\}.$$

If H is a subgroup of G, then $g^{-1}Hg$ is a subgroup of G isomorphic to H, and we say that $g^{-1}Hg$ is *conjugate* to H. Our next result shows that the stabilizers of two points in the same orbit of a group are conjugate.

Lemma 2.2.3 *Let G be a permutation group on the set V and let x be a point in V. If $g \in G$, then $g^{-1}G_x g = G_{x^g}$.*

Proof. Suppose that $x^g = y$. First we show that every element of $g^{-1}G_x g$ fixes y. Let $h \in G_x$. Then

$$y^{g^{-1}hg} = x^{hg} = x^g = y,$$

and therefore $g^{-1}hg \in G_y$. On the other hand, if $h \in G_y$, then ghg^{-1} fixes x, whence we see that $g^{-1}G_x g = G_y$. □

If g is a permutation of V, then $\text{fix}(g)$ denotes the set of points in V fixed by g. The following lemma is traditionally (and wrongly) attributed to Burnside; in fact, it is due to Cauchy and Frobenius.

Lemma 2.2.4 ("Burnside") *Let G be a permutation group on the set V. Then the number of orbits of G on V is equal to the average number of points fixed by an element of G.*

Proof. We count in two ways the pairs (g, x) where $g \in G$ and x is a point in V fixed by g. Summing over the elements of G we find that the number of such pairs is

$$\sum_{g \in G} |\text{fix}(g)|,$$

which, of course, is $|G|$ times the average number of points fixed by an element of G. Next we must sum over the points of V, and to do this we first note that the number of elements of G that fix x is $|G_x|$. Hence the number of pairs is

$$\sum_{x \in V} |G_x|.$$

Now, $|G_x|$ is constant as x ranges over an orbit, so the contribution to this sum from the elements in the orbit x^G is $|x^G|\,|G_x| = |G|$. Hence the total sum is equal to $|G|$ times the number of orbits, and the result is proved. □

2.3 Asymmetric Graphs

A graph is *asymmetric* if its automorphism group is the identity group. In this section we will prove that almost all graphs are asymmetric, i.e., the proportion of graphs on n vertices that are asymmetric goes to 1 as $n \to \infty$. Our main tool will be Burnside's lemma.

Let V be a set of size n and consider all the distinct graphs with vertex set V. If we let K_V denote a fixed copy of the complete graph on the vertex set V, then there is a one-to-one correspondence between graphs with vertex set V and subsets of $E(K_V)$. Since K_V has $\binom{n}{2}$ edges, the total number of different graphs is

$$2^{\binom{n}{2}}.$$

Given a graph X, the set of graphs isomorphic to X is called the *isomorphism class* of X. The isomorphism classes partition the set of graphs with vertex set V. Two such graphs X and Y are isomorphic if there is a permutation of $\text{Sym}(V)$ that maps the edge set of X onto the edge set of Y. Therefore, an isomorphism class is an orbit of $\text{Sym}(V)$ in its action on subsets of $E(K_V)$.

Lemma 2.3.1 *The size of the isomorphism class containing X is*

$$\frac{n!}{|\mathrm{Aut}(X)|}.$$

Proof. This follows from the orbit-stabilizer lemma. We leave the details as an exercise. □

Now we will count the number of isomorphism classes, using Burnside's lemma. This means that we must find the average number of subsets of $E(K_V)$ fixed by the elements of $\mathrm{Sym}(V)$. Now, if a permutation g has r orbits in its action on $E(K_V)$, then it fixes 2^r subsets in its action on the power set of $E(K_V)$. For any $g \in \mathrm{Sym}(V)$, let $\mathrm{orb}_2(g)$ denote the number of orbits of g in its action on $E(K_V)$. Then Burnside's lemma yields that the number of isomorphism classes of graphs with vertex set V is equal to

$$\frac{1}{n!} \sum_{g \in \mathrm{Sym}(V)} 2^{\mathrm{orb}_2(g)}. \tag{2.1}$$

If all graphs were asymmetric, then every isomorphism class would contain $n!$ graphs and there would be exactly

$$\frac{2^{\binom{n}{2}}}{n!}$$

isomorphism classes. Our next result shows that in fact, the number of isomorphism classes of graphs on n vertices is quite close to this, and we will deduce from this that almost all graphs are asymmetric. Recall that $o(1)$ is shorthand for a function that tends to 0 as $n \to \infty$.

Lemma 2.3.2 *The number of isomorphism classes of graphs on n vertices is at most*

$$(1 + o(1))\frac{2^{\binom{n}{2}}}{n!}.$$

Proof. We will leave some details to the reader. The *support* of a permutation is the set of points that it does not fix. We claim that among all permutations $g \in \mathrm{Sym}(V)$ with support of size an even integer $2r$, the maximum value of $\mathrm{orb}_2(g)$ is realized by the permutation with exactly r cycles of length 2.

Suppose $g \in \mathrm{Sym}(V)$ is such a permutation with r cycles of length two and $n-2r$ fixed points. Since $g^2 = e$, all its orbits on pairs of elements from V have length one or two. There are two ways in which an edge $\{x, y\} \in E(K_V)$ can be not fixed by g. Either both x and y are in the support of g, but $x^g \neq y$, or x is in the support of g and y is a fixed point of g. There are $2r(r-1)$ edges in the former category, and $2r(n-2r)$ is the latter category. Therefore the number of orbits of length 2 is $r(r-1)+r(n-2r) = r(n-r-1)$, and the total number of orbits of g on $E(K_V)$ is

$$\text{orb}_2(g) = \binom{n}{2} - r(n-r-1).$$

Now we are going to partition the permutations of $\text{Sym}(V)$ into 3 classes and make rough estimates for the contribution that each class makes to the sum (2.1) above.

Fix an even integer $m \le n-2$, and divide the permutations into three classes as follows: $C_1 = \{e\}$, C_2 contains the nonidentity permutations with support of size at most m, and C_3 contains the remaining permutations. We may estimate the sizes of these classes as follows:

$$|C_1| = 1, \qquad |C_2| \le \binom{n}{m} m! < n^m, \qquad |C_3| < n! < n^n.$$

An element $g \in C_2$ has the maximum number of orbits on pairs if it is a single 2-cycle, in which case it has $\binom{n}{2} - (n-2)$ such orbits. An element $g \in C_3$ has support of size at least m and so has the maximum number of orbits on pairs if it has $m/2$ 2-cycles, in which case it has

$$\binom{n}{2} - \frac{m}{2}\left(n - \frac{m}{2} - 1\right) \le \binom{n}{2} - \frac{nm}{4}$$

such orbits.

Therefore,

$$\begin{aligned}\sum_{g \in \text{Sym}(V)} 2^{\text{orb}_2(g)} &\le 2^{\binom{n}{2}} + n^m 2^{\binom{n}{2}-(n-2)} + n^n 2^{\binom{n}{2}-nm/4} \\ &= 2^{\binom{n}{2}}\left(1 + n^m 2^{-(n-2)} + n^n 2^{-nm/4}\right).\end{aligned}$$

The sum of the last two terms can be shown to be $o(1)$ by expressing it as

$$2^{m\log n - n + 2} + 2^{n \log n - nm/4}$$

and taking $m = \lfloor c \log n \rfloor$ for $c > 4$. □

Corollary 2.3.3 *Almost all graphs are asymmetric.*

Proof. Suppose that the proportion of isomorphism classes of graphs on V that are asymmetric is μ. Each isomorphism class of a graph that is not asymmetric contains at most $n!/2$ graphs, whence the average size of an isomorphism class is at most

$$n!\left(\mu + \frac{(1-\mu)}{2}\right) = \frac{n!}{2}(1+\mu).$$

Consequently,

$$\frac{n!}{2}(1+\mu)(1+o(1))\frac{2^{\binom{n}{2}}}{n!} > 2^{\binom{n}{2}},$$

from which it follows that μ tends to 1 as n tends to infinity. Since the proportion of asymmetric graphs on V is at least as large as the proportion of isomorphism classes (why?), it follows that the proportion of graphs on n vertices that are asymmetric goes to 1 as n tends to ∞. □

Although the last result assures us that most graphs are asymmetric, it is surprisingly difficult to find examples of graphs that are obviously asymmetric. We describe a construction that does yield such examples. Let T be a tree with no vertices of valency two, and with at least one vertex of valency greater than two. Assume that it has exactly m end-vertices. We construct a *Halin graph* by drawing T in the plane, and then drawing a cycle of length m through its end-vertices, so as to form a planar graph. An example is shown in Figure 2.1.

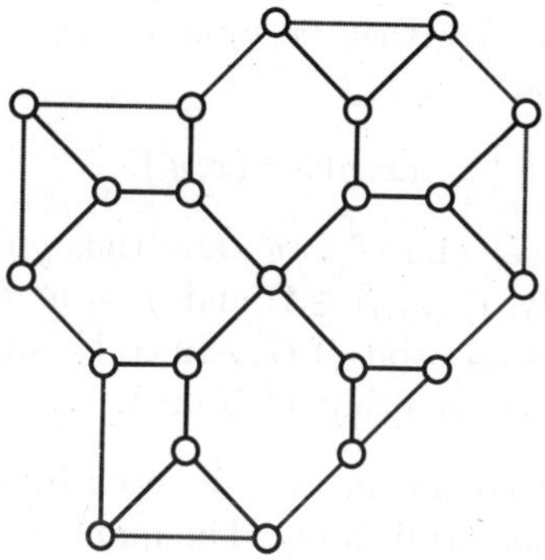

Figure 2.1. A Halin graph

Halin graphs have a number of interesting properties; in particular, it is comparatively easy to construct cubic Halin graphs with no nonidentity automorphisms. They all have the property that if we delete any two vertices, then the resulting graph is connected, but if we delete any edge, then we can find a pair of vertices whose deletion will disconnect the graph. (To use language from Section 3.4, they are 3-connected, but any proper subgraph is at most 2-connected.)

2.4 Orbits on Pairs

Let G be a transitive permutation group acting on the set V. Then G acts on the set of ordered pairs $V \times V$, and in this section we study its orbits on this set. It is so common to study G acting on both V and $V \times V$ that the orbits of G on $V \times V$ are often given the special name *orbitals*.

Since G is transitive, the set

$$\{(x, x) : x \in V\}$$

is an orbital of G, known as the *diagonal orbital*. If $\Omega \subseteq V \times V$, we define its *transpose* Ω^T to be

$$\{(y, x) : (x, y) \in \Omega\}.$$

It is a routine exercise to show that Ω^T is G-invariant if and only if Ω is. Since orbits are either equal or disjoint, it follows that if Ω is an orbital of G, either $\Omega = \Omega^T$ or $\Omega \cap \Omega^T = \emptyset$. If $\Omega = \Omega^T$, we call it a *symmetric* orbital.

Lemma 2.4.1 *Let G be a group acting transitively on the set V, and let x be a point of V. Then there is a one-to-one correspondence between the orbits of G on $V \times V$ and the orbits of G_x on V.*

Proof. Let Ω be an orbit of G on $V \times V$, and let Y_Ω denote the set $\{y : (x, y) \in \Omega\}$. We claim that the set Y_Ω is an orbit of G_x acting on V. If y and y' belong to Y_Ω, then (x, y) and (x, y') lie in Ω and there is a permutation g such that

$$(x, y)^g = (x, y').$$

This implies that $g \in G_x$ and $y^g = y'$, and thus y and y' are in the same orbit of G_x. Conversely, if $(x, y) \in \Omega$ and $y' = y^g$ for some $g \in G_x$, then $(x, y') \in \Omega$. Thus Y_Ω is an orbit of G_x. Since V is partitioned by the sets Y_Ω, where Ω ranges over the orbits of G on $V \times V$, the lemma follows. □

This lemma shows that for any $x \in V$, each orbit Ω of G on $V \times V$ is associated with a unique orbit of G_x. The number of orbits of G_x on V is called the *rank* of the group G. If Ω is symmetric, the corresponding orbit of G_x is said to be *self-paired*. Each orbit Ω of G on $V \times V$ may be viewed as a directed graph with vertex set V and arc set Ω. When Ω is symmetric this directed graph is a graph: (x, y) is an arc if and only if (y, x) is. If Ω is not symmetric, then the directed graph has the property that if (x, y) is an arc, then (y, x) is not an arc. Such directed graphs are often known as oriented graphs (see Section 8.3).

Lemma 2.4.2 *Let G be a transitive permutation group on V and let Ω be an orbit of G on $V \times V$. Suppose $(x, y) \in \Omega$. Then Ω is symmetric if and only if there is a permutation g in G such that $x^g = y$ and $y^g = x$.*

Proof. If (x, y) and (y, x) both lie in Ω, then there is a permutation $g \in G$ such that $(x, y)^g = (x^g, y^g) = (y, x)$. Conversely, suppose there is a permutation g swapping x and y. Since $(x, y)^g = (y, x) \in \Omega$, it follows that $\Omega \cap \Omega^T \neq \emptyset$, and so $\Omega = \Omega^T$. □

If a permutation g swaps x and y, then (xy) is a cycle in g. It follows that g has even order (and so G itself must have even order). A permutation group G on V is *generously transitive* if for any two distinct elements x and y from V there is a permutation that swaps them. All orbits of G on $V \times V$ are symmetric if and only if G is generously transitive.

We have seen that each orbital of a transitive permutation group G on V gives rise to a graph or an oriented graph. It is clear that G acts as a transitive group of automorphisms of each of these graphs. Similarly, the union of any set of orbitals is a directed graph (or graph) on which G acts transitively. We consider one example. Let V be the set of all 35 triples from a fixed set of seven points. The symmetric group Sym(7) acts on V as a transitive permutation group G, and it is not hard to show that G is generously transitive. Fix a particular triple x and consider the orbits of G_x on V. There are four orbits, namely x itself, the triples that meet x in 2 points, the triples that meet x in 1 point, and those disjoint from x. Hence these correspond to four orbitals, the first being the diagonal orbital, with the remaining three yielding the graphs $J(7,3,2)$, $J(7,3,1)$, and $J(7,3,0)$. It is clear that G is a subgroup of the automorphism group of each of these graphs, but although it can be shown that G is the full automorphism group of $J(7,3,2)$ and $J(7,3,0)$, it is not the full automorphism group of $J(7,3,1)$!

Lemma 2.4.3 *The automorphism group of* $J(7,3,1)$ *contains a group isomorphic to* Sym(8).

Proof. There are 35 partitions of the set $\{0,1,\ldots,7\}$ into two sets of size four. Let X be the graph with these partitions as vertices, where two partitions are adjacent if and only if the intersection of a 4-set from one with a 4-set from the other is a set of size two. Clearly, $\mathrm{Aut}(X)$ contains a subgroup isomorphic to Sym(8). However, X is isomorphic to $J(7,3,1)$. To see this, observe that a partition of $\{0,1,\ldots,7\}$ into two sets of size four is determined by the 4-set in it that contains 0, and this 4-set in turn is determined by its nonzero elements. Hence the partitions correspond to the triples from $\{1,2,\ldots,7\}$ and two partitions are adjacent in X if and only if the corresponding triples have exactly one element in common. □

2.5 Primitivity

Let G be a transitive group on V. A nonempty subset S of V is a *block of imprimitivity* for G if for any element g of G, either $S^g = S$ or $S \cap S^g = \emptyset$. Because G is transitive, it is clear that the translates of S form a partition of V. This set of distinct translates is called a *system of imprimitivity* for G.

An example of a system of imprimitivity is readily provided by the cube Q shown in Figure 2.2. It is straightforward to see that $\mathrm{Aut}(Q)$ acts transitively on Q (see Section 3.1 for more details).

For each vertex x there is a unique vertex x' at distance three from it; all other vertices in Q are at distance at most two. If $S = \{x, x'\}$ and $g \in \mathrm{Aut}(Q)$, then either $S^g = S$ or $S \cap S^g = \emptyset$, so S is a block of imprimitivity. There are four disjoint sets of the form S^g, as g ranges over

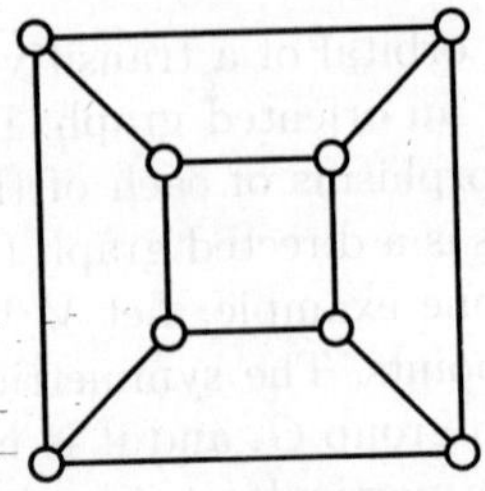

Figure 2.2. The cube Q

the elements of $\mathrm{Aut}(Q)$, and these sets are permuted among themselves by $\mathrm{Aut}(Q)$.

The partition of V into singletons is a system of imprimitivity as is the partition of V into one cell containing all of V. Any other system of imprimitivity is said to be nontrivial. A group with no nontrivial systems of imprimitivity is *primitive*; otherwise, it is *imprimitive*. There are two interesting characterizations of primitive permutation groups. We give one now; the second is in the next section.

Lemma 2.5.1 *Let G be a transitive permutation group on V and let x be a point in V. Then G is primitive if and only if G_x is a maximal subgroup of G.*

Proof. In fact, we shall be proving the contrapositive statement, namely that G has a nontrivial system of imprimitivity if and only if G_x is not a maximal subgroup of G. First some notation: We write $H \leq G$ if H is a maximal subgroup of G, and $H < G$ if H is a proper subgroup of G.

Suppose first that G has a nontrivial system of imprimitivity and that B is a block of imprimitivity that contains x. Then we will show that $G_x < G_B < G$ and therefore that G_x is not maximal. If $g \in G_x$, then $B \cap B^g$ is nonempty (for it contains x) and hence $B = B^g$. Thus $G_x \leq G_B$. To show that the inclusion is proper we find an element in G_B that is not in G_x. Let $y \neq x$ be another element of B. Since G is transitive it contains an element h such that $x^h = y$. But then $B = B^h$, yet $h \notin G_x$, and hence $G_x < G_B$.

Conversely, suppose that there is a subgroup H such that $G_x < H < G$. We shall show that the orbits of H form a nontrivial system of imprimitivity. Let B be the orbit of H containing x and let $g \in G$. To show that B is a block of imprimitivity it is necessary to show that either $B = B^g$ or $B \cap B^g = \emptyset$. Suppose that $y \in B \cap B^g$. Then because $y \in B$ there is an element $h \in H$ such that $y = x^h$. Moreover, because $y \in B^g$ there is some element $h' \in H$ such that $y = x^{h'g}$. Then $x^{h'gh^{-1}} = x$, and so $h'gh^{-1} \in G_x < H$. Therefore, $g \in H$, and because B is an orbit of H we have $B = B^g$. Because $G_x < H$, the block B does not consist of x alone, and because $H < G$, it does not consist of the whole of V, and hence it is a nontrivial block of imprimitivity. □

2.6 Primitivity and Connectivity

Our second characterization of primitive permutation groups uses the orbits of G on $V \times V$, and requires some preparation. A path in a directed graph D is a sequence $u_0, \ldots, u_r$ of distinct vertices such that (u_{i-1}, u_i) is an arc of D for $i = 1, \ldots, r$. A *weak path* is a sequence $u_0, \ldots, u_r$ of distinct vertices such that for $i = 1, \ldots, r$, either (u_{i-1}, u_i) or (u_i, u_{i-1}) is an arc. (We will use this terminology in this section only.) A directed graph is *strongly connected* if any two vertices can be joined by a path and is *weakly connected* if any two vertices can be joined by a weak path. A directed graph is weakly connected if and only if its "underlying" undirected graph is connected. (This is often used as a definition of weak connectivity.) A *strong component* of a directed graph is an induced subgraph that is maximal, subject to being strongly connected. Since a vertex is strongly connected, it follows that each vertex lies in a strong component, and therefore the strong components of D partition its vertices.

The *in-valency* of a vertex in a directed graph is the number of arcs ending on the vertex, the *out-valency* is defined analogously.

Lemma 2.6.1 *Let D be a directed graph such that the in-valency and out-valency of any vertex are equal. Then D is strongly connected if and only if it is weakly connected.*

Proof. The difficulty is to show that if D is weakly connected, then it is strongly connected. Assume by way of contradiction that D is weakly, but not strongly, connected and let $D_1, \ldots, D_r$ be the strong components of D. If there is an arc starting in D_1 and ending in D_2, then any arc joining D_1 to D_2 must start in D_1. Hence we may define a directed graph D' with the strong components of D as its vertices, and such that there is an arc from D_i to D_j in D' if and only if there is an arc in D starting in D_i and ending in D_j. This directed graph cannot contain any cycles. (Why?) It follows that there is a strong component, D_1 say, such that any arc that ends on a vertex in it must start at a vertex in it. Since D is weakly connected, there is at least one arc that starts in D_1 and ends on a vertex not in D_1. Consequently the number of arcs in D_1 is less than the sum of the out-valencies of the vertices in it. But on the other hand, each arc that ends in D_1 must start in it, and therefore the number of arcs in D_1 is equal to the sum of the in-valencies of its vertices. By our hypothesis on D, though, the sum of the in-valencies of the vertices in D_1 equals the sum of the out-valencies. Thus we have the contradiction we want. □

What does this lemma have to do with permutation groups? Let G act transitively on V and let Ω be an orbit of G on $V \times V$ that is not symmetric. Then Ω is an oriented graph, and G acts transitively on its vertices. Hence each point in V has the same out-valency in Ω and the same in-valency. As the in- and out-valencies sum to the number of arcs in Ω, the in- and out-valencies of any point of V in Ω are the same. Hence Ω is weakly connected if and only if it is strongly connected, and so we will refer to weakly or strongly connected orbits as connected orbits.

Lemma 2.6.2 *Let G be a transitive permutation group on V. Then G is primitive if and only if each nondiagonal orbit is connected.*

Proof. Suppose that G is imprimitive, and that $B_1, \ldots, B_r$ is a system of imprimitivity. Assume further that x and y are distinct points in B_1 and Ω is the orbit of G (on $V \times V$) that contains (x, y). If $g \in G$, then x^g and y^g must lie in the same block; otherwise B^g contains points from two distinct blocks. Therefore, each arc of Ω joins vertices in the same block, and so Ω is not connected.

Now suppose conversely that Ω is a nondiagonal orbit that is not connected, and let B be the point set of some component of Ω. If $g \in G$, then B and B^g are either equal or disjoint. Therefore, B is a nontrivial block and G is imprimitive. $\square$

Exercises

1. Show that the size of the isomorphism class containing X is
$$\frac{n!}{|\mathrm{Aut}(X)|}.$$

2. Prove that $|\mathrm{Aut}(C_n)| = 2n$. (You may assume that $2n$ is a lower bound on $|\mathrm{Aut}(C_n)|$.)

3. If G is a transitive permutation group on the set V, show that there is an element of G with no fixed points. (What if G has more than one orbit, but no fixed points?)

4. If g is a permutation of a set of n points with support of size s, show that $\mathrm{orb}_2(g)$ is maximal when all nontrivial cycles of g are transpositions.

5. The goal of this exercise is to prove Frobenius's lemma, which asserts that if the order of the group G is divisible by the prime p, then G contains an element of order p. Let Ω denote the set of all ordered p-tuples $(x_1, \ldots, x_p)$ of elements of G such that $x_1 \cdots x_p = e$. Let π denote the permutation of G^p that maps $(x_1, x_2, \ldots, x_p)$ to $(x_2, \ldots, x_p, x_1)$. Show that π fixes Ω as a set. Using the facts that π

fixes $(e, \dots, e)$ and $|\Omega| = |G|^{p-1}$, deduce that π must fix at least p elements of Ω and hence Frobenius's lemma holds.

6. Construct a cubic planar graph on 12 vertices with trivial automorphism group, and provide a proof that it has no nonidentity automorphism.

7. Decide whether the cube is a Halin graph.

8. Let X be a self-complementary graph with more than one vertex. Show that there is a permutation g of $V(X)$ such that:
   (a) $\{x, y\} \in E(X)$ if and only if $\{x^g, y^g\} \in E(\overline{X})$,
   (b) $g^2 \in \mathrm{Aut}(X)$ but $g^2 \neq e$,
   (c) the orbits of g on $V(X)$ induce self-complementary subgraphs of X.

9. If G is a permutation group on V, show that the number of orbits of G on $V \times V$ is equal to

$$\frac{1}{|G|} \sum_{g \in G} |\mathrm{fix}(g)|^2$$

and derive a similar formula for the number of orbits of G on the set of pairs of distinct elements from V.

10. If H and K are subsets of a group G, then HK denotes the subset

$$\{hk : h \in H,\ k \in K\}.$$

If H and K are subgroups and $g \in G$, then HgK is called a *double coset*. The double coset HgK is a union of right cosets of H and a union of left cosets of K, and G is partitioned by the distinct double cosets HgK, as g varies over the elements of G. Now (finally) suppose that G is a transitive permutation group on V and $H \leq G$. Show that each orbit of H on V corresponds to a double coset of the form $G_x gH$. Also show that the orbit of G_x corresponding to the double coset $G_x g G_x$ is self-paired if and only if $G_x g G_x = G_x g^{-1} G_x$.

11. Let G be a transitive permutation group on V. Show that it has a symmetric nondiagonal orbit on $V \times V$ if and only if $|G|$ is even.

12. Show that the only primitive permutation group on V that contains a transposition is $\mathrm{Sym}(V)$.

13. Let X be a graph such that $\mathrm{Aut}(X)$ acts transitively on $V(X)$ and let B be a block of imprimitivity for $\mathrm{Aut}(X)$. Show that the subgraph of X induced by B is regular.

14. Let G be a generously transitive permutation group on V and let B be a block for G. Show that $G \restriction B$ and the permutation group induced by G on the translates of B are both generously transitive.

15. Let G be a transitive permutation group on V such that for each element v in V there is an element of G with order two that has v as its only fixed point. (Thus $|V|$ must be odd.) Show that G is generously transitive.
16. Let X be a nonempty graph that is not connected. If $\mathrm{Aut}(X)$ is transitive, show that it is imprimitive.
17. Show that $\mathrm{Aut}(J(4n-1, 2n-1, n-1))$ contains a subgroup isomorphic to $\mathrm{Sym}(4n)$. Show further that $\omega(J(4n-1, 2n-1, n-1)) \leq 4n-1$, and characterize the cases where equality holds.

Notes

The standard reference for permutation groups is Wielandt's classic [5]. We also highly recommend Cameron [1]. For combinatorialists these are the best starting points. However, almost every book on finite group theory contains enough information on permutation groups to cover our modest needs. Neumann [3] gives an interesting history of Burnside's lemma.

The result of Exercise 15 is due to Shult. Exercise 17 is worth some thought, even if you do not attempt to solve it, because it appears quite obvious that $\mathrm{Aut}(J(4n-1, 2n-1, n-1))$ should be $\mathrm{Sym}(4n-1)$. The second part will be easier if you know something about Hadamard matrices.

Call a graph *minimal asymmetric* if it is asymmetric, but any proper induced subgraph with at least two vertices has a nontrivial automorphism. Sabidussi and Nešetřil [2] conjecture that there are finitely many isomorphism classes of minimal asymmetric graphs. In [4], Sabidussi verifies this for all graphs that contain an induced path of length at least 5, finding that there are only two such graphs. In [2], Sabidussi and Nešetřil show that there are exactly seven minimal asymmetric graphs in which the longest induced path has length four.

References

[1] P. J. CAMERON, *Permutation Groups*, Cambridge University Press, Cambridge, 1999.

[2] J. NEŠETŘIL AND G. SABIDUSSI, *Minimal asymmetric graphs of induced length 4*, Graphs Combin., 8 (1992), 343–359.

[3] P. M. NEUMANN, *A lemma that is not Burnside's*, Math. Sci., 4 (1979), 133–141.

[4] G. SABIDUSSI, *Clumps, minimal asymmetric graphs, and involutions*, J. Combin. Theory Ser. B, 53 (1991), 40–79.

[5] H. WIELANDT, *Finite Permutation Groups*, Academic Press, New York, 1964.

3

Transitive Graphs

We are going to study the properties of graphs whose automorphism group acts vertex transitively. A vertex-transitive graph is necessarily regular. One challenge is to find properties of vertex-transitive graphs that are not shared by all regular graphs. We will see that transitive graphs are more strongly connected than regular graphs in general. Cayley graphs form an important class of vertex-transitive graphs; we introduce them and offer some reasons why they are important and interesting.

3.1 Vertex-Transitive Graphs

A graph X is *vertex transitive* (or just *transitive*) if its automorphism group acts transitively on $V(X)$. Thus for any two distinct vertices of X there is an automorphism mapping one to the other.

An interesting family of vertex-transitive graphs is provided by the *k-cubes* Q_k. The vertex set of Q_k is the set of all 2^k binary k-tuples, with two being adjacent if they differ in precisely one coordinate position. We have already met the 3-cube Q_3, which is normally just called the cube (see Figure 2.2), and Figure 3.1 shows the 4-cube Q_4.

Lemma 3.1.1 *The k-cube Q_k is vertex transitive.*

Proof. If v is a fixed k-tuple, then the mapping

$$\rho_v : x \mapsto x + v$$

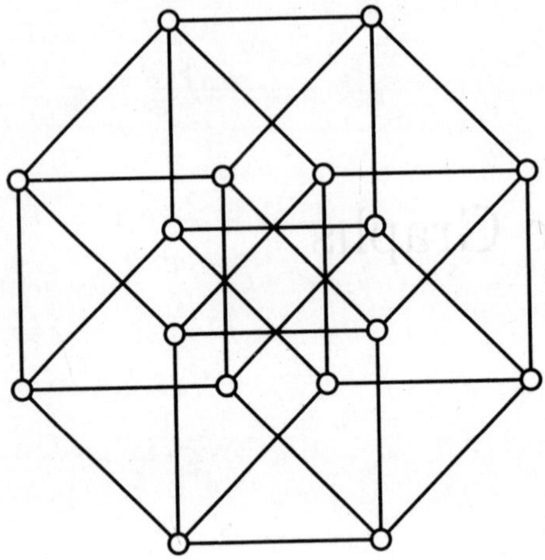

Figure 3.1. The 4-cube Q_4

(where addition is binary) is a permutation of the vertices of Q_k. This mapping is an automorphism because the k-tuples x and y differ in precisely one coordinate position if and only if $x + v$ and $y + v$ differ in precisely one coordinate position. There are 2^k such permutations, and they form a subgroup H of the automorphism group of Q_k. This subgroup acts transitively on $V(Q_k)$ because for any two vertices x and y, the automorphism ρ_{y-x} maps x to y. □

The group H of Lemma 3.1.1 is not the full automorphism group of Q_k. Any permutation of the k coordinate positions is an automorphism of Q_k, and the set of all these permutations forms a subgroup K of $\mathrm{Aut}(Q_k)$, isomorphic to $\mathrm{Sym}(k)$. Therefore, $\mathrm{Aut}(Q_k)$ contains the set HK. By standard group theory, the size of HK is given by

$$|HK| = \frac{|H||K|}{|H \cap K|}.$$

It is straightforward to see that $H \cap K$ is the identity subgroup, whence we conclude that $|\mathrm{Aut}(Q_k)| \geq 2^k k!$.

Another family of vertex-transitive graphs that we have met before are the circulants. Any vertex can be mapped to any other vertex by using a suitable power of the cyclic permutation described in Section 1.5.

The circulants and the k-cubes are both examples of a more general construction that produces many, but not all, vertex-transitive graphs.

Let G be a group and let C be a subset of G that is closed under taking inverses and does not contain the identity. Then the *Cayley graph* $X(G, C)$ is the graph with vertex set G and edge set

$$E(X(G, C)) = \{gh : hg^{-1} \in C\}.$$

If C is an arbitrary subset of G, then we can define a directed graph $X(G, C)$ with vertex set G and arc set $\{(g, h) : hg^{-1} \in C\}$. If C is inverse-closed and does not contain the identity, then this graph is undirected and has no loops, and the definition reduces to that of a Cayley graph. Most of the results for Cayley graphs apply in the more general directed case without modification, but we explicitly use directed Cayley graphs only in Section 3.8.

Theorem 3.1.2 *The Cayley graph $X(G, C)$ is vertex transitive.*

Proof. For each $g \in G$ the mapping

$$\rho_g : x \mapsto xg$$

is a permutation of the elements of G. This is an automorphism of $X(G, C)$ because

$$(yg)(xg)^{-1} = ygg^{-1}x^{-1} = yx^{-1},$$

and so $xg \sim yg$ if and only if $x \sim y$. The permutations ρ_g form a subgroup of the automorphism group of $X(G, C)$ isomorphic to G. This subgroup acts transitively on the vertices of $X(G, C)$ because for any two vertices g and h, the automorphism $\rho_{g^{-1}h}$ maps g to h. □

The k-cube is a Cayley graph for the elementary abelian group $(\mathbb{Z}_2)^k$, and a circulant on n vertices is a Cayley graph for the cyclic group of order n.

Most small vertex-transitive graphs are Cayley graphs, but there are also many families of vertex-transitive graphs that are not Cayley graphs. In particular, the graphs $J(v, k, i)$ are vertex transitive because $\mathrm{Sym}(v)$ contains permutations that map any k-set to any other k-set, but in general they are not Cayley graphs. We content ourselves with a single example.

Lemma 3.1.3 *The Petersen graph is not a Cayley graph.*

Proof. There are only two groups of order 10, the cyclic group $\mathbb{Z}_{10}$ and the dihedral group D_{10}. You may verify that none of the cubic Cayley graphs on these groups are isomorphic to the Petersen graph (Exercise 2). □

We will return to study Cayley graphs in more detail in Section 3.7.

3.2 Edge-Transitive Graphs

A graph X is *edge transitive* if its automorphism group acts transitively on $E(X)$. It is straightforward to see that the graphs $J(v, k, i)$ are edge transitive, but the circulants are not usually edge transitive.

An *arc* in X is an ordered pair of adjacent vertices, and X is *arc transitive* if $\mathrm{Aut}(X)$ acts transitively on its arcs. It is frequently useful to view an edge in a graph as a pair of oppositely directed arcs. An arc-transitive graph is necessarily vertex and edge transitive. In this section we will consider the relations between these various forms of transitivity.

The complete bipartite graphs $K_{m,n}$ are edge transitive, but not vertex transitive unless $m = n$, because no automorphism can map a vertex of valency m to a vertex of valency n. The next lemma shows that all graphs that are edge transitive but not vertex transitive are bipartite.

Lemma 3.2.1 *Let X be an edge-transitive graph with no isolated vertices. If X is not vertex transitive, then* $\text{Aut}(X)$ *has exactly two orbits, and these two orbits are a bipartition of X.*

Proof. Suppose X is edge but not vertex transitive. Suppose that $\{x, y\} \in E(X)$. If $w \in V(X)$, then w lies on an edge and there is an element of $\text{Aut}(X)$ that maps this edge onto $\{x, y\}$. Hence any vertex of X lies in either the orbit of x under $\text{Aut}(X)$, or the orbit of y. This shows that $\text{Aut}(X)$ has exactly two orbits. An edge that joins two vertices in one orbit cannot be mapped by an automorphism to an edge that contains a vertex from the other orbit. Since X is edge transitive and every vertex lies in an edge, it follows that there is no edge joining two vertices in the same orbit. Hence X is bipartite and the orbits are a bipartition for it. □

Figure 3.2 shows a regular graph that is edge transitive but not vertex transitive. The colouring of the vertices shows the bipartition.

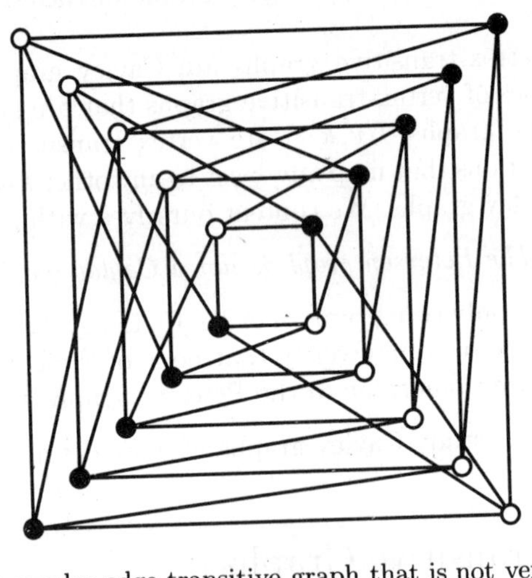

Figure 3.2. A regular edge-transitive graph that is not vertex transitive

An arc-transitive graph is, as we noted, always vertex and edge transitive. The converse is in general false; see the Notes at the end of the chapter for more. We do at least have the next result.

Lemma 3.2.2 *If the graph X is vertex and edge transitive, but not arc transitive, then its valency is even.*

Proof. Let $G = \text{Aut}(X)$, and suppose that $x \in V(X)$. Let y be a vertex adjacent to x and Ω be the orbit of G on $V \times V$ that contains (x, y). Since X is edge transitive, every arc in X can be mapped by an automorphism to either (x, y) or (y, x). Since X is not arc transitive, $(y, x) \notin \Omega$, and therefore Ω is not symmetric. Therefore, X is the graph with edge set $\Omega \cup \Omega^T$. Because the out-valency of x is the same in Ω and Ω^T, the valency of X must be even. □

A simple corollary to this result is that a vertex- and edge-transitive graph of odd valency must be arc transitive. Figure 3.3 gives an example of a vertex- and edge-transitive graph that is not arc-transitive.

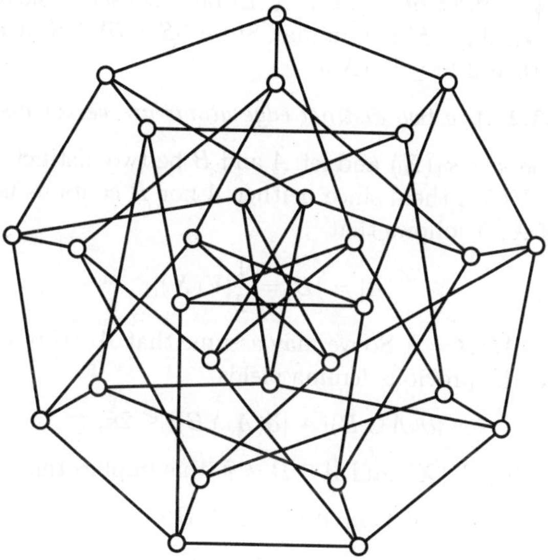

Figure 3.3. A vertex- and edge- but not arc-transitive graph

3.3 Edge Connectivity

An *edge cutset* in a graph X is a set of edges whose deletion increases the number of connected components of X. For a connected graph X, the *edge connectivity* is the minimum number of edges in an edge cutset, and will be denoted by $\kappa_1(X)$. If a single edge e is an edge cutset, then we call e a *bridge* or a *cut-edge*. As the set of edges adjacent to a vertex is an edge cutset, the edge connectivity of a graph cannot be greater than its minimum valency. Therefore, the edge connectivity of a vertex-transitive graph is at most its valency. In this section we will prove that the edge connectivity of a connected vertex-transitive graph is equal to its valency.

If $A \subseteq V(X)$, then we define ∂A to be the set of edges with one end in A and one end not in A. So if $A = \emptyset$ or $A = V(X)$, then $\partial A = \emptyset$, while the edge connectivity is the minimum size of ∂A as A ranges over the proper subsets of $V(X)$.

Lemma 3.3.1 *Let A and B be subsets of $V(X)$, for some graph X. Then*

$$|\partial(A \cup B)| + |\partial(A \cap B)| \leq |\partial A| + |\partial B|.$$

Proof. The details are left as an exercise. We simply note here that the difference between the two sides is twice the number of edges joining $A \setminus B$ to $B \setminus A$. □

Define an *edge atom* of a graph X to be a subset S such that $|\partial S| = \kappa_1(X)$ and, given this, $|S|$ is minimal. Since $\partial S = \partial V \setminus S$, it follows that if S is an atom, then $2|S| \leq |V(X)|$.

Corollary 3.3.2 *Any two distinct edge atoms are vertex disjoint.*

Proof. Assume $\kappa = \kappa_1(X)$ and let A and B be two distinct edge atoms in X. If $A \cup B = V(X)$, then, since neither A nor B contains more than half the vertices of X, it follows that

$$|A| = |B| = \frac{1}{2}|V(X)|$$

and hence that $A \cap B = \emptyset$. So we may assume that $A \cup B$ is a proper subset of $V(X)$. Now, the previous lemma yields

$$|\partial(A \cup B)| + |\partial(A \cap B)| \leq 2\kappa,$$

and, since $A \cup B \neq V(X)$ and $A \cap B \neq \emptyset$, this implies that

$$|\partial(A \cup B)| = |\partial(A \cap B)| = \kappa.$$

Since $A \cap B$ is a nonempty proper subset of the edge atom A, this is impossible. We are forced to conclude that A and B are disjoint. □

Our next result answers all questions about the edge connectivity of a vertex-transitive graph.

Lemma 3.3.3 *If X is a connected vertex-transitive graph, then its edge connectivity is equal to its valency.*

Proof. Suppose that X has valency k. Let A be an edge atom of X. If A is a single vertex, then $|\partial A| = k$ and we are finished. Suppose that $|A| \geq 2$. If g is an automorphism of X and $B = A^g$, then $|B| = |A|$ and $|\partial B| = |\partial A|$. From the previous lemma we see that either $A = B$ or $A \cap B = \emptyset$. Therefore, A is a block of imprimitivity for $\text{Aut}(X)$, and by Exercise 2.13 it follows that the subgraph of X induced by A is regular.

Suppose that the valency of this subgraph is ℓ. Then each vertex in A has exactly $k - \ell$ neighbours not in A, and so

$$|\partial A| = |A|(k - \ell).$$

Since X is connected, $\ell < k$, and so if $|A| \geq k$, then $|\partial A| \geq k$. Hence we may assume that $|A| < k$. Since $\ell \leq |A| - 1$, it follows that

$$|\partial A| \geq |A|(k + 1 - |A|).$$

The minimum value of the right side here occurs if $|A| = 1$ or k, when it is equal to k. Therefore, $|\partial A| \geq k$ in all cases. □

3.4 Vertex Connectivity

A *vertex cutset* in a graph X is a set of vertices whose deletion increases the number of connected components of X. The *vertex connectivity* (or just *connectivity*) of a connected graph X is the minimum number of vertices in a vertex cutset, and will be denoted by $\kappa_0(X)$. For any $k \leq \kappa_0(X)$ we say that X is *k-connected*. Complete graphs have no vertex cutsets, but it is conventional to define $\kappa_0(K_n)$ to be $n - 1$. The fundamental result on connectivity is Menger's theorem, which we state after establishing one more piece of terminology. If u and v are distinct vertices of X, then two paths P and Q from u to v are *openly disjoint* if $V(P \setminus \{u, v\})$ and $V(Q \setminus \{u, v\})$ are disjoint sets.

Theorem 3.4.1 (Menger) *Let u and v be distinct vertices in the graph X. Then the maximum number of openly disjoint paths from u to v equals the minimum size of a set of vertices S such that u and v lie in distinct components of $X \setminus S$.* □

We say that the subset S of the theorem *separates* u and v. Clearly, two vertices joined by m openly disjoint paths cannot be separated by any set of size less than m. The significance of this theorem is that it implies that two vertices that cannot be separated by fewer than m vertices must be joined by m openly disjoint paths. A simple consequence of Menger's theorem is that two vertices that cannot be separated by a single vertex must lie on a cycle. This is not too hard to prove directly. However, to prove that two vertices that cannot be separated by a set of size two are joined by three openly disjoint paths seems to be essentially as hard as the general case. Nonetheless, this is possibly the most important case. (It is the one that we make use of.)

There are a number of variations of Menger's theorem. In particular, two subsets A and B of $V(X)$ of size m cannot be separated by fewer than m vertices if and only if there are m disjoint paths starting in A and ending in B. This is easily deduced from the result stated; we leave it as an exercise.

We have a precise bound for the connectivity of a vertex-transitive graph, which requires much more effort to prove than determining its edge connectivity did.

Theorem 3.4.2 *A vertex-transitive graph with valency k has vertex connectivity at least* $\frac{2}{3}(k+1)$.

Figure 3.4 shows a 5-regular graph with vertex connectivity four, showing that equality can occur in this theorem.

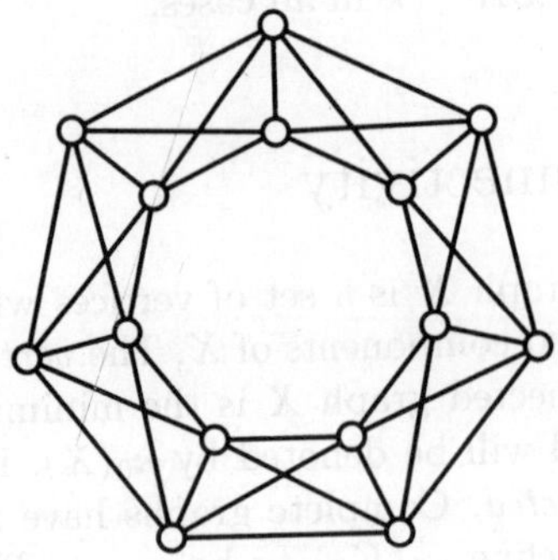

Figure 3.4. A 5-regular graph with vertex connectivity four

Before proving this result we need to develop some theory. If A is a set of vertices in X, let $N(A)$ denote the vertices in $V(X)\setminus A$ with a neighbour in A and let $\overline{A}$ be the complement of $A \cup N(A)$ in $V(X)$. A *fragment* of X is a subset A such that $\overline{A} \neq \emptyset$ and $|N(A)| = \kappa_0(X)$. An *atom* of X is a fragment that contains the minimum possible number of vertices. Note that an atom must be connected and that if X is a k-regular graph with an atom consisting of a single vertex, then $\kappa_0(X) = k$. It is not hard to show that if A is a fragment, then $N(A) = N(\overline{A})$ and $\overline{\overline{A}} = A$.

The following lemma records some further useful properties of fragments.

Lemma 3.4.3 *Let A and B be fragments in a graph X. Then*

(a) $N(A \cap B) \subseteq (A \cap N(B)) \cup (N(A) \cap B) \cup (N(A) \cap N(B))$.
(b) $N(A \cup B) = (\overline{A} \cap N(B)) \cup (N(A) \cap \overline{B}) \cup (N(A) \cap N(B))$.
(c) $\overline{A} \cup \overline{B} \subseteq \overline{A \cap B}$.
(d) $\overline{A \cup B} = \overline{A} \cap \overline{B}$.

Proof. Suppose first that $x \in N(A \cap B)$. Since $A \cap B$ and $N(A \cap B)$ are disjoint, if $x \in A$, then $x \notin B$, and therefore it must lie in $N(B)$. Similarly, if $x \in B$, then $x \in N(A)$. If x does not lie in A or B, then $x \in N(A) \cap N(B)$. Thus we have proved (a).

Analogous arguments show that $N(A \cup B)$ is contained in the union of $\overline{A} \cap N(B)$, $N(A) \cap \overline{B}$, and $N(A) \cap N(B)$. To obtain the reverse inclusion, note that if $x \in \overline{A} \cap N(B)$, then x does not lie in A or B. Since $x \in N(B)$, it follows that $x \in N(A \cup B)$. Similarly, we see that if $x \in N(A) \cap \overline{B}$ or $N(A) \cap N(B)$, then $x \in N(A \cup B)$.

Next, if $x \in \overline{A}$, then x does not lie in A or $N(A)$, and therefore does not lie in $A \cap B$ or $N(A \cap B)$. So $x \in \overline{A \cap B}$. This proves (c). We leave the proof of (d) as an exercise. □

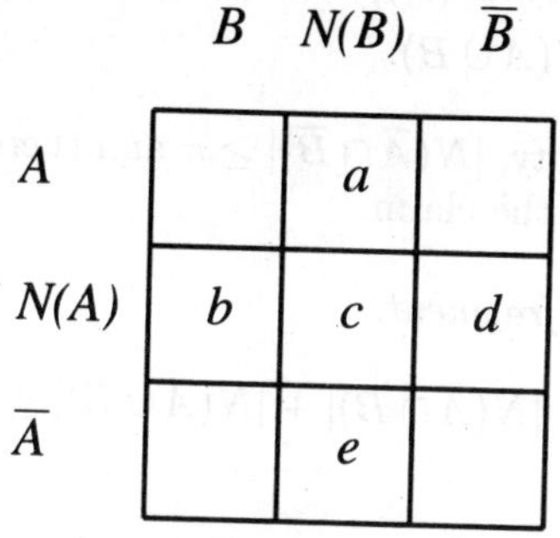

Figure 3.5. Intersection of fragments

Theorem 3.4.4 *Let X be a graph on n vertices with connectivity κ. Suppose A and B are fragments of X and $A \cap B \neq \emptyset$. If $|A| \leq |\overline{B}|$, then $A \cap B$ is a fragment.*

Proof. The intersections of A, $N(A)$, and $\overline{A}$ with the sets B, $N(B)$, and $\overline{B}$ partition $V(X)$ into nine pieces, as shown in Figure 3.5. The cardinalities of five of these pieces are also defined in this figure. We present the proof as a number of steps.

(a) $|A \cup B| < n - \kappa$.

Since $|F| + |\overline{F}| = n - \kappa$ for any fragment F of X,

$$|A| \leq |\overline{B}| = n - \kappa - |B|,$$

and therefore $|A| + |B| \leq n - \kappa$. Since $A \cap B$ is nonempty, the claim follows.

(b) $|N(A \cup B)| \leq \kappa$.

From Lemma 3.4.3 we find that $|N(A \cap B)| \leq a + b + c$ and $|N(A \cup B)| = c + d + e$. Consequently,

$$2\kappa = |N(A)| + |N(B)| = a+b+2c+d+e \geq |N(A \cap B)| + |N(A \cup B)|. \quad (3.1)$$

Since $|N(A \cap B)| \geq \kappa$, this implies that $|N(A \cup B)| \leq \kappa$.

(c) $\overline{A} \cap \overline{B} \neq \emptyset$.

From (a) and (b) we see that $|A \cup B| + |N(A \cup B)| < n$. Hence $\overline{A \cup B} \neq \emptyset$, and the claim follows from Lemma 3.4.3(d).

(d) $|N(A \cup B)| = \kappa$.

For any fragment F we have $N(F) = N(\overline{F})$. Using part (a) of Lemma 3.4.3 and (b) we obtain

$$\begin{aligned} N(\overline{A} \cap \overline{B}) &\subseteq (\overline{A} \cap N(\overline{B})) \cup (\overline{B} \cap N(\overline{A})) \cup (N(\overline{A}) \cap N(\overline{B})) \\ &= (\overline{A} \cap N(B)) \cup (\overline{B} \cap N(A)) \cup (N(A) \cap N(B)) \\ &= N(A \cup B). \end{aligned}$$

Since $\overline{A} \cap \overline{B}$ is nonempty, $|N(\overline{A} \cap \overline{B})| \geq \kappa$ and therefore $|N(A \cup B)| \geq \kappa$. Taken with (b), we get the claim.

(e) *The set $A \cap B$ is a fragment.*

From (3.1) we see that $|N(A \cap B)| + |N(A \cup B)| \leq 2\kappa$, whence (d) yields that $N(A \cap B) \leq \kappa$. □

Corollary 3.4.5 *If A is an atom and B is a fragment of X, then A is a subset of exactly one of B, $N(B)$, and $\overline{B}$.*

Proof. Since A is an atom, $|A| \leq |B|$ and $|A| \leq |\overline{B}|$. Hence the intersection of A with B or $\overline{B}$, if nonempty, would be a fragment. Since A is an atom, no proper subset of it can be a fragment. The result follows immediately. □

Now, we can prove Theorem 3.4.2. Suppose that X is a vertex-transitive graph with valency k, and let A be an atom in X. If A is a single vertex, then $|N(A)| = k$, and the theorem holds. Thus we may assume that $|A| \geq 2$. If $g \in \text{Aut}(X)$, then A^g is also an atom, and so by Corollary 3.4.5, either $A = A^g$ or $A \cap A^g = \emptyset$. Hence A is a block of imprimitivity for $\text{Aut}(X)$, and its translates partition $V(X)$. Corollary 3.4.5 now yields that $N(A)$ is partitioned by translates of A, and therefore

$$|N(A)| = t|A|$$

for some integer t. Suppose u is a vertex A. Then the valency of u is at most

$$|A| - 1 + |N(A)| = (t+1)|A| - 1,$$

and from this it follows that $k + 1 \leq (t+1)|A|$ and $\kappa_0(X) \geq \frac{t}{t+1}k$. To complete the proof we show that $t \geq 2$.

This is actually a consequence of Exercise 20, but since X is vertex transitive and the atoms are blocks of imprimitivity for $\text{Aut}(X)$, there is a shorter argument. Suppose for a contradiction that $t = 1$. By Corollary 3.4.5, $N(A)$ is a union of atoms, and so $N(A)$ is an atom. Since $\text{Aut}(X)$ acts transitively on the atoms of X, it follows that $|N(N(A))| = |A|$, and since $A \cap N(N(A))$ is nonempty, $A = N(N(A))$. This implies that $\overline{A} = \emptyset$, and so A is not a fragment.

3.5 Matchings

A *matching* M in a graph X is a set of edges such that no two have a vertex in common. The size of a matching is the number of edges in it. A vertex contained in an edge of M is *covered* by M. A matching that covers every vertex of X is called a *perfect matching* or a *1-factor*. Clearly, a graph that contains a perfect matching has an even number of vertices. A *maximum matching* is a matching with the maximum possible number of edges. (Without this convention there is a chance of confusion, since matchings can be partially ordered by inclusion or by size.)

We will prove the following result. It implies that a connected vertex-transitive graph on an even number of vertices has a perfect matching, and that each vertex in a connected vertex-transitive graph on an odd number of vertices is missed by a matching that covers all the remaining vertices.

Theorem 3.5.1 *Let X be a connected vertex-transitive graph. Then X has a matching that misses at most one vertex, and each edge is contained in a maximum matching.*

We first verify the claim about the maximum size of a matching. This requires some preparation, including two lemmas. If M is a matching in X and P is a path in X such that every second edge of P lies in M, we say that P is an *alternating path* relative to M. Similarly, an *alternating cycle* is a cycle with every second edge in M.

Suppose that M and N are matchings in X, and consider their symmetric difference $M \oplus N$. Since M and N are regular subgraphs with valency one, $M \oplus N$ is a subgraph with maximum valency two, and therefore each component of it is either a path or a cycle. Since no vertex in $M \oplus N$ lies in two edges of M or of N, these paths and cycles are alternating relative to both M and N. In particular, each cycle must have even length.

Suppose P is a path in $M \oplus N$ with odd length. We may assume without loss that P contains more edges of M than of N, in which case it follows that $N \oplus P$ is a matching in X that contains more edges than N. Hence, if M and N are maximum matchings, all paths in $M \oplus N$ have even length.

Lemma 3.5.2 *Let u and v be vertices in X such that no maximum matching misses both of them. Suppose that M_u and M_v are maximum matchings that miss u and v, respectively. Then there is a path of even length in $M_u \oplus M_v$ with u and v as its end-vertices.*

Proof. Our hypothesis implies that u and v are vertices of valency one in $M_u \oplus M_v$ so, by our ruminations above, both vertices are end-vertices of paths in $M_u \oplus M_v$. As M_u and M_v have maximum size, these paths have even length. If they are end-vertices of the same path, we are finished.

Assume that they lie on distinct paths and let P be the path on u. Then P is an alternating path relative to M_v with even length, and $M_v \oplus P$ is a matching in X that misses u and v and has the same size as M_v. This contradicts our choice of u and v. □

We are almost ready to prove the first part of our theorem. We call a vertex u in X *critical* if it is covered by every maximum matching. If X is vertex transitive and one vertex is critical, then all vertices are critical, whence X has a perfect matching. Given this, the next result implies the first claim in our theorem.

Lemma 3.5.3 *Let u and v be distinct vertices in X, and let P be a path from u to v. If no vertex in $V(P)\setminus\{u,v\}$ is critical, then no maximum matching misses both u and v.*

Proof. The proof is by induction on the length of P. If $u \sim v$, then no maximum matching misses both u and v; hence we may assume that P has length at least two.

Let x be a vertex on P distinct from u and v. Then u and x are joined in X by a path that contains no critical vertices. This path is shorter than P, so by induction, we conclude that no maximum matching misses both u and x. Similarly, no maximum matching misses both v and x.

Since x is not critical, there is a maximum matching M_x that misses it. Assume by way of contradiction that N is a maximum matching that misses u and v. Then, by Lemma 3.5.2 applied to the vertices u and x, there must be a path in $M_x \oplus N$ with u and x as its end-vertices. Applying the same argument to v and x, we find that $M_x \oplus N$ contains a path with v and x as its end-vertices. This implies that $u = v$, a contradiction. □

We noted above that a vertex-transitive graph that contains a critical vertex must have a perfect matching. By the above lemma, if X is vertex-transitive and does not contain a critical vertex, then no two vertices are missed by a maximum matching, and therefore a maximum matching covers all but one vertex of X.

It remains for us to show that every edge of X lies in a maximum matching. We assume inductively that this claim holds for all connected vertex-transitive graphs with fewer vertices or edges than X. If X is edge transitive, we are finished, so we assume it is not. Suppose e is an edge that does not lie in a maximum matching. Let Y be the subgraph of X with edge set consisting of the orbit of e under the action of $\mathrm{Aut}(X)$. Thus Y is a vertex-transitive spanning subgraph of X. Since X is not edge transitive, Y has fewer edges than X. We shall show that X has a matching containing an edge of Y that misses at most one vertex. This can be mapped to a matching containing e that misses at most one vertex.

If Y is connected, then by induction, each edge in it lies in a matching that misses at most one vertex. So suppose Y is not connected. The components of Y form a system of imprimitivity for $\mathrm{Aut}(X)$, and are pairwise isomorphic vertex-transitive graphs. If the number of vertices in a component of Y is even, then by induction, each component has a perfect matching and the union of these perfect matchings is a perfect matching in Y.

Assume then that the number of vertices in a component of Y is odd. Let $Y_1, \ldots, Y_r$ denote the distinct components of Y. Consider the graph Z with the components of Y as its vertices, with Y_i adjacent to Y_j if and only if there is an edge in X joining some vertex in Y_i to a vertex in Y_j. Then Z is a vertex-transitive graph and so, by induction, contains a matching N that misses at most one vertex. Suppose (Y_i, Y_j) is an edge in this matching. Since Y_i is adjacent to Y_j in Z, there are vertices y_i in Y_i and y_j in Y_j such that y_i is adjacent to y_j in X. Because Y_i and Y_j are vertex transitive and have an odd number of vertices, there is a matching in Y_i that misses only y_i and, similarly, a matching in Y_j that misses only y_j. The union of these two matchings, together with the edge y_iy_j, is a matching in X that covers all vertices in $Y_i \cup Y_j$. Thus each edge of N determines a matching that covers the vertices in two components of Y.

If the number of components of Y is even, it follows that X has a perfect matching. If the number is odd, we still have a matching in X that covers all the vertices outside one of the components, Y_1 say, of Y. Taken together with a matching of Y_1 that misses exactly one vertex of Y_1, we get a matching of X that misses exactly one vertex.

3.6 Hamilton Paths and Cycles

A *Hamilton path* in a graph is a path that meets every vertex, and a *Hamilton cycle* is a cycle that meets every vertex. A graph with a Hamilton cycle is called *hamiltonian*. All known vertex-transitive graphs have Hamilton paths, and only five are known that do not have Hamilton cycles. We consider these five graphs.

Clearly, K_2 is vertex transitive and does not have a Hamilton cycle. We pass on. A more interesting observation is that the Petersen graph $J(5,2,0)$ does not have a Hamilton cycle. This can be proved by a suitable case argument; in Chapter 8 we will offer an algebraic proof. The Coxeter graph, an arc-transitive cubic graph on 28 vertices that we will discuss in Section 4.6, also has no Hamilton cycle. For references to proofs of this, see the Notes at the end of this chapter.

The remaining two graphs are constructed from the Petersen and Coxeter graphs, by replacing each vertex with a triangle (see Figure 3.6).

We give a more formal definition of what this means, using subdivision graphs. The *subdivision graph* $S(X)$ of a graph X is obtained by putting one new vertex in the middle of each edge of X. Therefore, the vertex set of $S(X)$ is actually $V(X) \cup E(X)$, where two vertices of $S(X)$ are adjacent if they form an incident vertex/edge pair in X. The subdivision graph $S(X)$ is bipartite, with the two colour classes corresponding to $V(X)$ and $E(X)$.

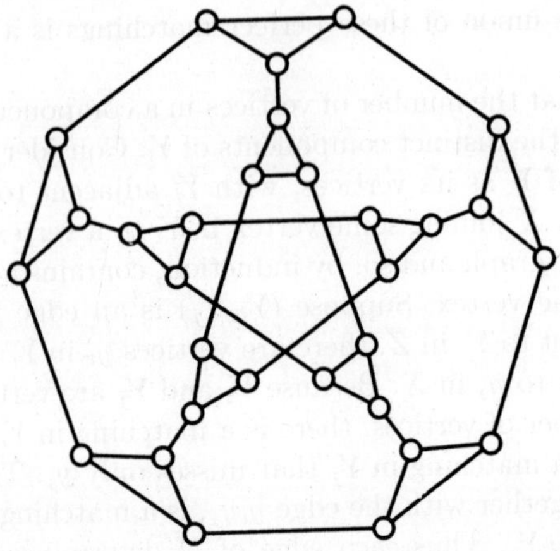

Figure 3.6. The Petersen graph with each vertex replaced by a triangle

The vertices in the "edge class" of $S(X)$ all have valency two. If X is regular with valency k, then the vertices in the "vertex class" of $S(X)$ all have valency k. Thus $S(X)$ is a semiregular bipartite graph.

The proof of the next result is left as an exercise.

Lemma 3.6.1 *Let X be a cubic graph. Then $L(S(X))$ has a Hamilton cycle if and only if X does.* □

If X is arc transitive and cubic, it follows that $L(S(X))$ is vertex transitive. Hence we obtain the last two of the five known vertex-transitive graphs without Hamilton cycles. Of these five graphs, only K_2 is a Cayley graph. Despite this somewhat limited evidence it has been conjectured that all Cayley graphs (other than K_2) are hamiltonian, and even more strongly that all vertex-transitive graphs other than these five are hamiltonian. The two conjectures have been quite intensively studied, and although some positive results are known, both conjectures seem to be wide open. (The conjecture is false for directed graphs.)

It is natural to look for sensible lower bounds on the length of a longest cycle in a vertex-transitive graph X. Some measure of our inadequacy is provided by the fact that the best known bound is of order $O(\sqrt{|V(X)|})$. Since this is all we have, we derive it here anyway. We need one further result about permutation groups.

Lemma 3.6.2 *Let G be a transitive permutation group on a set V, let S be a subset of V, and set c equal to the minimum value of $|S \cap S^g|$ as g ranges over the elements of G. Then $|S| \geq \sqrt{c|V|}$.*

Proof. We count the pairs (g, x) where $g \in G$ and $x \in S \cap S^g$. For each element of G there are at least c such points in S, and therefore there are at least $c|G|$ such pairs. On the other hand, the elements of G that map x

to y form a coset of G_x, and so there are exactly $|S|\,|G_x|$ elements g^{-1} of G such that $x^{g^{-1}} \in S$, i.e., $x \in S^g$. Hence

$$c|G| \leq |S|^2|G_x|,$$

and since G is transitive,

$$|G|/|G_x| = |V|$$

by the orbit stabilizer lemma. Consequently, $|S| \geq \sqrt{c|V|}$, as claimed. □

The proof of the next result depends on the fact that in a 3-connected graph any two maximum-length cycles must have at least three vertices in common.

Theorem 3.6.3 *A connected vertex-transitive graph on n vertices contains a cycle of length at least $\sqrt{3n}$.*

Proof. Let X be our graph and let G be its automorphism group. First we need to know that a connected vertex-transitive graph with valency at least three is at least 3-connected. This is a consequence of Theorem 3.4.2. Now we let C be a maximum-length cycle of X. Then by Exercise 19, $|C \cap C^g| \geq 3$ for any automorphism g of X, and the result follows from the previous lemma. □

In fact, we can find a cycle through all but one vertex in both the Petersen and Coxeter graphs (see Figure 3.7 for the Petersen graph).

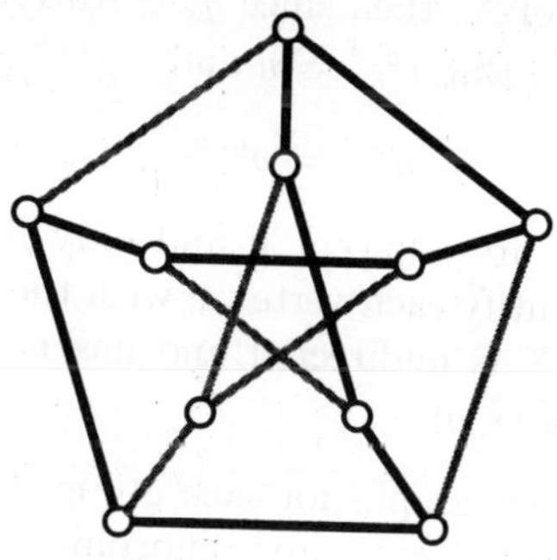

Figure 3.7. A cycle through nine vertices of the Petersen graph

3.7 Cayley Graphs

We now develop some of the basic properties of Cayley graphs. First we need some more terminology about permutation groups. A permutation group G acting on a set V is *semiregular* if no nonidentity element of G fixes a point of V. By the orbit-stabilizer lemma it follows that if G is semiregular, then all of its orbits have length equal to $|G|$. A permutation group is *regular* if it is semiregular and transitive. If G is regular on V, then $|G| = |V|$.

Given any group G we can always find a set on which it acts regularly—namely G itself. For each $g \in G$ recall that ρ_g is the permutation of the elements of G that maps x to xg. The mapping $g \mapsto \rho_g$ is a permutation representation of G (called the *right regular representation*). This group is isomorphic to G and acts transitively on G, hence is regular. Therefore, the proof of Theorem 3.1.2 implies the following result.

Lemma 3.7.1 *Let G be a group and let C be an inverse-closed subset of $G\backslash e$. Then $\mathrm{Aut}(X(G,C))$ contains a regular subgroup isomorphic to G.* □

There is a converse to this lemma.

Lemma 3.7.2 *If a group G acts regularly on the vertices of the graph X, then X is a Cayley graph for G relative to some inverse-closed subset of $G\backslash e$.*

Proof. Choose a fixed vertex u of X. Now, if v is any vertex of X, then since G acts regularly on $V(X)$, there is a unique element, g_v say, in G such that $u^{g_v} = v$. Define

$$C := \{g_v : v \sim u\}.$$

If x and y are vertices of X, then since $g_x \in \mathrm{Aut}(X)$, we see that $x \sim y$ if and only if $x^{g_x^{-1}} \sim y^{g_x^{-1}}$. But $x^{g_x^{-1}} = u$ and

$$y^{g_x^{-1}} = u^{g_y g_x^{-1}},$$

and therefore x and y are adjacent if and only if $g_y g_x^{-1} \in C$. It follows therefore that if we identify each vertex x with the group element g_x, then $X = X(G,C)$. Since X is undirected and has no loops, the set C is an inverse-closed subset of $G\backslash e$. □

There are many Cayley graphs for each group. It is natural to ask when Cayley graphs for the same group are isomorphic. The next lemma provides a partial answer to this question. If G is a group, then an *automorphism* of G is a bijection

$$\theta : G \to G$$

such that

$$\theta(gh) = \theta(g)\theta(h)$$

for all $g, h \in G$.

Lemma 3.7.3 *If θ is an automorphism of the group G, then $X(G,C)$ and $X(G,\theta(C))$ are isomorphic.*

Proof. For any two vertices x and y of $X(G,C)$ we have

$$\theta(y)\theta(x)^{-1} = \theta(yx^{-1}),$$

and so $\theta(y)\theta(x)^{-1} \in \theta(C)$ if and only if $yx^{-1} \in C$. Therefore, θ is an isomorphism from $X(G,C)$ to $X(G,\theta(C))$. □

The converse of this lemma is not true. Two Cayley graphs for a group G can be isomorphic even if there is no automorphism of G relating their connection sets.

A subset C of a group G is a *generating set* for G if every element of G can be written as a product of elements of C. Equivalently, the only subgroup of G that contains C is G itself. The proof of the following is left as an exercise.

Lemma 3.7.4 *The Cayley graph $X(G,C)$ is connected if and only if C is a generating set for G.* □

3.8 Directed Cayley Graphs with No Hamilton Cycles

In this section we show that it is relatively easy to find vertex-transitive directed graphs that are not hamiltonian, and in fact our examples are even directed Cayley graphs.

Theorem 3.8.1 *Suppose that distinct group elements α and β generate the finite group G, and that $X = X(G,\{\alpha,\beta\})$ is the directed Cayley graph of G with connection set $\{\alpha,\beta\}$. Suppose further that α and β have k and ℓ cycles, respectively, in their action by left multiplication on G. If $\beta^{-1}\alpha$ has odd order and $V(X)$ has a partition into r disjoint directed cycles, then r, k, and ℓ all have the same parity.*

Proof. Suppose that $V(X)$ has a partition into r directed cycles. Define a permutation π of G by $x^\pi = y$ if the arc (x,y) is in one of the directed cycles. If we define

$$P = \{x \in V(X) : x^\pi = \alpha x\}, \qquad Q = \{x \in V(X) : x^\pi = \beta x\},$$

then P and Q partition $V(X)$.

Let τ be the permutation of G defined by

$$x^\tau = \beta^{-1}x^\pi.$$

Clearly, τ fixes every element of Q, and thus it fixes P setwise. Moreover, for any element $x \in P$ we have $x^\tau = \beta^{-1}\alpha x$, and since $\beta^{-1}\alpha$ has odd order, so does τ. Therefore, τ is an even permutation. (An element of odd order is the square of some element in the cyclic group it generates, and so is even.)

Now, "recall" that the parity of a permutation of n elements with exactly r cycles equals the parity of $n+r$. Since left multiplication by β^{-1} followed by π is an even permutation, we see that $\ell+r$ is even. Exchanging α and β in the above argument yields that $k+r$ is even, and the result follows.□

The two permutations $\alpha = (1,2)$ and $\beta = (1,2,3,4)$ generate the symmetric group Sym(4); the Cayley graph $X = X(\mathrm{Sym}(4), \{\alpha, \beta\})$ is shown in Figure 3.8 (where undirected edges represent arcs in both directions). Now,

$$\beta^{-1}\alpha = (1,4,3),$$

which has odd order, and since Sym(4) has order 24, α and β have 12 and 6 cycles, respectively, in their action on Sym(4) by left multiplication. Therefore, $V(X)$ can only be partitioned into an even number of directed cycles, and so in particular does not have a directed Hamilton cycle.

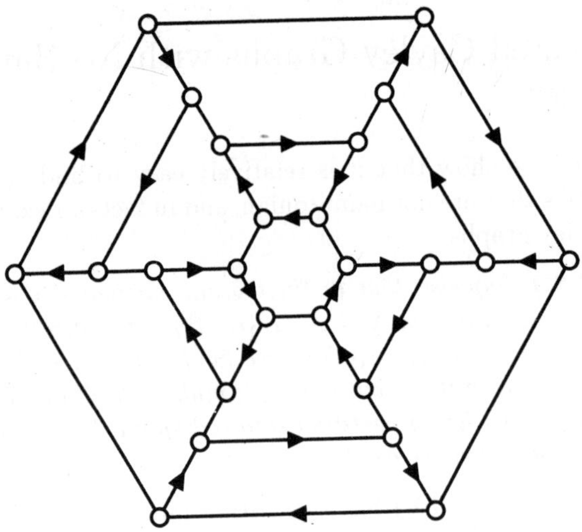

Figure 3.8. A nonhamiltonian directed Cayley graph

This example can be generalized to an infinite family $X(n)$ of directed Cayley graphs, where

$$X(n) = X(\mathrm{Sym}(n), \{(1,2), (1,2,3,\ldots,n)\}).$$

Corollary 3.8.2 *If n is even and $n \geq 4$, then the directed Cayley graph $X(n)$ is not hamiltonian.* □

It is known that $X(3)$ and $X(5)$ are hamiltonian, but it is unknown whether $X(n)$ is hamiltonian for odd $n \geq 7$.

3.9 Retracts

Recall that a subgraph Y of X is a retract if there is a homomorphism f from X to Y such that the restriction $f \restriction Y$ of f to Y is the identity map. In fact, it is enough to require that $f \restriction Y$ be a bijection, in which case it is an automorphism of Y. (See Exercise 1.5.)

The main result of this section is a proof that every vertex-transitive graph is a retract of a Cayley graph. Let G be a group acting transitively on the vertex set of X, and let x be a vertex of X. If y is a vertex of X, then by Lemma 2.2.1, the set of elements of G that map x to y is a right coset of G_x. Therefore, there is a bijection from $V(X)$ to the right cosets of G_x, and so we can identify each vertex of X with a right coset of G_x. The action of G on $V(X)$ coincides with the action of G by right multiplication on the cosets of G_x. (You ought to verify this.)

Theorem 3.9.1 *Any connected vertex-transitive graph is a retract of a Cayley graph.*

Proof. Suppose X is a connected vertex-transitive graph and let x be a vertex of X. Let C be the set

$$C := \{g \in G : x \sim x^g\}.$$

Then C is a union of right cosets of G_x, and since x is not adjacent to itself $C \cap G_x = \emptyset$. Furthermore, $x^a \sim x^b$ if and only if $x \sim x^{ba^{-1}}$, which is true if and only if $ba^{-1} \in C$.

If $g \in C$ and $h, h' \in G_x$, then

$$x = x^h \sim x^{gh} = x^{h'gh},$$

and thus $h'gh \in C$. Therefore, $G_xCG_x \subseteq C$, and since it is clear that $C \subseteq G_xCG_x$, we have $C = G_xCG_x$.

Let G be the subgroup of $\mathrm{Aut}(X)$ generated by the elements of C. An elementary induction argument on the diameter of X yields that G acts transitively on $V(X)$.

Now, let Y be the Cayley graph $X(G, C)$. The right cosets of G_x partition $V(Y)$, so we can express every element of G in the form ga for some $g \in G_x$. If g and h lie in G_x, then the two vertices ga and hb are adjacent if and only if

$$hb(ga)^{-1} = hba^{-1}g^{-1} \in C,$$

which happens if and only if $ba^{-1} \in C$. Therefore, any two distinct right cosets either have no edges between them or are completely joined, and since $e \notin C$, the subgraph of Y induced by each right coset is empty.

Thus the subgraph of Y induced by any complete set of coset representatives of G_x is isomorphic to X. The map sending the vertices in Y in a given right coset of G_x to the corresponding right coset, viewed as a vertex of X, is a homomorphism from Y to X. Its restriction to a complete set of coset representatives is a bijection, and thus we have the retraction we need. □

A careful reading of the above proof reveals that the Cayley graph Y can be obtained from X by replacing each vertex in X by an independent set of size $|G_x|$. The graph induced by a pair of these independent sets is empty when the vertices in X are not adjacent, or is a complete bipartite subgraph if they are adjacent. It follows that

$$\frac{|V(X)|}{\alpha(X)} = \frac{|V(Y)|}{\alpha(Y)}.$$

We will make use of this in Section 7.7.

3.10 Transpositions

We consider some special Cayley graphs for the symmetric groups. A set of transpositions from $\mathrm{Sym}(n)$ can be viewed as the edge set of a graph on n vertices (the transposition (ij) corresponding to the edge $\{i, j\}$). Every permutation in $\mathrm{Sym}(n)$ can be expressed as a product of transpositions, whence the transpositions form a generating set for $\mathrm{Sym}(n)$. A generating set C is minimal if $C \setminus g$ is not a generating set for any element g of C.

Lemma 3.10.1 *Let $\mathcal{T}$ be a set of transpositions from* $\mathrm{Sym}(n)$. *Then $\mathcal{T}$ is a generating set for* $\mathrm{Sym}(n)$ *if and only if its graph is connected.*

Proof. Let T be the graph of $\mathcal{T}$, which has vertex set $\{1, \ldots, n\}$. Let G be the group generated by $\mathcal{T}$. If $(1i)$ and (ij) are elements of $\mathcal{T}$, then

$$(1j) = (ij)(1i)(ij) \in G.$$

Consequently, a simple induction argument shows that if there is a path from 1 to i in T, then $(1i) \in G$. It follows that if k and ℓ lie in the same component, then $(k\ell) \in G$. (Repeat the above argument with k in place of 1.) Hence the transpositions belonging to a particular component of T generate the symmetric group on the vertices of that component.

Since no transposition can map a vertex in one component of T to a vertex in a second component, it follows that the components of T are the orbits of G. □

Lemma 3.10.2 *Let $\mathcal{T}$ be a set of transpositions from* $\mathrm{Sym}(n)$. *Then the following are equivalent:*

(a) *$\mathcal{T}$ is a minimal generating set for* $\mathrm{Sym}(n)$.

(b) *The graph of $\mathcal{T}$ is a tree.*

(c) *The product of the elements of $\mathcal{T}$ in any order is a cycle of length n.*

Proof. A connected graph on n vertices must have at least $n-1$ edges, and has exactly $n-1$ edges if and only if it is a tree. Thus (a) and (b) are equivalent. The equivalence of (b) and (c) is left as an exercise. □

There are $(n-1)!$ possible products of $n-1$ transpositions, and if (c) holds, then these will be distinct, i.e., every cycle of length n will arise exactly once.

If $\mathcal{T}$ is a set of transpositions, then the Cayley graph $X(\mathrm{Sym}(n), \mathcal{T})$ has no triangles: If $\{e, g, h\}$ were a triangle, then g, h, and hg would all be in $\mathcal{T}$, which is impossible because no transposition is a product of two transpositions. (In fact, it is almost as easy to prove that $X(\mathrm{Sym}(n), \mathcal{T})$ is bipartite.)

From Lemma 3.10.2 we see that each tree on n vertices determines a Cayley graph of $\mathrm{Sym}(n)$. We will use the next result to show that the Cayley graph is determined by the tree.

Lemma 3.10.3 *Let $\mathcal{T}$ be a set of transpositions from* $\mathrm{Sym}(n)$ *and let g and h be elements of $\mathcal{T}$. If the graph of $\mathcal{T}$ contains no triangles, then g and h have exactly one common neighbour in $X(\mathrm{Sym}(n), \mathcal{T})$ if $gh \neq hg$, and exactly two common neighbours otherwise.*

Proof. The neighbours of a vertex g of $X(\mathrm{Sym}(n), \mathcal{T})$ have the form xg, where $x \in \mathcal{T}$. So suppose $xg = yh$ is a common neighbour of g and h. Then $yx = hg$, and any solution to this equation yields a common neighbour.

If h and g commute, then they have disjoint support, and without loss of generality we may take $h = (12)$ and $g = (34)$. Then

$$hg = (12)(34) = (34)(12) = gh,$$

and there are two solutions to the above equation, yielding the two common neighbours e and hg.

If h and g do not commute, then they have overlapping support, and without loss of generality we may take $h = (12)$ and $g = (13)$. Then $hg = (123)$, and the only way in which this can be factored into transpositions is

$$(123) = (12)(13) = (13)(23) = (23)(12).$$

However, since both (12) and (13) lie in $\mathcal{T}$ and the graph of $\mathcal{T}$ contains no triangles, then (23) does not lie in $\mathcal{T}$, and hence there is only one possible factorization of hg, yielding e as the only common neighbour of g and h. □

Theorem 3.10.4 *Let $\mathcal{T}$ be a minimal generating set of transpositions for* $\mathrm{Sym}(n)$. *If the graph of $\mathcal{T}$ is asymmetric, then*

$$\mathrm{Aut}(X(\mathrm{Sym}(n), \mathcal{T})) \cong \mathrm{Sym}(n).$$

Proof. Let T be the graph of $\mathcal{T}$. Since $\mathcal{T}$ is a minimal generating set, T is a tree and hence contains no cycles. Then by Lemma 3.10.3 we can determine from $X(\mathrm{Sym}(n), \mathcal{T})$ which pairs of transpositions in $\mathcal{T}$ do not commute, or equivalently, which have overlapping support. Thus $X(\mathrm{Sym}(n), \mathcal{T})$ determines the line graph of $\mathcal{T}$. By Exercise 1.21, the tree T is determined by its line graph.

Any element g of $\mathrm{Aut}(X(\mathrm{Sym}(n), \mathcal{T}))_e$ induces a permutation of $\mathcal{T}$. Since automorphisms preserve paths of length two, it follows that the restriction of g to T is an automorphism of T. Therefore, it is trivial.

Now, suppose that g is an automorphism of $X(\mathrm{Sym}(n), \mathcal{T})$ fixing at least one vertex. We wish to show that g is the identity and hence that $\mathrm{Aut}(X(\mathrm{Sym}(n), \mathcal{T}))$ acts regularly. Suppose for a contradiction that g is not the identity. Then since $X(\mathrm{Sym}(n), \mathcal{T})$ is connected, there is a vertex v fixed by g adjacent to a vertex w that is not fixed by it. Then $\rho_v g \rho_v^{-1}$ fixes e and moves the adjacent vertex wv^{-1}. This is impossible, and so we are forced to conclude that g is the identity. Therefore, the automorphism group acts regularly. □

It is often quite difficult to determine the full automorphism group of a Cayley graph, which makes Theorem 3.10.4 more interesting.

Exercises

1. We describe a construction for the Folkman graph in Figure 3.2. Construct a multigraph by doubling each edge in $S(K_5)$, then replace each vertex of valency eight by two vertices of valency four with identical neighbourhoods. (This description is somewhat ambiguous, resolving this forms part of the problem.) Show that the result is the Folkman graph and prove that it is edge transitive but not vertex transitive.

2. Show that the Petersen graph is not a Cayley graph.

3. Show that the dodecahedron (Figure 1.4) is not a Cayley graph.

4. Prove that a Cayley graph $X(G, C)$ is connected if and only if C generates the group G.

5. If $\mathcal{T}$ is a set of transpositions from $\mathrm{Sym}(n)$, show that the Cayley graph $X(\mathrm{Sym}(n), \mathcal{T})$ is bipartite.

6. Prove that any vertex-transitive graph on a prime number of vertices is a Cayley graph. (Use Frobenius's lemma; see Exercise 2.5.)

7. Let G be a transitive permutation group on V, let S be a nonempty proper subset of V, and let c be the minimum value of $|S \cap S^g|$ as g ranges over the elements of G. Can $|V|$ be equal to $c^{-1}|S|^2$?

8. Prove that a transitive abelian permutation group is regular.

9. Let G be an abelian group and let C be an inverse-closed subset of $G \setminus e$. Show that if $|C| \geq 3$, then $X(G, C)$ has girth at most four.

10. Let C be an inverse-closed subset of $G \backslash e$. Show that if G is abelian and contains an element of order at least three, then $|\text{Aut}(X(G, C))| \geq 2|G|$.

11. Let G be a group. If $g \in G$, let λ_g be the permutation of G such that $\lambda_g(h) = gh$ for all h in G. Then $\{\lambda_g : g \in G\}$ is a subgroup of $\text{Sym}(G)$. Show that each element of this subgroup commutes with the group $\{\rho_g : g \in G\}$, and then determine when these two groups are equal.

12. Let a and b be two elements that generate the group G. Let A and B respectively denote the nonidentity elements of the cyclic subgroups generated by a and b. If $A \cap B = \emptyset$, show that the Cayley graph $X(G, A \cup B)$ is a line graph.

13. Show that $S(X)$ is edge transitive if and only if X is arc transitive and is vertex transitive if and only if X is a union of cycles of the same length.

14. If X is a cubic vertex-transitive graph with a triangle and is not K_4, show that X can be obtained by replacing each vertex of a cubic arc-transitive graph with a triangle. (This smaller graph may have multiple edges).

15. Let X be a connected arc-transitive graph with valency four and girth three. If X is not complete, show that it is the line graph of a cubic graph.

16. Prove that the vertex connectivity of a connected edge-transitive graph is equal to its minimum valency.

17. Let X be a graph. Show that two subsets A and B of $V(X)$ of size m cannot be separated by fewer than m vertices if and only if there are m disjoint paths starting in A and ending in B.

18. Show that any two paths of maximum length in a connected graph must have at least one vertex in common.

19. Show that any two cycles of maximum length in a 3-connected graph have at least three vertices in common.

20. If A is an atom and B is a fragment of X such that $A \subseteq N(B)$, show that $|A| \leq |N(B)|/2$.

21. Show that if a vertex-transitive graph with valency k has connectivity $\frac{2}{3}(k+1)$, then the atoms induce complete graphs.

22. Prove that if X is cubic, then $L(S(X))$ has a Hamilton cycle if and only if X does.

23. A Cayley graph $X(G, C)$ for the group G is minimal if C generates G but for any element c of C the set $C \setminus \{c, c^{-1}\}$ does not generate

G. Show that the connectivity of a minimal Cayley graph is equal to its valency.

24. The alternating group Alt(5) is generated by the two permutations

$$\alpha = (1, 2, 3), \quad \beta = (3, 4, 5).$$

Show that the directed Cayley graph $X(\text{Alt}(5), \{\alpha, \beta\})$ is not hamiltonian.

25. If G is a group of order $2^m k$ where k is odd, then G has a single conjugacy class of subgroups of order 2^m called the Sylow 2-subgroups of G. Suppose that G is generated by two elements α and β, where β has odd order. Show that if the Sylow 2-subgroups of G are not cyclic, then the directed Cayley graph

$$X(G, \{\alpha, \alpha\beta\})$$

is not hamiltonian.

Notes

The example in Figure 3.3 comes from Holt [6]. It follows from Alspach et al. [1], and earlier work reported there, that there are no smaller examples. (The smallest known examples with primitive automorphism group, found by Praeger and Xu [10], have 253 vertices and valency 24.)

The fact that the edge connectivity of a vertex-transitive graph equals its valency is due to Mader [9]. Our derivation of the lower bound on the connectivity of a vertex transitive graph follows the treatment in Chapter VI of Graver and Watkins [5]. The result is due independently to Mader [8] and Watkins [15]. For more details on the matching structure of vertex-transitive graphs, see [7, pp. 207–211].

Theorem 3.6.3 is due to Babai [2]. Exercise 19, which it uses, is due to Bondy. Our inability to improve on Babai's bound is regrettable evidence of our ignorance. Biggs [3] notes that there exactly six 1-factors in the Petersen graph, all equivalent under the action of its automorphism group. He also shows that the Coxeter graph has exactly 84 1-factors, all equivalent under its automorphism group. Deleting any one of them leaves $2C_{14}$, whence the Coxeter graph is not hamiltonian, but is 1-factorable. For the original proof that the Coxeter graph has no Hamilton cycle, see Tutte [14]. In Section 9.2, we will provide another proof that the Petersen graph does not have a Hamilton cycle.

Sabidussi [12] first noted that if G acts as a regular group of automorphisms of a graph X, then X must be a Cayley graph for G. The fact that each vertex-transitive graph is a retract of a Cayley graph is also due to him.

Exercise 23 is based on [4].

Lovász has conjectured that every connected vertex-transitive graph has a Hamilton path. This is open even for Cayley graphs. Witte [16] has shown that directed Cayley graphs of groups of prime-power order have Hamilton cycles.

There is a family of examples, due to Milnor, of directed Cayley graphs of metacyclic groups that do not have Hamilton paths. (A group G is metacyclic if it has a cyclic normal subgroup H such that G/H is cyclic.) These are described in Section 3.4 of Witte and Gallian's survey [17].

Our discussion of nonhamiltonian directed Cayley graphs in Section 3.8 follows Swan [13]. Ruskey et al. [11] prove that the directed Cayley graph $X(5)$ is hamiltonian.

References

[1] B. Alspach, D. Marušič, and L. Nowitz, *Constructing graphs which are 1/2-transitive*, J. Austral. Math. Soc. Ser. A, 56 (1994), 391–402.

[2] L. Babai, *Long cycles in vertex-transitive graphs*, J. Graph Theory, 3 (1979), 301–304.

[3] N. Biggs, *Three remarkable graphs*, Canad. J. Math., 25 (1973), 397–411.

[4] C. D. Godsil, *Connectivity of minimal Cayley graphs*, Arch. Math. (Basel), 37 (1981), 473–476.

[5] J. E. Graver and M. E. Watkins, *Combinatorics with Emphasis on the Theory of Graphs*, Springer-Verlag, New York, 1977.

[6] D. F. Holt, *A graph which is edge transitive but not arc transitive*, J. Graph Theory, 5 (1981), 201–204.

[7] L. Lovász and M. D. Plummer, *Matching Theory*, North-Holland Publishing Co., Amsterdam, 1986.

[8] W. Mader, *Über den Zusammenhang symmetrischer Graphen*, Arch. Math. (Basel), 21 (1970), 331–336.

[9] ——, *Minimale n-fach kantenzusammenhängende Graphen*, Math. Ann., 191 (1971), 21–28.

[10] C. E. Praeger and M. Y. Xu, *Vertex-primitive graphs of order a product of two distinct primes*, J. Combin. Theory Ser. B, 59 (1993), 245–266.

[11] F. Ruskey, M. Jiang, and A. Weston, *The Hamiltonicity of directed σ-τ Cayley graphs (or: A tale of backtracking)*, Discrete Appl. Math., 57 (1995), 75–83.

[12] G. Sabidussi, *On a class of fixed-point-free graphs*, Proc. Amer. Math. Soc., 9 (1958), 800–804.

[13] R. G. Swan, *A simple proof of Rankin's campanological theorem*, Amer. Math. Monthly, 106 (1999), 159–161.

[14] W. T. Tutte, *A non-Hamiltonian graph*, Canad. Math. Bull., 3 (1960), 1–5.

[15] M. E. Watkins, *Connectivity of transitive graphs*, J. Combinatorial Theory, 8 (1970), 23–29.

[16] D. WITTE, *Cayley digraphs of prime-power order are Hamiltonian*, J. Combin. Theory Ser. B, 40 (1986), 107–112.

[17] D. WITTE AND J. A. GALLIAN, *A survey: Hamiltonian cycles in Cayley graphs*, Discrete Math., 51 (1984), 293–304.

4

Arc-Transitive Graphs

An arc in a graph is an ordered pair of adjacent vertices, and so a graph is arc-transitive if its automorphism group acts transitively on the set of arcs. As we have seen, this is a stronger property than being either vertex transitive or edge transitive, and so we can say even more about arc-transitive graphs. The first few sections of this chapter consider the basic theory leading up to Tutte's remarkable results on cubic arc-transitive graphs. We then consider some examples of arc-transitive graphs, including three of the most famous graphs of all: the Petersen graph, the Coxeter graph, and Tutte's 8-cage.

4.1 Arc-Transitive Graphs

An *s-arc* in a graph is a sequence of vertices $(v_0, \ldots, v_s)$ such that consecutive vertices are adjacent and $v_{i-1} \neq v_{i+1}$ when $0 < i < s$. Note that an s-arc is permitted to use the same vertex more than once, although in all cases of interest this will not happen. A graph is *s-arc transitive* if its automorphism group is transitive on s-arcs. If $s \geq 1$, then it is both obvious and easy to prove that an s-arc transitive graph is also $(s-1)$-arc transitive. A 0-arc transitive graph is just another name for a vertex-transitive graph, and a 1-arc transitive graph is another name for an arc-transitive graph. A 1-arc transitive graph is also sometimes called a *symmetric* graph.

A cycle on n vertices is s-arc transitive for all s, which only shows that truth and utility are different concepts. A more interesting example is provided by the cube, which is 2-arc transitive. The cube is not 3-arc transitive because 3-arcs that form three sides of a four-cycle cannot be mapped to 3-arcs that do not (see Figure 4.1).

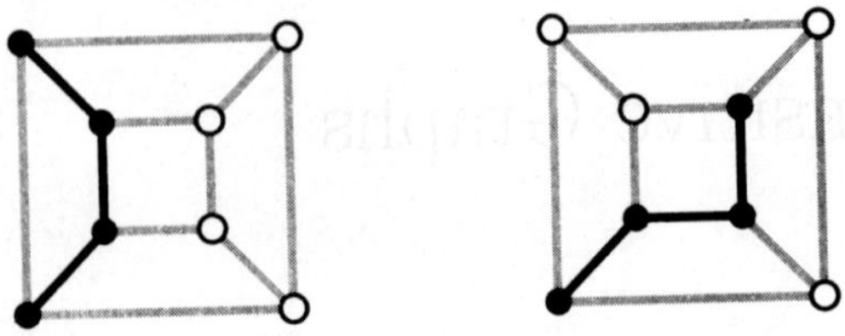

Figure 4.1. Inequivalent 3-arcs in the cube

A graph X is s-arc transitive if it has a group G of automorphisms such that G is transitive, and the stabilizer G_u of a vertex u acts transitively on the s-arcs with initial vertex u.

Lemma 4.1.1 *The graphs $J(v,k,i)$ are at least arc transitive.*

Proof. Consider the vertex $\{1,\ldots,k\}$. The stabilizer of this vertex contains $\mathrm{Sym}(k) \times \mathrm{Sym}(v-k)$. Clearly, any two k-sets meeting this initial vertex in an i-set can be mapped to each other by this group. □

Lemma 4.1.2 *The graphs $J(2k+1,k,0)$ are at least 2-arc transitive.* □

The *girth* of a graph is the length of the shortest cycle in it. Our first result implies that the subgraphs induced by s-arcs in s-arc transitive graphs are paths.

Lemma 4.1.3 (Tutte) *If X is an s-arc transitive graph with valency at least three and girth g, then $g \geq 2s-2$.*

Proof. We may assume that $s \geq 3$, since the condition on the girth is otherwise meaningless. It is easy to see that X contains a cycle of length g and a path of length g whose end-vertices are not adjacent. Therefore X contains a g-arc with adjacent end-vertices and a g-arc with nonadjacent end-vertices; clearly, no automorphism can map one to the other, and so $s < g$. Since X contains cycles of length g, and since these contain s-arcs, it follows that any s-arc must lie in a cycle of length g. Suppose that $v_0,\ldots,v_s$ is an s-arc. Denote it by α. Since v_{s-1} has valency at least three, it is adjacent to a vertex w other than v_{s-2} and v_s, and since the girth of X is at least s, this vertex cannot lie in α. Hence we may replace v_s by w, obtaining a second s-arc β that intersects α in an $(s-1)$-arc. Since β must lie in a circuit of length g, we thus obtain a pair of circuits of length g that have at least $s-1$ edges in common.

If we delete these $s-1$ edges from the graph formed by the edges of the two circuits of length g, the resulting graph still contains a cycle of length at most $2g-2s+2$. Hence $2g-2s+2 \geq g$, and the result follows. □

Given this lemma, it is natural to ask what can be said about the s-arc transitive graphs with girth $2s-2$. It follows from our next result that these graphs are, in the language of Section 5.6, generalized polygons. It is a consequence of results we state there that $s \leq 9$.

Lemma 4.1.4 (Tutte) *If X is an s-arc transitive graph with girth $2s-2$, it is bipartite and has diameter $s-1$.*

Proof. We first observe that if X has girth $2s-2$, then any s-arc lies in at most one cycle of length $2s-2$, and so if X is s-arc transitive, it follows that every s-arc lies in a unique cycle of length $2s-2$. Clearly, X has diameter at least $s-1$, because opposite vertices in a cycle of length $2s-2$ are at this distance. Now, let u be a vertex of X and suppose for a contradiction that v is a vertex at distance s from it. Then there is an s-arc joining u to v, which must lie in a cycle of length $2s-2$. Since a cycle of this length has diameter $s-1$, it follows that v cannot be of distance s from u. Therefore, the diameter of X is at most $s-1$ and hence equal to $s-1$.

If X is not bipartite, then it contains an odd cycle; suppose C is an odd cycle of minimal length. Because the diameter of X is $s-1$, the cycle must have length $2s-1$. Let u be a vertex of C, and let v and v' be the two adjacent vertices in C at distance $s-1$ from u. Then we can form an s-arc $(u, \dots, v, v')$. This s-arc lies in a cycle C' of length $2s-2$. The vertices of C and C' not internal to the s-arc form a cycle of length less than $2s-2$, which is a contradiction. □

In Section 4.5 we will use this lemma to show that s-arc transitive graphs with girth $2s-2$ are distance transitive.

4.2 Arc Graphs

If $s \geq 1$ and $\alpha = (x_0, \dots, x_s)$ is an arc in X, we define its *head* head(α) to be the $(s-1)$-arc $(x_1, \dots, x_s)$ and its *tail* tail(α) to be the $(s-1)$-arc $(x_0, \dots, x_{s-1})$. If α and β are s-arcs, then we say that β *follows* α if there is an $(s+1)$-arc γ such that head$(\gamma) = \beta$ and tail$(\gamma) = \alpha$. (Somewhat more colourfully, we say that α can be *shunted* onto β, and envisage pushing α one step onto β.) Let s be a nonnegative integer. We use $X^{(s)}$ to denote the directed graph with the s-arcs of X as its vertices, such that (α, β) is an arc if and only if α can be shunted onto β. Any automorphisms of X extend naturally to automorphisms of $X^{(s)}$, and so if X is s-arc transitive, then $X^{(s)}$ is vertex transitive.

Lemma 4.2.1 *Let X and Y be directed graphs and let f be a homomorphism from X onto Y such that every edge in Y is the image of an edge in X. Suppose $y_0, \dots, y_r$ is a path in Y. Then for each vertex x_0 in X such that $f(x_0) = y_0$, there is a path $x_0, \dots, x_r$ such that $f(x_i) = y_i$.*

Proof. Exercise. □

Define a "spindle" in X to be a subgraph consisting of two given vertices joined by three paths, with any two of these paths having only the given vertices in common. Define a "bicycle" to be a subgraph consisting either of two cycles with exactly one vertex in common, or two vertex-disjoint cycles and a path joining them having only its end-vertices in common with the cycles. We claim that if X is a spindle or a bicycle, then $X^{(1)}$ is strongly connected. We leave the proof of this as an easy exercise. Nonetheless, it is the key to the proof of the following result.

Theorem 4.2.2 *If X is a connected graph with minimum valency two that is not a cycle, then $X^{(s)}$ is strongly connected for all $s \geq 0$.*

Proof. First we shall prove the result for $s = 0$ and $s = 1$, and then by induction on s. If $s = 0$, then X_0 is the graph obtained by replacing each edge of X with a pair of oppositely directed arcs, so the result is clearly true. If $s = 1$, then we must show that any 1-arc can be shunted onto any other 1-arc. Since X is connected, we can shunt any 1-arc onto any edge of X, but not necessarily facing in the right direction. Therefore, it is necessary and sufficient to show that we can reverse the direction of any 1-arc, that is, shunt xy onto yx.

Since X has minimum valency at least two and is finite, it contains a cycle, C say. If C does not contain both x and y, then there is a (possibly empty) path in X joining y to C. It is now easy to shunt xy along the path, around C, then back along the path in the opposite direction to yx.

If x and y are in $V(C)$ but $xy \notin E(C)$, then C together with the edge xy is a spindle, and we are done.

Hence we may assume that $xy \in E(C)$. Since X is not a cycle, there is a vertex in C adjacent to a vertex not in C. Suppose w in $V(C)$ is adjacent to a vertex z not in C. Let P be a path with maximal length in X, starting with w and z, in this order. Then the last vertex of P is adjacent to a vertex in P or a vertex in C. If it is adjacent to a vertex in C other than w, then xy is an edge in a spindle. If it is adjacent to w or to a vertex of P not in C, then xy is an edge in a bicycle. In either case we are done.

Now, assume that $X^{(s)}$ is strongly connected for some $s \geq 1$. It is easy to see that the operation of taking the head of an $(s+1)$-arc is a homomorphism from $X^{(s+1)}$ to $X^{(s)}$. Since X has minimum valency at least two, each s-arc is the head of an $(s+1)$-arc, and it follows that every edge of $X^{(s)}$ is the image of an edge in $X^{(s+1)}$. Let α and β be any two $(s+1)$-arcs in X. Since $X^{(s)}$ is strongly connected, there is a path in it joining head(α) to tail(β). By the lemma above, this path lifts to a path in $X^{(s+1)}$ from α to a vertex γ where head(γ) = tail(β). Since $s \geq 1$ and X has minimum valency at least two, we see that γ can be shunted onto β. Thus α can be shunted to β via γ, and so there is a path in $X^{(s+1)}$ from α to β. □

In the next section we will use this theorem to prove that an arc-transitive cubic graph is s-arc regular, for some s. This is a crucial step in Tutte's work on arc-transitive cubic graphs.

4.3 Cubic Arc-Transitive Graphs

In 1947 Tutte showed that for any s-arc transitive cubic graph, $s \leq 5$. This was, eventually, the stimulus for a lot of work. One outcome of this was a proof, by Richard Weiss, that for any s-arc transitive graph, $s \leq 7$. This is a very deep result, the proof of which depends on the classification of the finite simple groups.

We used a form of the next result in proving Theorem 3.10.4.

Lemma 4.3.1 *Let X be a strongly connected directed graph, let G be a transitive subgroup of its automorphism group, and, if $u \in V(X)$, let $N(u)$ be the set of vertices v in $V(X)$ such that (u, v) is an arc of X. If there is a vertex u of X such that $G_u \restriction N(u)$ is the identity, then G is regular.*

Proof. Suppose $u \in V(X)$ and $G_u \restriction N(u)$ is the identity group. By Lemma 2.2.3, if $v \in V(X)$, then G_v is conjugate in G to G_u. Hence $G_v \restriction N(v)$ must be the identity for all vertices v of X.

Assume, by way of contradiction, that G_u is not the identity group. Since X is strongly connected, we may choose a directed path that goes from u to a vertex, w say, that is not fixed by G_u. Choose this path to have minimum possible length, and let v denote the second-last vertex on it. Thus v is fixed by G_u, and (v, w) is an arc in X. Since G_u fixes all vertices in $N(u)$, we see that $v \neq u$.

Since G_u fixes v, it fixes $N(v)$ but acts nontrivially on it, because it does not fix w. Hence $G_v \restriction N(v)$ is not the identity. This contradiction forces us to conclude that $G_u = \langle e \rangle$. □

A graph is *s-arc regular* if for any two s-arcs there is a unique automorphism mapping the first to the second.

Lemma 4.3.2 *Let X be a connected cubic graph that is s-arc transitive, but not $(s+1)$-arc transitive. Then X is s-arc regular.*

Proof. We note that if X is cubic, then $X^{(s)}$ has out-valency two. Now let G be the automorphism group of X, let α be an s-arc in X, and let H be the subgroup of G fixing each vertex in α. Then G acts vertex transitively on $X^{(s)}$, and H is the stabilizer in G of the vertex α in $X^{(s)}$. If the restriction of H to the out-neighbours of α is not trivial, then H must swap the two s-arcs that follow α. Now, any two $(s+1)$-arcs in X can be mapped by elements of G to $(s+1)$-arcs that have α as the "initial" s-arc; hence in this case we see that G is transitive on the $(s+1)$-arcs of X, which contradicts our initial assumption.

Hence the restriction of H to the out-neighbours of α is trivial, and it follows from Lemma 4.3.1 that H itself is trivial. Therefore, we have proved that $G_\alpha = \langle e \rangle$, and so G acts regularly on the s-arcs of X. □

If X is a regular graph with valency k on n vertices and $s \geq 1$, then there exactly $nk(k-1)^{s-1}$ s-arcs. It follows that if X is s-arc transitive then $|\mathrm{Aut}(X)|$ must be divisible by $nk(k-1)^{s-1}$, and if X is s-arc regular, then $|\mathrm{Aut}(X)| = nk(k-1)^{s-1}$. In particular, a cubic arc-transitive graph X is s-arc regular if and only if

$$|\mathrm{Aut}(X)| = (3n)2^{s-1}.$$

For an example, consider the cube. The alternative drawing of the cube in Figure 4.2 makes it clear that the stabilizer of a vertex contains Sym(3), and therefore its automorphism group has size at least 48. We observed earlier that the cube is not 3-arc transitive, so by Lemma 4.3.2 it must be precisely 2-arc regular, with full automorphism group of order 48.

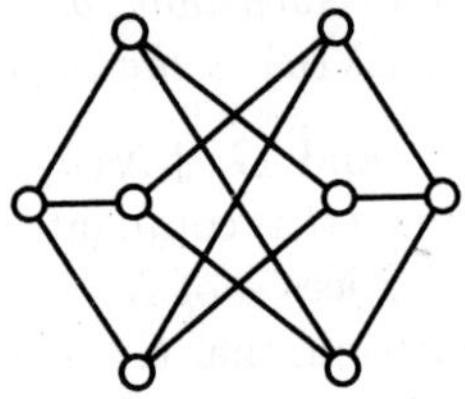

Figure 4.2. The cube redrawn

Finally, we state Tutte's theorem.

Theorem 4.3.3 *If X is an s-arc regular cubic graph, then $s \leq 5$.* □

The smallest 5-arc regular cubic graph is Tutte's 8-cage on 30 vertices, which we shall meet in Section 4.7.

Corollary 4.3.4 *If X is an arc-transitive cubic graph, $v \in V(X)$, and $G = \mathrm{Aut}(X)$, then $|G_v|$ divides 48 and is divisible by three.* □

4.4 The Petersen Graph

The Petersen graph is one of the most remarkable of all graphs. Despite having only 10 vertices, it plays a central role in so many different aspects of graph theory that almost any graph theorist will automatically be forced to give it special consideration when forming or testing new theorems. We have already met the Petersen graph in several guises: as $J(5,2,0)$ or $\overline{L(K_5)}$ in Section 1.5, as the dual of K_6 in the projective plane in Section 1.8, and as one of the five nonhamiltonian vertex-transitive graphs in Section 3.6. Two different drawings of it are shown in Figure 4.3.

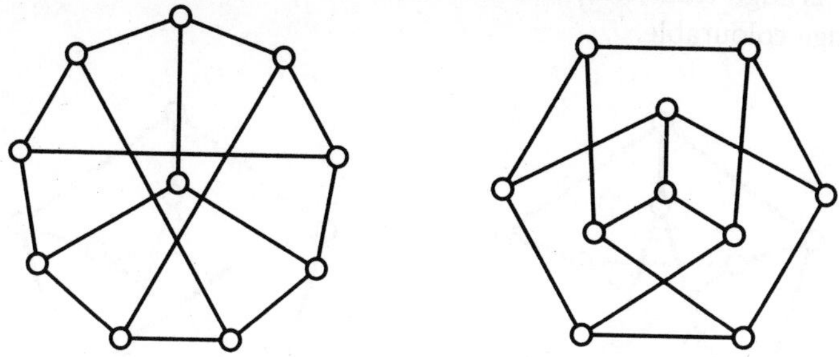

Figure 4.3. Two more drawings of the Petersen graph

The Petersen graph can also be constructed from the dodecahedron, which is shown in Figure 1.4. Every vertex v in the dodecahedron has a unique vertex v' at distance five from it. Consider the graph whose vertex set is the ten pairs of the form $\{v, v'\}$, where $\{u, u'\}$ is adjacent to $\{v, v'\}$ if and only if there is a perfect matching between them. The resulting graph is the Petersen graph. In this situation we say that the dodecahedron is a *2-fold cover* of the Petersen graph. We consider covers in more detail in Section 6.8.

Since $\mathrm{Sym}(5)$ acts on $J(5, 2, 0)$, we see that the automorphism group of the Petersen graph has order at least 120, and therefore it is at least 3-arc transitive (also see Exercise 5). Because the Petersen graph has girth five, by Lemma 4.1.8 it cannot be 4-arc transitive. Hence it is 3-arc regular, and its automorphism group has order exactly 120. Therefore, $\mathrm{Sym}(5)$ in its action on the 2-element subsets of a set of five elements is the full automorphism group of the Petersen graph.

The Petersen graph plays an important role in one of the most famous of all graph-theoretical problems. The four colour problem asks whether every plane graph can have its faces coloured with four colours such that faces with a common edge receive different colours. It can be shown that this is equivalent to the assertion that a cubic planar graph with edge connectivity at least two can have its edges coloured with three colours such that incident edges receive different colours (that is, has a proper *3-edge colouring*). The Petersen graph was the first cubic graph discovered that did not have a proper 3-edge colouring.

Theorem 4.4.1 *The Petersen graph cannot be 3-edge coloured.*

Proof. Let P denote the Petersen graph, and suppose for a contradiction that it can be 3-edge coloured. Since P is cubic, each colour class is a 1-factor of P. A simple case argument shows that each edge lies in precisely two 1-factors (Figure 4.4 shows the two 1-factors containing the vertical "spoke" edge). For each of these 1-factors, the remaining edges form a graph isomorphic to $2C_5$ that cannot be partitioned into two 1-factors. Since P is edge transitive, this is true for all 1-factors of P, and thus P is not 3-edge colourable. □

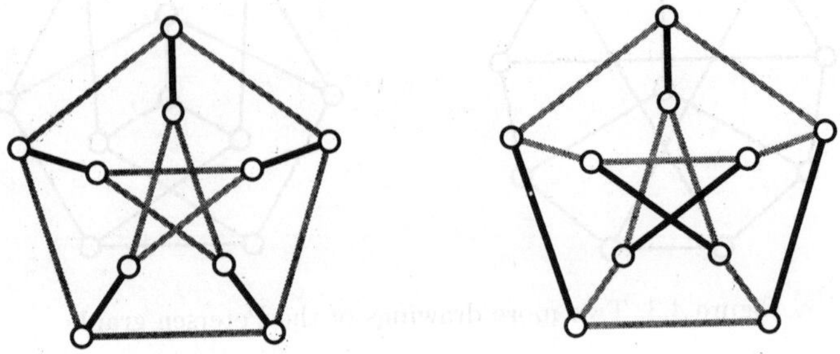

Figure 4.4. Two 1-factors through an edge of P

Thus the Petersen graph made the first of its many appearances as a counterexample. Since it is not planar, it is not a counterexample to the four colour problem, which was eventually proved in 1977, thus becoming the four colour theorem.

We have already observed that there is a cycle through any nine vertices in the Petersen graph. Let $X \setminus v$ denote the subgraph of X induced by $V(X) \setminus \{v\}$. A nonhamiltonian graph X such that $X \setminus v$ is hamiltonian for all v is called *hypohamiltonian*. The Petersen graph is the unique smallest hypohamiltonian graph; the next smallest have 13 and 15 vertices, respectively, and are closely related to the Petersen graph. We will see two more hypohamiltonian graphs in Section 13.6.

So far, we have only scratched the surface of the many ways in which the Petersen graph is special. It will reappear in several of the remaining sections of this book. In particular, the Petersen graph is a *distance-transitive* graph (Section 4.5), a *Moore graph* (Section 5.8), and a *strongly regular graph* (Chapter 10).

4.5 Distance-Transitive Graphs

A connected graph X is *distance transitive* if given any two ordered pairs of vertices (u, u') and (v, v') such that $d(u, u') = d(v, v')$, there is an automorphism g of X such that $(v, v') = (u, u')^g$. A distance-transitive graph is always at least 1-arc transitive. The complete graphs, the complete bipartite graphs with equal-sized parts, and the circuits are the cheapest examples available. A more interesting example is provided by the Petersen graph. It is not hard to see that this is distance transitive, since it is arc transitive and its complement, the line graph of K_5, is also arc transitive.

Another family of examples is provided by the k-cubes described in Section 3.7. If $d(u,u') = d(v,v') = i$, then by adding u to the first pair and v to the second pair, we can assume that $u = v = 0$. Then u' and v' are simply different vectors with i nonzero coordinates that can be mapped to one another by $\mathrm{Sym}(k)$ acting on coordinate positions.

Lemma 4.5.1 *The graph $J(v,k,k-1)$ is distance transitive.*

Proof. The key is to prove that two vertices u and v have distance i in $J(v,k,k-1)$ if and only if $|u \cap v| = k - i$ (viewing u and v as k-sets). We leave the details as an exercise. □

Lemma 4.5.2 *The graph $J(2k+1, k+1, 0)$ is distance transitive.* □

There is an alternative definition of a distance-transitive graph that often proves easier to work with. If u is a vertex of X, then let $X_i(u)$ denote the set of vertices at distance i from u. The partition $\{u, X_1(u), \ldots, X_d(u)\}$ is called the *distance partition* with respect to u. Figure 4.5 gives a drawing of the dodecahedron that displays the distance partition from a vertex.

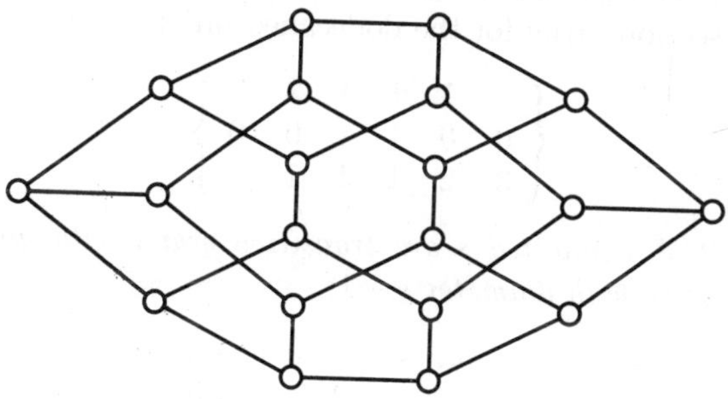

Figure 4.5. The dodecahedron

Suppose G acts distance transitively on X and $u \in V(X)$. If v and v' are two vertices at distance i from u, there is an element of G that maps (u,v) to (u,v'), i.e., there is an element of G_u that maps v to v', and so G acts transitively on $X_i(u)$. Therefore, the cells of the distance partition with respect to u are the orbits of G_u. If X has diameter d, then it follows that G acts distance transitively on X if and only if it acts transitively and, for any vertex u in X, the vertex stabilizer G_u has exactly $d+1$ orbits. In other words, the group G is transitive with rank $d+1$.

Since the cells of the distance partition are orbits of G_u, every vertex in $X_i(u)$ is adjacent to the same number of other vertices, say a_i, in $X_i(u)$. Similarly, every vertex in $X_i(u)$ is adjacent to the same number, say b_i, of vertices in $X_{i+1}(u)$ and the same number, say c_i, of vertices in $X_{i-1}(u)$. Equivalently, the graph induced by any cell is regular, and the graph induced by any pair of cells is semiregular. The graph X is regular, and its valency is given by b_0, so if the diameter of X is d, we have

$$c_i + a_i + b_i = b_0, \quad i = 0, 1, \ldots, d.$$

These numbers are called the parameters of the distance-transitive graph, and determine many of its properties. We can record these numbers in the $3 \times (d+1)$ *intersection array*

$$\left\{ \begin{array}{cccccc} - & c_1 & \cdots & c_{d-1} & c_d \\ a_0 & a_1 & \cdots & a_{d-1} & a_d \\ b_0 & b_1 & \cdots & b_{d-1} & - \end{array} \right\}.$$

Since each column sums to the valency of the graph, it is necessary to give only two rows of the matrix to determine it entirely. It is customary to use the following abbreviated version of the intersection array:

$$\{b_0, b_1, \ldots, b_{d-1}; c_1, c_2, \ldots, c_d\}.$$

For example, consider the dodecahedron. It is easy to see that every vertex in $X_2(u)$ is adjacent to one vertex in $X_1(u)$, one in $X_2(u)$, and one in $X_3(u)$, and therefore $a_2 = b_2 = c_2 = 1$. Continuing similarly we find that the intersection array for the dodecahedron is

$$\left\{ \begin{array}{cccccc} - & 1 & 1 & 1 & 2 & 3 \\ 0 & 0 & 1 & 1 & 0 & 0 \\ 3 & 2 & 1 & 1 & 1 & - \end{array} \right\}.$$

Lemma 4.5.3 *A connected s-arc transitive graph with girth $2s - 2$ is distance-transitive with diameter $s - 1$.*

Proof. Let X satisfy the hypotheses of the lemma and let (u, u') and (v, v') be pairs of vertices at distance i. Since X has diameter $s-1$ by Lemma 4.1.4, we see that $i \leq s-1$. The two pairs of vertices are joined by paths of length i, and since X is transitive on i-arcs, there is an automorphism mapping (u, u') to (v, v'). □

Distance transitivity is a symmetry property in that it is defined in terms of the existence of certain automorphisms of a graph. These automorphisms impose regularity properties on the graph, namely that the numbers a_i, b_i, and c_i are well-defined. There is an important combinatorial analogue to distance transitivity, which simply asks that the numerical regularity properties hold, whether or not the automorphisms exist. Given any graph X we can compute the distance partition from any vertex u, and it may occur "by accident" that every vertex in $X_i(u)$ is adjacent to a constant number of vertices in $X_{i-1}(u)$, $X_i(u)$, and $X_{i+1}(u)$, regardless of whether there are any automorphisms that force this to occur. (Looking forward to Section 9.3 this is saying that the distance partition is an equitable partition.) If the intersection array is well-defined and is the same for the distance partition from any vertex, then X is said to be *distance regular*.

It is immediate that any distance-transitive graph is distance regular, but the converse is far from true.

We provide one class of distance-regular graphs that includes many graphs that are not distance transitive. A *Latin square* of order n is an $n \times n$ matrix with entries from $\{1, \dots, n\}$ such that each integer i occurs exactly once in each row and exactly once in each column. Given an $n \times n$ Latin square L, we obtain a set of n^2 triples of the form

$$(i, j, L_{ij}).$$

Let $X(L)$ be the graph with these triples as vertices, and where two triples are adjacent if they agree on the first, second, or third coordinate. (In fact, two triples can agree on at most one coordinate.) Alternatively, we may view $X(L)$ as the graph whose vertices are the n^2 positions in L, and two "positions" are adjacent if they lie in the same row or column, or contain the same entry. The graph $X(L)$ has n^2 vertices, diameter two, and is regular with valency $3(n-1)$. It is distance regular, and is in general not distance transitive. The proof that $X(L)$ is distance regular is left as an exercise. We will encounter these graphs again in Chapter 10.

4.6 The Coxeter Graph

The Coxeter graph is a 28-vertex cubic graph with girth seven. It is shown in Figure 4.6. From the picture we see that it is constructed from circulants based on $\mathbb{Z}_7$. Start with the three circulants $X(\mathbb{Z}_7, \{1, -1\})$, $X(\mathbb{Z}_7, \{2, -2\})$, $X(\mathbb{Z}_7, \{3, -3\})$ and then add seven more vertices, joining each one to the same element of $\mathbb{Z}_7$ in each of the three circulants.

We describe another construction for the Coxeter graph, which identifies it as an induced subgraph of $J(7, 3, 0)$. The vertices of $J(7, 3, 0)$ are the 35 triples from the set $\Omega = \{1, \dots, 7\}$. Two triples are adjacent if they are disjoint, at distance two if they intersect in two points and at distance three if they have exactly one point in common. A *heptad* is a set of seven triples from Ω such that each pair of triples meet in exactly one point, and there is no point in all of them. In graph theoretic terms, a heptad is a set of seven vertices of $J(7, 3, 0)$ such that each pair of distinct vertices is at distance three. The following seven triples are an example of a heptad:

$$124,\ 235,\ 346,\ 457,\ 561,\ 672,\ 713.$$

This set of triples is invariant under the action of the 7-cycle (1234567); denote this permutation by σ. It is easy to verify that the four triples

$$357,\ 367,\ 567,\ 356$$

lie in distinct orbits under σ. The orbit of 356 is another heptad. The orbits of the first three triples are isomorphic, in the order given, to $X(\mathbb{Z}_7, \{1, -1\})$, $X(\mathbb{Z}_7, \{2, -2\})$, and $X(\mathbb{Z}_7, \{3, -3\})$, respectively. It is easy

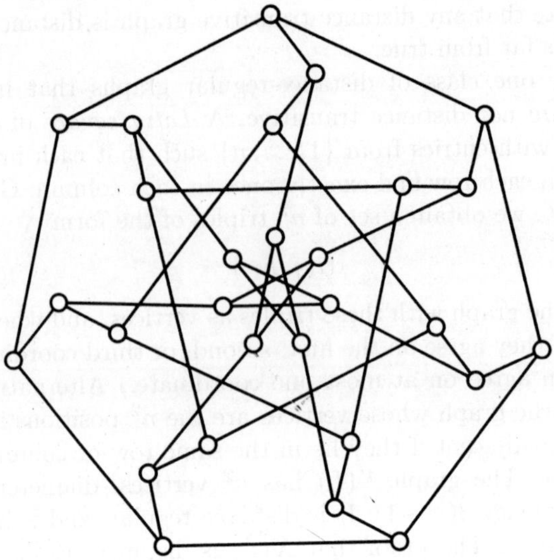

Figure 4.6. The Coxeter graph

to check that 357, 367 and 567 are the unique triples in their orbits that are disjoint from 124, and somewhat more tedious to see that no triple from one of these orbits is disjoint from a triple in one of the other two. Thus, by a minor miracle, we infer that the orbits of the triples

$$124, \ 357, \ 367, \ 567$$

induce a subgraph of $J(7,3,0)$ isomorphic to the Coxeter graph. We also see that the vertices not in this Coxeter graph form a heptad.

We use the embedding of the Coxeter graph in $J(7,3,0)$ to show that its girth is seven. First we determine the girth of $J(7,3,0)$.

Lemma 4.6.1 *The diameter of $J(7,3,0)$ is three and its girth is six.*

Proof. If we denote $J(7,3,0)$ by Y, then $Y_1(u)$ consists of the triples disjoint from u, while $Y_2(u)$ consists of the triples that meet u in two points, and $Y_3(u)$ consists of the triples that meet u in one point. Therefore, there are no edges in $Y_1(u)$ or $Y_2(u)$, so the girth of $J(7,3,0)$ is at least six. Since it is easy to find a six-cycle, its girth is exactly six. □

An easy argument shows that every triple from Ω that is not in a heptad is disjoint from precisely one triple of the heptad. Therefore, deleting a heptad from $J(7,3,0)$ results in a 28-vertex cubic graph X. To show that X has girth seven we must demonstrate that every heptad meets every six-cycle of $J(7,3,0)$. To this end, we characterize the six-cycles.

Lemma 4.6.2 *There is a one-to-one correspondence between six-cycles in $J(7,3,0)$ and partitions of Ω of the form $\{abc, de, fg\}$.*

Proof. The partition $\{abc, de, fg\}$ corresponds to the six-cycle ade, bfg, cde, afg, bde, cfg. To show that every six-cycle has this form, it suffices to consider six-cycles through 123. Without loss of generality we can assume that the neighbours of 123 in the six-cycle are 456 and 457. The vertex at distance three from 123 has one point in common with 123, say 1, and two in common with 456 and 457 and hence must be 145. This then determines the partition $\{167, 23, 45\}$, and it is straightforward to verify that the six-cycle must be of the type described above. □

Lemma 4.6.3 *Every heptad meets every six-cycle in* $J(7,3,0)$.

Proof. The seven triples of a heptad contain 21 pairs of points; since two distinct triples have only one point in common, these pairs must be distinct. Hence each pair of points from $\{1, \dots, 7\}$ lies in exactly one triple from the heptad. If the point i lies in r triples, then it lies in $2r$ pairs. Therefore, each point lies in exactly three triples.

Without loss of generality, we consider the six-cycle determined by the partition $\{123, 45, 67\}$. Each heptad has a triple of the form $a45$ and one of the form $b67$. At least one of a and b must be 1, 2, or 3, or else the two triples would meet in two points. Hence this six-cycle has a triple in common with every heptad. □

We will see in Section 5.9 that all heptads in $J(7,3,0)$ are equivalent under the action of Sym(7).

The automorphism group of the Coxeter graph is at least the size of the stabilizer in Sym(7) of the heptad. The heptad we used above is fixed by the permutations

$$(23)(47),\ (2347)(56),\ (235)(476),\ (1234567).$$

The first two permutations generate a group of order eight, so the group generated by all four permutations has order divisible by 8, 3, and 7, and therefore its order is at least 168.

This implies that the Coxeter graph is at least 2-arc transitive. In fact, there is an additional automorphism of order two (see Exercise 5.4), so its full automorphism group has size 336, and acts 3-arc regularly.

4.7 Tutte's 8-Cage

Another interesting cubic arc-transitive graph is Tutte's 8-cage on 30 vertices. In 1947 Tutte gave (essentially) the following two-sentence description of how to construct this graph. Take the cube and an additional vertex ∞. In each set of four parallel edges, join the midpoint of each pair of opposite edges by an edge, then join the midpoint of the two new edges by an edge, and finally join the midpoint of this edge to ∞. The resulting graph is shown in Figure 4.7.

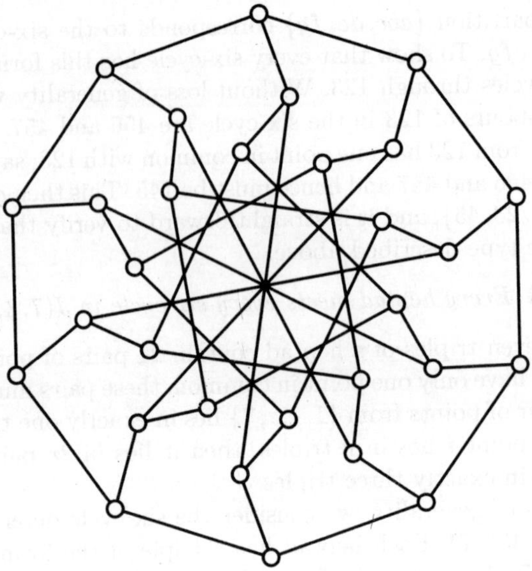

Figure 4.7. Tutte's 8-cage

An alternative description of this graph makes use of the edges and 1-factors of the complete graph K_6. There are fifteen edges in K_6. Each of these edges lies in three 1-factors, and as each 1-factor contains three edges, this implies that there are fifteen 1-factors.

Construct a bipartite graph T with the fifteen edges as one colour class and the fifteen 1-factors as the other, where each edge is adjacent to the three 1-factors that contain it. This is a cubic graph on 30 vertices, which is Tutte's 8-cage again. One advantage of this description is that is easy to see that Sym(6) acts as a group of automorphisms with the two parts of the bipartition as its two orbits.

However, we have not yet established why the two descriptions are equivalent. At the end of this section we sketch a proof that there is a unique bipartite cubic graph on 30 vertices with girth eight. First we will verify that T has girth eight. For reasons that we will reveal in Section 5.4, we do this by first establishing the following lemma.

Lemma 4.7.1 *Let F be a 1-factor of K_6 and let e be an edge of K_6 that is not contained in F. Then there is a unique 1-factor on e that contains an edge of F.*

Proof. Two edges of K_6 lie in a 1-factor if and only if they are disjoint, and two disjoint edges lie in a unique 1-factor. Since $e \notin F$, it meets two distinct edges of F, and hence is disjoint from precisely one edge of F, with which it lies in a unique 1-factor. □

Since T is bipartite, if its girth is less than eight, then it must be four or six. Since you should not just sit here reading, you may eliminate the possibility that the girth is four. Having completed that we shall now eliminate the possibility that the girth is six. Suppose to the contrary that there is a 6-cycle

$$e_1,\ F_1,\ e_2,\ F_2,\ e_3,\ F_3.$$

This implies that F_2 and F_3 are distinct 1-factors on e_3 that contain the edges e_2 and e_1, respectively, contradicting the previous lemma. Hence we are forced to conclude that the graph has girth at least eight. There are a number of ways by which you might show that the girth is equal to eight.

Now, we shall show that T is arc transitive. It is easy to see that if e_1 and e_2 are edges of K_6, then there is a permutation of Sym(6) mapping e_1 to e_2. It is also clear that the stabilizer of e_1 in Sym(6) is transitive on the three neighbours of e_1 in T. Therefore, we conclude that Sym(6) is transitive on 1-arcs starting at an "edge vertex." Similarly, Sym(6) is transitive on 1-arcs starting at a "1-factor vertex".

The sole remaining task is to show that there is an automorphism that exchanges the two classes of vertices, and this requires some preparation. Although this argument is quite long, it is not atypical for such computations done by hand.

A *1-factorization* of a graph is a partition of its edge set into 1-factors. Given a 1-factor F of K_6, there are six 1-factors that share an edge with F, and hence eight that are edge-disjoint from F. The union of two disjoint 1-factors is a 6-cycle, and hence the remaining edges of K_6 form a 3-prism (see Figure 4.8). It is straightforward to check that the 3-prism has four 1-factors and a unique 1-factorization (see Figure 4.8). Therefore, any two disjoint 1-factors lie in a unique 1-factorization. Counting triples $(F, G, \mathcal{F})$ where F and G are 1-factors contained in the 1-factorization $\mathcal{F}$, we see that there are six 1-factorizations of K_6. Since each 1-factor lies in the same number of 1-factorizations, this implies that each 1-factor lies in two 1-factorizations. There are fifteen pairs of distinct 1-factorizations, and so any two distinct 1-factorizations have a unique 1-factor in common.

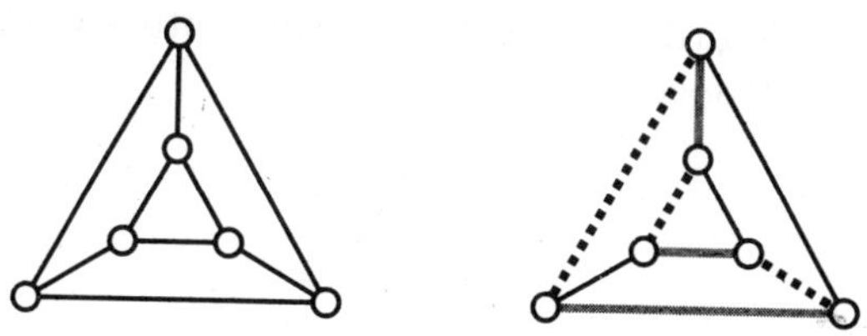

Figure 4.8. The 3-prism together with its unique 1-factorization

We will use the six 1-factorizations to determine a bijection between the edges of K_6 and the 1-factors of K_6, and show that this bijection is an automorphism of T. Arbitrarily label the six 1-factorizations of K_6 as $\mathcal{F}_1, \ldots, \mathcal{F}_6$. Then define a map ψ as follows. If $e = ij$ is an edge of K_6, then let $\psi(e)$ be the 1-factor that $\mathcal{F}_i$ and $\mathcal{F}_j$ have in common. The five edges of K_6 containing i are mapped by ψ to the five 1-factors contained in $\mathcal{F}_i$. If e and f are incident edges of K_6, then $\psi(e)$ and $\psi(f)$ are edge-disjoint 1-factors. Since there are only eight 1-factors disjoint from a given one, this shows that if e and f are not incident, then $\psi(e)$ and $\psi(f)$ have an edge in common.

Therefore, three independent edges of K_6 are mapped by ψ to three 1-factors, any two of which have an edge in common. Any such set must consist of the three 1-factors on a single edge. So, if $F = \{e, f, g\}$ is a 1-factor, then define $\psi(F)$ to be the edge of K_6 common to $\psi(e)$, $\psi(f)$, and $\psi(g)$.

All that remains is to show that ψ is an automorphism of T. Suppose that the edge e is adjacent to the 1-factor $F = \{e, f, g\}$. Then $\psi(F)$ is the edge that $\psi(e)$, $\psi(f)$, and $\psi(g)$ have in common. In particular, $\psi(F)$ is an edge in $\psi(e)$, and so $\psi(e) \sim \psi(F)$ in T. Consequently, ψ is an automorphism of T that swaps its two colour classes. Therefore, T is vertex transitive and hence arc transitive with an automorphism group of order at least $2 \times 6! = 1440$. If T is s-arc transitive, then $s \geq 5$, and Lemma 4.1.3 (or Theorem 4.3.3) yields that $s = 5$. From Lemma 4.5.3 we conclude that T is distance transitive with diameter four.

Now we sketch a proof that there is a unique cubic bipartite graph on 30 vertices with girth eight, thus showing that both descriptions above are equivalent. So let X be a cubic bipartite graph on 30 vertices with girth eight. Let v be any vertex of X and consider the graph induced by $X_3(v) \cup X_4(v)$. The eight vertices of $X_4(v)$ have valency three, and the twelve vertices of $X_3(v)$ all have valency two and join two vertices of $X_4(v)$. Therefore, this graph is a subdivision of a cubic graph Y, that has girth four. The fact that X has girth eight implies that it must be possible to partition $E(Y)$ into pairs of edges at distance three. The cube is the unique graph on eight vertices with these properties, and therefore $X_3(v) \cup X_4(v)$ is the subdivision graph of the cube. It is now straightforward to check that the only way to extend a subdivision of the cube to a bipartite cubic graph on 30 vertices with girth eight is by following Tutte's original description of the 8-cage (see Exercise 15).

Exercises

1. Show that the graph $J(2k+1, k, 0)$ is at least 2-arc transitive.

2. Prove that if X is a spindle or a bicycle, then $X^{(1)}$ is strongly connected.

3. Define the *directed line graph* $DL(X)$ of a directed graph X to be the directed graph with the arcs of X as its vertices. If α and β are arcs in X, then (α, β) is an arc in $DL(X)$ if and only if $\text{head}(\alpha) = \text{tail}(\beta)$ and $\text{tail}(\alpha) \neq \text{head}(\beta)$. (Thus (α, β) is an arc if and only if α is the tail and β the head of a 2-arc in X.) Prove that if $s \geq 1$, then $X^{(s+1)}$ is the directed line graph of $X^{(s)}$.

4. Let X be a vertex-transitive cubic graph on n vertices and let G be its automorphism group. If 3 divides the order of the stabilizer G_u of a vertex u, show that X is arc transitive.

5. Show directly (without using Tutte's theorem) that the automorphism group of the Petersen graph has one orbit on 3-arcs, and hence that P is 3-arc transitive.

6. Show that each edge of the Petersen graph lies in exactly two 1-factors. Conclude that it contains precisely six 1-factors, and that these 1-factors are equivalent under the action of the automorphism group. Deduce from this that the Petersen graph does not have a Hamilton cycle.

7. Find the intersection arrays of the Petersen graph, the Coxeter graph, and Tutte's 8-cage.

8. Prove that $J(v, k, k-1)$ is distance transitive, and determine its intersection array.

9. Prove that $J(2k+1, k, 0)$ is distance transitive, and determine its intersection array.

10. The multiplication table of a group is a Latin square. Show that its Latin square graph is a Cayley graph.

11. There are two distinct groups of order four, and their multiplication tables can be viewed as Latin squares. Show that the graphs of these Latin squares are not isomorphic. (One approach is to determine the value of $\chi(X)$ for both graphs.)

12. Show that an automorphism of Coxeter's graph that fixes two vertices at distance three is necessarily the identity, and conclude that Coxeter's graph is not 4-arc transitive.

13. Show that a distance-transitive graph with girth at least five is 2-arc transitive. Determine a relation between the girth and degree of arc transitivity.

14. Show that an s-arc transitive graph with girth $2s-1$ has diameter s and is distance transitive.

15. Let the edges of the graph K_6 be given as pairs (i, j) where $1 \leq i < j \leq 6$. By labelling the vertices of a cube with the edges $(1,3)$, $(1,4)$,

$(1,5)$, $(1,6)$, $(2,3)$, $(2,4)$, $(2,5)$, $(2,6)$ and labelling the point ∞ with $(1,2)$, reconcile Tutte's one-sentence description of the 8-cage with the description in terms of edges and 1-factors of K_6.

Notes

Theorem 4.2.2 is based on unpublished notes by D. G. Wagner.

Biggs [1] shows that there are exactly six 1-factors in the Petersen graph, all equivalent under the action of its automorphism group. He also shows that the Coxeter graph has exactly 84 1-factors, all equivalent under its automorphism group. Deleting any one of them leaves $2C_{14}$, whence the Coxeter graph is not hamiltonian, but is 1-factorable. Coxeter gave a geometric construction for Tutte's graph, and so it is sometimes referred to as the Tutte–Coxeter graph.

Biggs [2] provides a proof of Tutte's theorem on arc-transitive cubic graphs, which loosely follows Tutte's treatment. Weiss provides a more succinct proof of a slightly more general result in [3]. (The advantages of his approach will be lost if you do not read German, unfortunately.)

References

[1] N. Biggs, *Three remarkable graphs*, Canad. J. Math., 25 (1973), 397–411.

[2] ——, *Algebraic Graph Theory*, Cambridge University Press, Cambridge, second edition, 1993.

[3] R. M. Weiss, *Über s-reguläre Graphen*, J. Combinatorial Theory Ser. B, 16 (1974), 229–233.

5

Generalized Polygons and Moore Graphs

A graph with diameter d has girth at most $2d+1$, while a bipartite graph with diameter d has girth at most $2d$. While these are very simple bounds, the graphs that arise when they are met are particularly interesting. Graphs with diameter d and girth $2d+1$ are known as Moore graphs. They were introduced by Hoffman and Singleton in a paper that can be viewed as one of the prime sources of algebraic graph theory. After considerable development, the tools they used in this paper led to a proof that a Moore graph has diameter at most two. They themselves proved that a Moore graph of diameter two must be regular, with valency 2, 3, 7, or 57. We will provide the machinery to prove this last result in our work on strongly regular graphs in Chapter 10.

Bipartite graphs with diameter d and girth $2d$ are known as generalized polygons. They were introduced by Tits in fundamental work on the classification of finite simple groups. The complete bipartite graphs, with diameter two and girth four, are the only examples we have met already. Surprisingly, generalized polygons are related to classical geometry; in fact, a generalized polygon with diameter three is another manifestation of a projective plane. When $d=4$ they are known as generalized quadrangles, and many of the known examples are related to quadrics in projective space.

In this chapter we consider these two classes of graphs. We develop some of the basic theory of generalized polygons proving that "nondegenerate" generalized polygons are necessarily semiregular bipartite graphs. We present the classical examples of generalized triangles and generalized quadrangles, and the smallest generalized hexagons. We show that the Moore graphs are distance regular, which is surprising, because it is not even immediate that they are regular. We give a construction of the Hoffman–Singleton graph, the unique Moore graph of diameter two and valency seven, which along with the Petersen graph and the 5-cycle completes the list of known Moore graphs of diameter two. The chapter concludes with a brief introduction to designs, which provide another source of highly structured graphs.

5.1 Incidence Graphs

An *incidence structure* consists of a set $\mathcal{P}$ of points, a set $\mathcal{L}$ of lines (disjoint from $\mathcal{P}$), and a relation

$$I \subseteq \mathcal{P} \times \mathcal{L}$$

called *incidence*. If $(p, L) \in I$, then we say that the point p and the line L are *incident*. If $\mathcal{I} = (\mathcal{P}, \mathcal{L}, I)$ is an incidence structure, then its *dual* incidence structure is given by $\mathcal{I}^* = (\mathcal{L}, \mathcal{P}, I^*)$, where $I^* = \{(L,p) \mid (p, L) \in I\}$. Informally, this simply corresponds to interchanging the names of "points" and "lines."

The *incidence graph* $X(\mathcal{I})$ of an incidence structure $\mathcal{I}$ is the graph with vertex set $\mathcal{P} \cup \mathcal{L}$, where two vertices are adjacent if and only if they are incident. The incidence graph of an incidence structure is a bipartite graph. Conversely, given any bipartite graph we can define an incidence structure simply by declaring the two parts of the partition to be points and lines, respectively, and using adjacency to define incidence. Since we can choose either half of the partition to be the points, any bipartite graph determines a dual pair of incidence structures. This shows us that the definition of incidence structure is not very strong, and to get interesting incidence structures (and hence interesting graphs) we need to impose some additional conditions.

A *partial linear space* is an incidence structure in which any two points are incident with at most one line. This implies that any two lines are incident with at most one point.

Lemma 5.1.1 *The incidence graph X of a partial linear space has girth at least six.*

Proof. If X contains a four-cycle p, L, q, M, then p and q are incident to two lines. Since the girth of X is even and not four, it is at least six. □

When referring to partial linear spaces we will normally use geometric terminology. Thus two points are said to be joined by a line, or to be *collinear*, if they are incident to a common line. Similarly, two lines meet at a point, or are *concurrent*, if they are incident to a common point.

An automorphism of an incidence structure $(\mathcal{P}, \mathcal{L}, I)$ is a permutation σ of $\mathcal{P} \cup \mathcal{L}$ such that $\mathcal{P}^\sigma = \mathcal{P}$, $\mathcal{L}^\sigma = \mathcal{L}$, and

$$(p^\sigma, L^\sigma) \in I \Longleftrightarrow (p, L) \in I.$$

This yields an automorphism of the incidence graph that preserves the two parts of the bipartition. An incidence-preserving permutation σ of $\mathcal{P} \cup \mathcal{L}$ such that $\mathcal{P}^\sigma = \mathcal{L}$ and $\mathcal{L}^\sigma = \mathcal{P}$ is called a *duality*. An incidence structure with a duality is isomorphic to its dual, and called *self-dual*.

5.2 Projective Planes

One of the most interesting classes of incidence structures is that of projective planes. A *projective plane* is a partial linear space satisfying the following three conditions:

(1) Any two lines meet in a unique point.
(2) Any two points lie in a unique line.
(3) There are three pairwise noncollinear points (a *triangle*).

The first two conditions are duals of each other, while the third is self-dual, so the dual of a projective plane is again a projective plane.

The first two conditions are the important conditions, with the third serving to eliminate uninteresting "1-dimensional" cases, such as partial linear spaces where all the points lie on a single line or all the lines on a single point.

Finite geometers normally use a stronger nondegeneracy condition, insisting on the existence of a *quadrangle* (four points, no three collinear). Figure 5.1 shows a projective plane that a geometer would regard as degenerate. The reasons for this will become apparent in Section 5.6.

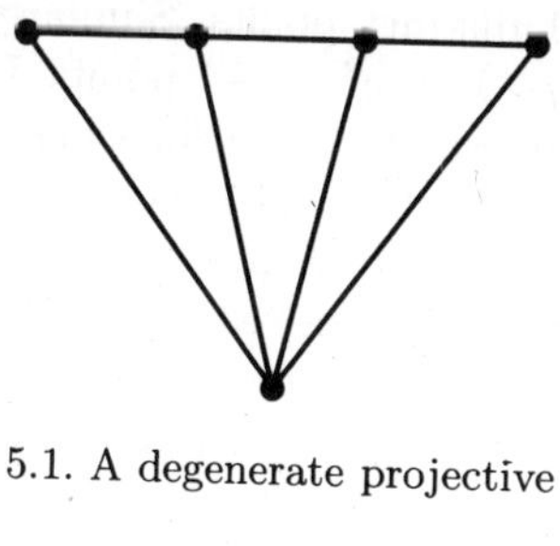

Figure 5.1. A degenerate projective plane

Theorem 5.2.1 *Let $\mathcal{I}$ be a partial linear space that contains a triangle. Then $\mathcal{I}$ is a (possibly degenerate) projective plane if and only if its incidence graph $X(\mathcal{I})$ has diameter three and girth six.*

Proof. Let $\mathcal{I}$ be a projective plane that contains a triangle. Any two points lie at distance two in $X(\mathcal{I})$, similarly for any two lines. Now, consider a line L and a point p not on L. Any line M through p must meet L in a point p', and thus L, p', M, p is a path of length three from L to p. Hence any two vertices are at distance at most three, and the existence of the triangle guarantees one pair at distance exactly three, so the diameter of $X(\mathcal{I})$ is three. Since $\mathcal{I}$ is a partial linear space, the girth of $X(\mathcal{I})$ is at least six, and the existence of the triangle guarantees that it is exactly six.

Conversely, let $X(\mathcal{I})$ be the incidence graph of an incidence structure and suppose that it has diameter three and girth six. Then one half of the bipartition corresponds to the points of $\mathcal{I}$ and the other to the lines of $\mathcal{I}$. Any two points are at an even distance from each other, and since this distance is at most three, it must be two. There must be a unique path of length two between the two points; else there would be a four-cycle in $X(\mathcal{I})$. Hence there is a unique line between any two points. A dual argument shows that any two lines meet in a unique point, and hence we have a projective plane. □

5.3 A Family of Projective Planes

Let V be the three-dimensional vector space over the field $\mathbb{F}$ with q elements. We can define an incidence structure $PG(2, q)$ as follows: The *points* of $PG(2, q)$ are the 1-dimensional subspaces of V, and the *lines* are the 2-dimensional subspaces of V. We say that a point p is *incident* with a line L if the 1-dimensional subspace p is contained in the 2-dimensional subspace L. A k-dimensional subspace of V contains $q^k - 1$ nonzero vectors. Therefore, a line L contains $q^2 - 1$ nonzero vectors, while each 1-dimensional subspace contains $q - 1$ nonzero vectors. Therefore, each line contains $(q^2 - 1)/(q - 1) = q + 1$ distinct points. Similarly, the entire projective plane contains $(q^3 - 1)/(q - 1) = q^2 + q + 1$ points. It is also not hard to see that there are $q^2 + q + 1$ lines, with $q + 1$ lines passing through each point.

Each point may be represented by a vector a in V, where a and λa represent the same point if $\lambda \neq 0$. A line can be represented by a pair of linearly independent vectors, or by a vector a^T. Here the understanding is that a line is the subspace of dimension two formed by the vectors x such $a^T x = 0$. Of course, if $\lambda \neq 0$, then λa^T and a^T determine the same line. Then the point represented by a vector b lies on the line represented by a^T if and only if $a^T b = 0$.

Two one-dimensional subspaces of V lie in a unique two-dimensional subspace of V, so there is a unique line joining two points. Two two-dimensional subspaces of V intersect in a one-dimensional subspace, so any two lines meet in a unique point. Therefore, $PG(2, q)$ is a projective plane.

By Theorem 5.2.1, the incidence graph X of $PG(2, q)$ is a bipartite graph with diameter three and girth six. It has $2(q^2+q+1)$ vertices and is regular of valency $q+1$. However, we can say more. We aim to prove that it is 4-arc transitive. To start with we must find some automorphisms of it.

We denote the group of all invertible 3×3 matrices over $\mathbb{F}$ by $GL(3, q)$. It is called the 3-dimensional *linear group* over $\mathbb{F}$. Each element of it permutes the nonzero vectors in V and maps subspaces to subspaces, therefore giving rise to an automorphism of X. By elementary linear algebra, there is an invertible matrix that maps any ordered basis to any other ordered basis, so $GL(3, q)$ acts transitively on the set of all ordered bases of V.

Let $p \vee q$ denote the unique line joining the points p and q. If p, q, and r are three noncollinear points, then

$$p,\ p\vee q,\ q,\ q\vee r,\ r,\ p\vee r$$

is a hexagon in X. The sequence

$$(p,\ p\vee q,\ q,\ q\vee r,\ r)$$

is a 4-arc in X, and so it follows that $\mathrm{Aut}(X)$ acts transitively on the 4-arcs that start at a "point-vertex" of X. The same argument shows that $\mathrm{Aut}(X)$ acts transitively on 4-arcs starting at a "line-vertex" of X. Therefore, to show that $\mathrm{Aut}(X)$ is 4-arc transitive, it remains only to prove that there is an automorphism of X that swaps point-vertices and line-vertices of X.

This automorphism is easy to describe. For each vector a, it swaps the point represented by a with the line represented by a^T. Since $a^Tb = 0$ if and only if $b^Ta = 0$, this maps adjacent vertices to adjacent vertices, and hence is an example of a duality.

Given this, it follows that X is a 4-arc transitive graph. In addition, from Lemma 4.5.3, X is distance transitive.

5.4 Generalized Quadrangles

A second interesting class of incidence structures is provided by generalized quadrangles. A *generalized quadrangle* is a partial linear space satisfying the following two conditions:

(1) Given any line L and a point p not on L there is a unique point p' on L such that p and p' are collinear.

(2) There are noncollinear points and nonconcurrent lines.

These conditions are self-dual, so the dual of a generalized quadrangle is again a generalized quadrangle.

Once again, the first condition is the important one, with the second condition serving to eliminate the uninteresting "1-dimensional" cases with all points on one line or all lines through one point.

We have already seen a generalized quadrangle. Lemma 4.7.1 showed that the incidence structure defined on the edges and 1-factors of K_6 is a generalized quadrangle with Tutte's 8-cage as its incidence graph.

Two simple generalized quadrangles, called the *grid* and its dual are shown in Figure 5.2. In a grid, every point is on two lines, while in a dual grid, every line contains two points. For reasons that will become apparent in Section 5.6, finite geometers also sometimes regard these as degenerate.

Figure 5.2. A grid and a dual grid

Theorem 5.4.1 *Let $\mathcal{I}$ be a partial linear space that contains noncollinear points and nonconcurrent lines. Then $\mathcal{I}$ is a generalized quadrangle if and only if its incidence graph $X(\mathcal{I})$ has diameter four and girth eight.*

Proof. Let $\mathcal{I}$ be a generalized quadrangle, and consider the distances in $X(\mathcal{I})$ from a point p. A line is distance one from p if it contains p, and at distance three otherwise (by the condition defining a generalized quadrangle). A point is at distance two from p if it is collinear with p, and at distance four otherwise. The existence of noncollinear points guarantees the existence of a pair of points at distance four. A dual argument for lines completes the argument, showing that the diameter of $X(\mathcal{I})$ is indeed four.

The girth of $X(\mathcal{I})$ is at least six. If it were exactly six, then the point and line opposite each other in a six-cycle would violate the condition defining a generalized quadrangle. To show that there is an 8-cycle we let p and q be two noncollinear points. Then there is a line L_p on p that does not contain q, and a line L_q on q that does not contain p. But then there is a unique point on L_p incident to q and a unique point on L_q incident to p. These eight elements form a cycle of length eight in the incidence graph, and hence the girth is eight.

Conversely, suppose $X(\mathcal{I})$ is the incidence graph of some partial linear space, and that it has diameter four and girth eight. Then one part of the bipartition corresponds to the points of $\mathcal{I}$, and the other part to the lines of $\mathcal{I}$. Consider a line L and point p at distance three. Since the girth is eight, there is a unique path L, p', L', p from L to p. This provides the unique point p' satisfying the condition defining a generalized quadrangle. □

5.5 A Family of Generalized Quadrangles

In this section we describe an infinite class of generalized quadrangles. The smallest member of this family has Tutte's graph as its incidence graph.

Let V be the vector space of dimension four over the field $\mathbb{F}$ with order q. The *projective space* $PG(3, q)$ is the system of one-, two- and three-dimensional subspaces of V. We will refer to these as the points, lines, and planes, respectively of $PG(3, q)$. There are $q^4 - 1$ nonzero vectors in V, and each 1-dimensional subspace contains $q - 1$ nonzero vectors, so there are exactly $(q^4 - 1)/(q - 1) = (q + 1)(q^2 + 1)$ points. We will construct an incidence structure using all of these points, but just some of the lines of $PG(3, q)$.

Let H be the matrix defined by

$$H = \begin{pmatrix} 0 & 1 & 0 & 0 \\ -1 & 0 & 0 & 0 \\ 0 & 0 & 0 & 1 \\ 0 & 0 & -1 & 0 \end{pmatrix}.$$

(If the field has characteristic 2 or, equivalently, q is even, then $-1 = 1$.) A subspace S of V is *totally isotropic* if $u^T H v = 0$ for all u and v in S. It is easy to see that $u^T H u = 0$ for all u, so all the 1-dimensional subspaces of V are totally isotropic. We will be concerned with the 2-dimensional totally isotropic subspaces of V. Our first task is to count them. A 2-dimensional subspace of V spanned by u and v is totally isotropic if and only if $u^T H v = 0$. For a nonzero vector u, define the set $u^\perp$ as follows:

$$u^\perp = \{v \in V \mid u^T H v = 0\}.$$

The determinant of H is one, so H is invertible, and the vector $u^T H$ is nonzero. Since $u^\perp$ consists of all the vectors orthogonal to $u^T H$, it is a 3-dimensional subspace of V that contains u. We count the number of pairs of vectors u, v such that $\langle u, v\rangle$ is a 2-dimensional totally isotropic subspace. There are $q^4 - 1$ choices for the vector u and then $q^3 - q$ choices for a vector v that it is in $u^\perp$ but not in the span of u. Therefore, there are $(q^4 - 1)(q^3 - q)$ pairs of vectors spanning 2-dimensional totally isotropic subspaces. Each 2-dimensional subspace is spanned by $(q^2 - 1)(q^2 - q)$ pairs of vectors, so the total number of 2-dimensional totally isotropic subspaces is $(q^2 + 1)(q + 1)$.

Using the language of geometry, we say that $PG(3, q)$ contains $(q^2+1)(q+1)$ totally isotropic points and $(q^2 + 1)(q + 1)$ totally isotropic lines. A 2-dimensional space contains $q+1$ subspaces of dimension one, so each totally isotropic line contains $q + 1$ totally isotropic points. Because the numbers of points and lines are equal, this implies that each totally isotropic point is contained in $q+1$ totally isotropic lines. Now, let $W(q)$ be the incidence structure whose points and lines are the totally isotropic points and totally isotropic lines of $PG(3, q)$.

Lemma 5.5.1 *Let $W(q)$ be the point/line incidence structure whose points and lines are the totally isotropic points and totally isotropic lines of $PG(3,q)$. Then $W(q)$ is a generalized quadrangle.*

Proof. We need to prove that given a point p and a line L not containing p, there is a unique point on L that is collinear with p. Suppose that the point p is spanned by the vector u. Any point collinear with p is spanned by a vector in $u^\perp$. The 3-dimensional subspace $u^\perp$ intersects the 2-dimensional subspace L in a subspace of dimension one, and hence there is a unique point on L that is collinear with p. □

If X is the incidence graph of $W(q)$, then it is a bipartite graph on $2(q^2+1)(q+1)$ vertices that is regular with valency $q+1$. By Theorem 5.4.1, it has diameter four and girth eight. As will become apparent in Section 5.6, it is also distance regular.

Applying the construction to the field of order two, we obtain a generalized quadrangle with fifteen points and fifteen lines; this is the same as the generalized quadrangle on the edges and 1-factors of K_6.

The matrix H we used can be replaced by any invertible 4×4 matrix over $\mathbb{F}$ with all diagonal entries zero such that $H^T = -H$. However, this does not change the generalized quadrangle that results.

Although the construction described in this section yields generalized quadrangles that are regular, it should be noted that there are many that are not regular. We will not present any general constructions, although an example will arise later.

5.6 Generalized Polygons

In Section 5.2 and Section 5.4 we saw that two classes of interesting (and highly studied) incidence structures are equivalent to bipartite graphs with diameter d and girth $2d$, for $d = 3$ and $d = 4$. This motivates us to define a *generalized polygon* to be a finite bipartite graph with diameter d and girth $2d$. When it is important to specify the diameter, a generalized polygon of diameter d is called a generalized d-gon, and the normal names for small polygons (triangle for 3-gon, quadrangle for 4-gon, etc.) are used.

A vertex in a generalized polygon is called *thick* if its valency is at least three. Vertices that are not thick are *thin*. A generalized polygon is called *thick* if all its vertices are thick. Although on the face of it the definition of a generalized polygon is not very restrictive, we will show that the thick generalized polygons are regular or semiregular, and that the generalized polygons that are not thick arise purely as subdivisions of generalized polygons.

The argument proceeds by a series of simple structural lemmas. The first such lemma is a trivial observation, but we will use it repeatedly.

Lemma 5.6.1 *If $d(v, w) = m < d$, then there is a unique path of length m from v to w.* □

Lemma 5.6.2 *If $d(v, w) = d$, then v and w have the same valency.*

Proof. Since X is bipartite of diameter d, any neighbour v' of v has distance $d-1$ from w. Therefore, there is a unique path of length $d-1$ from v' to w that contains precisely one neighbour of w. Each such path contains a different neighbour of w, and therefore w has at least as many neighbours as v. Similarly, v has at least as many neighbours as w, and hence they have equal valency. □

Lemma 5.6.3 *Every vertex in X has valency at least two.*

Proof. Let C be a cycle of length $2d$ in X. Clearly, the vertices of C have valency at least two. Let x be a vertex not on C, let P be the shortest path joining x to C, and denote the length of P by i. Travelling around the cycle for $d - i$ steps we arrive at a vertex x' at distance d from x. Then x has the same valency as x', which is at least two. □

Lemma 5.6.4 *Any two vertices lie in a cycle of length $2d$.*

Proof. Let v and w be any two vertices of X. Let P be the shortest path between them. By repeatedly choosing any neighbour of an endpoint of P not already in P, we can extend P to a geodesic path of length d with endpoints x and y. Then x has a neighbour x' not in P, and hence it has distance $d - 1$ from y, and by following the unique path of length $d - 1$ from x' to y we extend P into a cycle of length $2d$ as required. □

The next series of lemmas shows that generalized polygons that are not thick are largely trivial modifications of those that are thick.

Lemma 5.6.5 *Let C be a cycle of length $2d$. Then any two vertices at the same distance in C from a thick vertex in C have the same valency.*

Proof. Let v be a thick vertex contained in C and let w be its antipode in C (that is, the unique vertex in C at distance d from v). Now, because v is thick, it has at least one further neighbour v', and hence there is a path P from v' to w that is disjoint from C except at w. Therefore, C together with P forms three internally vertex-disjoint paths of length d from v to w. Consider two vertices v_1, v_2 in C both at distance h from v. Let x be the vertex in P at distance $d - h$ from v. Then x is at distance d from both v_1 and v_2, and hence v_1 and v_2 both have the same valency as x, and so they have equal valencies. □

Lemma 5.6.6 *The minimum distance k between any pair of thick vertices in X is a divisor of d. If d/k is odd, then all the thick vertices have the same valency; if it is even, then the thick vertices share at most two valencies. Moreover, any vertex at distance k from a thick vertex is itself thick.*

Proof. Let v and w be two thick vertices of X such that $d(v, w) = k$. and let x be any other thick vertex of X. By extending the path from x to the closer of v and w, we can form a cycle C of length $2d$ containing v, w and x. Repeatedly applying the previous lemma to the thick vertices of C starting at v, we see that every kth vertex of C is thick. Since the antipode v' of v in C is thick, k must divide d. Again using the previous lemma we see that every second thick vertex in C has the same valency, and therefore every thick vertex in C has the same valency as either v or w. Thus our arbitrarily chosen thick vertex x has one of these two valencies, and the set of all thick vertices shares at most two valencies. If d/k is odd, then v' has the same valency as w, so the valency of v is equal to the valency of w. However, if d/k is even, then v and w may have different valencies. Finally, consider any vertex x' at distance k from x. If $x' \in C$, then the argument above shows that it is thick. If $x' \notin C$, then we can form a new cycle of length $2d$ that includes x', x, and one of the vertices of C at distance k from x. Then repeating the above argument with the new cycle yields that x' is itself thick. □

We have already defined the subdivision graph $S(X)$ as being the graph obtained from X by putting a vertex in the middle of each edge. We could also regard this as replacing each edge by a path of length 2. Taking this point of view we define the k-fold subdivision of a graph X to be the graph obtained from X by replacing each edge by a path of length k.

Theorem 5.6.7 *A generalized polygon X that is not thick is either a cycle, the k-fold subdivision of a multiple edge, or the k-fold subdivision of a thick generalized polygon.*

Proof. If X has no thick vertices at all, then it is a cycle. Otherwise, the previous lemma shows that any path between two thick vertices of X has length a multiple of k with every kth vertex being thick and the remainder thin. Therefore, we can define a graph X' whose vertices are the thick vertices of X, and where two vertices are adjacent in X' if they are joined by a path of length k in X. Clearly, X is the k-fold subdivision of X'. If $k = d$, then two thick vertices at maximum distance are joined by a collection of k-vertex paths of thin vertices. This collection of paths contains all the vertices of X, so X contains only two thick vertices and is just a subdivided multiple edge. (If we are willing to accept a multiple edge as a thick generalized 2-gon, then we eliminate the necessity for this case altogether.) If $k < d$, then X' has diameter $d' := d/k$ because a path of length d between two thick vertices in X is a k-fold subdivision of a path of length d' between two vertices of X'. Similarly, a cycle of length $2d$ in X is a k-fold subdivision of a cycle of length $2d/k$ in X'. Therefore, X' has diameter d' and girth $2d'$. It is clear that X' must be bipartite, for if it contained an odd cycle, then any k-fold subdivision of such a cycle would have a thin vertex at distance at least $kd' + 1$ from some thick vertex, contradicting the fact that X has diameter d. Therefore, X' is a thick generalized polygon. □

Therefore, the study of generalized polygons reduces to the study of thick generalized polygons, with the remainder being considered the degenerate cases. The degenerate projective plane of Figure 5.1 is a 3-fold subdivision of a multiple edge. The grids and dual grids are 2-fold subdivisions of the complete bipartite graph, which is a generalized 2-gon.

Although the proofs of the main results about thick generalized polygons are beyond our scope, the results themselves are easy to state. The following famous theorem shows that in a thick generalized polygon, the diameter d is severely restricted.

Theorem 5.6.8 (Feit and Higman) *If a generalized d-gon is thick, then $d \in \{3, 4, 6, 8\}$.* □

We have already seen examples of thick generalized triangles ($d = 3$) and thick generalized quadrangles ($d = 4$). In fact generalized triangles and generalized quadrangles exist in great profusion. Generalized hexagons and octagons do exist, but only a few families are known. Unfortunately, even the simplest of these families are difficult to describe.

Since a projective plane is a thick generalized triangle, it is necessarily regular. If all the vertices have valency $s+1$, then we say that the projective plane has order s. The other thick generalized polygons may be regular or semiregular. If the valencies of the vertices of a thick generalized polygon X are $s+1$ and $t+1$, then X is said to have order (s, t) (where s may equal t).

We leave as an exercise the task of establishing the following result.

Lemma 5.6.9 *If a generalized polygon is regular, then it is distance regular.* □

The order of a thick generalized polygon satisfies certain inequalities due to Higman and Haemers. We will prove the first of these later, as Lemma 10.8.3.

Theorem 5.6.10 *Let X be a thick generalized d-gon of order (s, t).*

(a) *If $d = 4$, then $s \le t^2$ and $t \le s^2$.*
(b) *If $d = 6$, then st is a square and $s \le t^3$ and $t \le s^3$.*
(c) *If $d = 8$, then $2st$ is a square and $s \le t^2$ and $t \le s^2$.* □

Note that it is possible to take a generalized polygon of order (s, s) and subdivide each edge exactly once to form a generalized polygon of order $(1, s)$. Therefore, it is possible to have a generalized 12-gon that is neither thick nor a cycle.

5.7 Two Generalized Hexagons

Although it is known that an infinite number of generalized hexagons exist, it is not straightforward to present an elementary construction of an infinite family. Therefore, we content ourselves with a construction of the smallest thick generalized hexagon.

The smallest thick generalized hexagon has order $(2,2)$, and hence is a cubic graph with girth 6 and diameter 12. By Exercise 5 it is distance regular with intersection array

$$\{3,2,2,2,2,2;1,1,1,1,1,3\}.$$

Given the intersection array we can count the number of vertices in each cell of the distance partition from any vertex u. For example, it is clear that $|X_1(u)| = 3$. Therefore, there are six edges between $X_1(u)$ and $X_2(u)$, and since each vertex of $X_2(u)$ is adjacent to one vertex in $X_1(u)$, we conclude that $|X_2(u)| = 6$. Continuing similarly we see that the cells of the distance partition from u have 1, 3, 6, 12, 24, 48, and 32 vertices, respectively.

Lemma 5.7.1 *If X is a generalized hexagon of order $(2,2)$, then the graph $X_5(u) \cup X_6(u)$ is the subdivision $S(Y)$ of a cubic graph Y on 32 vertices.*

Proof. It is straightforward to confirm that the 48 vertices of $X_5(u)$ are each adjacent to two vertices of $X_6(u)$, and each vertex of $X_6(u)$ is adjacent to three from $X_5(u)$. □

We will describe the generalized hexagon by giving the cubic graph Y on 32 vertices, and explaining which vertex of $X_5(u)$ subdivides each edge of Y. First we give a simple encoding of the 48 vertices in $X_5(u)$. Let the three vertices adjacent to u be called r, g, and b. Then each of these has two neighbours in $X_2(u)$: We call them $r0$, $r1$, $b0$, $b1$, $g0$, and $g1$. Similarly, we denote the two neighbours of $r0$ in $X_3(u)$ by $r00$ and $r01$. Continuing in this fashion, every vertex of $X_5(u)$ is labelled by a word of length 5 with first entry r, g, or b and whose remaining four entries are binary.

Lemma 5.7.2 *For $c \in \{r,g,b\}$, the 16 edges of Y subdivided by the 16 vertices of $X_5(u)$ with first entry c form a one-factor of Y.*

Proof. The distance from $c \in \{r,g,b\}$ to any vertex of $X_5(u)$ with first entry c is four, and so there is a path of length at most eight between any two such vertices. Since X has no cycles of length 10, two such vertices cannot subdivide incident edges of Y. □

Figure 5.3 shows a bipartite cubic graph with 32 vertices, along with a 1-factorization given by the three different edge colours. This graph is drawn on the torus, but in an unusual manner. Rather than identifying points on the opposite sides of a square, this diagram identifies points on the opposite sides of a hexagon.

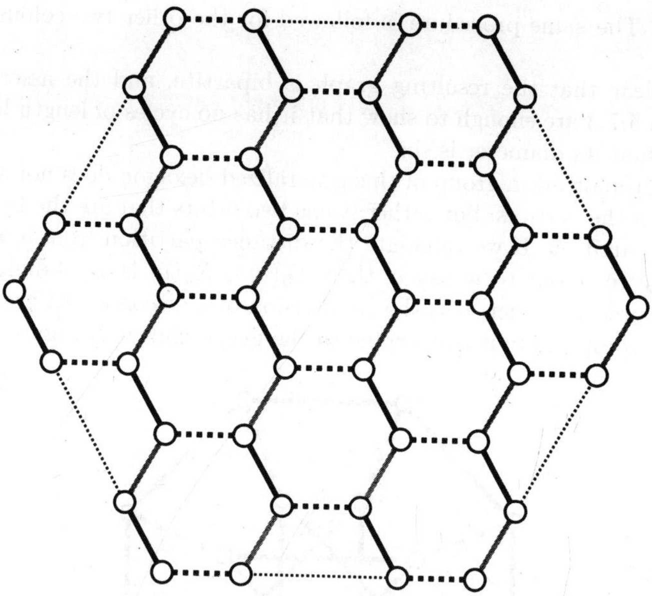

Figure 5.3. Building block for the generalized hexagon

Temporarily define the distance between two edges of a graph Z to be the distance that the corresponding vertices have in the subdivision $S(Z)$. (Thus incident edges of Y are deemed to have distance two.)

Theorem 5.7.3 *Let Y be the graph of Figure 5.3 and let R be the set of edges in one of the colour classes. Then for every edge $e \in R$, there is a unique edge $e' \in R$ at distance* 10 *from* e. *Moreover,*

(a) *There is a unique partition of the eight pairs $\{e, e'\}$ into four quartets of edges with pairwise distance at least eight, and*

(b) *There is a unique partition of these four quartets into two octets of edges with pairwise distance at least six.*

Proof. A few minutes with a photocopy of Y and a pencil will be far more convincing than any written proof, so this is left as an exercise. □

Now, it should be clear how we will subdivide the edges of Y to form a generalized hexagon. The edges in R are assigned to the vertices of $X_5(u)$ with first element r. The two octets of edges are assigned to the two octets of vertices whose codes agree in the first two positions, the four quartets of edges are assigned to the quartets of vertices whose codes agree in the first three positions, and the eight pairs of edges are assigned to the pairs of vertices whose codes agree in the first four positions. Then the two edges of a pair are subdivided arbitrarily by the two vertices to which the pair is assigned. The same procedure is followed for the other two colour classes of edges.

It is clear that the resulting graph is bipartite, and the assertions of Theorem 5.7.3 are enough to show that it has no cycles of length less than 12, and that its diameter is six.

The automorphism group of this generalized hexagon does not act transitively on the vertices, but rather it has two orbits that are the two halves of the bipartition. If we calculate the distance partition from a vertex of the opposite colour to u, say v, then $X_5(v) \cup X_6(v)$ is a subdivision of a different graph. In fact, it is the subdivision of a disconnected graph, each of whose components is isomorphic to the graph shown in Figure 5.4.

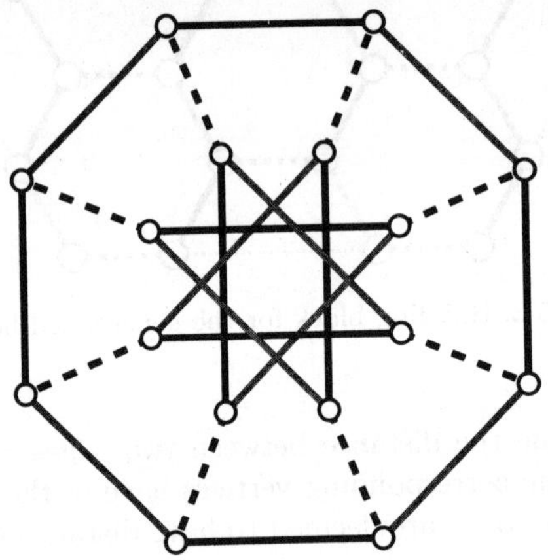

Figure 5.4. Building block for the dual generalized hexagon

Although we shall not do so here, it can be shown that the two graphs of Figure 5.3 and Figure 5.4 are the only possibilities for Y, and hence there is a unique dual pair of generalized hexagons of order $(2, 2)$.

5.8 Moore Graphs

A *Moore graph* is a graph with diameter d and girth $2d+1$. We already know two examples: C_5 and the Petersen graph. Unfortunately, there are at most two more Moore graphs. (The proof of this is one of the major achievements in algebraic graph theory.) In this section we prove that a Moore graph must be distance regular, and in the next section we provide the third known example.

Lemma 5.8.1 *Let X be a graph with diameter d and girth $2d+1$. Then X is regular.*

Proof. First we shall show that any two vertices at distance d have the same valency, and then we shall show that this implies that all vertices have the same valency. Let v and w be two vertices of X such that $d(v, w) = d$. Let P be the path of length d joining them. Consider any neighbour v' of v that is not on P. Then the distance from v' to w is exactly d; hence there is a unique path from v' to w that contains one neighbour of w. Each such path uses a different neighbour of w, and hence w has at least as many neighbours as v. Similarly, v has at least as many neighbours as w, and so they have equal valency. Let C be a cycle of length $2d + 1$. Starting with any given vertex v and taking two d-step walks around C shows that the neighbours of v have the same valency as v. Therefore, all vertices of C have the same valency. Given any vertex x not on C, form a path of length i from x to C. The vertex x' that is $d - i$ further steps around C has distance d from x, and hence x has the same valency as x'. Therefore, all the vertices of X have the same valency, and X is regular. □

Theorem 5.8.2 *A Moore graph is distance regular.*

Proof. Let X be a Moore graph of diameter d. By the previous lemma, X is regular, so denote its valency by k. In order to show that X is distance regular it is sufficient to show that the intersection numbers a_i, b_i, and c_i of Section 4.5 are well-defined. Let v be a vertex of the Moore graph and let $X_1(v), \ldots, X_d(v)$ be the cells of the distance partition. The arguments are straightforward, relying on the simple fact that for each vertex $w \in X_i(v)$ there is a unique path of length i from v to w.

For any $1 \le i \le d$ a vertex w in $X_i(v)$ cannot have two neighbours in $X_{i-1}(v)$ because if so, there would be a cycle of length at most $2i$ containing v and w. On the other hand, w must have at least one neighbour in $X_{i-1}(v)$, and so $c_i = 1$ for all $1 \le i \le d$.

For any $1 \le i \le d-1$ a vertex w in $X_i(v)$ cannot have a neighbour w' in the same cell, because if so, there would be a cycle of length at most $2i+1$ containing v, w, and w'. Therefore, $a_i = 0$ for all $1 \le i \le d-1$.

By the previous lemma X is regular, and hence this is enough to show that $b_0 = k$, $b_i = k-1$ for $1 \le i \le d-1$ and $a_d = k-1$. Therefore, the intersection numbers are well-defined and hence X is distance regular. □

The theory of distance-regular graphs can be used to show that Moore graphs of diameter greater than two do not exist, and that if a Moore graph of diameter two does exist, then its valency is either 2, 3, 7, or 57. We will develop enough of this theory in our work on strongly regular graphs in Chapter 10 to determine the possible valencies of a Moore graph of diameter two. (In fact, it will become a fairly routine exercise: Exercise 10.7.) We construct a Moore graph of valency seven in the next section; the existence of a Moore graph of valency 57 is a long-standing and famous open problem in graph theory.

5.9 The Hoffman–Singleton Graph

In this section we show that there is a Moore graph of diameter two and valency seven and study some of its properties. This graph is known as the *Hoffman–Singleton graph* after its discoverers. By counting the number of vertices at distance one and two from a fixed vertex, we find that it has $1+7+42=50$ vertices.

Lemma 5.9.1 *An independent set C in a Moore graph of diameter two and valency seven contains at most 15 vertices. If $|C|=15$, then every vertex not in C has exactly three neighbours in C.*

Proof. Let X be a Moore graph of diameter two and valency seven. Suppose that C is an independent set in X with c vertices in it. Without loss of generality we may assume that the vertices are labelled so that the vertices $\{1,\ldots,50-c\}$ are the ones not in C. If i is a vertex not in C, let k_i denote the number of its neighbours that lie in C. Since no two vertices in C are joined by an edge, we have

$$7c = \sum_{i=1}^{50-c} k_i.$$

Now, consider the paths of length two joining two vertices in C. Since every pair of nonadjacent vertices in X has exactly one common neighbour, counting these in two ways yields

$$\binom{c}{2} = \sum_{i=1}^{50-c} \binom{k_i}{2}.$$

From these last two equations it follows that for any real number μ,

$$\sum_{i=1}^{50-c} (k_i-\mu)^2 = (50-c)\mu^2 - 14c\mu + c^2 + 6c. \qquad (5.1)$$

The right side here must be nonnegative for all values of μ, so regarding it as a quadratic in μ, we see that it must have at most one zero. Therefore, the discriminant

$$196c^2 - 4(50-c)(c^2+6c) = 4c(c-15)(c+20)$$

of the quadratic must be less than or equal to 0. It follows that $c \le 15$.

If $c=15$, then the right side of (5.1) becomes

$$35\mu^2 - 210\mu + 315 = 35(\mu-3)^2,$$

and so setting μ equal to three in (5.1) yields that

$$\sum_{i=1}^{35} (k_i-3)^2 = 0.$$

Therefore, $k_i = 3$ for all i, as required. □

We will now describe a construction of the Hoffman–Singleton graph, using the heptads of Section 4.6. Once again we consider the 35 triples from the set $\Omega = \{1, \ldots, 7\}$. A set of triples is *concurrent* if there is some point common to them all, and the intersection of any two of them is this common point. A *triad* is a set of three concurrent triples. The remainder of the argument is broken up into a number of separate claims.

(a) *No two distinct heptads have three nonconcurrent triples in common.*

It is enough to check that for one set of three nonconcurrent triples, there is a unique heptad containing them.

(b) *Each triad is contained in exactly two heptads.*

Without loss of generality we may take our triad to be 123, 145, and 167. By a routine calculation one finds that there are two heptads containing this triad:

123	123
145	145
167	167
246	247
257	256
347	346
356	357

Note that the second of these heptads can be obtained from the first by applying the permutation (67) to each of its triples.

(c) *There are exactly* 30 *heptads.*

There are 15 triads on each point, thus we obtain 210 pairs consisting of a triad and a heptad containing it. Since each heptad contains exactly 7 triads, it follows that there must be 30 heptads.

(d) *Any two heptads have* 0, 1, *or* 3 *triples in common.*

If two heptads have four (or more) triples in common, then they have three nonconcurrent triples in common. Hence two heptads can have at most three triples in common. If two triples meet in precisely one point, there is a unique third triple concurrent with them. Any heptad containing the first two triples must contain the third. (Why?)

(e) *The automorphism group of a heptad has order* 168, *and consists of even permutations.*

Firstly, we note that Sym(7) acts transitively on the set of heptads, as it acts transitively on the set of triads and there are permutations mapping the two heptads on a triad to each other. Since there are 30 heptads, we deduce that the subgroup of Sym(7) fixing a heptad has order 168, and in Section 4.6 we exhibited such a group consisting of even permutations.

(f) *The heptads form two orbits of length* 15 *under the action of the alternating group* Alt(7). *Any two heptads in the same orbit have exactly one triple in common.*

Since the subgroup of Alt(7) fixing a heptad has order 168, the number of heptads in an orbit is 15. Let Π denote the first of the heptads above. The permutations (123) and (132) lie in Alt(7) and map Π onto two distinct heptads having exactly one triple in common with Π. (Check it!) From each triple in Π we obtain two 3-cycles in Alt(7); hence we infer that there are 14 heptads in the same orbit as Π under Alt(7) each with exactly one triple in common with Π. Since there are only 15 heptads in an Alt(7) orbit, and since all heptads in an Alt(7) orbit are equivalent, it follows that any two heptads in such an orbit have exactly one triple in common.

(g) *Each triple from* Ω *lies in exactly six heptads, three from each* Alt(7) *orbit.*

Simple counting.

We can now construct the Hoffman–Singleton graph. Choose an Alt(7) orbit of heptads from Ω. Take the vertices of our graph to be these heptads, together with the 35 triples in Ω. We join a heptad to a triple if and only if it contains the triple. Two triples are adjacent if and only if they are disjoint. The resulting graph is easily seen to have valency seven and diameter two. Since it has $50 = 7^2 + 1$ vertices, it is a Moore graph. The collection of 15 heptads forms an independent set of size 15 as considered in Lemma 5.9.1.

5.10 Designs

Another important class of incidence structures is the class of t-designs. In general, t-designs are not partial linear spaces, and design theorists tend to use the word "block" rather than "line", and to identify a block with the subset of points to which it is incident.

In this language, a t-(v, k, λ_t) design is a set $\mathcal{P}$ of v points, together with a collection $\mathcal{B}$ of k-subsets of points, called *blocks* such that every t-set of points lies in precisely λ_t blocks. The projective planes $PG(2, q)$ have the property that every two points lie in a unique block, and so they are 2-$(q^2 + q + 1, q + 1, 1)$ designs.

Now, suppose that $\mathcal{D}$ is a t-(v, k, λ_t) design and let S be an s-set of points for some $s < t$. We will count the number of blocks λ_s of $\mathcal{D}$ containing S. We will do this by counting in two ways the pairs (T, B) where T is a t-set containing S and B is a block containing T. Firstly, S lies in $\binom{v-s}{t-s}$ t-subsets T, each of which lies in λ_t blocks. Secondly, for each block containing S

there are $\binom{k-s}{t-s}$ possible choices for T. Hence

$$\lambda_s \binom{k-s}{t-s} = \lambda_t \binom{v-s}{t-s}, \tag{5.2}$$

and since the number of blocks does not depend on the particular choice of S, we see that $\mathcal{D}$ is also an s-(v, k, λ_s) design. This yields a necessary condition for the existence of a t-design in that the values of λ_s must be integers for all $s < t$.

The parameter λ_0 is the total number of blocks in the design, and is normally denoted by b. Putting $s = 0$ into (5.2), we get

$$b\binom{k}{t} = \lambda_t \binom{v}{t}.$$

The parameter λ_1 is the number of blocks containing each point, and it is normally called the *replication number* and denoted by r. Putting $t = 1$ in the previous equation yields that

$$bk = vr.$$

If $\lambda_t = 1$, then the design is called a *Steiner system*, and a 2-design with $\lambda_2 = 1$ and $k = 3$ is called a *Steiner triple system*. The projective plane $PG(2, 2)$ is a 2-$(7, 3, 1)$ design, so is a Steiner triple system. It is usually called the *Fano plane* and drawn as shown in Figure 5.5, where the blocks are the straight lines and the central circle.

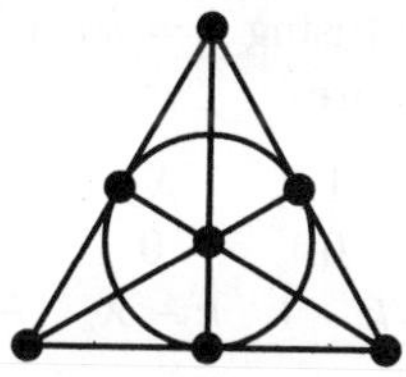

Figure 5.5. The Fano plane

The *incidence matrix* of a design is the matrix N with rows indexed by points and columns by blocks such that $N_{ij} = 1$ if the ith point lies in the jth block, and $N_{ij} = 0$ otherwise. Then the matrix N has constant row sum r and constant column sum k, and satisfies the equation

$$NN^T = (r - \lambda_2)I + \lambda_2 J,$$

where J is the all-ones matrix. Conversely, any 01-matrix with constant row sum and constant column sum satisfying this equation yields a 2-design.

The proof of the next result relies on some results from linear algebra that will be covered in Chapter 8.

Lemma 5.10.1 *In a 2-design with $k < v$ we have $b \geq v$.*

Proof. Putting $t = 2$ and $s = 1$ into (5.2) we get that $r(k-1) = (v-1)\lambda_2$, and so $k < v$ implies that $r - \lambda_2 > 0$. By the remark at the end of Section 8.6, it follows that NN^T is invertible. It follows that the rows of N are linearly independent, and therefore that $b \geq v$. □

A 2-design with $b = v$ is called *symmetric*. The dual of a 1-design is a 1-design, but in general the dual of a 2-design is not a 2-design. The next result shows that symmetric designs are exceptional.

Lemma 5.10.2 *The dual $\mathcal{D}^*$ of a symmetric design $\mathcal{D}$ is a symmetric design with the same parameters.*

Proof. If N is the incidence matrix of $\mathcal{D}$, then N^T is the incidence matrix of $\mathcal{D}^*$. Since $\mathcal{D}$ is a 2-design, we have $NN^T = (r - \lambda_2)I + \lambda_2 J$, and thus $N^T = N^{-1}((r - \lambda_2)I + \lambda_2 J)$. Since $\mathcal{D}$ is symmetric, $r = k$, and so N commutes with both I and J. Therefore, $N^T N = (r - \lambda_2)I + \lambda_2 J$, showing that $\mathcal{D}^*$ is a 2-design with the same parameters as $\mathcal{D}$. □

Theorem 5.10.3 *A bipartite graph is the incidence graph of a symmetric 2-design if and only if it is distance regular with diameter three.*

Proof. Let $\mathcal{D}$ be a symmetric 2-(v, k, λ_2) design with incidence graph X. Any two points lie at distance two in X, and similarly for blocks. Therefore, a block lies at distance three from a point not on the block, and this is the diameter of X. Now, consider the distance partition from a point. Clearly, X is bipartite, so we have $a_1 = a_2 = a_3 = 0$. Since two points lie in λ_2 blocks, we have $c_2 = \lambda_2$, and (using $r = k$) it is straightforward to verify that the intersection numbers are

$$\left\{\begin{matrix} - & 1 & \lambda_2 & k \\ 0 & 0 & 0 & 0 \\ k & k-1 & k-\lambda_2 & - \end{matrix}\right\}.$$

Since the dual of $\mathcal{D}$ is a design with the same parameters, the distance partition from a line yields the same intersection numbers.

Conversely, suppose that X is a bipartite distance-regular graph with diameter three. Declare one part of the bipartition to be points, and the other to be blocks. Considering the distance partition from a point we see that every point lies in b_0 blocks, and every two points lie in c_2 blocks, hence we have a 2-design with $r = b_0$ and $\lambda_2 = c_2$. Considering the distance partition from a block, we see that every block contains b_0 points and every two blocks meet in c_2 blocks. Thus we have a 2-design with $k = b_0 = r$ and hence $b = v$. □

Since projective planes are symmetric designs, this provides another proof of Lemma 5.6.9 for the case of generalized polygons with diameter three. The incidence graph of the Fano plane is called the Heawood graph, and shown in Figure 5.6.

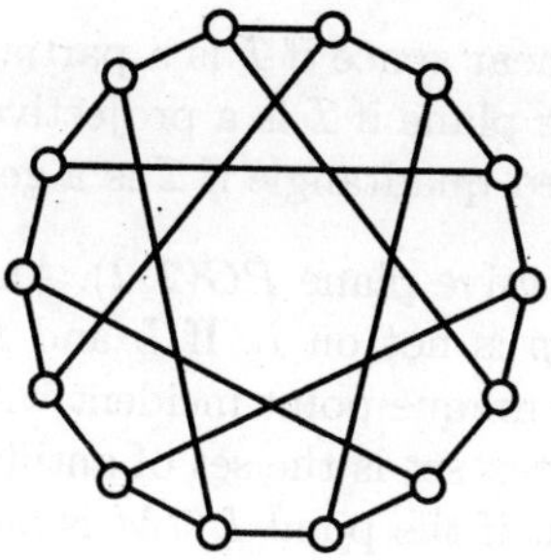

Figure 5.6. The Heawood graph

Another way to form a graph from a design $\mathcal{D}$ is to consider the *block graph* whose vertices are the blocks of $\mathcal{D}$, where two vertices are adjacent if the corresponding blocks intersect. More generally, blocks in a design may meet in differing numbers of points, and interesting graphs can often be found by taking two blocks to be adjacent if they meet in some fixed number of points.

Theorem 5.10.4 *The block intersection graph of a Steiner triple system with $v > 7$ is distance regular with diameter two.*

Proof. Let $\mathcal{D}$ be a 2-$(v, 3, 1)$ design, and let X be the block intersection graph of $\mathcal{D}$. Every point lies in $(v-1)/2$ blocks, and so X is regular with valency $3(v-3)/2$. If we consider two blocks that intersect, then there are $(v-5)/2$ further blocks through that point of intersection, and four blocks containing a pair of points, one from each block, other than the point of intersection. Therefore,

$$a_1 = (v-5)/2 + 4 = (v+3)/2.$$

If we now consider two blocks that do not intersect, then we see that there are nine blocks containing a pair of points, one from each block, and so $c_2 = 9$. This also shows that the diameter of X is two. From these it is straightforward to compute the remaining intersection numbers and hence show that X is distance regular. □

Exercises

1. Find the degenerate projective planes (those that do not contain a triangle).

2. Determine the degenerate generalized quadrangles (those without noncollinear points and nonconcurrent lines).

3. Let G be a group of automorphisms of an incidence structure $\mathcal{I}$ and consider the set of points and lines fixed by G. Show that this is:

(1) A partial linear space if $\mathcal{I}$ is a partial linear space.
(2) A projective plane if $\mathcal{I}$ is a projective plane.
(3) A generalized quadrangle if $\mathcal{I}$ is a generalized quadrangle.

4. Consider the projective plane $PG(2,2)$. An *antiflag* in $PG(2,2)$ is a pair (p, L) where p is not on L. If L and M are two lines, then let $L \cap M$ denote the unique point incident with both L and M. Define a graph whose vertex set is the set of antiflags, and where (p, L) and (q, M) are adjacent if the point $L \cap M$ is on the line $p \vee q$. Show that this graph is the Coxeter graph and that the duality gives rise to an automorphism of order two additional to the automorphism group of $PG(2,2)$. (We discussed the Coxeter graph at length in Section 4.6.)

5. Show that a generalized hexagon of order $(2,2)$ is distance regular with intersection array $\{3,2,2,2,2,2;1,1,1,1,1,3\}$.

6. Determine how to subdivide the edges of two copies of the graph of Figure 5.4 to form a generalized hexagon.

7. The Shrikhande graph can be embedded as a triangulation on the torus as shown in Figure 5.7. Show that the graph of Figure 5.3 is the dual of the Shrikhande graph.

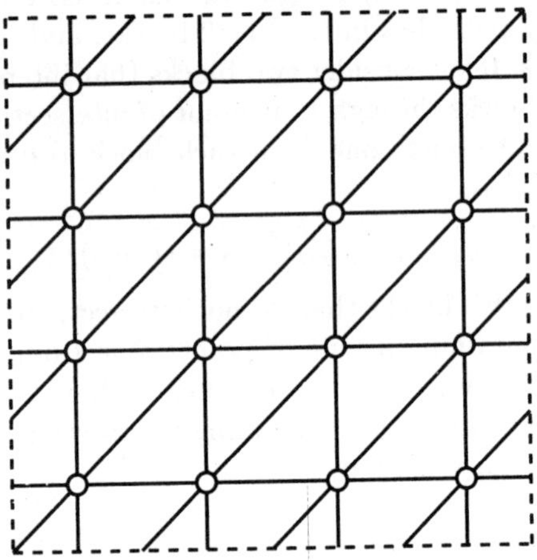

Figure 5.7. The Shrikhande graph on the torus

8. Let Y denote the graph of Figure 5.4. Show that every vertex $v \in V(Y)$ has a unique vertex v' at distance 4 from it. Define a graph Y' whose vertex set is the eight pairs of vertices, and where two pairs are adjacent if there are any edges between them in Y. What graph is Y'? (It will follow from our work in Section 6.8 that Y is a cover of Y'.)

9. Show that the graph of Figure 5.4 is the unique cubic graph of girth six on 16 vertices. Show further that the Heawood graph is the only smaller cubic graph of girth six.

10. Show that if a Moore graph of valency 57 and diameter two exists, then an independent set in it has size at most 400.

11. Suppose that X is a Moore graph with diameter two and valency k. Show that there are exactly $k(k-1)^2/2$ 5-cycles through a given vertex of X. Deduce from this that 5 divides $(k^2+1)k(k-1)^2$, and hence that $k \not\equiv 4 \bmod 5$.

12. Show that a Moore graph with diameter two and valency seven contains a subgraph isomorphic to the Petersen graph with an edge deleted. Show further that the subgraph induced by this is isomorphic to the Petersen graph.

13. Let X be the incidence graph of a projective plane of order n. (See Section 5.3 for details.) Let Y be the graph obtained from X by deleting an adjacent pair of vertices and all their neighbours. Show that the resulting graph is distance regular.

14. Let X be a Moore graph with diameter two and valency seven. If $u \in V(X)$, show that the graph induced by the vertices at distance two from u is distance regular.

15. Let X be a Moore graph with diameter two and valency seven, and let Y be an induced subgraph of X isomorphic to the Petersen graph. Show that each vertex in $V(X) \setminus V(Y)$ has exactly one neighbour in Y. (It follows that the partition $(V(Y), V(X) \setminus V(Y))$ of $V(X)$ is an example of an equitable partition. We will study these in Section 9.3.)

16. Let X be a Moore graph with diameter two and suppose G is a group of automorphisms of X. Let Y be the subgraph of X induced by the fixed points of G. Show that Y is isomorphic to either K_1, $K_{1,r}$ for some r, or a Moore graph of diameter two.

17. In Section 1.8 we saw that the Petersen graph is the dual of K_6 embedded in the projective plane. This embedding determines a set of 5-cycles such that each edge lies in exactly two of them. We build an incidence structure as follows. Let V be the set $\{0, 1, \ldots, 10\}$ and let $V \setminus 0$ be the vertices of a copy of the Petersen graph. The first six blocks of the incidence structure are the vertex sets of the six 5-cycles given by the embedding of this graph in the projective plane. There are five further blocks, consisting of a set of four independent vertices in the Petersen graph, together with 0. (You might wish to verify that the Petersen graph has exactly five independent sets of size four.) Show that the 11 points in V and the 11 blocks just described form a 2-(11, 5, 2) design.

Notes

Feit and Higman proved their theorem in [8]. A number of alternative proofs of this result are known now. For one of these, and more details, see Section 6.5 of [3]. The study of Moore graphs began with the work of Hoffman and Singleton [9]. Biggs [2] presents a proof that a Moore graph has diameter at most two. His treatment follows Damerell [7]; this result was proved independently by Bannai and Ito [1].

Although generalized triangles and generalized quadrangles had previously been studied, the concept of a generalized polygon is due to Tits [11]. Our proof that a generalized polygon is semiregular is based on Yanushka [12], while our proof that a Moore graph is necessarily regular follows Singleton [10]. The construction of the generalized hexagon in Section 5.7 is essentially the same as that offered by Cohen and Tits in [6]. There they also prove that there is a unique dual pair of generalized hexagons with order $(2,2)$.

The construction of the Hoffman–Singleton graph we presented is related to the projective geometry $PG(3,2)$ of dimension three over GF(2). We may take the heptads in one Alt(7) orbit to be the points, the triples as the lines, and the remaining Alt(7) orbit of heptads as the planes. The resulting collection of points, lines, and planes is $PG(3,2)$.

An independent set C in a Moore graph of valency 57 and diameter two has size at most 400. (See Exercise 10.) If such a set C exists, then each of 2850 vertices not in it is adjacent to exactly eight vertices in C. This gives us a 2-$(400,8,1)$ design. The projective geometry $PG(3,7)$ has 400 points and 2850 lines. It is perhaps tempting to use this to construct a Moore graph of valency 57, with these points and lines as its vertices. A point will be adjacent to a line if it is on it; the unresolved difficulty is to decide how to define a suitable adjacency relation on the lines of $PG(3,7)$.

G. Higman [4] showed that if a Moore graph with valency 57 and diameter two exists, it cannot be vertex transitive. This improved on earlier work of Aschbacher, who showed that the automorphism group of such a graph could not be a rank-three group.

A solution to Exercise 12 will be found in Chapter 6 of [5].

We do not seem to be at all close to deciding whether there is a Moore graph with diameter two and valency 57. This is one of the most famous open problems in graph theory.

References

[1] E. Bannai and T. Ito, *On finite Moore graphs*, J. Fac. Sci. Univ. Tokyo Sect. IA Math., 20 (1973), 191–208.

[2] N. Biggs, *Algebraic Graph Theory*, Cambridge University Press, Cambridge, second edition, 1993.

[3] A. E. Brouwer, A. M. Cohen, and A. Neumaier, *Distance-Regular Graphs*, Springer-Verlag, Berlin, 1989.

[4] P. J. Cameron, *Permutation Groups*, Cambridge University Press, Cambridge, 1999.

[5] P. J. Cameron and J. H. van Lint, *Designs, Graphs, Codes and their Links*, Cambridge University Press, Cambridge, 1991.

[6] A. M. Cohen and J. Tits, *On generalized hexagons and a near octagon whose lines have three points*, European J. Combin., 6 (1985), 13–27.

[7] R. M. Damerell, *On Moore graphs*, Proc. Cambridge Philos. Soc., 74 (1973), 227–236.

[8] W. Feit and G. Higman, *The nonexistence of certain generalized polygons*, J. Algebra, 1 (1964), 114–131.

[9] A. J. Hoffman and R. R. Singleton, *On Moore graphs with diameters* 2 *and* 3, IBM J. Res. Develop., 4 (1960), 497–504.

[10] R. Singleton, *There is no irregular Moore graph*, Amer. Math. Monthly, 75 (1968), 42–43.

[11] J. Tits, *Buildings of spherical type and finite BN-pairs*, Springer-Verlag, Berlin, 1974.

[12] A. Yanushka, *On order in generalized polygons*, Geom. Dedicata, 10 (1981), 451–458.

6

Homomorphisms

Although any isomorphism between two graphs is a homomorphism, the study of homomorphisms between graphs has quite a different flavour to the study of isomorphisms. In this chapter we support this claim by introducing a number of topics involving graph homomorphisms. We consider the relationship between homomorphisms and graph products, and in particular a famous unsolved conjecture of Hedetniemi, which asserts that if two graphs are not n-colourable, then neither is their product. Our second major topic is the exploration of the core of a graph, which is the minimal subgraph of a graph that is also a homomorphic image of the graph. Studying graphs that are equal to their core leads us to an interesting class of graphs first studied by Andrásfai. We finish the chapter with an exploration of the cores of vertex-transitive graphs.

6.1 The Basics

If X, Y, and Z are graphs and there are homomorphisms f from X to Y and g from Y to Z, then the composition $g \circ f$ is a homomorphism from X to Z. (This needs a line of proof, which is a task for the reader.) Note that we have $g \circ f$ and not $f \circ g$, an unfortunate consequence of the fact that it is traditional to write homomorphisms on the left rather than the right. Now, define a relation "$\to$" on the class of all graphs by $X \to Y$ if there is a homomorphism from X to Y. (It may help if you read "$\to$" as "has a homomorphism into.") Since the composition of two homomorphisms is a homomorphism, "$\rightarrow$" is a transitive relation. Since the identity map is a homomorphism, we have $X \rightarrow X$ for any graph X, and therefore $\rightarrow$ is reflexive as well. Most reflexive transitive relations you have met have been partial orders such as:

(a) $\leq$ on the reals,

(b) m divides n on the integers, or

(c) $\subseteq$ on the subsets of a set.

Our new relation is not a partial order because it is not antisymmetric, that is to say, if $X \rightarrow Y$ and $Y \rightarrow X$, it does not necessarily follow that $X = Y$. (Take X to be any bipartite graph and Y to be an edge.)

If X and Y are graphs such that there is a homomorphism from X to Y and a homomorphism from Y to X, we say they are *homomorphically equivalent*. A homomorphism from X to Y is *surjective* if every vertex of Y is the image of a vertex of X. If there is a surjective homomorphism from X to Y and from Y to X, then X and Y are isomorphic (this implicitly uses the fact that X and Y are finite).

If f is a homomorphism from X to Y, then the preimages $f^{-1}(y)$ of each vertex y in Y are called the *fibres* of f. The fibres of f determine a partition π of $V(X)$ called the *kernel* of f. If Y has no loops, then the kernel is a partition into independent sets. Given a graph X together with a partition π of $V(X)$, define a graph X/π with vertex set the cells of π and with an edge between two cells if there is an edge of X with an endpoint in each cell (and a loop if there is an edge within a cell). There is a natural homomorphism from X to X/π with kernel π.

Although it is generally a hard task to show that there is no homomorphism from one graph to another, there are two parameters that can be useful. Recall from Lemma 1.4.1 that a graph Y can be properly coloured with r colours if and only if there is a homomorphism from Y to K_r. Therefore, if there is a homomorphism from X to Y, we have $X \rightarrow Y \rightarrow K_r$, and so $\chi(X) \leq \chi(Y)$. Hence if $\chi(X) > \chi(Y)$, then there can be no homomorphism from X to Y. Second, if X has an induced odd cycle of length ℓ and any induced odd cycle in Y has length greater than ℓ, then there cannot be a homomorphism from X to Y, because the homomorphic image of an odd cycle must be an odd cycle of no greater length. We call the length of a shortest odd cycle in X the *odd girth* of X; the odd girth of X is an upper bound on the odd girth of any homomorphic image of X.

6.2 Cores

A graph X is a *core* if any homomorphism from X to itself is a bijection or, equivalently, if its endomorphism monoid equals its automorphism group. The simplest examples of cores are the complete graphs. A subgraph Y of

X is a core of X if Y is a core and there is a homomorphism from X to Y. We will see below that every graph has a core, and that all its cores are isomorphic. We denote the core of X by $X^\bullet$. If Y is a core of X and f is a homomorphism from X to Y, then $f \restriction Y$ must be an automorphism of Y. The composition of f with the inverse of this automorphism is the identity mapping on Y; hence any core of X is a retract (see Exercise 1.5).

A graph X is χ-*critical* (or just *critical*) if the chromatic number of any proper subgraph is less than $\chi(X)$. A χ-critical graph cannot have a homomorphism to any proper subgraph, and hence must be its own core. This provides a wide class of cores, including all complete graphs and odd cycles.

The next lemma implies that $\rightarrow$ is a partial order on isomorphism classes of cores.

Lemma 6.2.1 *Let X and Y be cores. Then X and Y are homomorphically equivalent if and only if they are isomorphic.*

Proof. Suppose X and Y are homomorphically equivalent and that $f : X \rightarrow Y$ and $g : Y \rightarrow X$ are the homomorphisms between them. Then because both $f \circ g$ and $g \circ f$ must be surjective, we see that both f and g are surjective, so X and Y are isomorphic. □

Lemma 6.2.2 *Every graph X has a core, which is an induced subgraph and is unique up to isomorphism.*

Proof. Since X is finite and the identity mapping is a homomorphism, the family of subgraphs of X to which X has a homomorphism is finite and nonempty and hence has a minimal element with respect to inclusion. Since a core is a retract, it is clearly an induced subgraph. Now, suppose that Y_1 and Y_2 are cores of X and let f_i be a homomorphism from X to Y_i. Then $f_1 \restriction Y_2$ is a homomorphism from Y_2 to Y_1, and $f_2 \restriction Y_1$ is a homomorphism from Y_1 to Y_2. Therefore, by the previous lemma, Y_1 and Y_2 are isomorphic. □

Lemma 6.2.3 *Two graphs X and Y are homomorphically equivalent if and only if their cores are isomorphic.*

Proof. If there is a homomorphism $f : X \rightarrow Y$, then we have a sequence of homomorphisms

$$X^\bullet \rightarrow X \rightarrow Y \rightarrow Y^\bullet,$$

which composes to give a homomorphism from $X^\bullet$ to $Y^\bullet$. Hence, if X and Y are homomorphically equivalent, so are $X^\bullet$ and $Y^\bullet$.

On the other hand, if $f : X^\bullet \rightarrow Y^\bullet$ is a homomorphism, then we have a sequence of homomorphisms

$$X \rightarrow X^\bullet \rightarrow Y^\bullet \rightarrow Y,$$

which composes to yield a homomorphism from X to Y, and so X and Y are homomorphically equivalent if $X^\bullet$ and $Y^\bullet$ are.

Hence two graphs are homomorphically equivalent if and only if their cores are. By Lemma 6.2.1, two cores are homomorphically equivalent if and only if they are isomorphic. Hence the proof is complete. □

If we view "$\rightarrow$" as a relation on the set of isomorphism classes of cores, then the above results have the following consequence.

Corollary 6.2.4 *The relation "$\rightarrow$" is a partial order on the set of isomorphism classes of cores.*

Proof. We have already seen that "$\rightarrow$" is a transitive and reflexive relation on the set of isomorphism classes of graphs, whence it follows that it is transitive and reflexive on isomorphism classes of cores. By Lemma 6.2.1, if X and Y are cores and $X \rightarrow Y$ and $Y \rightarrow X$, then X and Y are isomorphic. Hence "$\rightarrow$" is antisymmetric, and a transitive, reflexive, antisymmetric relation is a partial order. □

We will learn more about this partial order in the next section.

6.3 Products

If X and Y are graphs, then their *product* $X \times Y$ has vertex set $V(X) \times V(Y)$, and $(x, y) \sim (x', y')$ if and only if $x \sim x'$ and $y \sim y'$. The map that sends (x, y) to (y, x) is an isomorphism from $X \times Y$ to $Y \times X$, and it is no harder to describe an isomorphism from $(X \times Y) \times Z$ to $X \times (Y \times Z)$, so this product behaves in much the way we might expect. However,

$$K_2 \times 2K_3 \cong 2C_6 \cong K_2 \times C_6$$

(as you are invited to verify), and so if $X \times Y_1 \cong X \times Y_2$, it does not follow that $Y_1 \cong Y_2$. The product of connected graphs is connected if and only if at least one of the factors is not bipartite. (Another exercise.) We also point out that the product $X \times K_1$ is the empty graph, which is possibly not what you expected.

For fixed x in $V(X)$, the vertices of the form (x, y) in $X \times Y$ form an independent set. Therefore, the mapping

$$p_X : (x, y) \mapsto x$$

is a homomorphism from $X \times Y$ to X. It is dignified by calling it the *projection* from $X \times Y$ to X. Similarly, there is a projection p_Y from $X \times Y$ to Y.

Theorem 6.3.1 *Let X, Y, and Z be graphs. If $f : Z \rightarrow X$ and $g : Z \rightarrow Y$, then there is a unique homomorphism ϕ from Z to $X \times Y$ such that $f = p_X \circ \phi$ and $g = p_Y \circ \phi$.*

Proof. Assume that we are given homomorphisms $f : Z \to X$ and $g : Z \to Y$. The map

$$\phi : z \mapsto (f(z), g(z))$$

is readily seen to be a homomorphism from Z to $X \times Y$. Clearly, $p_X \circ \phi = f$ and $p_Y \circ \phi = g$, and furthermore, ϕ is uniquely determined by f and g. □

If X and Y are graphs, we use $\text{Hom}(X, Y)$ to denote the set of all homomorphisms from X to Y.

Corollary 6.3.2 *For any graphs X, Y, and Z,*

$$|\text{Hom}(Z, X \times Y)| = |\text{Hom}(Z, X)|\,|\text{Hom}(Z, Y)|. \qquad \square$$

Our last theorem allows us to derive another property of the set of isomorphism classes of cores. Recall that a partially ordered set is a *lattice* if each pair of elements has a least upper bound and a greatest lower bound.

Lemma 6.3.3 *The set of isomorphism classes of cores, partially ordered by "$\to$", is a lattice.*

Proof. We start with the least upper bound. Let X and Y be cores. For any core Z, if $X \to Z$ and $Y \to Z$, then $X \cup Y \to Z$. Hence $(X \cup Y)^\bullet$ is the least upper bound of X and Y.

For the greatest lower bound we note that by the previous theorem, if $Z \to X$ and $Z \to Y$, then $Z \to X \times Y$. Hence $(X \times Y)^\bullet$ is the greatest lower bound of X and Y.

□

It is probably a surprise that the greatest lower bound $(X \times Y)^\bullet$ normally has more vertices than the least upper bound. Life can be surprising.

If X is a graph, then the vertices (x, x), where $x \in V(X)$, induce a subgraph of $X \times X$ isomorphic to X. We call it the *diagonal* of the product. In general, $X \times Y$ need not contain a copy of X; consider the product $K_2 \times K_3$, which is isomorphic to C_6 and thus contains no copy of K_3.

To conclude this section we describe another construction closely related to the product. Suppose that X and Y are graphs with homomorphisms f and g, respectively, to a graph F. The *subdirect product* of (X, f) and (Y, g) is the subgraph of $X \times Y$ induced by the set of vertices

$$\{(x, y) \in V(X) \times V(Y) : f(x) = g(y)\}.$$

(The proof is left as an exercise.) If X is a connected bipartite graph, then it has exactly two homomorphisms f_1 and f_2 to K_2. Suppose Y is connected and g is a homomorphism from Y to K_2. Then the two subdirect products of (X, f_i) with (Y, g) form the components of $X \times Y$. (Yet another exercise.)

6.4 The Map Graph

Let F and X be graphs. The *map graph* F^X has the set of functions from $V(X)$ to $V(F)$ as its vertices; two such functions f and g are adjacent in F^X if and only if whenever u and v are adjacent in X, the vertices $f(u)$ and $g(v)$ are adjacent in F. A vertex in F^X has a loop on it if and only if the corresponding function is a homomorphism. Even if there are no homomorphisms from X to F, the map graph F^X can still be very useful, as we will see.

Now, suppose that ψ is a homomorphism from X to Y. If f is a function from $V(Y)$ to $V(F)$, then the composition $f \circ \psi$ is a function from $V(X)$ to $V(F)$. Hence ψ determines a map from the vertices of F^Y to F^X, which we call the *adjoint* map to ψ.

Theorem 6.4.1 *If F is a graph and ψ is a homomorphism from X to Y, then the adjoint of ψ is a homomorphism from F^Y to F^X.*

Proof. Suppose that f and g are adjacent vertices of F^Y and that x_1 and x_2 are adjacent vertices in X. Then $\psi(x_1) \sim \psi(x_2)$, and therefore $f(\psi(x_1)) \sim g(\psi(x_2))$. Hence $f \circ \psi$ and $g \circ \psi$ are adjacent in F^X. □

Theorem 6.4.2 *For any graphs F, X, and Y, we have $F^{X \times Y} \cong (F^X)^Y$.*

Proof. It is immediate that $F^{X \times Y}$ and $(F^X)^Y$ have the same number of vertices. We start by defining the natural bijection between these sets, and then we will show that it is an isomorphism.

Suppose that g is a map from $V(X \times Y)$ to F. For any fixed $y \in V(Y)$ the map

$$g_y : x \mapsto g(x, y)$$

is an element of F^X. Therefore, the map

$$\Phi_g : y \mapsto g_y$$

is an element of $(F^X)^Y$. The mapping $g \mapsto \Phi_g$ is the bijection that we need.

Now, we must show that this bijection is in fact an isomorphism. So let f and g be adjacent vertices of $F^{X \times Y}$. We must show that Φ_f and Φ_g are adjacent vertices of $(F^X)^Y$. Let y_1 and y_2 be adjacent vertices in Y. For any two vertices $x_1 \sim x_2$ in X we have

$$(x_1, y_1) \sim (x_2, y_2),$$

and since $f \sim g$,

$$f(x_1, y_1) \sim g(x_2, y_2),$$

and so

$$\Phi_f(y_1) \sim \Phi_g(y_2).$$

A similar argument shows that if $f \not\sim g$, then $\Phi_f \not\sim \Phi_g$, and hence the result follows. □

Corollary 6.4.3 *For any graphs F, X, and Y, we have*

$$|\mathrm{Hom}(X \times Y, F)| = |\mathrm{Hom}(Y, F^X)|.$$

Proof. We have just seen that $F^{X \times Y} \cong (F^X)^Y$, and so they have the same number of loops, which are precisely the homomorphisms. □

Since there is a homomorphism from $X \times F$ to F, the last result implies that there is a homomorphism from F into F^X. We can be more precise, although we leave the proof as an exercise.

Lemma 6.4.4 *If X has at least one edge, the constant functions from $V(X)$ to $V(F)$ induce a subgraph of F^X isomorphic to F.* □

6.5 Counting Homomorphisms

By counting homomorphisms we will derive another interesting property of the map graph.

Lemma 6.5.1 *Let X and Y be fixed graphs. Suppose that for all graphs Z we have*

$$|\mathrm{Hom}(Z, X)| = |\mathrm{Hom}(Z, Y)|.$$

Then X and Y are isomorphic.

Proof. Let $\mathrm{Inj}(A, B)$ denote the set of injective homomorphisms from a graph A to a graph B. We aim to show that for all Z we have $|\mathrm{Inj}(Z, X)| = |\mathrm{Inj}(Z, Y)|$. By taking Z equal to X and then Y, we see that there are injective homomorphisms from X to Y and Y to X. Since X and Y must have the same number of vertices, an injective homomorphism is surjective, and thus X is isomorphic to Y.

We prove that $|\mathrm{Inj}(Z, X)| = |\mathrm{Inj}(Z, Y)|$ by induction on the number of vertices in Z. It is clearly true if Z has one vertex, because any homomorphism from a single vertex is injective.

We can partition the homomorphisms from Z into any graph W according to the kernel, so we get

$$|\mathrm{Hom}(Z, W)| = \sum_{\pi} |\mathrm{Inj}(Z/\pi, W)|,$$

where π ranges over all partitions. A homomorphism is an injection if and only if its kernel is the discrete partition, which we shall denote by δ. Therefore,

$$|\mathrm{Inj}(Z, W)| = |\mathrm{Hom}(Z, W)| - \sum_{\pi \neq \delta} |\mathrm{Inj}(Z/\pi, W)|.$$

Now, by the induction hypothesis, all the terms on the right hand side of this sum are the same for $W = X$ and $W = Y$. Therefore, we conclude that

$$|\mathrm{Inj}(Z, X)| = |\mathrm{Inj}(Z, Y)|,$$

and the result follows. □

Lemma 6.5.2 *For any graphs F, X, and Y we have*

$$F^{X \cup Y} \cong F^X \times F^Y.$$

Proof. For any graph Z, we have

$$\begin{aligned} |\mathrm{Hom}(Z, F^{X \cup Y})| &= |\mathrm{Hom}(Z \times (X \cup Y), F)| \\ &= |\mathrm{Hom}((Z \times X) \cup (Z \times Y), F)| \\ &= |\mathrm{Hom}(Z \times X, F)||\mathrm{Hom}(Z \times Y, F)| \\ &= |\mathrm{Hom}(Z, F^X)||\mathrm{Hom}(Z, F^Y)|. \end{aligned}$$

By Corollary 6.3.2, the last product equals the number of homomorphisms from Z to $F^X \times F^Y$. Now, the previous lemma completes the argument. □

It is not hard to find a direct proof of the last result, but the argument we have given has its own charm.

6.6 Products and Colourings

We recall that if $X \to Y$, then $\chi(X) \leq \chi(Y)$. Since both X and Y are homomorphic images of $X \times Y$ (using the projection homomorphisms), we have that

$$\chi(X \times Y) \leq \min\{\chi(X), \chi(Y)\}.$$

S. Hedetniemi has conjectured that for all graphs X and Y equality occurs in the above bound and hence that $\chi(X \times Y) = \min\{\chi(X), \chi(Y)\}$.

An equivalent formulation of Hedetniemi's conjecture is that if X and Y are graphs that are not n-colourable, then the product $X \times Y$ is not n-colourable. When $n = 2$ we can prove this by showing that the product of two odd cycles contains an odd cycle. For $n = 3$, the conjecture was proved by El-Zahar and Sauer in 1985. The remaining cases are still open.

Our first result uses the map graph to simplify the study of Hedetniemi's conjecture.

Theorem 6.6.1 *Suppose $\chi(X) > n$. Then K_n^X is n-colourable if and only if $\chi(X \times Y) > n$ for all graphs Y such that $\chi(Y) > n$.*

Proof. By Corollary 6.4.3,

$$|\mathrm{Hom}(X \times K_n^X, K_n)| = |\mathrm{Hom}(K_n^X, K_n^X)| > 0,$$

and therefore $X \times K_n^X$ is n-colourable. Consequently, if $\chi(X) > n$ and $\chi(X \times Y) > n$ whenever $\chi(Y) > n$, then K_n^X must be n-colourable.

Assume conversely that $\chi(K_n^X) \leq n$ and let Y be a graph such that $\chi(Y) > n$. Then there are no homomorphisms from Y into any n-colourable graph, and therefore

$$0 = |\text{Hom}(Y, K_n^X)| = |\text{Hom}(X \times Y, K_n)|.$$

Hence $\chi(X \times Y) > n$. □

This theorem tells us that we can prove Hedetniemi's conjecture by proving that $\chi(K_n^X) \leq n$ if $\chi(X) > n$. The next few results summarize the limited number of cases where the conjecture is known to be true.

Theorem 6.6.2 *The map graph $K_n^{K_{n+1}}$ is n-colourable.*

Proof. We construct a proper n-colouring ϕ of $K_n^{K_{n+1}}$. For any $f \in K_n^{K_{n+1}}$, there are two distinct vertices i and j such that $f(i) = f(j)$. Define $\phi(f)$ to be the least value in the range of f that is the image of at least two vertices. If $\phi(f) = \phi(g)$, then for some distinct vertices i' and j' we have

$$f(i) = f(j) = g(i') = g(j').$$

Because i is not equal to both i' and j', this implies that $f \not\sim g$. Therefore, ϕ is a proper n-colouring of $K_n^{K_{n+1}}$. □

Corollary 6.6.3 *Suppose that the graph X contains a clique of size $n+1$. Then K_n^X is n-colourable.*

Proof. Since $K_{n+1} \to X$, by Theorem 6.4.1

$$K_n^X \to K_n^{K_{n+1}}.$$

By the theorem, $K_n^{K_{n+1}}$ is n-colourable, and so K_n^X is n-colourable. □

Theorem 6.6.4 *All loops in $K_n^{K_n}$ are isolated vertices. The subgraph of $K_n^{K_n}$ induced by the vertices without loops is n-colourable.*

Proof. Suppose $f \in K_n^{K_n}$ and f is a proper n-colouring of K_n. If g is adjacent to f, then $g(i) \neq f(j)$ for j in $V(K_n)\backslash i$. This implies that $g(i) = f(i)$ and hence that $g = f$.

For any f in the loopless part of $K_n^{K_n}$, there are at least two distinct vertices i and j such that $f(i) = f(j)$, and we can define a proper n-colouring of this part of $K_n^{K_n}$ as in Theorem 6.6.2. □

The next result is remarkably useful.

Theorem 6.6.5 *If X is connected and not n-colourable, then K_n^X contains a unique n-clique, namely the constant functions.*

Proof. By Lemma 6.4.4, the subgraph of K_n^X induced by the constant functions is an n-clique. We need to prove this is the only n-clique.

If $\chi(X) > n$ and f is a homomorphism from X to $K_n^{K_n}$, then, by the previous theorem, f must map each vertex of X onto the same loop of $K_n^{K_n}$. Since K_n has exactly $n!$ proper n-colourings, $K_n^{K_n}$ has exactly $n!$ loops and therefore

$$\begin{aligned} n! &= |\mathrm{Hom}(X, K_n^{K_n})| \\ &= |\mathrm{Hom}(K_n \times X, K_n)| \\ &= |\mathrm{Hom}(X \times K_n, K_n)| \\ &= |\mathrm{Hom}(K_n, K_n^X)|. \end{aligned}$$

Thus there are exactly $n!$ homomorphisms from K_n into K_n^X, and therefore K_n^X contains a unique n-clique. □

The above proof shows that if $\chi(X) > n$, then there are exactly $n!$ homomorphisms from $X \times K_n$ to K_n. Hence $X \times K_n$ is uniquely n-colourable. (For more about this, see the next section.)

Theorem 6.6.6 *Suppose $n \geq 2$ and let X and Y be connected graphs, each containing an n-clique. If X and Y are not n-colourable, neither is $X \times Y$.*

Proof. Let $x_1, \ldots, x_n$ and $y_1, \ldots, y_n$ be the respective n-cliques in X and Y and suppose, by way of contradiction, that there is a homomorphism f from $X \times Y$ into K_n. Consider the induced homomorphism from Y into K_n^X. By Theorem 6.6.5, the image of $y_1, \ldots, y_n$ in K_n^X consists of the constant maps. In other words, $f(x, y_j)$, viewed as a function of x, is constant for each y_j. A similar argument yields that $f(x_i, y)$ is constant as a function of y. Then $f(x_1, y_1) = f(x_1, y_2) = f(x_2, y_2)$, so the adjacent vertices (x_1, y_1) and (x_2, y_2) are mapped to the same vertex of K_n, contradicting the assumption that f is a homomorphism. □

One consequence of this theorem is that if X and Y are not bipartite, then neither is $X \times Y$.

Corollary 6.6.7 *Let X be a graph such that every vertex lies in an n-clique and $\chi(X) > n$. If Y is a connected graph with $\chi(Y) > n$, then $\chi(X \times Y) > n$.*

Proof. Suppose by way of contradiction that there is a homomorphism f from $X \times Y$ into K_n. Then consider the induced mapping Φ_f from X into K_n^Y. Because K_n^Y has no loops, every n-clique in X is mapped injectively onto the unique n-clique in K_n^Y. Every vertex of X lies in an n-clique, and so every vertex of X is mapped to this n-clique. Therefore, Φ_f is a homomorphism from X into K_n, which is a contradiction. □

6.7 Uniquely Colourable Graphs

If X is a graph with chromatic number n, then each n-colouring of X determines a partition of $V(X)$ into n independent sets; conversely, each partition of $V(X)$ into n independent sets gives rise to exactly $n!$ proper n-colourings. We say that a graph is *uniquely n-colourable* if it has chromatic number n, and there is a unique partition of its vertex set into n independent sets. It is not hard to see that if a graph X has at least n vertices, then it is uniquely n-colourable if and only if there are exactly $n!$ homomorphisms from X to K_n. The simplest examples are the connected bipartite graphs with at least one edge, which are uniquely 2-colourable.

There are a number of conjectures concerning uniquely colourable graphs related to Hedetniemi's conjecture. The connections arise because of our next result, which is implicit in the proof of Theorem 6.6.5, as we noted earlier.

Theorem 6.7.1 *If X is a connected graph with $\chi(X) > n$, then $X \times K_n$ is uniquely n-colourable.* □

We have the following generalization of the first part of Theorem 6.6.4.

Lemma 6.7.2 *If X is uniquely n-colourable, then each proper n-colouring of X is an isolated vertex in K_n^X.*

Proof. Let f be a proper n-colouring of X and let x be a vertex in X. Since X is uniquely n-colourable, each of the $n-1$ colours other than $f(x)$ must occur as the colour of a vertex in the neighbourhood of x. It follows that if $g \sim f$, then $g(x) = f(x)$, and so the only vertex of K_n^X adjacent to f is f itself. □

Let $\lambda(K_n^X)$ denote the subgraph of K_n^X induced by its loopless vertices. We can state the following conjectures:

(B_n) If X is uniquely n-colourable and Y is a connected graph that is not n-colourable, then $X \times Y$ is uniquely n-colourable.
(D_n) If X is uniquely n-colourable, then the subgraph of K_n^X induced by its loopless vertices is n-colourable.
(H_n) If $\chi(X) = \chi(Y) = n+1$, then $\chi(X \times Y) = n+1$.

The conjecture that (H_n) holds for all positive integers n is equivalent to Hedetniemi's conjecture. We will show that

$$(B_n) \Leftrightarrow (D_n) \Rightarrow (H_n).$$

Suppose that (B_n) holds, and let X be uniquely n-colourable. If Y is any subgraph of $\lambda(K_n^X)$, then there are more than $n!$ homomorphisms from Y into K_n^X (there is one homomorphism for each of the $n!$ loops, along with the identity map), and so

$$|\mathrm{Hom}(X \times Y, K_n)| = |\mathrm{Hom}(Y, K_n^X)| \geq n! + 1.$$

This shows that $X \times Y$ is not uniquely n-colourable, whence (B_n) implies that $\chi(Y) \leq n$. Hence (B_n) implies (D_n).

But (D_n) implies (B_n) too. For if Y is connected and $\chi(Y) > n$ and $\lambda(K_n^X)$ is n-colourable, then the only homomorphisms from Y to K_n^X are the maps onto the loops. Therefore,

$$n! = |\mathrm{Hom}(Y, K_n^X)| = |\mathrm{Hom}(X \times Y, K_n)|,$$

with the implication that $X \times Y$ is uniquely n-colourable.

We will use the next lemma to show that (D_n) implies (H_n) (which, we recall, is Hedetniemi's conjecture).

Lemma 6.7.3 *If $\chi(X) > n$, then there is a homomorphism from K_n^X to the subgraph of $K_n^{X \times K_n}$ induced by the loopless vertices.*

Proof. Let p_X be the projection homomorphism from $X \times K_n$ to X, and let φ be the induced mapping from K_n^X to $K_n^{X \times K_n}$. (See Theorem 6.4.1, where this was introduced.) If $g \in K_n^X$, then g is not a proper colouring of X, and so there are adjacent vertices u and v in X such that $g(u) = g(v)$. Now, $\varphi(g) = g \circ p_X$, whence $\varphi(g)$ maps (u, i) and (v, j) to $g(u)$, for any vertices i and j in K_n. Hence $\varphi(g)$ is not a proper colouring of $X \times K_n$, which means that it is not a loop. □

So now suppose that (D_n) holds and let X be a graph with $\chi(X) > n$. By Theorem 6.7.1, $X \times K_n$ is uniquely n-colourable, and so (D_n) implies that $\lambda(K_n^{X \times K_n})$ is n-colourable. Hence, by the lemma, K_n^X is n-colourable, and Hedetniemi's conjecture holds.

6.8 Foldings and Covers

We call a homomorphism from X to Y a *simple folding* if it has one fibre consisting of two vertices at distance two, and all other fibres are singletons. For example, the two homomorphisms from the path on three vertices to K_2 are simple foldings. A homomorphism is a *folding* if it is the composition of a number of simple foldings.

Lemma 6.8.1 *If f is a retraction from a connected graph X to a proper subgraph Y, then it is a folding.*

Proof. We proceed by induction on the number of vertices in X. Suppose f is a retraction from X to Y, that is, f is a homomorphism from X to Y and $f \upharpoonright Y$ is the identity. If $X = Y$, we have nothing to prove. Otherwise, since X is connected, there is a vertex y in Y adjacent to a vertex x not in Y. Now, f fixes y and maps x to some neighbour, z say, of y in Y.

Let π be the partition of $V(X)$ with $\{x, z\}$ as one cell and with all other cells singletons. There is a homomorphism f_1 from X to a graph X_1 with kernel π. Since the kernel of f_1 is a refinement of the kernel of f, there is a homomorphism f_2 from X_1 to Y such that $f_2 \circ f_1 = f$. Since f_1 maps each vertex in Y to itself, it follows that Y is a subgraph of X_1, and f_2 is a retraction from X_1 to Y. Finally, f_1 is a simple folding, and by induction, we may assume that f_2 is a folding. This proves the lemma. □

We call a homomorphism a *local injection* if the minimum distance between two vertices in the same fibre is at least three. Clearly, any automorphism is a local injection.

Lemma 6.8.2 *Every homomorphism h from X to Y can be expressed as the composition $f \circ g$, where g is a folding and f a local injection.*

Proof. Let π be the kernel of h. If u and v are vertices of X, write $u \approx v$ if u and v lie in the same cell of π and are equal or at distance two in X. This is a symmetric and reflexive relation on the vertices of X. Hence its transitive closure is an equivalence relation, which determines a partition π' of $V(X)$. There is a homomorphism g from X to X/π' with kernel π' and a homomorphism f from X/π' to Y such that $h = f \circ g$.

Clearly, g is a folding. We complete the proof by showing that f is a local injection. Assume by way of contradiction that α and β are vertices in X/π' at distance two such that $f(\alpha) = f(\beta)$. Let γ be a common neighbour of α and β in X/π'. There must be a vertex u of X in $g^{-1}(\alpha)$ adjacent (in X) to a vertex u' in $g^{-1}(\gamma)$, and a vertex v of X in $g^{-1}(\beta)$ adjacent (in X) to a vertex v' in $g^{-1}(\gamma)$. But u' and v' are joined in X by a walk of even length, hence this holds true for u and v as well. This implies that u and v must lie in the same cell of π', a contradiction that completes the proof. □

A homomorphism f from X to Y is a *local isomorphism* if for each vertex y in Y, the induced mapping from the set of neighbours of a vertex in $f^{-1}(y)$ to the neighbours of y is bijective. We call f a *covering map* if it is a surjective local isomorphism, in which case we say that X *covers* Y. If f is a local isomorphism, then each fibre is an independent set of vertices in X, and between each fibre there are either no edges or there is a matching. If the image of X is connected, then each fibre has the same size. This number is called the *index* r of the cover, and X is said to be an *r-fold cover* of Y. There may be more than one covering map from a graph X to a graph Y, so we define a *covering graph* X of Y to be a pair (X, f), where f is a local isomorphism from X to Y.

If (X, f) is a cover of Y and Y_1 is an induced subgraph of Y, then $f^{-1}(Y_1)$ covers Y_1. This means that questions about covers of Y can be reduced to questions about the covers of its components. If Y is a connected graph and (X, f) is a cover of Y, then each component of X covers Y. (We leave the proof of this as an exercise.)

Our next result is a simple but fundamental property of covering maps.

Lemma 6.8.3 *If X covers Y and Y is a tree, then X is the disjoint union of copies of Y.*

Proof. Suppose f is a covering map from X to Y. Since f is a local isomorphism, if $x \in V(X)$, then the valency of $f(x)$ in Y equals the valency of x in X. This implies that the image of any cycle in X is a cycle in Y, and hence the girth of X cannot be less than the girth of Y. Thus, since Y is acyclic, so is X.

A local isomorphism is locally surjective; hence if $f(x) = y$, then each edge on y is the image under f of an edge on x. It follows that any path in Y that starts at y is the image under f of a path in X that starts at x (and this is also true for walks). Therefore, there is a tree T in X such that f is an isomorphism from T to Y. Hence Y is a retract of X, and as each component of X covers Y, it follows from Lemma 6.8.1 that X is the disjoint union of copies of Y. □

We say that a cover (X, f) of index r over Y is *trivial* if X is isomorphic to r vertex disjoint copies of Y and the restriction of f to any copy of Y is an isomorphism. The previous lemma implies that any cover of a tree is trivial.

Interesting covers are surprisingly common. The cube Q has the property that for each vertex x there is a unique vertex in Q at distance three from x. Thus $V(Q)$ can partitioned into four pairs, and these pairs are the fibres of a covering map from Q onto K_4 (see Figure 6.1).

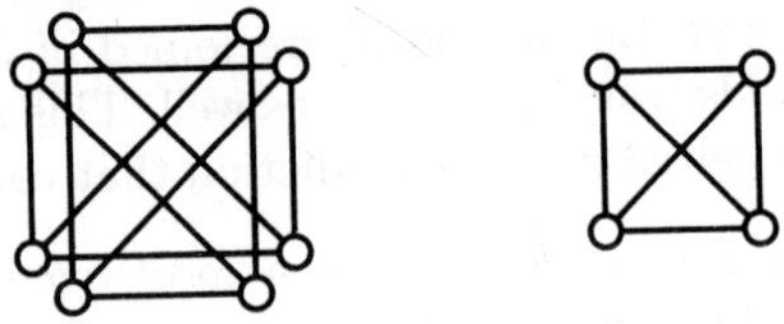

Figure 6.1. The cube is a 2-fold cover of K_4

Similarly, the dodecahedron covers the Petersen graph and the line graph of the Petersen graph covers K_5. The 42 vertices at distance two from a fixed vertex in the Hoffman–Singleton graph form a 6-fold cover of K_7. For any graph X, the product $X \times K_2$ is a 2-fold cover of X. In Chapter 11 we will study two-graphs, which can be defined as 2-fold covers of complete graphs.

If (X, f) and (Y, g) are covers of F, then so is their subdirect product. (The proof is left as an exercise. We defined the subdirect product in Section 6.3.)

6.9 Cores with No Triangles

We showed in Section 6.2 that every graph has a core, but despite this, it is not trivial to provide examples of cores. Critical graphs provide one such class. So far, the only critical graphs we have identified are the odd cycles and the complete graphs. There are many critical graphs known that are more interesting, but we do not consider them here, since we have nothing to say about them from an algebraic viewpoint. Since any homomorphism must map triangles to triangles, it would seem comparatively easy to construct examples of cores that contain many triangles. In this section we therefore take a more difficult route, and construct examples of cores without triangles.

We begin by deriving a simple sufficient condition for a graph to be a core.

Lemma 6.9.1 *Let X be a connected nonbipartite graph. If every 2-arc lies in a shortest odd cycle of X, then X is a core.*

Proof. Let f be a homomorphism from X to X. This necessarily maps a shortest odd cycle of X onto an odd cycle of the same length, so any two vertices in the cycle have different images under f. Since every 2-arc lies in a shortest odd cycle, this shows that f is a local injection, and hence by Lemma 6.8.1, it cannot map X onto a proper subgraph of itself. □

If two vertices u and v in a graph X have identical neighbourhoods, then X is certainly not a core, for there is a retraction from X to $X \setminus u$. This motivates the following definition: A graph is *reduced* if it has no isolated vertices and the neighbourhoods of distinct vertices are distinct.

Now, suppose that X is a triangle-free graph. If u and v are two vertices in X at distance at least three, then the graph obtained by adding the edge uv is also triangle free. Continuing this process, we see that any triangle-free graph X is a spanning subgraph of a triangle-free graph with diameter two.

We note a useful property of reduced triangle-free graphs with diameter two.

Lemma 6.9.2 *Let X be a reduced triangle-free graph with diameter two. For any pair of distinct nonadjacent vertices u and v, there is a vertex adjacent to u but not to v.*

Proof. Suppose for a contradiction that $N(u) \subseteq N(v)$. Since X is reduced, there is some vertex w adjacent to v but not to u. Since X has no triangles, w is not adjacent to any neighbour of u, which implies that the distance between u and w is at least three. □

This last result enables us to characterize a class of cores.

Lemma 6.9.3 *Let X be a triangle-free graph with diameter two. Then X is a core if and only if it is reduced.*

Proof. Our comments above establish that a graph that is not reduced is not a core. So we assume that X is reduced and show that each 2-arc in X lies in a 5-cycle, whence the result follows from Lemma 6.9.1.

Assume that (u, v, w) is a 2-arc. Then w is at distance two from u. Since $N(w)$ is not contained in $N(u)$, there is a neighbour, w' say, of w at distance two from u. Now, w' must have a neighbour, v' say, adjacent to u. Since X has no triangles, $v' \neq v$, and therefore (u, v, w, w', v') is a 5-cycle. □

It follows immediately that any reduced triangle-free graph of diameter two is a core. The graph obtained by deleting a vertex from the Petersen graph is triangle free with diameter three, but any proper homomorphic image of it contains a triangle, thus showing that the condition in the lemma is not necessary.

6.10 The Andrásfai Graphs

We define a family of Cayley graphs And(k), each of which is a reduced triangle-free graph with diameter two. For any integer $k \geq 1$, let $G = \mathbb{Z}_{3k-1}$ denote the additive group of integers modulo $3k - 1$ and let C be the subset of $\mathbb{Z}_{3k-1}$ consisting of the elements congruent to 1 modulo 3. Then we denote the Cayley graph $X(G, C)$ by And(k). The graph And(2) is isomorphic to the 5-cycle, And(3) is known as the *Möbius ladder* (see Exercise 44), and And(4) is depicted in Figure 6.2.

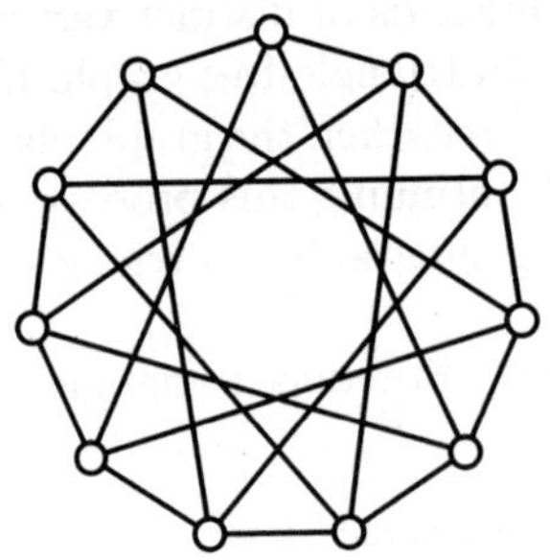

Figure 6.2. $\text{And}(4) = X(\mathbb{Z}_{11}, \{1, 4, 7, 10\})$

Lemma 6.10.1 *For $k \geq 2$, the Cayley graph* And(k) *is a reduced triangle-free graph with diameter two.*

Proof. First we show that And(k) is reduced. If And(k) has two distinct vertices with the same neighbours, then there must be an element g in G such that $g \neq 0$ and $g + C = C$. It follows that both $g + 1$ and $g - 1$ lie in C, which is impossible, since they are not both congruent to 1 modulo 3.

Next we show that And(k) has no triangles containing the vertex 0. Let g and h be two neighbours of 0. Then g and h are in C, and so $g - h$ is zero modulo 3. Thus $g - h \notin C$, and so g is not adjacent to h. Since And(k) is transitive, this suffices to show that it is triangle-free.

Finally, we show that And(k) has diameter two, by showing that there is a path of length at most two from 0 to any other vertex. If $g = 3i$, then the path $(0, 3i + 1, 3i)$ has length two, and if $g = 3i + 2$, then the path $(0, 3i + 1, 3i + 2)$ has length two. Every other vertex is adjacent to 0, and so the result follows. □

Let X be a reduced triangle-free graph with diameter two, and let S be an independent set in X. If S is an independent set that is maximal under inclusion, then every vertex of X is adjacent to at least one vertex in S. The graph we get by taking a new vertex and joining it to each vertex of S is triangle-free with diameter two. Provided that S is not the neighbourhood of a vertex, this new graph is also reduced. This gives us a procedure for embedding each reduced triangle-free graph with diameter two as an induced subgraph of a reduced triangle-free graph with diameter two with one more vertex—unless each independent set in X is contained in the neighbourhood of a vertex. This observation should make the following result more interesting.

Lemma 6.10.2 *Each independent set of vertices of* And(k) *is contained in the neighbourhood of a vertex.*

Proof. Consider the $k-1$ pairs of adjacent vertices of the form $\{3i, 3i-1\}$ for $1 \leq i < k$. Now, any vertex $g \in C$ has the form $3q + 1$. If $q \geq i$, then g is adjacent to $3i$, but not to $3i-1$. If $q < i$, then g is adjacent to $3i-1$, but not to $3i$. Therefore, every vertex in C is adjacent to precisely one vertex from each pair.

Suppose now that there is an independent set not contained in the neighbourhood of a vertex. Then we can find an independent set S and an element x such that $S \cup x$ is independent, S is in the neighbourhood of a vertex, but $S \cup x$ is not in the neighbourhood of a vertex. By the transitivity of And(k) we can assume that S is in the neighbourhood of 0. Since x is not adjacent to 0, either $x = 3i$ or $x = 3i - 1$ for some i. If $x = 3i$, then every vertex of S is not adjacent to x, and so is adjacent to $x - 1$, which implies that $S \cup x$ is in the neighbourhood of $x - 1$. Finally, if $x = 3i - 1$, then every vertex of S is not adjacent to x and so is adjacent to $x + 1$, which implies that $S \cup x$ is in the neighbourhood of $x + 1$. Therefore, $S \cup x$ is in the neighbourhood of some vertex, which is a contradiction. □

6.11 Colouring Andrásfai Graphs

In the next section we will use the following properties of Andrásfai graphs to characterize them.

Lemma 6.11.1 *If $k \geq 2$, then the number of 3-colourings of* And(k) *is* $6(3k-1)$*, and they are all equivalent under its automorphism group.*

Proof. In any 3-colouring of And(k), the average size of a colour class is $k - \frac{1}{3}$. Since the maximum size of a colour class is k, two colour classes have size k, and the third has size $k - 1$. By Lemma 6.10.2, the two big colour classes are neighbourhoods of vertices.

Suppose that we have a 3-colouring of And(k) and that one of the big colour classes consists of the neighbours of 0; this colour class may be any of the three colours.

Now, consider the set of vertices not adjacent to 0. This can be partitioned into the two sets

$$A := \{2, 5, \ldots, 3k - 4\}, \quad B := \{3, 6, \ldots, 3k - 3\}.$$

It is immediate that A and B are independent sets and that the ith vertex of A is adjacent to the last $k - i$ vertices of B (Figure 6.3 shows this for And(4)). Hence $A \cup B$ induces a connected bipartite graph, and can therefore be coloured in exactly two ways with two colours.

There are now two choices for the colour assigned to 0, and so in total there are 12 distinct colourings with the neighbours of 0 as a big colour class. Since And(k) is transitive with $3k - 1$ vertices, and each 3-colouring has two big colour classes, the first claim follows.

The permutation that exchanges i and $-i$ modulo $3k - 1$ is an automorphism of And(k) that exchanges A and B, and hence the second claim follows. □

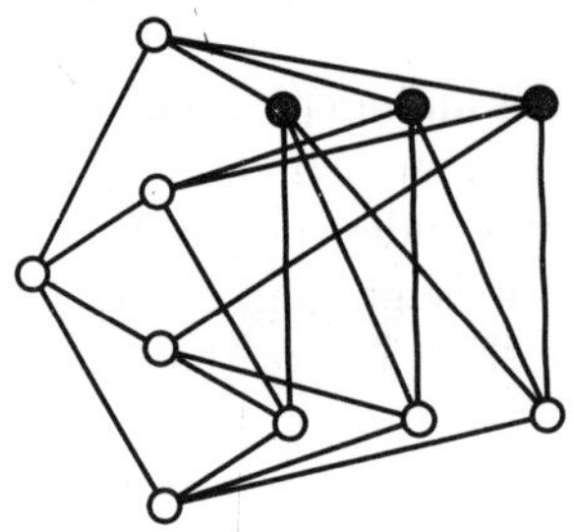

Figure 6.3. Another view of And(4)

We note another property of the Andrásfai graphs. The subgraph of And(k) induced by $\{0, 1, \ldots, 3(k-1) - 2\}$ is And($k-1$). Therefore, we can get And($k - 1$) from And(k) by deleting the path $(3k - 4, 3k - 3, 3k - 2)$.

Lemma 6.11.2 *Let X be a triangle-free regular graph with valency $k \geq 2$, and suppose that P is a path of length two in X. If $X \setminus P \cong$ And$(k - 1)$, then $X \cong$ And(k).*

Proof. Let P be the path (u, v, w) and let Y denote $X \setminus P$. Since X is triangle-free and regular, the neighbours of u, v, and w that are in Y form independent sets of size $k - 1$, $k - 2$, and $k - 1$ respectively. Since Y is regular of valency $k-1$, each vertex of Y is adjacent to precisely one vertex of P. Therefore, these independent sets are a 3-colouring of Y. Since all 3-colourings of Y are equivalent under $\mathrm{Aut}(Y)$, the result follows. □

6.12 A Characterization

The condition that each independent set lies in the neighbourhood of a vertex is very strong: We show that a reduced triangle-free graph with this property must be one of the Cayley graphs $\mathrm{And}(k)$. This is surprising, because we know very few interesting cases where a simple combinatorial condition implies so much symmetry.

Theorem 6.12.1 *If X is a reduced triangle-free graph such that each independent set in X is contained in the neighbourhood of a vertex, then X is an Andrásfai graph.*

Proof. We break the proof into a number of steps.

(a) *If u and v are distinct nonadjacent vertices, then there is a unique vertex $\sigma_u(v)$ adjacent to v but not u and such that*

$$X_1(u) \cap X_1(\sigma_u(v)) = X_1(u) \setminus (X_1(u) \cap X_1(v)).$$

The set of vertices

$$\{v\} \cup (X_1(u) \setminus X_1(v))$$

is independent and, by Lemma 6.9.2, contains at least one vertex of $X_1(u)$. Therefore, it is contained in the neighbourhood of some vertex, w say, not adjacent to u. We will take $\sigma_u(v)$ to be w, but must show that it is unique. Suppose for a contradiction that there is another vertex w' such that w' is adjacent to all the vertices in the above set. Then w and w' are not adjacent, and so there is a vertex x that is adjacent to w but not w'. However, this implies that $\{u, x, w'\}$ is an independent set. Any vertex adjacent to u and w' is adjacent to w, and therefore it cannot be adjacent to x (because w is). Thus we have an independent set that is not contained in the neighbourhood of a vertex, which is the required contradiction. This implies that σ_u is a fixed-point-free involution on the set of vertices at distance two from u.

(b) *X is a k-regular graph on $3k-1$ vertices.*

Let u be a vertex of X, and consider the edges between $X_1(u)$ and $X_2(u)$. Every vertex of $X_1(u)$ is adjacent to exactly one vertex from each pair $\{v, \sigma_u(v)\}$. Therefore, every vertex in $X_1(u)$ has the same valency, which implies that every pair of vertices at distance two has the same valency. Consequently, either X is bipartite or it is regular of valency k. If X has two distinct nonadjacent vertices u and v, then u, v, and $\sigma_u(v)$ lie in a 5-cycle, and so X is not bipartite. Therefore, if X is bipartite, it is complete and thus equal to K_2, which is And(1). If it is regular of valency k, then for any vertex u, there are $k-1$ pairs $\{v, \sigma_u(v)\}$ in $X_2(u)$, and hence $|X_2(u)| = 2(k-1)$.

(c) *If $k \geq 2$, then for each vertex u in X there is a vertex w such that u and w have a unique common neighbour.*

First we show that there is a vertex v such that u and v have $k-1$ common neighbours. Let v be a vertex with the largest number of common neighbours with u, and suppose that they have $k-s$ common neighbours, where $s \geq 1$. Let U denote $X_1(u) \backslash X_1(v)$ and V denote $X_1(v) \backslash X_1(u)$. Then both U and V contain s vertices; moreover, U contains $\sigma_v(u)$ and V contains $\sigma_u(v)$.

If $s > 1$, then V contains a vertex v' other than $\sigma_u(v)$. Since $\sigma_u(v)$ is the unique vertex adjacent to v and everything in U, it follows that there is some vertex $u' \in U$ not adjacent to v'. Then

$$\{u', v'\} \cup (X_1(u) \cap X_1(v))$$

is independent and hence contained in the neighbourhood of a vertex $w \neq u$. Therefore, u and w have at least $k-s+1$ common neighbours. By the choice of v, this cannot occur, and therefore $s = 1$, and u and v have $k-1$ common neighbours.

If u and v have $k-1$ common neighbours, then u and $w = \sigma_u(v)$ have one common neighbour, and the claim is proved.

(d) *Let $k \geq 2$, and suppose that $P = (u, v, w)$ is a path in X such that v is the unique common neighbour of u and w. Then $X \setminus P$ is a reduced triangle-free graph such that every independent set is in the neighbourhood of a vertex.*

If Y denotes $X \backslash P$, then it is immediate that Y is a triangle-free graph, and so we must show that every independent set of Y lies in the neighbourhood of a vertex, and that it is reduced.

Let U, V, and W be the neighbours of u, v, and w, respectively, in Y. Since no vertex of Y is in two of these sets, and they have sizes $k-1$, $k-2$, and $k-1$, respectively, these three sets partition $V(Y)$.

Suppose that S is an independent set in Y, and hence an independent set in X. Then S lies in the neighbourhood of a vertex, x say, in X. If x is not in $\{u, v, w\}$, then it is a vertex of Y, and there is nothing to prove. If $x = u$, then $S \subseteq U$ and since the vertex $\sigma_u(v)$ is adjacent to everything in U, the set S is in the neighbourhood of $\sigma_u(v)$. An analogous argument deals with the case where $x = w$. For the final case, where $x = v$, note that v and $\sigma_u(w)$ have w as their unique common neighbour, and therefore $\sigma_v(\sigma_u(w))$ lies in U and is adjacent to everything in V. Therefore, we conclude that in every case S lies in the neighbourhood of a vertex in Y.

Finally, we prove that Y is reduced, by showing that any two vertices x and y have different neighbourhoods. If x and y are both in U, then since they have different neighbourhoods in X, they have different neighbourhoods in Y. The same argument applies if they are both in V or both in W. So suppose that x and y are in different sets from $\{U, V, W\}$, and without loss of generality we assume that $x \in U$. Then $\sigma_w(u)$ is in U and is adjacent to everything in W, and $\sigma_v(\sigma_u(w))$ is in U and is adjacent to everything in V. Therefore, every vertex in $V(Y)\backslash U$ is adjacent to a vertex in U and so cannot have the same neighbourhood as x.

(e) *X is an Andrásfai graph.*

It is easy to check that And(2) is the unique reduced triangle-free graph with valency two such that every independent set is in the neighbourhood of a vertex. The result follows by induction using (d) and Lemma 6.11.2. □

6.13 Cores of Vertex-Transitive Graphs

In this section we consider some further simple techniques that allow us to identify classes of cores. Although these techniques use the theory of homomorphisms that we have developed in a relatively elementary way, we can get some quite strong results that would be difficult to prove without using homomorphisms. The first result is quite surprising, as it provides a somewhat unexpected connection between homomorphisms and automorphisms.

Theorem 6.13.1 *If X is a vertex-transitive graph, then its core $X^\bullet$ is vertex transitive.*

Proof. Let x and y be two distinct vertices of $X^\bullet$. Then there is an automorphism of X that maps x to y. The composition of this automorphism with a retraction from X to $X^\bullet$ is a homomorphism f from X to $X^\bullet$. The restriction $f\restriction X^\bullet$ is an automorphism of $X^\bullet$ mapping x to y. □

The graph of Figure 6.4 is an example of a quartic vertex-transitive graph whose core is the vertex-transitive graph C_5.

Theorem 6.13.2 *If X is a vertex-transitive graph, then $|V(X^\bullet)|$ divides $|V(X)|$.*

Proof. We show that the fibres of any homomorphism from X to $X^\bullet$ have the same size. Let f be a homomorphism from X to X whose image Y is a core of X. For any element g of $\text{Aut}(X)$, the translate Y^g is mapped onto Y by f, and therefore Y^g has one vertex in each fibre of f.

Now, suppose $v \in V(X)$ and let F be the fibre of f that contains v. Since X is vertex transitive, the number of automorphisms g such that Y^g

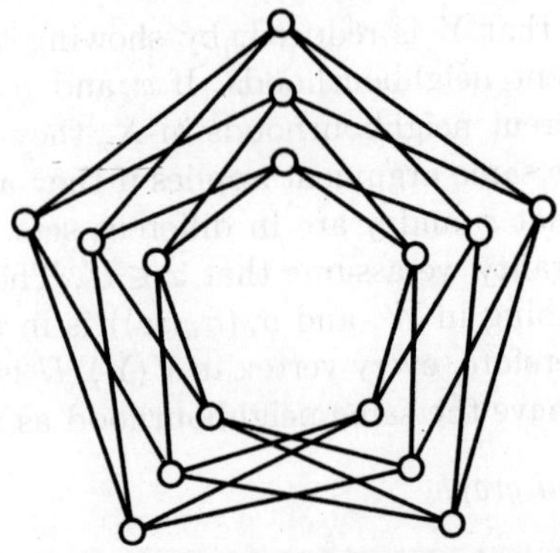

Figure 6.4. Quartic vertex-transitive graph with core C_5

contains v is independent of our choice of v. If we denote this number by N, then since every image Y^g of Y meets F,

$$|\mathrm{Aut}(X)| = |F|N.$$

Since N does not depend on F, this implies that all fibres of f have the same size. □

This result has an immediate corollary that provides us with further large classes of cores.

Corollary 6.13.3 *If X is a nonempty vertex-transitive graph with a prime number of vertices, then X is a core.* □

More surprisingly, it also yields an elegant proof of a result in graph colouring theory.

Corollary 6.13.4 *Let X be a vertex-transitive graph on n vertices with chromatic number three. If n is not a multiple of three, then X is triangle-free.*

Proof. Since X is 3-colourable, it has a homomorphism onto K_3. If X contained a triangle, then the core of X would be a triangle and n would be a multiple of three, contradicting the hypothesis. Therefore, X has no triangles. □

This result can easily be generalized to other chromatic numbers, as you are asked to show in Exercise 41.

To complete this section, we note another application of Lemma 6.9.1.

Theorem 6.13.5 *If X is a connected 2-arc transitive nonbipartite graph, then X is a core.*

Proof. Since X is not bipartite, it contains an odd cycle; since X is 2-arc transitive, each 2-arc lies in a shortest odd cycle. □

This provides a simple proof that the Petersen graph and the Coxeter graph are cores; alternative proofs seem to require tedious case arguments. By Lemma 4.1.2 and the previous theorem, we see that the Kneser graphs $J(2k+1, k, 0)$ are cores; in Chapter 7 we will show that all Kneser graphs are cores.

6.14 Cores of Cubic Vertex-Transitive Graphs

The cycles are the only connected vertex-transitive graphs of valency two, so cubic vertex-transitive graphs are the first interesting vertex-transitive graphs, and as such, they have been widely studied. In this section we consider cores of cubic vertex-transitive graphs, strengthen some of the results of Section 6.13, and provide some interesting examples.

We start by showing that a connected cubic graph is a core if it is arc transitive, thus strengthening Theorem 6.13.5.

Theorem 6.14.1 *If X is a connected arc-transitive nonbipartite cubic graph, then X is a core.*

Proof. Let C be a shortest odd cycle in X, and let x be a vertex in C with three neighbours x_1, x_2, and x_3, where x_1 and x_2 are in C. If G is the automorphism group of X, then the vertex stabilizer G_x contains an element g of order three, which can be taken without loss of generality to contain the cycle $(x_1x_2x_3)$. The 2-arc $\beta = (x_1, x, x_2)$ is in a shortest odd cycle, and therefore so are $\beta^g = (x_2, x, x_3)$ and $\beta^{gg} = (x_3, x, x_1)$. Hence any 2-arc with x as middle vertex lies in a shortest odd cycle, and because X is vertex transitive, the same is true for every 2-arc. Thus by Lemma 6.9.1, X is a core. □

We note in passing that there are cubic graphs that satisfy the condition of Lemma 6.9.1 that are not arc transitive. For example, Figure 6.5 shows two such graphs that are not even vertex transitive.

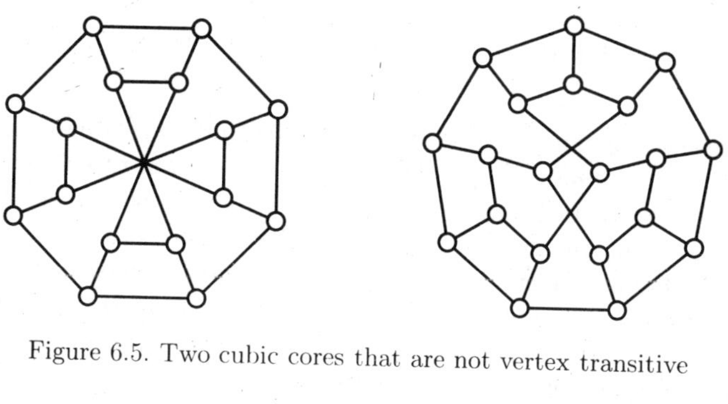

Figure 6.5. Two cubic cores that are not vertex transitive

It is easy to show that a graph with maximum valency Δ can be properly coloured with $\Delta + 1$ colours. The following useful strengthening of this observation is a standard result from graph theory, known as Brooks's theorem.

Theorem 6.14.2 (Brooks) *If X is a connected graph of maximum valency Δ that is neither complete nor an odd cycle, then the chromatic number of X is at most Δ.* □

Theorem 6.14.3 *If X is a connected vertex-transitive cubic graph, then $X^{\bullet}$ is K_2, an odd cycle, or X itself.*

Proof. The proof of this is left as Exercise 42. □

This theorem raises the question as to whether we can identify cubic vertex-transitive graphs whose cores are odd cycles. We content ourselves with presenting an interesting example, which is the smallest cubic vertex-transitive graph after the 10-vertex ladder that has core C_5. First some notation: Given a graph X, a *truncation* of X is a graph Y obtained by replacing each vertex v of valency k with k new vertices, one for each edge incident to v. Pairs of vertices corresponding to the edges of X are adjacent in Y, and the k vertices of Y corresponding to a single vertex of X are joined in a cycle of length k. If $k = 3$, then there is only one way to do this, but otherwise the order in which the k vertices are joined must be specified. If the graph X is embedded in a surface, then there is a "natural" truncation obtained by joining the k vertices in the cyclic order given by the embedding (see Figure 6.6).

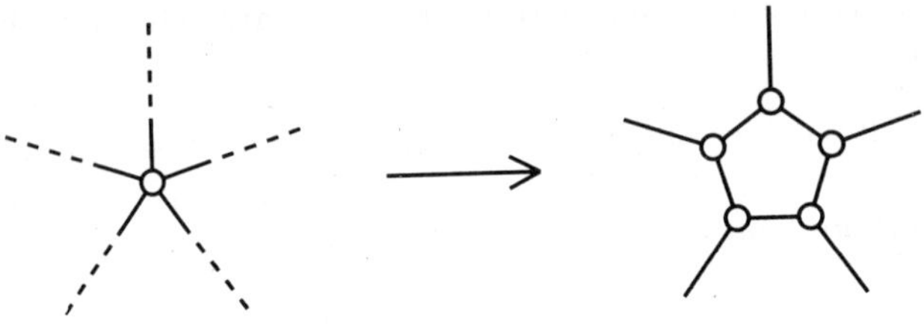

Figure 6.6. Truncating a vertex of valency k

The graph K_6 can be embedded in the real projective plane as we saw in Figure 1.13. Truncating this graph yields the vertex-transitive graph on 30 vertices shown in Figure 6.7 (also drawn in the real projective plane). The odd girth of this graph is five, so by Theorem 6.14.3, it is either a core or has a homomorphism onto C_5. In fact, it can be shown that it has a homomorphism onto C_5, as given by the colouring of the vertices in the figure. It is the second-smallest cubic vertex-transitive graph with core C_5 after the 10-vertex ladder (see Exercise 44).

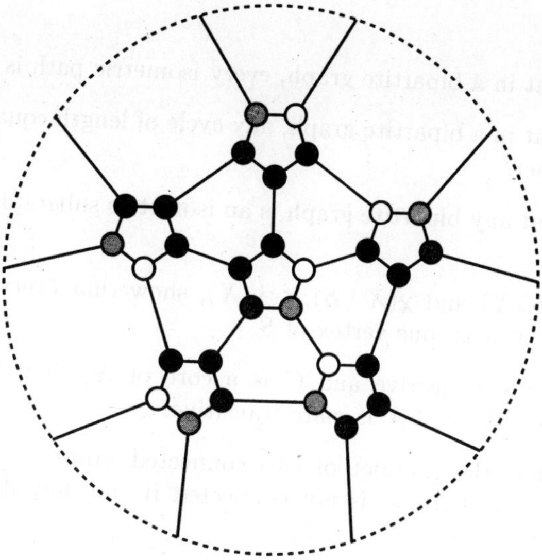

Figure 6.7. Cubic vertex-transitive graph with core C_5

Truncating the *icosahedron* embedded in the plane (Figure 6.8) yields a cubic vertex-transitive graph on 60 vertices (see Figure 9.5). The truncated icosahedron is well known because it describes the structure of the molecule C_{60} (here C stands for carbon not cycle!) known as buckminsterfullerene. Like the cube, the truncated icosahedron is antipodal; it is a 2-fold cover of the graph of Figure 6.7. One consequence of this is that the truncated icosahedron also has core C_5. We will meet the truncated icosahedron once again when we study fullerenes in Section 9.8.

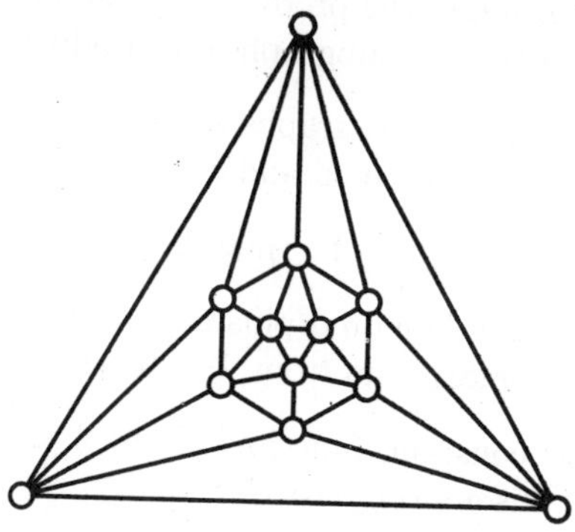

Figure 6.8. The icosahedron

Exercises

1. Show that in a bipartite graph, every isometric path is a retract.

2. Show that in a bipartite graph, any cycle of length equal to the girth is a retract.

3. Show that any bipartite graph is an isometric subgraph of a product of paths.

4. If $S \subseteq V(X)$ and $\chi(X \setminus S) < \chi(X)$, show that every retract of X contains at least one vertex of S.

5. If X is arc transitive and C is a core of X, show that C is arc transitive. What if X is s-arc transitive?

6. Show that the product of two connected graphs X and Y (with at least two vertices) is not connected if and only if X and Y are bipartite.

7. Let X and Y be two graphs. Show that $\omega(X \times Y)$ is the minimum of $\omega(X)$ and $\omega(Y)$. Show that the odd girth of $X \times Y$ is the maximum of the odd girths of X and Y. (This implies that if X and Y are not bipartite, then neither is $X \times Y$.)

8. Show that $K_2 \times J(2k-1, k-1, 0) \cong J(2k, k, k-1)$. (See Section 1.6 if you have forgotten the notation.)

9. For $i = 1, 2$, let X_i and Y_i be graphs and let f_i be a homomorphism from X_i to Y_i. Show that the mapping that sends a vertex (x_1, x_2) in $X_1 \times X_2$ to $(f_1(x_1), f_2(x_2))$ is a homomorphism to $Y_1 \times Y_2$.

10. Suppose that for each pair of distinct vertices u and v in X, there is an r-colouring of X where u and v have different colours. Show that X is a subgraph of the product of some number of copies of K_r. Deduce that $\mathrm{And}(k)$ is a subgraph of a product of copies of K_3.

11. Let X and Y be fixed graphs. Show that if $|\mathrm{Hom}(X, Z)| = |\mathrm{Hom}(Y, Z)|$ for all Z, then X and Y are isomorphic.

12. Show that if $X \times X \cong Y \times Y$, then $X \cong Y$.

13. Show that there is a homomorphism from X into $X \times Y$ if and only if there is a homomorphism from X into Y.

14. Show that the constant functions from $V(X)$ to $V(F)$ induce a subgraph of F^X isomorphic to F.

15. If X is not bipartite, show that K_2^X is the disjoint union of K_2 with some (usually large) number of isolated vertices. Using this deduce that if $X \times Y$ is bipartite, then X or Y is bipartite.

16. Show that there is bijection between the arcs of F^X and the set of homomorphisms from $X \times K_2$ to F.

17. Show that for any graph X, the product $X \times K_n^X$ is n-colourable. (In fact, construct an explicit n-colouring, then prove the result again by counting homomorphisms.)

18. Suppose that there are graphs X and Y, neither n-colourable, such that $\chi(X \times Y) = n$. Show there are subgraphs X' and Y' of X and Y, respectively, such that $\chi(X') = \chi(Y') = n+1$ and $\chi(X' \times Y') = n$.

19. If X and Y are connected graphs, then show that K_n is a retract of $X \times Y$ only if it is a retract of X or a retract of Y.

20. Show that Hedetniemi's conjecture is equivalent to the statement that K_n is a retract of the product $X \times Y$ of two graphs only if it is a retract of X or a retract of Y.

21. Show that the subdirect product of two covers of Y is a cover of Y.

22. If X and Y are connected bipartite graphs, show that the two components of $X \times Y$ are subdirect products of X and Y.

23. Show that if (X, f) is a cover of the connected graph Y, then each component of X covers Y.

24. Show that a nontrivial automorphism of a connected cover that fixes each fibre must be fixed-point free.

25. Let X be a Moore graph of diameter two and valency m. Show that the $m^2 - m$ vertices at distance two from a fixed vertex in X form an $(m-1)$-fold cover of K_m.

26. Let X be the incidence graph of a projective plane of order n. (See Section 5.3 for details.) Let Y be the graph obtained from X by deleting an adjacent pair of vertices and all their neighbours. Show that the resulting graph is an n-fold cover of $K_{n,n}$.

27. A homomorphism f from a graph X to a graph Y determines a map, f' say, from $L(X)$ to $L(Y)$. Show that the following are equivalent:

    (a) f is a local injection,
    (b) f' is a homomorphism,
    (c) f' is a local injection.

28. Show that if f is a local injection from X to itself, then f is an automorphism.

29. Suppose the Cayley graph $X(G, C)$ for the group G is triangle-free. Show that there is a subset D of $G \backslash e$ such that $C \subseteq D$ and $X(G, D)$ is triangle-free and has diameter two.

30. The complement of $\text{And}(k)$ (see Section 6.9) has the property that the neighbourhood of each vertex is covered by two vertex-disjoint cliques. Prove or disprove that it is a line graph.

31. Show that the graph we get by deleting a vertex from the Petersen graph has the property that each proper homomorphic image contains a triangle.

32. Let X be a reduced triangle-free graph with diameter two. If u and v are nonadjacent vertices in X, show that there is a unique vertex v' adjacent to v such that each neighbour of u is adjacent to v or v'.

33. Show that the Cayley graph $\text{And}(k)$ does not contain an induced copy of C_6.

34. Show that a triangle-free graph X on n vertices that contains a subgraph isomorphic to the Möbius ladder on eight vertices (see Exercise 44 for the definition of Möbius ladder) has minimum valency at most $3n/8$. (Hint: Any independent set in $\text{And}(3)$ contains at most three vertices. Hence if $\text{And}(3)$ is a subgraph of X, then any vertex not in this subgraph has at most three neighbours in it.)

35. Show that a triangle-free graph on n vertices with minimum valency greater than $[3n/8]$ has a homomorphism into C_5.

36. Let r and k be integers such that $r \geq 3$ and $k \geq 2$. Let $\text{And}_r(k)$ denote the Cayley graph for $\mathbb{Z}_{(k-1)r+2}$ with connection set

$$C = \{1, r+1, \dots, (k-1)r+1\}.$$

Show that $\{0, 1, 2, \dots, r-1\}$ induces a path and

$$\text{And}_r(k) \setminus \{0, 1, 2, \dots, r-1\} \cong \text{And}_r(k-1).$$

37. Suppose $X = \text{And}(k)$, with vertex set $\{0, 1, \dots, 3k-2\}$. Show that $\sigma_i(i+2) = i+3$, where addition is modulo $3k-1$.

38. Prove or disprove: If $k \geq r$, then $\chi(\text{And}_r(k)) = r$.

39. Let H_k be the graph defined as follows. The vertices of H_k are the $3k-1$ vertices of a regular $(3k-1)$-gon inscribed in a circle. Two vertices are adjacent if and only if their distance is greater than the side of the regular triangle that can be inscribed in the circle. Show that H_k is a Cayley graph for $\mathbb{Z}_{3k-1}$, and then show that H_k is isomorphic to $\text{And}(k)$.

40. Show that the icosahedron is a core.

41. We saw (in Corollary 6.13.4) that if X is a vertex-transitive graph with $\chi(X) = 3$ and $|V(X)|$ is not divisible by three, then X is triangle-free. Find and prove an analogue of this result for graphs with chromatic number k.

42. Prove that if X is cubic and vertex transitive, then $X^\bullet$ is K_2, an odd cycle, or X itself. (Hint: Consider the odd girth of X).

43. Show that if X is a quartic vertex-transitive graph on an odd number of vertices, then its core is either complete, an odd cycle, or itself. What about quartic graphs on 2^n vertices?

44. The *ladder* $L(2n)$ is the cubic graph constructed as follows: Take two copies of the cycle C_n on disjoint vertex sets $\{a_1, \ldots, a_n\}$ and $\{b_1, \ldots, b_n\}$, and join the corresponding vertices $a_i b_i$ for $1 \le i \le n$. The Möbius ladder $M(2n)$ is obtained from the ladder by deleting the edges a_1a_2 and b_1b_2 and then inserting edges a_1b_2 and a_2b_1. Find the cores of $L(2n)$ and $M(2n)$ for all n.

45. Consider the cubic graph obtained by subdividing every edge of the cube and joining pairs of vertices corresponding to opposite edges (the first step in the construction of Tutte's 8-cage). Show that this graph is a core.

46. Let X be a graph and let C be a subset of $E(X)$. Construct a graph Y with vertex set

$$V(X) \times \{0, 1\}$$

as follows. If $uv \in C$, then $(u, 0) \sim (v, 1)$ and $(u, 1) \sim (v, 0)$ (in Y). If $uv \in E(X) \setminus C$, then $(u, 0) \sim (v, 0)$ and $(u, 1) \sim (v, 1)$. Show that Y is a double cover of X, and that the girth of Y is greater than r if and only if each cycle of X with length at most r contains an odd number of edges from C.

47. Let X be a graph and let C be a subset of $E(X)$. If $u \in V(X)$, let $S(u)$ denote the set of edges in X that are incident with u. Show that the double cover determined by C is isomorphic to the double cover determined by the symmetric difference of C and $S(u)$.

48. The aim of this exercise is to show that if $n \ge 2$, then there is a unique double cover of the n-cube with girth six. For $n = 2$, this is immediate, so assume $n > 2$. View the n-cube Q_n as consisting of a top and bottom copy of Q_{n-1} with a perfect matching consisting of vertical edges joining the two copies. Now, proceed as follows:

    (a) If $C \subseteq E(Q_n)$, then there is a subset C' of $E(Q_n)$ that contains no vertical edges, but determines a double cover isomorphic to the one given by C.

    (b) Suppose C is a subset of $E(Q_n)$ that contains no vertical edges. For any edge e in the top copy of $E(Q_{n-1})$, let e' denote the corresponding edge in the bottom copy. Show that the double cover determined by C has girth at least six if and only if C contains precisely one edge from each pair $\{e, e'\}$.

(c) Prove that up to isomorphism there is a unique double cover of Q_n with girth six.

Notes

The survey by Hahn and Tardif [11] is an excellent source of information on graph homomorphisms, and has had a strong influence on our treatment.

Many of the questions about homomorphisms we have considered can be extended naturally to directed graphs. See, e.g., [10].

The theory of graph homomorphisms can be presented naturally in terms of the category with graphs as its objects and homomorphisms as its maps. Then $X \times Y$ is the categorical product; the subdirect product is the natural product for the category formed by the covers of a fixed graph Y, with local isomorphisms as mappings. Imrich and Izbicki enumerate all the natural graph products in [15]. The products used most frequently in graph theory are the "product," the strong product and the Cartesian product, all of which we have considered in this chapter. Imrich and Klavžar [16] provide an extensive treatment of graph products.

The map graph K_n^X was introduced by El-Zahar and Sauer and extended to F^X by Häggkvist et al. in [10]. Our treatment has been influenced by Duffus and Sauer's treatment in [6]. Lemma 6.5.1 is due to Lovász, as is Exercise 12. (See Section 5 of [18].)

The strongest result concerning Hedetniemi's conjecture is due to El-Zahar and Sauer [7], who proved that if X and Y are not 3-colourable, then neither is their product $X \times Y$. Their paper is elegant and accessible. Greenwell and Lovász first proved that $K_n \times X$ is uniquely n-colourable when $\chi(X) > n$ or X is uniquely n-colourable. Burr, Erdős, and Lovász [4] proved that $\chi(X \times Y) = n + 1$ if $\chi(X) = \chi(Y) = n + 1$ and each vertex of X lies in an n-clique. Welzl [21] and, independently, Duffus, Sands, and Woodrow [5] proved that if $\chi(X) = \chi(Y) = n + 1$ and both X and Y contain an n-clique, then $\chi(X \times Y) = n + 1$.

The comparison between the conclusions in Exercise 19 and Exercise 20 is very surprising. Most of the interesting results in graph theory that hold for connected graphs hold for all graphs, but the essential difficulty of Hedetniemi's conjecture lies in establishing it for graphs that are not necessarily connected! For further information related to this, see Larose and Tardif [17].

Covers play a significant role in the theory of graph embeddings, disguised as "voltage graphs". (See [8].) The theory of covering graphs can be viewed a special case of the theory of covering spaces in topology. If this approach is taken, it is more natural to allow our graphs to have multiple edges and loops, and our definition of a covering map needs adjustment.

It would be useful to have a combinatorial development of the theory and applications of covering graphs, but we do not know of one.

Hell and Nešetřil [14] prove that if Y is not bipartite, then the problem of deciding whether there is a homomorphism from a given graph X into Y is NP-complete.

The Cayley graphs And(k) were first used by Andrásfai in [1], and also appear in his book [2]. Pach [20] showed that a reduced triangle-free graph of diameter two such that each independent set lies in the neighbourhood of a vertex must be isomorphic to one of the Cayley graphs And(k). Brouwer [3] has a second proof. Exercise 39 provides the original description of the Andrásfai graphs.

Exercise 1 and Exercise 3 are from Hell [12] and [13], respectively, while Exercise 2 is an unpublished observation due to Sabidussi. Exercise 27 and Exercise 28 are due to Nešetřil [19]. Exercise 34 and Exercise 35 come from [9].

Without a doubt, Hedetniemi's conjecture remains one of the most important unsolved problems in this area.

References

[1] B. Andrásfai, *Graphentheoretische Extremalprobleme*, Acta Math. Acad. Sci. Hungar, 15 (1964), 413–438.

[2] B. Andrásfai, *Introductory Graph Theory*, Pergamon Press Inc., Elmsford, N.Y., 1977.

[3] A. E. Brouwer, *Finite graphs in which the point neighbourhoods are the maximal independent sets*, in From universal morphisms to megabytes: a Baayen space odyssey, Math. Centrum Centrum Wisk. Inform., Amsterdam, 1994, 231–233.

[4] S. A. Burr, P. Erdős, and L. Lovasz, *On graphs of Ramsey type*, Ars Combinatoria, 1 (1976), 167–190.

[5] D. Duffus, B. Sands, and R. E. Woodrow, *On the chromatic number of the product of graphs*, J. Graph Theory, 9 (1985), 487–495.

[6] D. Duffus and N. Sauer, *Lattices arising in categorial investigations of Hedetniemi's conjecture*, Discrete Math., 152 (1996), 125–139.

[7] M. El-Zahar and N. W. Sauer, *The chromatic number of the product of two 4-chromatic graphs is* 4, Combinatorica, 5 (1985), 121–126.

[8] J. L. Gross and T. W. Tucker, *Topological Graph Theory*, John Wiley & Sons Inc., New York, 1987.

[9] R. Häggkvist, *Odd cycles of specified length in nonbipartite graphs*, in Graph Theory, North-Holland, Amsterdam, 1982, 89–99.

[10] R. Häggkvist, P. Hell, D. J. Miller, and V. Neumann Lara, *On multiplicative graphs and the product conjecture*, Combinatorica, 8 (1988), 63–74.

[11] G. Hahn and C. Tardif, *Graph homomorphisms: structure and symmetry*, in Graph symmetry (Montreal, PQ, 1996), Kluwer Acad. Publ., Dordrecht, 1997, 107–166.

[12] P. Hell, *Absolute retracts in graphs*, Lecture Notes in Math., 406 (1974), 291–301.

[13] P. Hell, *Subdirect products of bipartite graphs*, Colloq. Math. Soc. János Bolyai, 10 (1975), 857–866.

[14] P. Hell and J. Nešetřil, *On the complexity of H-coloring*, J. Combin. Theory Ser. B, 48 (1990), 92–110.

[15] W. Imrich and H. Izbicki, *Associative products of graphs*, Monatsh. Math., 80 (1975), 277–281.

[16] W. Imrich and S. Klavžar, *Product Graphs: Structure and Recognition*, Wiley, 2000.

[17] B. Larose and C. Tardif, *Hedetniemi's conjecture and the retracts of a product of graphs*, Combinatorica, 20 (2000), 531–544.

[18] L. Lovász, *Combinatorial Problems and Exercises*, North-Holland Publishing Co., Amsterdam, 1979.

[19] J. Nešetřil, *Homomorphisms of derivative graphs*, Discrete Math., 1 (1971/72), 257–268.

[20] J. Pach, *Graphs whose every independent set has a common neighbour*, Discrete Math., 37 (1981), 217–228.

[21] E. Welzl, *Symmetric graphs and interpretations*, J. Combin. Theory Ser. B, 37 (1984), 235–244.

7

Kneser Graphs

The Kneser graph $K_{v:r}$ is the graph with the r-subsets of a fixed v-set as its vertices, with two r-subsets adjacent if they are disjoint. We have already met the complete graphs $K_{v:1}$, while $K_{v:2}$ is the complement of the line graph of K_v. The first half of this chapter is devoted to fractional versions of the chromatic number and clique number of a graph. We discover that for the fractional chromatic number, the Kneser graphs play a role analogous to that played by the complete graphs for the ordinary chromatic number. We use this setting to provide a proof of the Erdős–Ko–Rado theorem, which is a famous result from extremal set theory. In the remainder of the chapter, we determine the chromatic number of the Kneser graphs, which surprisingly uses a nontrivial result from topology, and study homomorphisms between Kneser graphs.

7.1 Fractional Colourings and Cliques

We will use $\mathcal{I}(X)$ to denote the set of all independent sets of X, and $\mathcal{I}(X, u)$ to denote the independent sets that contain the vertex u.

A *fractional colouring* of a graph X is a nonnegative real-valued function f on $\mathcal{I}(X)$ such that for any vertex x of X,

$$\sum_{S \in \mathcal{I}(X,x)} f(S) \geq 1.$$

The *weight* of a fractional colouring is the sum of all its values, and the *fractional chromatic number* $\chi^*(X)$ of the graph X is the minimum possible weight of a fractional colouring. (We address the question of why this minimum exists in a later section.) We call a fractional colouring *regular* if, for each vertex x of X, we have

$$\sum_{S \in \mathcal{I}(X,x)} f(S) = 1.$$

The colour classes of a proper k-colouring of X form a collection of k pairwise disjoint independent sets $V_1, \ldots, V_k$ whose union is $V(X)$. The function f such that $f(V_i) = 1$ and $f(S) = 0$ for all other independent sets S is a fractional colouring of weight k. Therefore, it is immediate that

$$\chi^*(X) \le \chi(X).$$

Conversely, suppose that X has a 01-valued fractional colouring f of weight k. Then the support of f consists of k independent sets $V_1, \ldots, V_k$ whose union is $V(X)$. If we colour a vertex x with the smallest i such that $x \in V_i$, then we have a proper k-colouring of X. Thus the chromatic number of X is the minimum weight of a 01-valued fractional colouring.

The five-cycle C_5 has exactly five independent sets of size two, and each vertex lies in two of them. Thus if we define f to take the value $\frac{1}{2}$ on each of these independent sets and 0 on all others, then f is a fractional colouring of C_5 with weight $\frac{5}{2}$. Of course, $\chi(C_5) = 3$, and thus we see that $\chi^*(X)$ can be strictly less than $\chi(X)$. (Despite this, $\chi^*(X)$ is no easier to compute than $\chi(X)$, in general.)

For a second example, consider the Kneser graph $K_{v:r}$. The r-sets that contain a given point i form an independent set of size $\binom{v-1}{r-1}$, and each vertex lies in exactly r of these independent sets. The function with value $1/r$ on each of these sets, and zero elsewhere, is a fractional colouring with weight v/r, and so $\chi^*(K_{v:r}) \le v/r$.

The empty set is an independent set, and so if f is a fractional colouring, then $f(\emptyset)$ is defined. However, if $f(\emptyset) \ne 0$, then we may adjust f by declaring it to be zero on $\emptyset$ (and leaving its value on all nonempty independent sets unaltered). The resulting function is a still a fractional colouring, but its weight is less than that of f. Thus we can usually assume without loss that $f(\emptyset) = 0$.

7.2 Fractional Cliques

A *fractional clique* of a graph X is a nonnegative real-valued function on $V(X)$ such that the sum of the values of the function on the vertices of any independent set is at most one. The *weight* of a fractional clique is the sum of its values. The *fractional clique number* of X is the maximum possible weight of a fractional clique, and it is denoted by $\omega^*(X)$. The characteristic function of any clique of size k in X is a 01-valued fractional clique of weight k, and thus

$$\omega(X) \le \omega^*(X).$$

The function with value $\frac{1}{2}$ on each vertex of C_5 is a fractional clique with weight $\frac{5}{2}$, and thus we see that $\omega^*(X)$ can be strictly greater than $\omega(X)$. More generally, if $\alpha(X)$ denotes the maximum size of an independent set in X, then $g := \alpha(X)^{-1}\mathbf{1}$ is a fractional clique. Hence we have the following.

Lemma 7.2.1 *For any graph X,*

$$\omega^*(X) \ge \frac{|V(X)|}{\alpha(X)}. \qquad \square$$

There is one important case where we can determine the fractional clique number.

Lemma 7.2.2 *If X is vertex transitive, then*

$$\omega^*(X) = \frac{|V(X)|}{\alpha(X)}$$

and $\alpha(X)^{-1}\mathbf{1}$ is a fractional clique with this weight.

Proof. Suppose g is a nonzero fractional clique of X. Then g is a function on $V(X)$. If $\gamma \in \mathrm{Aut}(X)$, define the function g^γ by

$$g^\gamma(x) = g(x^\gamma).$$

Then g^γ is again a fractional clique, with the same weight as g. It follows that

$$\hat{g} := \frac{1}{|\mathrm{Aut}(X)|} \sum_{\gamma \in \mathrm{Aut}(X)} g^\gamma$$

is also a fractional clique with the same weight as g. If X is vertex transitive, then it is easy to verify that $\hat{g}$ is constant on the vertices of X. Now, $c\mathbf{1}$ is a fractional clique if and only if $c \le \alpha(X)^{-1}$, and so the result follows. $\square$

So far we have not indicated why the fractional chromatic number and fractional clique number are well-defined, that is, why fractional colourings of minimum weight and fractional cliques of maximum weight must exist. We remedy this in the next section.

7.3 Fractional Chromatic Number

Let B be the 01-matrix with rows indexed by the vertices of X and with the characteristic vectors of the independent sets of X as columns. Then a nonnegative vector f such that $Bf \ge \mathbf{1}$ (that is, each coordinate of Bf is at least one) is the same thing as a fractional colouring, and a nonnegative vector g such that $g^T B \leq 1$ is a fractional clique. Our first lemma shows that if f is a fractional colouring, then there is a regular fractional colouring f' of no greater weight than f.

Lemma 7.3.1 *If a graph X has a fractional colouring f of weight w, then it has a fractional colouring f' with weight no greater than w such that $Bf' = \mathbf{1}$.*

Proof. If $Bf \neq \mathbf{1}$, then we will show that we can perturb f into a function f' of weight no greater than f such that Bf' has fewer entries not equal to one. The result then follows immediately by induction.

Suppose that some entry of Bf is greater than 1; say $(Bf)_j = b > 1$. Let $S_1, \ldots, S_t$ be the independent sets in the support of f that contain x_j. Choose values $a_1, \ldots, a_t$ such that

$$a_i \leq f(S_i) \qquad \text{and} \qquad \sum_{i=1}^{i=t} a_i = b - 1.$$

Then define f' by

$$f'(S) = \begin{cases} f(S) - a_i, & \text{if } S = S_i; \\ f(S) + a_i, & \text{if } S = S_i \setminus x_j \text{ and } S \neq \emptyset; \\ f(S), & \text{otherwise.} \end{cases}$$

Then f' is a fractional colouring with weight no greater than w such that $(Bf')_j = 1$ and $(Bf')_i = (Bf)_i$ for all $i \neq j$. □

The next result is extremely important, as it asserts that the fractional chromatic number of a graph is a well-defined rational number. It is possible to provide a reasonably short and direct proof of this result. However, it is a direct consequence of the basic theory of linear programming, and so we give only the statement, leaving the details of the direct proof as Exercise 1.

Theorem 7.3.2 *Any graph X has a regular rational-valued fractional colouring with weight $\chi^*(X)$.* □

Similarly, the fractional clique number is also well-defined and rational valued.

7.4 Homomorphisms and Fractional Colourings

We have already seen two graph parameters, namely the chromatic number and the odd girth, that can be used to demonstrate that there is no homomorphism from one graph to another. We show that the fractional chromatic number can also fill this role.

First, we make some remarks about the preimages of independent sets in Y. If φ is a homomorphism from X to Y and S is an independent set in Y, then the preimage $\varphi^{-1}(S)$ is an independent set in X, as is easily verified. If T is a second independent set in Y and

$$S \cap \varphi(X) = T \cap \varphi(X),$$

then $\varphi^{-1}(S) = \varphi^{-1}(T)$. It follows that the preimage of an independent set S of Y is determined by its intersection with $\varphi(X)$.

Now, suppose that φ is a homomorphism from X to Y and f is a fractional colouring of Y. We define a function $\hat{f}$ on $\mathcal{I}(X)$ by

$$\hat{f}(S) = \sum_{T:\varphi^{-1}(T)=S} f(T),$$

and say that $\hat{f}$ is obtained by *lifting* f. The support of $\hat{f}$ consists of independent sets in X of the form $\varphi^{-1}(S)$, where $S \in \mathcal{I}(Y)$. If two or more independent sets in Y have the same intersection with $\varphi(X)$, then they have the same preimage S, and so all contribute to the value $\hat{f}(S)$. As every independent set in Y makes a contribution to $\hat{f}$, the weight of $\hat{f}$ is the same as the weight of f.

Theorem 7.4.1 *If there is a homomorphism from X to Y and f is a fractional colouring of Y, then the lift $\hat{f}$ of f is a fractional colouring of X with weight equal to the weight of f. The support of $\hat{f}$ consists of the preimages of the independent sets in the support of f.*

Proof. If $u \in V(X)$, then

$$\begin{aligned}\sum_{T\in\mathcal{I}(X,u)} \hat{f}(T) &= \sum_{S:u\in\varphi^{-1}(S)} f(S)\\ &= \sum_{S\in\mathcal{I}(Y,\varphi(u))} f(S).\end{aligned}$$

It follows that $\hat{f}$ is a fractional colouring. □

Corollary 7.4.2 *If there is a homomorphism from X to Y, then $\chi^*(X) \leq \chi^*(Y)$.* □

If there is an independent set in the support of f that does not intersect $\varphi(X)$, then its preimage is the empty set. In this situation $\hat{f}(\emptyset) \neq 0$, and there is a fractional colouring that agrees with $\hat{f}$ on all nonempty independent sets and vanishes on $\emptyset$. Hence we have the following:

Corollary 7.4.3 *Let X and Y be two graphs with the same fractional chromatic number. If φ is a homomorphism from X to Y and f is a fractional colouring of Y with weight $\chi^*(Y)$, then the image of X in Y must meet every independent set in the support of f.* □

Lemma 7.4.4 *If X is vertex transitive, then $\chi^*(X) \le |V(X)|/\alpha(X)$.*

Proof. We saw in Section 7.1 that $\chi^*(K_{v:r}) \le v/r$. If X is vertex transitive, then by Theorem 3.9.1 and the remarks following its proof, it is a retract of a Cayley graph Y where $|V(Y)|/\alpha(Y) = |V(X)|/\alpha(X)$. By Corollary 7.4.2 we see that $\chi^*(X) = \chi^*(Y)$. If $n = |V(Y)|$ and $\alpha = \alpha(Y)$, then we will show that there is a homomorphism from Y into $K_{n:\alpha}$.

Thus suppose that Y is a Cayley graph $X(G, C)$ for some group G of order n. As in Section 3.1, we take the vertex set of Y to be G. Let S be an independent set of size $\alpha(Y)$ in Y, and define a map

$$\varphi : g \mapsto (S^{-1})g,$$

where $S^{-1} = \{s^{-1} : s \in S\}$. Now, suppose that $g \sim h$ and consider $\varphi(g) \cap \varphi(h)$. If $y \in \varphi(g) \cap \varphi(h)$, then $y = a^{-1}g = b^{-1}h$ where $a, b \in S$. But then $ba^{-1} = hg^{-1} \in C$, and so $a \sim b$, contradicting the fact that S is an independent set. Thus $\varphi(g)$ is disjoint from $\varphi(h)$, and φ is a homomorphism from Y to $K_{n:\alpha}$. □

In the previous proof we used the existence of a homomorphism into $K_{v:r}$ to bound $\chi^*(X)$. Our next result shows that in fact the Kneser graphs play the same central role in fractional graph colouring as the complete graphs in graph colouring.

Theorem 7.4.5 *For any graph X we have*

$$\chi^*(X) = \min\{v/r : X \to K_{v:r}\}.$$

Proof. We have already seen that $\chi^*(K_{v:r}) \le v/r$, and so by Corollary 7.4.2, it follows that if X has a homomorphism into $K_{v:r}$, then it has a fractional colouring with weight at most v/r.

Conversely, suppose that X is a graph with fractional chromatic number $\chi^*(X)$. By Theorem 7.3.2, $\chi^*(X)$ is a rational number, say m/n, and X has a regular fractional colouring f of this weight. Then there is a least integer r such that the function $g = rf$ is integer valued. The weight of g is an integer v, and since f is regular, the sum of the values of g on the independent sets containing x is r.

Now, let A be the $|V(X)| \times v$ matrix with rows indexed by $V(X)$, such that if S is an independent set in X, then A has $g(S)$ columns equal to the characteristic vector of S. Form a copy of the Kneser graph $K_{v:r}$ by taking Ω to be the set of columns of A. Each vertex x of X determines a set of r columns of A—those that have a 1 in the row corresponding to x—and since no independent set contains an edge, the sets of columns corresponding to adjacent vertices of X are disjoint. Hence the map from vertices of X to sets of columns is a homomorphism from X into $K_{v:r}$. □

The proof of Lemma 7.4.4 emphasized the role of Kneser graphs, but there are several alternative proofs. One of the shortest is to observe that if X is vertex transitive and S is an independent set of maximum size

$\alpha(X)$, then the translates of S under the action of Aut(X) are the support of a fractional colouring with weight $|V(X)|/\alpha(X)$. We will offer another argument in the next section.

7.5 Duality

We give an alternative description of $\chi^*(X)$ and $\omega^*(X)$, which will prove very useful.

Let B be the 01-matrix whose columns are the characteristic vectors of the independent sets in X. The fractional chromatic number $\chi^*(X)$ is equal to the value of the following linear optimization problem:

$$\begin{gathered} \min\ \mathbf{1}^T f \\ Bf \geq 1 \\ f \geq 0. \end{gathered}$$

Similarly, $\omega^*(X)$ is the value of the optimization problem

$$\begin{gathered} \max\ g^T \mathbf{1} \\ g^T B \leq 1 \\ g \geq 0. \end{gathered}$$

These are both linear programming problems; in fact, they form a dual pair.

We use the formulations just given to prove the following result. Since $\omega(X) \leq \omega^*(X)$ and $\chi^*(X) \leq \chi(X)$, this could be viewed as a strengthening of the simple inequality $\omega(X) \leq \chi(X)$.

Lemma 7.5.1 *For any graph X we have $\omega^*(X) \leq \chi^*(X)$.*

Proof. Suppose that f is a fractional colouring and g a fractional clique of X. Then

$$\begin{aligned} \mathbf{1}^T f - g^T \mathbf{1} &= \mathbf{1}^T f - g^T Bf + g^T Bf - g^T \mathbf{1} \\ &= (\mathbf{1}^T - g^T B)f + g^T(Bf - \mathbf{1}). \end{aligned}$$

Since g is a fractional clique, $\mathbf{1}^T - g^T B \geq 0$. Since f is a fractional colouring, $f \geq 0$, and consequently $(\mathbf{1}^T - g^T B)f \geq 0$. Similarly, g and $Bf - \mathbf{1}$ are nonnegative, and so $g^T(Bf - \mathbf{1}) \geq 0$. Hence we have that $\mathbf{1}^T f - g^T \mathbf{1}$ is the sum of two nonnegative numbers, and therefore $\mathbf{1}^T f \geq g^T \mathbf{1}$ for any fractional colouring f and fractional clique g. □

The above argument is essentially the proof of the weak duality theorem from linear programming. We point out that the strong duality theorem from linear programming implies that $\chi^*(X) = \omega^*(X)$ for any graph X. For vertex-transitive graphs we can prove this now.

Corollary 7.5.2 *If X is a vertex-transitive graph, then*

$$\omega^*(X) = \chi^*(X) = \frac{|V(X)|}{\alpha(X)}.$$

Proof. Lemma 7.2.1, Lemma 7.5.1, and Lemma 7.4.4 yield that

$$\frac{|V(X)|}{\alpha(X)} \leq \omega^*(X) \leq \chi^*(X) \leq \frac{|V(X)|}{\alpha(X)},$$

and so the result follows. □

The odd circuit C_{2m+1} is vertex transitive and $\alpha(C_{2m+1}) = m$, so we see that $\chi^*(C_{2m+1}) = 2 + \frac{1}{m}$.

We extract two consequences of the proof of Lemma 7.5.1, for later use.

Corollary 7.5.3 *For any graph X we have*

$$\chi^*(X) \geq \frac{|V(X)|}{\alpha(X)}.$$

Proof. Use Lemma 7.2.1. □

Lemma 7.5.4 *Let X and Y be vertex-transitive graphs with the same fractional chromatic number, and suppose φ is a homomorphism from X to Y. If S is a maximum independent set in Y, then $\varphi^{-1}(S)$ is a maximum independent set in X.*

Proof. Since X and Y are vertex transitive,

$$\frac{|V(X)|}{\alpha(X)} = \chi^*(X) = \chi^*(Y) = \frac{|V(Y)|}{\alpha(Y)}.$$

Let f be a fractional colouring of weight $\chi^*(X)$ and let $g = \alpha(X)^{-1}\mathbf{1}$. By Lemma 7.2.2 we have that g is a fractional clique of maximum weight. From the proof of Lemma 7.5.1

$$(\mathbf{1}^T - g^T B)f = 0.$$

Since the sum of the values of g on any independent set of size less than $\alpha(X)$ is less than 1, this implies that $f(S) = 0$ if S is an independent set with size less than $\alpha(X)$. On the other hand, Theorem 7.4.1 yields that X has a fractional colouring of weight $\chi^*(X)$ with $\varphi^{-1}(S)$ in its support. Therefore, $|\varphi^{-1}(S)| = \alpha(X)$. □

7.6 Imperfect Graphs

It is a trivial observation that $\chi(X) \geq \omega(X)$. We call a graph X *perfect* if for any induced subgraph Y of X we have $\chi(Y) = \omega(Y)$. A graph that is not perfect is called *imperfect*. The simplest examples of perfect graphs are the bipartite graphs, while the simplest examples of imperfect graphs are the odd cycles.

A much larger class of perfect graphs, known as *comparability graphs*, arise from partially ordered sets. If S is a set, partially ordered by "$\leq$", then we say that two elements a and b are comparable if $a \leq b$ or $b \leq a$. The comparability graph of S is the graph with vertex set S, where two vertices are adjacent if they are distinct and comparable. An induced subgraph of a comparability graph is also a comparability graph. A clique in a comparability graph corresponds to a chain in the partially ordered set, and an independent set to an antichain. A famous theorem of Dilworth asserts that the minimum number of antichains needed to cover the elements of a partially ordered set equals the maximum size of a chain. Equivalently, comparability graphs are perfect. Every bipartite graph is a comparability graph, and so this result generalizes the observation that bipartite graphs are perfect.

Dilworth also proved that the minimum number of chains needed to cover the elements of a poset equals the maximum size of an antichain. Expressed graph-theoretically, this result states that the complement of a comparability graph is perfect. Lovász settled a long-standing open problem by proving that the complement of any perfect graph is perfect, and we will present a short proof of this fact.

A graph is *minimally imperfect* if it is not perfect but each induced proper subgraph is perfect. The odd cycles are the simplest examples of minimally imperfect graphs. If X is minimally imperfect, then $\chi(X) = \omega(X) + 1$ and $\chi(X \setminus v) = \omega(X)$, for each vertex v of X. Let us say that an independent set S in a graph X is *big* if $|S| = \alpha(X)$, and that a clique is big if it has size $\omega(X)$.

Lemma 7.6.1 *Let X be a minimally imperfect graph. Then any independent set is disjoint from at least one big clique.*

Proof. Let S be an independent set in the minimally imperfect graph X. Then $X \setminus S$ is perfect, and therefore $\chi(X \setminus S) = \omega(X \setminus S)$. If S meets each big clique in at least one vertex, it follows that $\omega(X \setminus S) \leq \omega(X) - 1$. Consequently,

$$\chi(X) = 1 + \chi(X \setminus S) = \omega(X),$$

which is impossible. □

Suppose X is a minimally imperfect graph on n vertices, and let α and ω denote $\alpha(X)$ and $\omega(X)$, respectively. If $v \in V(X)$, then $X \setminus v$ has a partition into $\omega(X)$ independent sets. Each of these sets contains a neighbour of v, for otherwise we could extend the colouring to a proper colouring of X with ω colours. Thus these independent sets are maximal in X. We now define a collection $\mathcal{S}$ of independent sets in X. First choose an independent set S_0 of size α. For each vertex v in S_0, take ω independent sets in $X \setminus v$

that form an ω-colouring of $X \setminus v$. This gives a collection of $N = 1 + \alpha\omega$ independent sets $S_0, \dots, S_{N-1}$.

Lemma 7.6.2 *Each vertex of X lies in exactly α members of $\mathcal{S}$, and any big clique of X is disjoint from exactly one member of $\mathcal{S}$.*

Proof. We leave the first claim as an exercise.

Let K be a big clique of X, let v be an arbitrary vertex of X, and suppose that $X \setminus v$ is coloured with ω colours. Then K has at most one vertex in each colour class, and so either $v \notin K$ and K meets each colour class in one vertex, or $v \in K$ and K is disjoint from exactly one colour class.

We see now that if K is disjoint from S_0, then it must meet each of $S_1, \dots, S_{N-1}$ in one vertex. If K is not disjoint from S_0, then it meets it in a single vertex, u say. If $v \in S_0 \setminus u$, then K meets each of the independent sets we chose in $X \setminus v$. However, K misses exactly one of the independent sets from $X \setminus u$. □

Let A be the $N \times n$ matrix whose rows are the characteristic vectors of the independent sets in $\mathcal{S}$. By Lemma 7.6.1 we may form a collection $\mathcal{C}$ of big cliques C_i such that $C_i \cap S_i = \emptyset$ for each i. Let B be the $N \times n$ matrix whose rows are the characteristic vectors of these big cliques. Lemma 7.6.2 implies that S_i is the only member of $\mathcal{S}$ disjoint from C_i. Accordingly, the following result is immediate.

Lemma 7.6.3 $AB^T = J - I$. □

Theorem 7.6.4 *The complement of a perfect graph is perfect.*

Proof. For any graph X we have the trivial bound $|V(X)| \le \chi(X)\alpha(X)$, and so for perfect graphs we have $|V(X)| \le \alpha(X)\omega(X)$.

Since $J-I$ is invertible, the previous lemma implies that the rows of A are linearly independent, and thus $N \le n$. On the other hand, $|V(X \setminus v)| \le \alpha\omega$, and therefore $n \le N$. This proves that $N = n$, and so

$$n = |V(\overline{X})| = 1 + \alpha(X)\omega(X) = 1 + \omega(\overline{X})\alpha(\overline{X}).$$

Therefore, $\overline{X}$ cannot be perfect.

If X is imperfect, then it contains a minimally imperfect induced subgraph Z. The complement $\overline{Z}$ is then an induced subgraph of $\overline{X}$ that is not perfect, and so $\overline{X}$ is imperfect. Therefore, the complement of a perfect graph is perfect. □

We extract further consequences from the above proof. If X is minimally imperfect and $n = 1 + \alpha\omega$, the ω independent sets that partition a subgraph $X \setminus v$ must all have size α. Therefore, all members of $\mathcal{S}$ have size α.

We can define a function f on the independent sets of X by declaring f to have value $1/\alpha$ on each element of $\mathcal{S}$ and to be zero elsewhere. By the first part of Lemma 7.6.2, this is a fractional colouring. Since $|\mathcal{S}| = n$, its weight is n/α, and therefore $\chi^*(X) \leq n/\alpha$. On the other hand, Corollary 7.5.3 asserts that $|V(X)|/\alpha(X) \leq \chi^*(X)$, for any graph X. Hence we find that

$$\chi^*(X) = \frac{n}{\alpha(X)} = \omega(X) + \frac{1}{\alpha(X)}.$$

For any graph,

$$\omega(X) \leq \omega^*(X) \leq \chi^*(X) \leq \chi(X),$$

and so whenever Y is an induced subgraph of a perfect graph we have $\chi^*(Y) = \chi(Y)$. Therefore, we deduce that a graph X is perfect if and only if $\chi^*(Y) = \chi(Y)$ for any induced subgraph Y of X.

We note one final corollary of these results. Suppose that K is a big clique of X, and let x be its characteristic vector. By the second part of Lemma 7.6.2, we see that $b = Ax$ has one entry zero and all other entries equal to one. Hence b is a column of $J - I$. Since $AB^T = J - I$ and A is invertible, this implies that x must be a column of B^T, and therefore $K \in \mathcal{C}$. Thus $\mathcal{C}$ contains all big cliques of X. It follows similarly that $\mathcal{S}$ contains all big independent sets of X.

7.7 Cyclic Interval Graphs

The *cyclic interval graph* $C(v, r)$ is the graph whose vertex set is the set of all cyclic shifts, modulo v, of the subset $\{1, \ldots, r\}$ of $\Omega = \{1, \ldots, v\}$, and where two vertices are adjacent if the corresponding r-sets are disjoint. It is immediate that $C(v, r)$ is an induced subgraph of the Kneser graph $K_{v:r}$ and that $C(v, r)$ is a circulant, and hence vertex transitive.

If $v < 2r$, then every two vertices intersect and $C(v, r)$ is empty, and therefore we usually insist that $v \geq 2r$. In this case we can determine the maximum size of an independent set in $C(v, r)$ and characterize the independent sets of this size.

Lemma 7.7.1 *For $v \geq 2r$, an independent set in $C(v, r)$ has size at most r. Moreover, an independent set of size r consists of the vertices that contain a given element of $\{1, \ldots, v\}$.*

Proof. Suppose that S is an independent set in $C(v, r)$. Since $C(v, r)$ is vertex transitive, we may assume that S contains the r-set $\beta = \{1, \ldots, r\}$. Let S_1 and S_r be the r-sets in S that contain the points 1 and r, respectively. Let j be the least integer that lies in all the r-sets in S_r. The least element of each set in S_r is thus at most j, and since distinct sets in S_r have distinct least elements, it follows that $|S_r| \leq j$. On the other hand, each element of S_1 has a point in common with each element of S_r. Hence each element of S_1 contains j, and consequently, $|S_1| \leq r - j + 1$. Since $v \geq 2r$ this implies that $S_1 \cap S_r = \{\beta\}$, and so we have

$$|S| = |S_1| + |S_r| - 1 \leq (r - j + 1) + j - 1 = r.$$

If equality holds, then S consists of the vertices in $C(v,r)$ that contain j.□

Corollary 7.7.2 *If $v \geq 2r$, then $\chi^*(C(v,r)) = v/r$.* □

Corollary 7.7.3 *For $v \geq 2r$, the fractional chromatic number of the Kneser graph $K_{v:r}$ is v/r.*

Proof. Since $C(v,r)$ is a subgraph of $K_{v:r}$, it follows that

$$\frac{v}{r} = \chi^*(C(v,r)) \leq \chi^*(K_{v:r}),$$

and we have already seen that $\chi^*(K_{v:r}) \leq v/r$. □

Corollary 7.7.4 *If $v > 2r$, then the shortest odd-cycle in $K_{v:r}$ has length at least $v/(v-2r)$.*

Proof. If the odd cycle C_{2m+1} is a subgraph of $K_{v:r}$, then

$$2 + \frac{1}{m} = \chi^*(C_{2m+1}) \leq \frac{v}{r},$$

which implies that $m \geq r/(v-2r)$, and hence that $2m+1 \geq v/(v-2r)$.□

The bound of this lemma is tight for the odd graphs $K_{2r+1:r}$.

7.8 Erdős–Ko–Rado

We apply the theory we have developed to the Kneser graphs. The following result is one of the fundamental theorems in extremal set theory.

Theorem 7.8.1 (Erdős–Ko–Rado) *If $v > 2r$, then $\alpha(K_{v:r}) = \binom{v-1}{r-1}$. An independent set of size $\binom{v-1}{r-1}$ consists of the r-subsets of $\{1,\ldots,v\}$ that contain a particular point.*

Proof. From Corollary 7.7.3 and Corollary 7.5.2, it follows that

$$\alpha(K_{v:r}) = \binom{v-1}{r-1}.$$

Suppose that S is an independent set in $K_{v:r}$ with size $\binom{v-1}{r-1}$. Given any cyclic ordering of $\{1,\ldots,v\}$, the graph C induced by the cyclic shifts of the first r elements of the ordering is isomorphic to $C(v,r)$. The inclusion mapping from C to $K_{v:r}$ is a homomorphism, and $S \cap V(C)$ is the preimage of S under this homomorphism. Therefore, by Lemma 7.5.4 we have that $|S \cap V(C)| = r$ and $S \cap V(C)$ consists of the r cyclic shifts of some set of r consecutive elements in this ordering.

First consider the natural (numerical) ordering $\{1,\ldots,v\}$. Relabelling if necessary, we can assume that S contains the sets

$$\{1,2,\ldots,r\}, \{2,3,\ldots,r,r+1\},\ldots,\{r,r+1,\ldots,2r-1\},$$

that is, all the r-subsets that contain the element r. To complete the proof we need to show that by varying the cyclic ordering appropriately, we can conclude that S contains every r-subset containing r. First note that since S contains precisely r cyclic shifts from any cyclic ordering, it does not contain $\{x, 1, \ldots, r-1\}$ for any $x \in \{2r, \ldots, v\}$.

Now, let g be any element of $\mathrm{Sym}(v)$ that fixes $\{1, \ldots, r-1\}$ setwise and consider any cyclic ordering that starts

$$\{x, 1^g, 2^g, \ldots, (r-1)^g, r, \ldots\},$$

where $x \in \{2r, \ldots, v\}$. Then S contains $\beta = \{1^g, 2^g, \ldots, (r-1)^g, r\}$ but not $\{x, 1^g, \ldots, (r-1)^g\}$, and so S must contain the r right cyclic shifts of β. For any r-subset α containing r, there is some cyclic ordering of this form that has α as one of these r cyclic shifts unless α contains all of the elements $\{2r, \ldots, v\}$ (for then there is no suitable choice for x).

For any $y \in \{r+1, \ldots, 2r-1\}$, the same argument applies if we consider the natural cyclic ordering with $2r$ interchanged with y (with v as x). Therefore, every r-subset containing r is in S except possibly those containing all the elements of $\{y, 2r, \ldots, v\}$. By varying y, it follows that if there is any r-subset containing r that is not in S, then it contains all of the elements of $\{r+1, \ldots, v\}$. Since $v > 2r$, there are no such subsets and the result follows. □

Corollary 7.8.2 *The automorphism group of $K_{v:r}$ is isomorphic to the symmetric group* $\mathrm{Sym}(v)$.

Proof. Let X denote $K_{v:r}$ and let $X(i)$ denote the maximum independent set consisting of all the r-sets containing the point i from the underlying set Ω. Any automorphism of X must permute the maximum independent sets of X, and by the Erdős–Ko–Rado theorem, all the maximum independent sets are of the form $X(i)$ for some $i \in \Omega$. Thus any automorphism of X permutes the $X(i)$, and thus determines a permutation in $\mathrm{Sym}(v)$. It is straightforward to check that no nonidentity permutation can fix all the $X(i)$, and therefore $\mathrm{Aut}(X) \cong \mathrm{Sym}(v)$. □

The bound in Theorem 7.8.1 is still correct when $v = 2r$; the maximum size of an independent set is

$$\binom{2r-1}{r-1} = \frac{1}{2}\binom{2r}{r}.$$

But $K_{2r:r}$ is isomorphic to $\binom{2r-1}{r-1}$ vertex-disjoint copies of K_2, and therefore it has

$$2^{\binom{2r-1}{r-1}}$$

maximum independent sets, not just the $2r$ described in the theorem.

7.9 Homomorphisms of Kneser Graphs

In this section we consider homomorphisms between Kneser graphs. Our first result shows that Kneser graphs are cores.

Theorem 7.9.1 *If $v > 2r$, then $K_{v:r}$ is a core.*

Proof. Let X denote $K_{v:r}$, and let $X(i)$ denote the maximum independent set consisting of all the r-sets containing the point i from the underlying set Ω. Let φ be a homomorphism from X to X. We will show that it is onto. If $\beta = \{1, \dots, r\}$, then β is the unique element of the intersection

$$X(1) \cap X(2) \cap \dots \cap X(r).$$

By Lemma 7.5.4, the preimage $\varphi^{-1}(X(i))$ is an independent set of maximum size. By the Erdős–Ko–Rado theorem, this preimage is equal to $X(i')$, for some element i' of Ω. We have

$$\varphi^{-1}\{\beta\} = \varphi^{-1}(X(1)) \cap \varphi^{-1}(X(2)) \cap \dots \cap \varphi^{-1}(X(r)),$$

from which we see that $\varphi^{-1}\{\beta\}$ is the intersection of at most r distinct sets of the form $X(i')$. This implies that $\varphi^{-1}\{\beta\} \neq \emptyset$, and hence φ is onto. □

Our next result implies that $K_{v:r} \to K_{v-2r+2:1}$; since $K_{v-2r+2:1}$ is the complete graph on $v-2r+2$ vertices, this implies that $\chi(K_{v:r}) \le v-2r+2$. In the next section we will see that equality holds.

Theorem 7.9.2 *If $v \ge 2r$ and $r \ge 2$, there is a homomorphism from $K_{v:r}$ to $K_{v-2:r-1}$.*

Proof. If $v = 2r$, then $K_{v:r} = \binom{2r-1}{r-1}K_2$, which admits a homomorphism into any graph with an edge. So we assume $v > 2r$, and that the underlying set Ω is equal to $\{1, \dots, v\}$. We can easily find a homomorphism φ from $K_{v-1:r}$ to $K_{v-2:r-1}$: Map each r-set to the $(r-1)$-subset we get by deleting its largest element. We identify $K_{v-1:r}$ with the subgraph of $K_{v:r}$ induced by the vertices that do not contain v, and try to extend φ into a homomorphism from $K_{v:r}$ into $K_{v-2:r-1}$.

We note first that the vertices in $K_{v:r}$ that are not in our chosen $K_{v-1:r}$ all contain v, and thus they form an independent set in $K_{v:r}$. Denote this set of vertices by S and let S_i denote the subset of S formed by the r-sets that contain $v, v-1, \dots, v-i+1$, but not $v-i$. The sets $S_1, \dots, S_r$ form a partition of S. If $\alpha \in S_1$, define $\varphi(\alpha)$ to be $\alpha \setminus v$. If $i > 1$ and $\alpha \in S_i$, then $v - i \notin \alpha$. In this case let $\varphi(\alpha)$ be obtained from α by deleting v and replacing $v-1$ by $v-i$. It is now routine to check that φ is a homomorphism from $K_{v:r}$ into $K_{v-2:r-1}$. □

Lemma 7.9.3 *Suppose that $v > 2r$ and $v/r = w/s$. There is a homomorphism from $K_{v:r}$ to $K_{w:s}$ if and only if r divides s.*

Proof. Suppose r divides s; we may assume $s = mr$ and $w = mv$. Let W be a fixed set of size w and let π be a partition of it into v cells of size

m. Then the s-subsets of W that are the union of r cells of π induce a subgraph of $K_{w:s}$ isomorphic to $K_{v:r}$.

Assume that the vertices of X are the r-subsets of the v-set V, and for i in V, let $X(i)$ be the maximum independent set formed by the r-subsets that contain i. Similarly, assume that the vertices of Y are the s-subsets of W, and for j in W, let $Y(j)$ be the maximum independent set formed by the s-subsets that contain j. The preimage $\varphi^{-1}(Y(j))$ is equal to $X(i)$, for some i.

Let ν_i denote the number of elements j of W such that $\varphi^{-1}(Y(j)) = X(i)$. Let α be an arbitrary vertex of X. Then $\varphi^{-1}(Y(j)) = X(i)$ for some $i \in \alpha$ if and only if $\varphi^{-1}(Y(j))$ contains α if and only if $j \in \varphi(\alpha)$. Therefore,

$$\sum_{i\in\alpha} \nu_i = s. \tag{7.1}$$

Moreover, ν_i is independent of i. Suppose α and β are vertices of X such that $|\alpha \cap \beta| = r-1$. Then, if $\alpha \setminus \beta = \{k\}$ and $\beta \setminus \alpha = \{\ell\}$, we have

$$0 = s - s = \sum_{i\in\alpha} \nu_i - \sum_{i\in\beta} \nu_i = \nu_k - \nu_\ell.$$

Thus $\nu_k = \nu_\ell$, and therefore ν_i is constant. By (7.1) it follows that r divides s, as required. □

7.10 Induced Homomorphisms

We continue our study of homomorphisms between Kneser graphs by showing that in many cases a homomorphism from $K_{v:r}$ to $K_{w:\ell}$ induces a homomorphism from $K_{v-1:r}$ to $K_{w-2:\ell}$.

For this we need to consider the independent sets of the Kneser graphs. We have already seen that the maximum independent sets of $X = K_{v:r}$ are the sets $X(1), \ldots, X(v)$ where $X(i)$ comprises all the subsets that contain the point i. More generally, let S be an independent set in $K_{v:r}$. An element of the underlying set Ω is called a *centre* of S if it lies in each r-set in S. If an independent set in $K_{v:r}$ has a centre i, then it is a subset of $X(i)$. Let $\mathrm{h}_{v,r}$ denote the maximum size of an independent set in $K_{v:r}$ that does not have a centre.

Theorem 7.10.1 (Hilton–Milner) *If $v \geq 2r$, the maximum size of an independent set in $K_{v:r}$ with no centre is*

$$\mathrm{h}_{v,r} = 1 + \binom{v-1}{r-1} - \binom{v-r-1}{r-1}.$$ □

Lemma 7.10.2 *Suppose there is a homomorphism from $K_{v:r}$ to $K_{w:\ell}$. If*

$$\ell\binom{v}{r} > v\binom{v-1}{r-1} + (w-v)\,\mathrm{h}_{v,r},$$

then there is a homomorphism from $K_{v-1:r}$ to $K_{w-2:\ell}$.

Proof. Suppose that f is a homomorphism from $X = K_{v:r}$ to $Y = K_{w:\ell}$. Consider the preimages $f^{-1}(Y(i))$ of all the maximum independent sets of Y, and suppose that two of them, say $f^{-1}(Y(i))$ and $f^{-1}(Y(j))$, have the same centre c. Then f maps any r-set that does not contain c to an ℓ-set that does not contain i or j, and so its restriction to the r-sets not containing c is a homomorphism from $K_{v-1:r}$ to $K_{w-2:\ell}$.

Counting the pairs $(\alpha, Y(i))$ where $\alpha \in V(K_{v:r})$ and $Y(i)$ contains the vertex $f(\alpha)$ we find that

$$\sum_i |f^{-1}(Y(i))| = \ell\binom{v}{r}.$$

If no two of the preimages $f^{-1}(Y(i))$ have the same centre, then at most v of them have centres and the remaining $w - v$ do not have centres. In this case, it follows that

$$\sum_i |f^{-1}(Y(i))| \le v\binom{v-1}{r-1} + (w-v)\,\mathrm{h}_{v,r},$$

and thus the result holds. □

By way of illustration, suppose that there were a homomorphism from $K_{7:2}$ to $K_{11:3}$. The inequality of the lemma holds, and so this implies the existence of a homomorphism from $K_{6:2}$ to $K_{9:3}$. This can be ruled out either by using Lemma 7.9.3 or by applying the lemma one more time, and seeing that there would be an induced homomorphism from $K_{5:2}$ to $K_{7:3}$, which can be directly eliminated as

$$\chi^*(K_{5:2}) = \frac{5}{2} > \frac{7}{3} = \chi^*(K_{7:3}).$$

This argument can be extended to show that there is a homomorphism from $K_{v:2}$ to $K_{w:3}$ if and only if $w \ge 2v - 2$, but we leave the proof of this as an exercise.

7.11 The Chromatic Number of the Kneser Graph

We will use the following theorem from topology, known as Borsuk's theorem, to determine $\chi(K_{v:r})$. A pair of points $\{x, y\}$ on the unit sphere in $\mathbb{R}^n$ is *antipodal* if $y = -x$.

Theorem 7.11.1 *If the unit sphere in $\mathbb{R}^n$ is expressed as the union of n open sets, then one of the sets contains an antipodal pair of points.* □

At first (and second) glance, this bears no relation to colouring graphs. We therefore present Borsuk's theorem in an alternative form. If a is a nonzero vector, the *open half-space* $H(a)$ is the set of vectors x such that $a^T x > 0$.

Lemma 7.11.2 *Let $\mathcal{C}$ be a collection of closed convex subsets of the unit sphere in $\mathbb{R}^n$. Let X be the graph with the elements of $\mathcal{C}$ as its vertices, with two elements adjacent if they are disjoint. If for each unit vector a the open half-space $H(a)$ contains an element of $\mathcal{C}$, then X cannot be properly coloured with n colours.*

Proof. Suppose X has been coloured with the n colours $\{1,\ldots,n\}$. For $i \in \{1,\ldots,n\}$, let C_i be the set of vectors a on the unit sphere such that $H(a)$ contains a vertex of colour i. If $S \in V(X)$, then the set of vectors a such that $a^T x > 0$ for all x in S is open, and C_i is the union of these sets for all vertices of X with colour i. Hence C_i is open.

By our constraint on $\mathcal{C}$, we see that $\cup_{i=1}^n C_i$ is the entire unit sphere. Hence Borsuk's theorem implies that for some i, the set C_i contains an antipodal pair of points, a and $-a$ say. Then both $H(a)$ and $H(-a)$ contain a vertex of colour i; since these vertices must be adjacent, our colouring cannot be proper. □

To apply this result we need a further lemma. The proof of this involves only linear algebra, and will be presented in the next section.

Theorem 7.11.3 *There is a set Ω of v points in $\mathbb{R}^{v-2r+1}$ such that each open half-space $H(a)$ contains at least r points from Ω.* □

Theorem 7.11.4 $\chi(K_{v:r}) = v - 2r + 2$.

Proof. We have already seen that $v-2r+2$ is an upper bound on $\chi(K_{v:r})$; we must show that it is also a lower bound.

Assume that $\Omega = \{x_1,\ldots,x_v\}$ is a set of v points in $\mathbb{R}^{v-2r+1}$ such that each open half-space $H(a)$ contains at least r points of Ω. Call a subset S of Ω *conical* if it is contained in some open half-space. For each conical r-subset α of Ω, let $S(\alpha)$ be the intersection with the sphere with the cone generated by α. (In other words, let $S(\alpha)$ be the set of all unit vectors that are nonnegative linear combinations of the elements of α.) Then let X be the graph with the sets $S(\alpha)$ as vertices, and with two such vertices adjacent if they are disjoint. If $S(\alpha)$ is disjoint from $S(\beta)$, then clearly α is disjoint from β, and so the map

$$\varphi : S(\alpha) \to \alpha$$

is an injective homomorphism from X to $K_{v:r}$. Thus by Lemma 7.11.2, the chromatic number of $K_{v:r}$ is at least $v - 2r + 2$. □

Since the fractional chromatic number of $K_{v:r}$ is only v/r, this shows that the difference between the chromatic number and the fractional chromatic number can be arbitrarily large.

7.12 Gale's Theorem

We have already used the following to determine $\chi(K_{v:r})$; now we prove it.

Theorem 7.12.1 *If $v \geq 2r$, then there is a set Ω of v points in $\mathbb{R}^{v-2r+1}$ such that each open half-space $H(a)$ contains at least r points from Ω.*

Proof. Let $a_1, \ldots, a_v$ be any v distinct real numbers, and let G be the $(2m+1) \times v$ matrix

$$G = \begin{pmatrix} 1 & 1 & \cdots & 1 \\ a_1 & a_2 & \cdots & a_v \\ \vdots & \vdots & & \vdots \\ a_1^{2m} & a_2^{2m} & \cdots & a_v^{2m} \end{pmatrix}$$

for any integer m such that $v \geq 2m+2$.

We claim that the rank of G is $2m+1$. We shall show that for any vector $f = (f_0, \ldots, f_{2m})^T$ we have $f^T G \neq 0$, and hence the $2m+1$ rows of G are linearly independent. If $f(t)$ is the polynomial of degree at most $2m$ given by

$$f(t) = \sum_{i=0}^{2m} f_i t^i,$$

then

$$f^T G = (f(a_1), \ldots, f(a_v)),$$

and so $f^T G$ has at most $2m$ entries equal to 0; thus $f^T G \neq 0$.

Now, consider the null space of G. We shall show that any vector $x \neq 0$ such that $Gx = 0$ has at least $m+1$ negative entries and at least $m+1$ positive entries. Suppose for a contradiction that $Gx = 0$ and that x has at most m negative entries. Then

$$g(t) := \prod_{x_i < 0} (x - a_i)$$

is a polynomial of degree at most m, and $f(t) := g(t)^2$ is a polynomial of degree at most $2m$. Then $y = f^T G$ is a vector in the row space of G such that $y \geq 0$ and $y_i = 0$ if and only if $x_i < 0$. Since $y^T x = 0$, it follows that x can have no positive entries, and since $Gx = 0$, we have found a set of at most m linearly dependent columns, contradicting the fact that G has rank $2m+1$. Hence x has at least $m+1$ negative entries, and because $-x$ is also in the null space of G, we see that x has at least $m+1$ positive entries.

Now, let N be the $(v-2m-1) \times v$ matrix whose rows are a basis for the null space of G. The columns of N form a set of v vectors in $\mathbb{R}^{v-2m-1}$ such that for any vector a there are at least $m+1$ positive entries in $a^T N$. Therefore, the open half-space $H(a)$ in $\mathbb{R}^{v-2m-1}$ contains at least $m+1$ columns of N. Taking m equal to $r-1$, the theorem follows. □

7.13 Welzl's Theorem

The rational numbers have the property that between any two distinct rational numbers there is a third one. More formally, the usual order on the rationals is dense. It is amazing that the lattice of cores of nonbipartite graphs is also dense.

Theorem 7.13.1 (Welzl) *Let X be a graph such that $\chi(X) \geq 3$, and let Z be a graph such that $X \to Z$ but $Z \not\to X$. Then there is a graph Y such that $X \to Y$ and $Y \to Z$, but $Z \not\to Y$ and $Y \not\to X$.*

Proof. Since X is not empty or bipartite, any homomorphism from X to Z must map X into a nonbipartite component of Z. If we have homomorphisms $X \to Y$ and $Y \to Z$, it follows that the image of Y must be contained in a nonbipartite component of Z. Since Y cannot be empty, there is a homomorphism from any bipartite component of Z into Y. Hence it will suffice if we prove the theorem under the assumption that no component of Z is bipartite.

Let m be the maximum value of the odd girths of the components of Z and let n be the chromatic number of the map-graph X^Z. Let L be a graph with no odd cycles of length less than or equal to m and with chromatic number greater than n. (For example, we may take L to be a suitable Kneser graph.) Let Y be the disjoint union $X \cup (Z \times L)$.

Clearly, $X \to Y$. If there is a homomorphism from Y to X, then there must be one from $Z \times L$ to X, and therefore, by Corollary 6.4.3, a homomorphism from L to X^Z. Given the chromatic number of L, this is impossible.

Since there are homomorphisms from X to Z and from $Z \times L$ to Z, there is a homomorphism from Y to Z. Given the value of the odd girth of L, there cannot be a homomorphism that maps a component of Z into L. Therefore, there is no homomorphism from Z to L, and so there cannot be one from Z into $Z \times L$. □

An elegant illustration of this theorem is provided by the Andrásfai graphs. Each Andrásfai graph is 3-colourable, and so $\text{And}(k) \to K_3$, but $\text{And}(k)$ is triangle-free, and so $K_3 \not\to \text{And}(k)$. The theorem implies the existence of a graph Y such that $\text{And}(k) \to Y \to K_3$, and from our work in Section 6.11 we see that we can take Y to be $\text{And}(k+1)$. Therefore, we get an infinite sequence

$$\text{And}(2) \to \text{And}(3) \to \cdots \to K_3.$$

The fractional chromatic number of $\text{And}(k)$ is $(3k-1)/k$, and so the fractional chromatic numbers of the graphs in this sequence form an increasing sequence tending to 3.

7.14 The Cartesian Product

We introduce the Cartesian product of graphs, and show how it can be used to provide information about the size of r-colourable induced subgraphs of a graph.

If X and Y are graphs, their *Cartesian product* $X \square Y$ has vertex set $V(X) \times V(Y)$, where (x_1, y_1) is adjacent to (x_2, y_2) if and only if $x_1 = y_1$ and $x_2 \sim y_2$, or $x_1 \sim y_1$ and $x_2 = y_2$. Roughly speaking, we construct the Cartesian product of X and Y by taking one copy of Y for each vertex of X, and joining copies of Y corresponding to adjacent vertices of X by matchings of size $|V(Y)|$. For example, $K_m \square K_n = L(K_{m,n})$.

Let $\alpha_r(X)$ denote the maximum number of vertices in an r-colourable induced subgraph of X.

Lemma 7.14.1 *For any graph X, we have $\alpha_r(X) = \alpha(X \square K_r)$.*

Proof. Suppose that S is an independent set in $X \square K_r$. If $v \in V(K_r)$, then the set S_v, defined by

$$S_v = \{u \in V(X) : (u, v) \in S\},$$

is an independent set in X. Any two distinct vertices of $X \square K_r$ with the same first coordinate are adjacent, which implies that if v and w are distinct vertices of K_r, then $S_v \cap S_w = \emptyset$. Thus an independent set in $X \square K_r$ corresponds to a set of r pairwise-disjoint independent sets in X. The subgraph induced by such a set is an r-colourable subgraph of X. For the converse, suppose that X' is an r-colourable induced subgraph of X. Consider the set of vertices

$$S = \{(x, i) : x \in V(X') \text{ and } x \text{ has colour } i\}$$

in $X \square K_r$. All vertices in S have distinct first coordinates so can be adjacent only if they share the same second coordinate. However, if both (x, i) and (y, i) are in S, then x and y have the same colour in the r-colouring of X', so are not adjacent in X. Therefore, S is an independent set in $X \square K_r$. □

In Section 9.7 we will use this result to bound the size of the largest bipartite subgraphs of certain Kneser graphs.

If X and Y are vertex transitive, then so is their Cartesian product (as you are invited to prove). In particular, if X is vertex transitive, then so is $X \square K_r$.

Lemma 7.14.2 *If Y is vertex transitive and there is a homomorphism from X to Y, then*

$$\frac{|V(X)|}{\alpha_r(X)} \leq \frac{|V(Y)|}{\alpha_r(Y)}.$$

Proof. If there is a homomorphism from X to Y, then there is a homomorphism from $X \square K_r$ to $Y \square K_r$. Therefore, $\chi^*(X \square K_r) \leq \chi^*(Y \square K_r)$.

Using Corollary 7.5.3 and Corollary 7.5.2 in turn, we see that

$$\frac{|V(X \square K_r)|}{\alpha(X \square K_r)} \leq \chi^*(X \square K_r) \leq \chi^*(Y \square K_r) = \frac{|V(Y \square K_r)|}{\alpha(Y \square K_r)},$$

and then by the previous lemma

$$\frac{|V(X \square K_r)|}{\alpha_r(X)} \leq \frac{|V(Y \square K_r)|}{\alpha_r(Y)},$$

which immediately yields the result. □

We offer a generalization of this result in Exercise 25.

7.15 Strong Products and Colourings

The *strong product* $X * Y$ of two graphs X and Y is a graph with vertex set $X \times Y$; two distinct pairs (x_1, y_1) and (x_2, y_2) are adjacent in $X * Y$ if x_1 is equal or adjacent to x_2, and y_1 is equal or adjacent to y_2.

In the strong product $X * Y$ any set of vertices of the form

$$\{(x, y) : x \in V(X)\}$$

induces a subgraph isomorphic to X. Similarly, the sets

$$\{(x, y) : y \in V(Y)\}$$

induce copies of Y, and so it follows that

$$\chi(X * Y) \geq \max\{\chi(X), \chi(Y)\}.$$

This bound is not tight if both X and Y have at least one edge, as will be proved later.

We define the *n-colouring graph* $\mathcal{C}_n(X)$ of X to be the graph whose vertices are the n-colourings of X, with two vertices f and g adjacent if and only if there is an n-colouring of $X * K_2$ whose restrictions to the subsets $V(X) \times \{1\}$ and $V(X) \times \{2\}$ of $V(X * K_2)$ are f and g respectively. Notice that unlike the map graph K_n^X, the vertices of $\mathcal{C}_n(X)$ are restricted to be proper colourings of X.

Lemma 7.15.1 *For graphs X and Y,*

$$|\mathrm{Hom}(X * Y, K_n)| = |\mathrm{Hom}(Y, \mathcal{C}_n(X))|.$$

Proof. Exercise. □

Applying this lemma with $X = K_r$ we discover that

$$|\mathrm{Hom}(K_r * Y, K_n)| = |\mathrm{Hom}(Y, \mathcal{C}_n(K_r))|.$$

Recall that the *lexicographic product* $X[Y]$ of two graphs X and Y is the graph with vertex set $V(X) \times V(Y)$ and where

$$(x_1, y_1) \sim (x_2, y_2) \quad \text{if} \quad \begin{cases} x_1 = x_2 \text{ and } y_1 \sim y_2, & \text{or} \\ x_1 \sim x_2 \text{ and } y_1 = y_2, & \text{or} \\ x_1 \sim x_2 \text{ and } y_1 \sim y_2. \end{cases}$$

Theorem 7.15.2 $\mathcal{C}_n(K_r) = K_{n:r}[\overline{K_{r!}}]$.

Proof. Each vertex of $\mathcal{C}_n(K_r)$ is an n-colouring of K_r, and so its image (as a function) is a set of r distinct colours. Partitioning the vertices of $\mathcal{C}_n(K_r)$ according to their images gives $\binom{n}{r}$ cells each containing $r!$ pairwise nonadjacent vertices. Any two cells of this partition induce a complete bipartite graph if the corresponding r-sets are disjoint, and otherwise induce an empty graph. It is straightforward to see that this is precisely the description of the graph $K_{n:r}[\overline{K_{r!}}]$. □

Corollary 7.15.3 *$K_{n:r}$ and $\mathcal{C}_n(K_r)$ are homomorphically equivalent.* □

Corollary 7.15.4 *There is an n-colouring of $K_r * X$ if and only if there is a homomorphism from X into $K_{n:r}$.* □

A more unusual application of these results is the determination of the number of proper colourings of the lexicographic product $C_n[K_r]$.

Lemma 7.15.5 *The number of v-colourings of the graph $C_n[K_r]$ is equal to*

$$|\mathrm{Hom}(C_n, K_{v:r}[\overline{K_{r!}}])|.$$

Proof. The lexicographic product $C_n[K_r]$ is equal to the strong product $K_r * C_n$, and therefore we have

$$\begin{aligned} |\mathrm{Hom}(C_n[K_r], K_v)| &= |\mathrm{Hom}(K_r * C_n, K_v)| \\ &= |\mathrm{Hom}(C_n, \mathcal{C}_v(K_r))| \\ &= |\mathrm{Hom}(C_n, K_{v:r}[\overline{K_{r!}}])|. \end{aligned}$$

□

In Exercise 8.1 we will see how this last expression can be evaluated in terms of the eigenvalues of a suitable matrix.

Exercises

1. Let f be a fractional colouring of the graph X, and let B be the matrix with the characteristic vectors of the independent sets of X as its columns. Show that if the columns in $\operatorname{supp} f$ are linearly dependent, there is a fractional colouring f' such that $\operatorname{supp}(f') \subset \operatorname{supp}(f)$ and

the weight of f' is no greater than that of f. Deduce that there is a fractional colouring f with weight $\chi^*(X)$, and that $\chi^*(X)$ must be a rational number.

2. Prove that $\omega^*(X)$ is rational.

3. Show that $K_{v:r}$ is isomorphic to a subgraph of the product of enough copies of $K_{v-2:r-1}$. (Hint: First, if f is a homomorphism from $K_{v:r}$ to $K_{v-2:r-1}$ and $\gamma \in \mathrm{Aut}(K_{v:r})$, then $f \circ \gamma$ is a homomorphism from $K_{v:r}$ to $K_{v-2:r-1}$; second, the number of copies needed is at most $|\mathrm{Aut}(K_{v:r})|$.)

4. Prove that $C(v,r)$ is a core. (Hint: It is possible to use the proof of Theorem 7.9.1 as a starting point.)

5. Show that $\chi(C(v,r)) = \lceil \frac{v}{r} \rceil$.

6. Suppose $v \geq 2r$, $w \geq 2s$, and $v/r \leq w/s$. Show that there is a homomorphism from $C(v,r)$ into $C(w,s)$.

7. The *circular chromatic number* of a graph X is defined to be the minimum possible value of the ratio v/r, given that there is a homomorphism from X to $C(v,r)$. Denote this by $\chi^\circ(X)$. Show that this is well-defined and that for any X we have $\chi(X) - 1 < \chi^\circ(X) \leq \chi(X)$.

8. If X is vertex transitive, show that $\alpha(X)\omega(X) \leq |V(X)|$. Deduce that if X is vertex transitive but not complete and $|V(X)|$ is prime, then $\omega(X) < \chi(X)$.

9. Let X be a minimally imperfect graph and let A and B respectively be the incidence matrices for the big independent sets and big cliques of X, as defined in Section 7.6. Show that $BJ = JB = \omega J$ and that A and B^T commute. Deduce that each vertex of X lies in exactly $\omega(X)$ big cliques.

10. Let X be a minimally imperfect graph, and let $\mathcal{S}$ and $\mathcal{C}$ be the collections of big independent sets and big cliques of X, as defined in Section 7.6. Let S_i and S_j be two members of $\mathcal{S}$, and let C_i be the corresponding members of $\mathcal{C}$. Show that if neither $S_i \cap S_j$ nor $C_i \cap C_j$ is empty, then some entry of $B^T A$ is greater than 1. Deduce that for each vertex v in X there is a partition of $X \setminus v$ into α cliques from $\mathcal{C}$.

11. Let S be a subset of a set of v elements with size $2\ell - 1$. Show that the k-sets that contain at least ℓ elements of S form a maximal independent set in $K_{v:r}$, although the intersection of this family of k sets is empty.

12. Show that there are no homomorphisms from $K_{6:2}$ or $K_{9:3}$ to $K_{15:5}$, but there is a homomorphism from $K_{6:2} \times K_{9:3}$ to $K_{15:5}$.

13. Show that $\alpha(X * Y) \geq \alpha(X)\alpha(Y)$ and $\omega(X * Y) = \omega(X)\omega(Y)$.

14. Show that $\chi(X * Y) \leq \chi(X)\chi(Y)$, and that this bound is sharp for graphs whose chromatic and clique numbers are equal.

15. Show that $\chi(C_5 * C_5) = 5$.

16. If X is a graph on n vertices, show that $\alpha(X * \overline{X}) \geq n$.

17. Show that C_5 is homomorphically equivalent to $C_5(C_5)$. (Hence $\chi(X * C_5) = 5$ if and only if there is a homomorphism from X to C_5.)

18. Let X^* be the graph obtained from X by replacing each edge of X by a path of length three. (So X^* is a double subdivision of X, but not a subdivision of $S(X)$.) Show that there is a homomorphism from X^* into C_5 if and only if there is a homomorphism from X into K_5.

19. Convince yourself that if L_1 denotes the loop on one vertex, then $X \times L_1 \cong X$. Show that the subgraph of $F^{X \cup Y}$ induced by the homomorphisms is, essentially, the strong product of the subgraphs of F^X and F^Y induced respectively by the homomorphisms from X and Y into F. (And explain why we wrote "essentially" above.)

20. The usual version of Borsuk's theorem asserts that if the unit sphere in $\mathbb{R}^n$ is covered by n closed sets, then one of the sets contains an antipodal pair of points. The aim of this exercise is to show that this version implies the one we used. Suppose that $S_1, \ldots, S_k$ are open sets covering the unit sphere in $\mathbb{R}^n$. Show that there is an open set T_1 whose closure is contained in S_1 such that T_1, together with $S_2, \ldots, S_k$, covers the unit sphere; using this deduce the version of Borsuk's theorem from Section 7.11. [Hint: Let R_1 be the complement of $S_2 \cup \cdots \cup S_k$; this is a closed set contained in S_1. The boundary of R_1 is a compact set, hence can be covered by a finite number of open disks on the sphere, each of which is contained in S_1.]

21. Show that if X and Y are vertex transitive, then so is their Cartesian product.

22. Show that if there is a homomorphism from X to Y, then there is a homomorphism from $X \,\square\, K_r$ to $Y \,\square\, K_r$.

23. Show that if X has n vertices, then $\alpha(X \,\square\, C_5) \leq 2n$ and equality holds if and only if there is a homomorphism from X into C_5. Hence deduce that if X has an induced subgraph Y on m vertices such that there is a homomorphism from Y into C_5, then $2|V(Y)| \leq \alpha(X \square C_5)$.

24. Show that $K_{7:3} \,\square\, C_5$ contains an independent set of size 61.

25. Let $\nu(X, K)$ denote the maximum number of vertices in a subgraph of X that admits a homomorphism to K. If Y is vertex transitive and there is a homomorphism from X to Y, show that

$$\frac{|V(X)|}{\nu(X,K)} \leq \frac{|V(Y)|}{\nu(Y,K)}.$$

(Hint: Do not use the Cartesian product.)

26. If Y and X are graphs, let $Y[X]$ be the graph we get by replacing each vertex of Y by a copy of X, and each edge of Y by a complete bipartite graph joining the two copies of X. (For example, the complement of m vertex-disjoint copies of X is isomorphic to $K_m[\overline{X}]$.) If $v \geq 2r+1$ and $Y = K_{v:r}$, show that $\chi(X) \leq r$ if and only if $\chi(Y[X]) \leq v$. If $\chi(X) > r$, show that any n-colouring of Y determines a homomorphism from $K_{v:r}$ to $K_{n:r+1}$.

27. Show that there is a homomorphism from $K_{v:2}$ to $K_{w:3}$ if and only if $w \geq 2v - 2$.

28. Using the Hilton–Milner theorem (Theorem 7.10.1), prove that for $v \geq 7$ there is a homomorphism from $K_{v:3}$ to $K_{w:4}$ if and only if $w \geq 2v - 4$.

29. Let V be a set of size v, let α be a k-subset of V, and suppose $1 \in V \backslash \alpha$. Let $\mathcal{H}$ denote the set of all k-subsets of V that contain 1 and at least one point from α, together with the set α. Show that any two elements of $\mathcal{H}$ have at least one point in common, but the intersection of the elements of $\mathcal{H}$ is empty. (Note that $|\mathcal{H}| = 1 + \binom{v-1}{k-1} - \binom{v-k-1}{k-1}$.)

30. Let V be a set of size v that contains $\{1, 2, 3\}$. Show that the set of triples that contain at least two elements from $\{1, 2, 3\}$ has size $1 + 3(v - 3)$. (Note that $3v - 9 = \binom{v-1}{2} - \binom{v-4}{2}$.)

Notes

The theory of the fractional chromatic number provides a convincing application of linear programming methods to graph theory.

The idea of using the fractional chromatic number to restrict the existence of homomorphisms is apparently due to Perles. It is an extension of the "no homomorphism lemma" of Albertson and Collins, which appears in [1]. The results in Section 7.14 are also based on this paper.

The study of perfect graphs has been driven by two conjectures, due to Berge. The first, the so-called weak perfect graph conjecture, asserted that the complement of a perfect graph is perfect. This was first proved by Lovász, although it was subsequently realized that Fulkerson had come within easy distance of it. The proof we give is due to Gasparian [7]. The second conjecture, the strong perfect graph conjecture, asserts that a minimally imperfect graph is either an odd cycle or its complement. This is still open. Inventing new classes of perfect graphs has been a growth industry for many years.

The circular chromatic number of a graph, which we introduced in Exercise 7, was first studied by Vince [12], who called it the star chromatic number. See Bondy and Hell [4] and Zhu [14] for further work on this parameter.

For a treatment of the Erdős–Ko–Rado theorem from a more traditional viewpoint, see [3].

The chromatic number of the Kneser graphs was first determined by Lovász [8], thus verifying a 23-year-old conjecture due to Kneser. A shorter proof was subsequently found by Bárány [2], and this is what we followed.

The proof of Theorem 7.13.1 is due independently to Perles and Nešetřil. It is a significant simplification of Welzl's original argument, in [13].

The strong product $X * K_r$ is isomorphic to the lexicographic product $X[K_r]$ (see Exercise 1.26). Theorem 7.9.2 and Lemma 7.9.3 are due to Stahl [10]. Section 7.15 is based on Vesztergombi [11], which in turn is strongly influenced by [10].

Reinfeld [9] uses the result of Lemma 7.15.5 together with the spectrum of the Kneser graph to find the chromatic polynomial of the graphs $C_n[K_r]$ (see Chapter 15 for the definition of chromatic polynomial).

The truths expressed in Exercise 23 and Exercise 24 were pointed out to us by Tardif. In Section 9.7 we will present a technique that will allow us to prove that $\alpha(K_{7:3} \square C_5) = 61$. (See Exercise 9.18.) Tardif observes that this implies that an induced subgraph of $K_{7:3}$ with a homomorphism into C_5 must have at most 30 vertices; he has an example of such a subgraph with 29 vertices, and this can be shown by computer to be the largest such subgraph. For the solution to Exercise 25, see Bondy and Hell [4].

Exercise 26 is based on Garey and Johnson [6]. They use it to show that if a polynomial-time approximate algorithm for graph colouring exists, then there is a polynomial-time algorithm for graph colouring. (The expert consensus is that this is unlikely.) For information related to the Hilton–Milner theorem (Theorem 7.10.1), see Frankl [5]. Exercise 29 and Exercise 30 give all the families of k-sets without centres that realize the Hilton–Milner bound.

It would be interesting to find more tools for determining the existence of homomorphisms between pairs of Kneser graphs. We personally cannot say whether there is a homomorphism from $K_{10:4}$ to $K_{13:5}$. The general problem is clearly difficult, since it contains the problem of determining the chromatic numbers of the Kneser graphs.

References

[1] M. O. Albertson and K. L. Collins, *Homomorphisms of 3-chromatic graphs*, Discrete Math., 54 (1985), 127–132.

[2] I. Bárány, *A short proof of Kneser's conjecture*, J. Combin. Theory Ser. A, 25 (1978), 325–326.

[3] B. Bollobás, *Combinatorics*, Cambridge University Press, Cambridge, 1986.

[4] J. A. BONDY AND P. HELL, *A note on the star chromatic number*, J. Graph Theory, 14 (1990), 479–482.

[5] P. FRANKL, *Extremal set systems*, in Handbook of Combinatorics, Vol. 1, 2, Elsevier, Amsterdam, 1995, 1293–1329.

[6] M. R. GAREY AND D. S. JOHNSON, *The complexity of near-optimal graph coloring*, J. Assoc. Comp. Mach., 23 (1976), 43–49.

[7] G. S. GASPARIAN, *Minimal imperfect graphs: a simple approach*, Combinatorica, 16 (1996), 209–212.

[8] L. LOVÁSZ, *Kneser's conjecture, chromatic number, and homotopy*, J. Combin. Theory Ser. A, 25 (1978), 319–324.

[9] P. REINFELD, *Chromatic polynomials and the spectrum of the Kneser graph*, tech. rep., London School of Economics, 2000. LSE-CDAM-2000-02.

[10] S. STAHL, *n-tuple colorings and associated graphs*, J. Combinatorial Theory Ser. B, 20 (1976), 185–203.

[11] K. VESZTERGOMBI, *Chromatic number of strong product of graphs*, in Algebraic methods in graph theory, Vol. I, II (Szeged, 1978), North-Holland, Amsterdam, 1981, 819–825.

[12] A. VINCE, *Star chromatic number*, J. Graph Theory, 12 (1988), 551–559.

[13] E. WELZL, *Symmetric graphs and interpretations*, J. Combin. Theory Ser. B, 37 (1984), 235–244.

[14] X. ZHU, *Graphs whose circular chromatic number equals the chromatic number*, Combinatorica, 19 (1999), 139–149.

8
Matrix Theory

There are various matrices that are naturally associated with a graph, such as the adjacency matrix, the incidence matrix, and the Laplacian. One of the main problems of algebraic graph theory is to determine precisely how, or whether, properties of graphs are reflected in the algebraic properties of such matrices.

Here we introduce the incidence and adjacency matrices of a graph, and the tools needed to work with them. This chapter could be subtitled "Linear Algebra for Graph Theorists," because it develops the linear algebra we need from fundamental results about symmetric matrices through to the Perron–Frobenius theorem and the spectral decomposition of symmetric matrices.

Since many of the matrices that arise in graph theory are 01-matrices, further information can often be obtained by viewing the matrix over the finite field $GF(2)$. We illustrate this with an investigation into the binary rank of the adjacency matrix of a graph.

8.1 The Adjacency Matrix

The *adjacency matrix* $A(X)$ of a directed graph X is the integer matrix with rows and columns indexed by the vertices of X, such that the uv entry of $A(X)$ is equal to the number of arcs from u to v (which is usually 0 or 1). If X is a graph, then we view each edge as a pair of arcs in opposite directions, and $A(X)$ is a symmetric 01-matrix. Because a graph has no loops, the diagonal entries of $A(X)$ are zero. Different directed graphs on the same vertex set have different adjacency matrices, even if they are isomorphic. This is not much of a problem, and in any case we have the following consolation, the proof of which is left as an exercise.

Lemma 8.1.1 *Let X and Y be directed graphs on the same vertex set. Then they are isomorphic if and only if there is a permutation matrix P such that $P^T A(X)P = A(Y)$.* □

Since permutation matrices are orthogonal, $P^T = P^{-1}$, and so if X and Y are isomorphic directed graphs, then $A(X)$ and $A(Y)$ are similar matrices. The *characteristic polynomial* of a matrix A is the polynomial

$$\phi(A, x) = \det(xI - A),$$

and we let $\phi(X, x)$ denote the characteristic polynomial of $A(X)$. The *spectrum* of a matrix is the list of its eigenvalues together with their multiplicities. The spectrum of a graph X is the spectrum of $A(X)$ (and similarly we refer to the eigenvalues and eigenvectors of $A(X)$ as the eigenvalues and eigenvectors of X). Lemma 8.1.1 shows that $\phi(X, x) = \phi(Y, x)$ if X and Y are isomorphic, and so the spectrum is an invariant of the isomorphism class of a graph.

However, it is not hard to see that the spectrum of a graph does not determine its isomorphism class. Figure 8.1 shows two graphs that are not isomorphic but share the characteristic polynomial

$$(x + 2)(x + 1)^2(x - 1)^2(x^2 - 2x - 6),$$

and hence have spectrum

$$\{ -2,\ -1^{(2)},\ 1^{(2)},\ 1 \pm \sqrt{7} \}$$

(where the superscripts give the multiplicities of eigenvalues with multiplicity greater than one). Two graphs with the same spectrum are called *cospectral*.

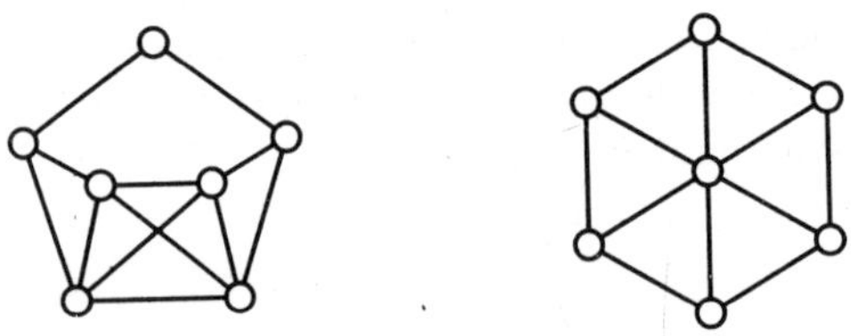

Figure 8.1. Two cospectral graphs

The graphs of Figure 8.1 show that the valencies of the vertices are not determined by the spectrum, and that whether a graph is planar is not determined by the spectrum. In general, if there is a cospectral pair of graphs, only one of which has a certain property $\mathcal{P}$, then $\mathcal{P}$ cannot be determined by the spectrum. Such cospectral pairs have been found for a large number of graph-theoretical properties.

However, the next result shows that there is some useful information that can be obtained from the spectrum. A *walk* of length r in a directed graph X is a sequence of vertices

$$v_0 \sim v_1 \sim \cdots \sim v_r.$$

A walk is *closed* if its first and last vertices are the same. This definition is similar to that of a path (Section 1.2), with the important difference being that a walk is permitted to use vertices more than once.

Lemma 8.1.2 *Let X be a directed graph with adjacency matrix A. The number of walks from u to v in X with length r is $(A^r)_{uv}$.*

Proof. This is easily proved by induction on r, as you are invited to do. □

The *trace* of a square matrix A is the sum of its diagonal entries and is denoted by $\operatorname{tr} A$. The previous result shows that the number of closed walks of length r in X is $\operatorname{tr} A^r$, and hence we get the following corollary:

Corollary 8.1.3 *Let X be a graph with e edges and t triangles. If A is the adjacency matrix of X, then*

(a) $\operatorname{tr} A = 0$,
(b) $\operatorname{tr} A^2 = 2e$,
(c) $\operatorname{tr} A^3 = 6t$.

□

Since the trace of a square matrix is also equal to the sum of its eigenvalues, and the eigenvalues of A^r are the rth powers of the eigenvalues of A, we see that $\operatorname{tr} A^r$ is determined by the spectrum of A. Therefore, the spectrum of a graph X determines at least the number of vertices, edges, and triangles in X. The graphs $K_{1,4}$ and $K_1 \cup C_4$ are cospectral and do not have the same number of 4-cycles, so it is difficult to extend these observations.

8.2 The Incidence Matrix

The *incidence matrix* $B(X)$ of a graph X is the 01-matrix with rows and columns indexed by the vertices and edges of X, respectively, such that the uf-entry of $B(X)$ is equal to one if and only if the vertex u is in the edge f. If X has n vertices and e edges, then $B(X)$ has order $n \times e$.

The rank of the adjacency matrix of a graph can be computed in polynomial time, but we do not have a simple combinatorial expression for it. We do have one for the rank of the incidence matrix.

Theorem 8.2.1 *Let X be a graph with n vertices and c_0 bipartite connected components. If B is the incidence matrix of X, then its rank is given by* $\operatorname{rk} B = n - c_0$.

Proof. We shall show that the null space of B has dimension c_0, and hence that $\operatorname{rk} B = n - c_0$. Suppose that z is a vector in $\mathbb{R}^n$ such that $z^T B = 0$. If uv is an edge of X, then $z_u + z_v = 0$. It follows by an easy induction that if u and v are vertices of X joined by a path of length r, then $z_u = (-1)^r z_v$. Therefore, if we view z as a function on $V(X)$, it is identically zero on any component of X that is not bipartite, and takes equal and opposite values on the two colour classes of any bipartite component. The space of such vectors has dimension c_0. □

The inner product of two columns of $B(X)$ is nonzero if and only if the corresponding edges have a common vertex, which immediately yields the following result.

Lemma 8.2.2 *Let B be the incidence matrix of the graph X, and let L be the line graph of X. Then $B^T B = 2I + A(L)$.* □

If X is a graph on n vertices, let $\Delta(X)$ be the diagonal $n \times n$ matrix with rows and columns indexed by $V(X)$ with uu-entry equal to the valency of vertex u. The inner product of any two distinct rows of $B(X)$ is equal to the number of edges joining the corresponding vertices. Thus it is zero or one according as these vertices are adjacent or not, and we have the following:

Lemma 8.2.3 *Let B be the incidence matrix of the graph X. Then $BB^T = \Delta(X) + A(X)$.* □

When X is regular the last two results imply a simple relation between the eigenvalues of $L(X)$ and those of X, but to prove this we also need the following result.

Lemma 8.2.4 *If C and D are matrices such that CD and DC are both defined, then* $\det(I - CD) = \det(I - DC)$.

Proof. If

$$X = \begin{pmatrix} I & C \\ D & I \end{pmatrix}, \qquad Y = \begin{pmatrix} I & 0 \\ -D & I \end{pmatrix},$$

then

$$XY = \begin{pmatrix} I - CD & C \\ 0 & I \end{pmatrix}, \qquad YX = \begin{pmatrix} I & C \\ 0 & I - DC \end{pmatrix},$$

and since $\det XY = \det YX$, it follows that $\det(I - CD) = \det(I - DC)$. □

This result implies that $\det(I - x^{-1}CD) = \det(I - x^{-1}DC)$, from which it follows that CD and DC have the same nonzero eigenvalues with the same multiplicities.

Lemma 8.2.5 *Let X be a regular graph of valency k with n vertices and e edges and let L be the line graph of X. Then*

$$\phi(L, x) = (x + 2)^{e-n}\phi(X, x - k + 2).$$

Proof. Substituting $C = x^{-1}B^T$ and $D = B$ into the previous lemma we get

$$\det\left(I_e - x^{-1}B^T B\right) = \det\left(I_n - x^{-1}BB^T\right),$$

whence

$$\det\left(xI_e - B^T B\right) = x^{e-n}\det\left(xI_n - BB^T\right).$$

Noting that $\Delta(X) = kI$ and using Lemma 8.2.2 and Lemma 8.2.3, we get

$$\det((x-2)I_e - A(L)) = x^{e-n}\det((x-k)I_n - A(X)),$$

and so

$$\phi(L, x-2) = x^{e-n}\phi(X, x-k),$$

whence our claim follows. □

8.3 The Incidence Matrix of an Oriented Graph

An *orientation* of a graph X is the assignment of a direction to each edge; this means that we declare one end of the edge to be the *head* of the edge and the other to be the *tail*, and view the edge as oriented from its tail to its head. Although this definition should be clear, we occasionally need a more formal version. Recall that an arc of a graph is an ordered pair of adjacent vertices. An orientation of X can then be defined as a function σ from the arcs of X to $\{-1, 1\}$ such that if (u, v) is an arc, then

$$\sigma(u, v) = -\sigma(v, u).$$

If $\sigma(u, v) = 1$, then we will regard the edge uv as oriented from tail u to head v.

An *oriented graph* is a graph together with a particular orientation. We will sometimes use X^σ to denote the oriented graph determined by the specific orientation σ. (You may, if you choose, view oriented graphs as a special class of directed graphs. We tend to view them as graphs with extra structure.) Figure 8.2 shows an example of an oriented graph, using arrows to indicate the orientation.

The *incidence matrix* $D(X^\sigma)$ of an oriented graph X^σ is the $\{0, \pm 1\}$-matrix with rows and columns indexed by the vertices and edges of X, respectively, such that the uf-entry of $D(X^\sigma)$ is equal to 1 if the vertex u is the head of the edge f, -1 if u is the tail of f, and 0 otherwise. If X

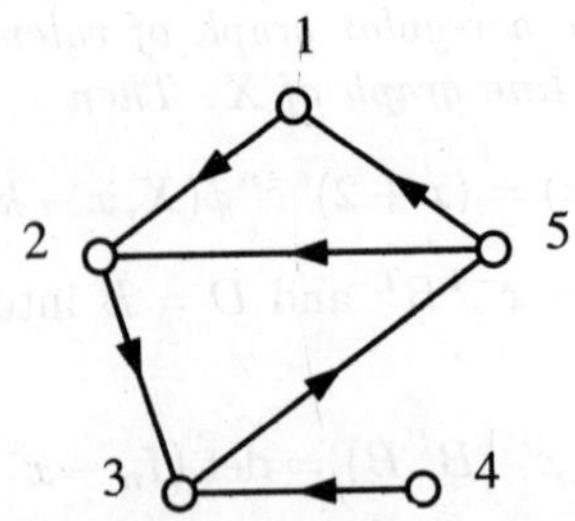

Figure 8.2. An oriented graph

has n vertices and e edges, then $D(X^\sigma)$ has order $n \times e$. For example, the incidence matrix of the graph of Figure 8.2 is

$$\begin{pmatrix} -1 & 1 & 0 & 0 & 0 & 0 \\ 1 & 0 & -1 & 1 & 0 & 0 \\ 0 & 0 & 1 & 0 & 1 & -1 \\ 0 & 0 & 0 & 0 & -1 & 0 \\ 0 & -1 & 0 & -1 & 0 & 1 \end{pmatrix}.$$

Although there are many different ways to orient a given graph, many of the results about oriented graphs are independent of the choice of orientation. For example, the next result shows that the rank of the incidence matrix of an oriented graph depends only on X, rather than on the particular orientation given to X.

Theorem 8.3.1 *Let X be a graph with n vertices and c connected components. If σ is an orientation of X and D is the incidence matrix of X^σ, then* $\operatorname{rk} D = n - c$.

Proof. We shall show that the null space of D has dimension c, and hence that $\operatorname{rk} D = n - c$. Suppose that z is a vector in $\mathbb{R}^n$ such that $z^T B = 0$. If uv is an edge of X, then $z_u - z_v = 0$. Therefore, if we view z as a function on $V(X)$, it is constant on any connected component of X. The space of such vectors has dimension c. □

We note the following analogue to Lemma 8.2.3.

Lemma 8.3.2 *If σ is an orientation of X and D is the incidence matrix of X^σ, then $DD^T = \Delta(X) - A(X)$.* □

If X is a plane graph, then each orientation of X determines an orientation of its dual. This orientation is obtained by viewing each edge of X^* as arising from rotating the corresponding edge of X through 90° clockwise (as in Figure 8.3). We will use σ to denote the orientation of both X and X^*.

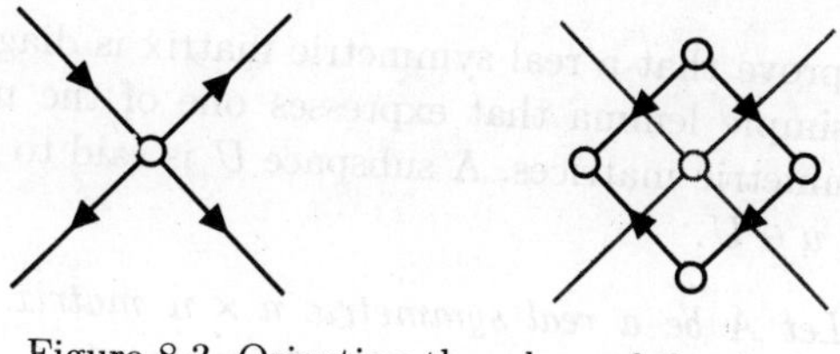

Figure 8.3. Orienting the edges of the dual

Lemma 8.3.3 *Let X and Y be dual plane graphs, and let σ be an orientation of X. If D and E are the incidence matrices of X^σ and Y^σ, then $DE^T = 0$.*

Proof. If u is an edge of X and F is a face, there are exactly two edges on u and in F. Denote them by g and h and assume, for convenience, that g precedes h as we go clockwise around F. Then the uF-entry of DE^T is equal to

$$D_{ug}E_{gF}^T + D_{uh}E_{hF}^T.$$

If the orientation of the edge g is reversed, then the value of the product $D_{ug}E_{gF}^T$ does not change. Hence the value of the sum is independent of the orientation σ, and so we may assume that g has head u and that f has tail u. This implies that the edges in Y corresponding to g and h both have head F, and a simple computation now yields that the sum is zero. □

8.4 Symmetric Matrices

In this section we review the main results of the linear algebra of symmetric matrices over the real numbers, which form the basis for the remainder of this chapter.

Lemma 8.4.1 *Let A be a real symmetric matrix. If u and v are eigenvectors of A with different eigenvalues, then u and v are orthogonal.*

Proof. Suppose that $Au = \lambda u$ and $Av = \tau v$. As A is symmetric, $u^T Av = (v^T Au)^T$. However, the left-hand side of this equation is $\tau u^T v$ and the right-hand side is $\lambda u^T v$, and so if $\tau \neq \lambda$, it must be the case that $u^T v = 0$. □

Lemma 8.4.2 *The eigenvalues of a real symmetric matrix A are real numbers.*

Proof. Let u be an eigenvector of A with eigenvalue λ. Then by taking the complex conjugate of the equation $Au = \lambda u$ we get $A\bar{u} = \bar{\lambda}\bar{u}$, and so $\bar{u}$ is also an eigenvector of A. Now, by definition an eigenvector is not zero, so $u^T\bar{u} > 0$. By the previous lemma, u and $\bar{u}$ cannot have different eigenvalues, so $\lambda = \bar{\lambda}$, and the claim is proved. □

We shall now prove that a real symmetric matrix is diagonalizable. For this we need a simple lemma that expresses one of the most important properties of symmetric matrices. A subspace U is said to be *A-invariant* if $Au \in U$ for all $u \in U$.

Lemma 8.4.3 *Let A be a real symmetric $n \times n$ matrix. If U is an A-invariant subspace of $\mathbb{R}^n$, then $U^\perp$ is also A-invariant.*

Proof. For any two vectors u and v, we have

$$v^T(Au) = (Av)^T u.$$

If $u \in U$, then $Au \in U$; hence if $v \in U^\perp$, then $v^T Au = 0$. Consequently, $(Av)^T u = 0$ whenever $u \in U$ and $v \in U^\perp$. This implies that $Av \in U^\perp$ whenever $v \in U^\perp$, and therefore $U^\perp$ is A-invariant. □

Any square matrix has at least one eigenvalue, because there must be at least one solution to the polynomial equation $\det(xI - A) = 0$. Hence a real symmetric matrix A has at least one real eigenvalue, θ say, and hence at least one real eigenvector (any vector in the kernel of $A - \theta I$, to be precise). Our next result is a crucial strengthening of this fact.

Lemma 8.4.4 *Let A be an $n \times n$ real symmetric matrix. If U is a nonzero A-invariant subspace of $\mathbb{R}^n$, then U contains a real eigenvector of A.*

Proof. Let R be a matrix whose columns form an orthonormal basis for U. Then, because U is A-invariant, $AR = RB$ for some square matrix B. Since $R^T R = I$, we have

$$R^T AR = R^T RB = B,$$

which implies that B is symmetric, as well as real. Since every symmetric matrix has at least one eigenvalue, we may choose a real eigenvector u of B with eigenvalue λ. Then $ARu = RBu = \lambda Ru$, and since $u \neq 0$ and the columns of R are linearly independent, $Ru \neq 0$. Therefore, Ru is an eigenvector of A contained in U. □

Theorem 8.4.5 *Let A be a real symmetric $n \times n$ matrix. Then $\mathbb{R}^n$ has an orthonormal basis consisting of eigenvectors of A.*

Proof. Let $\{u_1, \dots, u_m\}$ be an orthonormal (and hence linearly independent) set of $m < n$ eigenvectors of A, and let M be the subspace that they span. Since A has at least one eigenvector, $m \geq 1$. The subspace M is A-invariant, and hence $M^\perp$ is A-invariant, and so $M^\perp$ contains a (normalized) eigenvector u_{m+1}. Then $\{u_1, \dots, u_m, u_{m+1}\}$ is an orthonormal set of $m+1$ eigenvectors of A. Therefore, a simple induction argument shows that a set consisting of one normalized eigenvector can be extended to an orthonormal basis consisting of eigenvectors of A. □

Corollary 8.4.6 *If A is an $n \times n$ real symmetric matrix, then there are matrices L and D such that $L^T L = LL^T = I$ and $LAL^T = D$, where D is the diagonal matrix of eigenvalues of A.*

Proof. Let L be the matrix whose rows are an orthonormal basis of eigenvectors of A. We leave it as an exercise to show that L has the stated properties. □

8.5 Eigenvectors

Most introductory linear algebra courses impart the belief that the way to compute the eigenvalues of a matrix is to find the zeros of its characteristic polynomial. For matrices with order greater than two, this is false. Generally, the best way to obtain eigenvalues is to find eigenvectors: If $Ax = \theta x$, then θ is an eigenvalue of A.

When we work with graphs there is an additional refinement. First, we stated in Section 8.1 that the rows and columns of $A(X)$ are indexed by the vertices of X. Formally, this means we are viewing $A(X)$ as a linear mapping on $\mathbb{R}^{V(X)}$, the space of real functions on $V(X)$ (rather than on the isomorphic vector space $\mathbb{R}^n$, where $n = |V(X)|$). If $f \in \mathbb{R}^{V(X)}$ and $A = A(X)$, then the image Af of f under A is given by

$$(Af)(u) = \sum A_{uv} f(v);$$

since A is a 01-matrix, it follows that

$$(Af)(u) = \sum_{v \sim u} f(v).$$

In words, the value of Af at u is the sum of the values of f on the neighbours of u. If we suppose that f is an eigenvector of A with eigenvalue θ, then $Af = \theta f$, and so

$$\theta f(u) = \sum_{v \sim u} f(v).$$

In words, the sum of the values of f on the neighbours of u is equal to θ times the value of f at u. Conversely, any function f that satisfies this condition is an eigenvector of X. Figure 8.4 shows an eigenvector of the Petersen graph. It can readily be checked that the sum of the values on the neighbours of any vertex is equal to the value on that vertex; hence we have an eigenvector with eigenvalue one. (The viewpoint expressed in this paragraph is very fruitful, and we will make extensive use of it.)

Now, we will find the eigenvalues of the cycle C_n. Take the vertex set of C_n to be $\{0, 1, \ldots, n-1\}$. Let τ be an nth root of unity (so τ is probably

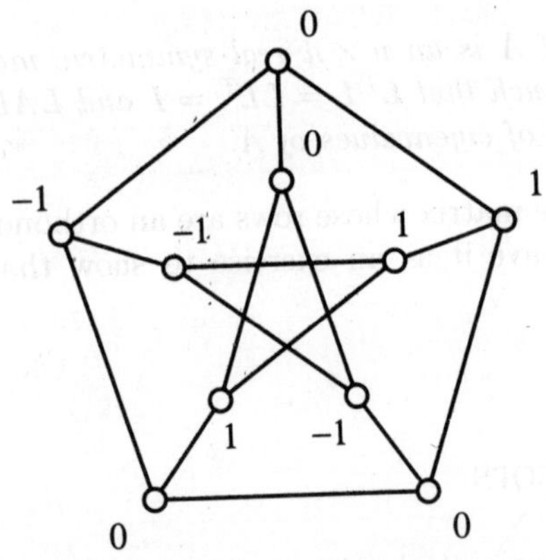

Figure 8.4. An eigenvector of P with eigenvalue 1

not a real number) and define $f(u) := \tau^u$. Then for all vertices u,

$$\sum_{v \sim u} f(v) = (\tau^{-1} + \tau)\tau^u,$$

and therefore $\tau^{-1} + \tau$ is an eigenvalue of C_n. Note that this is real, even if τ is not. By varying our choice of τ we find the n eigenvalues of C_n. This argument is easily extended to any circulant graph.

By taking $\tau = 1$ we see that the vector with all entries equal to one is an eigenvector of C_n with eigenvalue two. We shall denote this eigenvector by $\mathbf{1}$. It is clear that $\mathbf{1}$ is an eigenvector of a graph X with eigenvalue k if and only if X is regular with valency k. We can say more about regular graphs.

Lemma 8.5.1 *Let X be a k-regular graph on n vertices with eigenvalues $k, \theta_2, \dots, \theta_n$. Then X and its complement $\overline{X}$ have the same eigenvectors, and the eigenvalues of $\overline{X}$ are $n-k-1, -1-\theta_2, \dots, -1-\theta_n$.*

Proof. The adjacency matrix of the complement $\overline{X}$ is given by

$$A(\overline{X}) = J - I - A(X),$$

where J is the all-ones matrix. Let $\{\mathbf{1}, u_2, \dots, u_n\}$ be an orthonormal basis of eigenvectors of $A(X)$. Then $\mathbf{1}$ is an eigenvector of $\overline{X}$ with eigenvalue $n-k-1$. For $2 \le i \le n$, the eigenvector u_i is orthogonal to $\mathbf{1}$, and so

$$A(\overline{X})u_i = (J - I - A(X))u_i = (-1-\theta_i)u_i.$$

Therefore, u_i is an eigenvector of $A(\overline{X})$ with eigenvalue $-1-\theta_i$. □

Finally, suppose that X is a semiregular bipartite graph with bipartition $V(X) = V_1 \cup V_2$, and let k and ℓ be the valencies of the vertices in V_1 and V_2, respectively. Assume that u_1 is a vertex with valency k, and u_2 is a vertex with valency ℓ. We look for an eigenvector f that is constant on the two parts of the bipartition. If f is such an eigenvector and has eigenvalue

θ, then

$$\theta f(u_1) = kf(u_2), \qquad \theta f(u_2) = \ell f(u_1).$$

Because an eigenvector is a nonzero vector, we can multiply the two equations just given to obtain

$$\theta^2 = k\ell.$$

Thus, if $\theta = \pm\sqrt{k\ell}$, then defining f by

$$f(u) = \begin{cases} 1, & \text{if } u \in V_1, \\ \theta/k, & \text{if } u \in V_2, \end{cases}$$

yields two eigenvectors of X.

We comment on a feature of the last example. If A is the adjacency matrix of a graph X, and f is a function on $V(X)$, then so is Af. If X is a semiregular bipartite graph, then the space of functions on $V(X)$ that are constant on the two parts of the bipartition is A-invariant. (Indeed, this is equivalent to the fact that X is bipartite and semiregular.) By Lemma 8.4.4, an A-invariant subspace must contain an eigenvector of A; in the above example this subspace has dimension two, and the eigenvector is easy to find. In Section 9.3 we introduce and study equitable partitions, which provide many further examples of A-invariant subspaces.

8.6 Positive Semidefinite Matrices

A real symmetric matrix A is *positive semidefinite* if $u^T Au \geq 0$ for all vectors u. It is *positive definite* if it is positive semidefinite and $u^T Au = 0$ if and only if $u = 0$. (These terms are used only for symmetric matrices.) Observe that a positive semidefinite matrix is positive definite if and only if it is invertible.

There are a number of characterizations of positive semidefinite matrices. The first we offer involves eigenvalues. If u is an eigenvector of A with eigenvalue θ, then

$$u^T Au = \theta u^T u,$$

and so we see that a real symmetric matrix is positive semidefinite if and only if its eigenvalues are nonnegative.

Our second characterization involves a factorization. If $A = B^T B$ for some matrix B, then

$$u^T Au = u^T B^T Bu = (Bu)^T Bu \geq 0,$$

and therefore A is positive semidefinite. The *Gram matrix* of vectors $u_1, \ldots, u_n$ from $\mathbb{R}^m$ is the $n \times n$ matrix G such that $G_{ij} = u_i^T u_j$. Note that $B^T B$ is the Gram matrix of the columns of B, and that any Gram matrix is positive semidefinite. The next result shows that the converse is true.

Lemma 8.6.1 *If A is a positive semidefinite matrix, then there is a matrix B such that $A = B^T B$.*

Proof. Since A is symmetric, there is a matrix L such that

$$A = L^T \Lambda L,$$

where Λ is the diagonal matrix with ith entry equal to the ith eigenvalue of A. Since A is positive semidefinite, the entries of Λ are nonnegative, and so there is a diagonal matrix D such that $D^2 = \Lambda$. If $B = L^T DL$, then $B = B^T$ and $A = B^2 = B^T B$, as required. □

We can now establish some interesting results about the eigenvalues of graphs, the first being about line graphs.

Let $\theta_{\max}(X)$ and $\theta_{\min}(X)$ respectively denote the largest and smallest eigenvalues of $A(X)$.

Lemma 8.6.2 *If L is a line graph, then $\theta_{\min}(L) \geq -2$.*

Proof. If L is the line graph of X and B is the incidence matrix of X, we have

$$A(L) + 2I = B^T B.$$

Since $B^T B$ is positive semidefinite, its eigenvalues are nonnegative and all eigenvalues of $B^T B - 2I$ are at least -2. □

What is surprising about this lemma is how close it comes to characterizing line graphs. We will study this question in detail in Chapter 12.

Lemma 8.6.3 *Let Y be an induced subgraph of X. Then*

$$\theta_{\min}(X) \leq \theta_{\min}(Y) \leq \theta_{\max}(Y) \leq \theta_{\max}(X).$$

Proof. Let A be the adjacency matrix of X and abbreviate $\theta_{\max}(X)$ to θ. The matrix $\theta I - A$ has only nonnegative eigenvalues, and is therefore positive semidefinite. Let f be any vector that is zero on the vertices of X not in Y, and let f_Y be its restriction to $V(Y)$. Then

$$0 \leq f^T(\theta I - A)f = f_Y^T(\theta I - A(Y))f_Y,$$

from which we deduce that $\theta I - A(Y)$ is positive semidefinite. Hence $\theta_{\max}(Y) \leq \theta$. A similar argument applied to $A - \theta_{\min}(X)I$ yields the second claim of the lemma. □

It is actually true that if Y is any subgraph of X, and not just an induced subgraph, then $\theta_{\max}(Y) \leq \theta_{\max}(X)$. Furthermore, when Y is a proper subgraph, equality can hold only when X is not connected. We return to this when we discuss the Perron–Frobenius theorem in the next two sections.

Finally, we clear a debt incurred in Section 5.10. There we claimed that the matrix

$$(r - \lambda)I + \lambda J$$

is invertible when $r > \lambda \geq 0$. Note that $(r - \lambda)I$ is positive definite: All its eigenvalues are positive and $\lambda J = \lambda \mathbf{1}\mathbf{1}^T$ is positive semidefinite. But the sum of a positive definite and a positive semidefinite matrix is positive definite, and therefore invertible.

8.7 Subharmonic Functions

In this section we introduce subharmonic functions, and use them to develop some properties of nonnegative matrices. We will use similar techniques again in Section 13.9, when we show how linear algebra can be used to construct drawings of planar graphs.

If A is a square matrix, then we say that a nonnegative vector x is λ-*subharmonic* for A if $x \neq 0$ and $Ax \geq \lambda x$. When the value of λ is irrelevant, we simply say that x is subharmonic. We note one way that subharmonic vectors arise. Let $|A|$ denote the matrix obtained by replacing each entry of A with its absolute value. If x is an eigenvector for A with eigenvalue θ, then

$$|\theta|\,|x_i| = |\theta x_i| = |(Ax)_i| = \Big|\sum_j A_{ij}x_j\Big| \leq \sum_j |A_{ij}|\,|x_j|,$$

from which we see that $|x|$ is $|\theta|$-subharmonic for $|A|$.

Let A be an $n \times n$ real matrix. The *underlying directed graph* of A has vertex set $\{1, \ldots, n\}$, with an arc from vertex i to vertex j if and only if $A_{ij} \neq 0$. (Note that this directed graph may have loops.) A square matrix is *irreducible* if its underlying graph is strongly connected.

Lemma 8.7.1 *Let A be an $n \times n$ nonnegative irreducible matrix. Then there is a maximum real number ρ such that there is a ρ-subharmonic vector for A. Moreover, any ρ-subharmonic vector x is an eigenvector for A with eigenvalue ρ, and all entries of x are positive.*

Proof. Let

$$F(x) = \min_{i : x_i \neq 0} \frac{(Ax)_i}{x_i}$$

be a function defined on the set of nonnegative vectors, and consider the values of F on the vectors in the set

$$S = \left\{x : x \geq 0,\ \mathbf{1}^T x = 1\right\}.$$

It is clear that any nonnegative vector x is $F(x)$-subharmonic, and so we wish to show that there is some vector $y \in S$ such that F attains its maximum on y. Since S is compact, this would be immediate if F were continuous on S, but this is not the case at the boundary of S. As A is irreducible, Lemma 8.1.2 shows that the matrix $(I + A)^{n-1}$ is positive. Therefore, the set

$$T = (I + A)^{n-1} S$$

contains only positive vectors, and F is continuous on T. Since T is also compact, it follows that F attains its maximum value ρ at a point $z \in T$. If we set

$$y = \frac{z}{\mathbf{1}^T z},$$

then $y \in S$ and $F(y) = F(z) = \rho$. Moreover, for any vector x we have

$$F((I + A)^{n-1}(x)) \geq F(x),$$

and therefore by the choice of z, there is no vector $x \in S$ with $F(x) > \rho$.

We now prove that any ρ-subharmonic vector is an eigenvector for A, necessarily with eigenvalue ρ. If x is ρ-subharmonic, define $\sigma(x)$ by

$$\sigma(x) = \{i : (Ax)_i > \rho x_i\}.$$

Clearly, x is an eigenvector if and only if $\sigma(x) = \emptyset$. Assume by way of contradiction that $\sigma(x) \neq \emptyset$. The *support* of a vector v is the set of nonzero coordinates of v and is denoted by $\text{supp}(v)$. Let h be a nonnegative vector with support equal to $\sigma(x)$ and consider the vector $y = x + \epsilon h$.

We have

$$(Ay)_i - \rho y_i = (Ax)_i - \rho x_i + \epsilon (Ah)_i - \epsilon \rho h_i.$$

If $i \in \sigma(x)$, then $(Ax)_i > \rho x_i$, and so for all sufficiently small values of ϵ, the right side of (8.7) is positive. Hence

$$(Ay)_i > \rho y_i.$$

If $i \notin \sigma(x)$, then $(Ax)_i = \rho x_i$ and $h_i = 0$, so (8.7) yields that

$$(Ay)_i - \rho y_i = \epsilon (Ah)_i.$$

Provided that $\epsilon > 0$, the right side here is nonnegative. Since A is irreducible, there is at least one value of i not in $\sigma(x)$ such that $(Ah)_i > 0$, and hence $\sigma(y)$ properly contains $\sigma(x)$.

If $|\sigma(y)| = n$, it follows that y is ρ'-subharmonic, where $\rho' > \rho$, and this is a contradiction to our choice of ρ. Otherwise, y is ρ-subharmonic but $|\sigma(y)| > |\sigma(x)|$, and we may repeat the above argument with y in place of x. After a finite number of iterations we will arrive at a ρ'-subharmonic vector, with $\rho' > \rho$, again a contradiction.

Finally, we prove that if x is ρ-subharmonic, then $x > 0$. Suppose instead that $x_i = 0$ for some i. Because $\sigma(x) = \emptyset$, it follows that $(Ax)_i = 0$, but

$$(Ax)_i = \sum_j A_{ij} x_j,$$

and since $A \geq 0$, this implies that $x_j = 0$ if $A_{ij} \neq 0$. Since A is irreducible, a simple induction argument yields that all entries of x must be zero, which is the required contradiction. Therefore, x must be positive. □

The *spectral radius* $\rho(A)$ of a matrix A is the maximum of the moduli of its eigenvalues. (If A is not symmetric, these eigenvalues need not be real numbers.) The spectral radius of a matrix need not be an eigenvalue of it, e.g., if $A = -I$, then $\rho(A) = 1$. One consequence of our next result is that the real number ρ from the previous lemma is the spectral radius of A.

Lemma 8.7.2 *Let A be an $n \times n$ nonnegative irreducible matrix and let ρ be the greatest real number such that A has a ρ-subharmonic vector. If B is an $n \times n$ matrix such that $|B| \leq A$ and $Bx = \theta x$, then $|\theta| \leq \rho$. If $|\theta| = \rho$, then $|B| = A$ and $|x|$ is an eigenvector of A with eigenvalue ρ.*

Proof. If $Bx = \theta x$, then

$$|\theta||x| = |\theta x| = |Bx| \leq |B|\,|x| \leq A|x|.$$

Hence $|x|$ is $|\theta|$-subharmonic for A, and so $|\theta| \leq \rho$. If $|\theta| = \rho$, then $A|x| = |B||x| = \rho|x|$, and by the previous lemma, $|x|$ is positive. Since $A - |B| \geq 0$ and $(A - |B|)|x| = 0$, it follows that $A = |B|$. □

Lemma 8.7.3 *Let A be a nonnegative irreducible $n \times n$ matrix with spectral radius ρ. Then ρ is a simple eigenvalue of A, and if x is an eigenvector with eigenvalue ρ, then all entries of x are nonzero and have the same sign.*

Proof. The ρ-eigenspace of A is 1-dimensional, for otherwise we could find a ρ-subharmonic vector with some entry equal to zero, contradicting Lemma 8.7.1. If x is an eigenvector with eigenvalue ρ, then by the previous lemma, $|x|$ is a positive eigenvector with the same eigenvalue. Thus $|x|$ is a multiple of x, which implies that all the entries of x have the same sign.

Since the geometric multiplicity of ρ is 1, we see that $K = \ker(A - \rho I)$ has dimension 1 and the column space C of $A - \rho I$ has dimension $n - 1$. If C contains x, then we can find a vector y such that $x = (A - \rho I)y$. For any k, we have $(A - \rho I)(y + kx) = x$, and so by taking k sufficiently large, we may assume that y is positive. But then y is ρ-subharmonic and hence is a multiple of x, which is impossible. Therefore, we conclude that $K \cap C = 0$, and that $\mathbb{R}^n$ is the direct sum of K and C. Since K and C are A-invariant, this implies that the characteristic polynomial $\varphi(A, t)$ of A is the product of $t - \rho$ and the characteristic polynomial of A restricted to C. As x is not in C, all eigenvectors of A contained in C have eigenvalue different from ρ, and so we conclude that ρ is a simple root of $\varphi(A, t)$, and hence has algebraic multiplicity one. □

8.8 The Perron–Frobenius Theorem

The Perron–Frobenius theorem is the most important result on the eigenvalues and eigenvectors of nonnegative matrices.

Theorem 8.8.1 *Suppose A is a real nonnegative $n \times n$ matrix whose underlying directed graph X is strongly connected. Then:*

(a) *$\rho(A)$ is a simple eigenvalue of A. If x is an eigenvector for ρ, then no entries of x are zero, and all have the same sign.*

(b) *Suppose A_1 is a real nonnegative $n \times n$ matrix such that $A - A_1$ is nonnegative. Then $\rho(A_1) \leq \rho(A)$, with equality if and only if $A_1 = A$.*

(c) *If θ is an eigenvalue of A and $|\theta| = \rho(A)$, then $\theta/\rho(A)$ is an mth root of unity and $e^{2\pi ir/m}\rho(A)$ is an eigenvalue of A for all r. Further, all cycles in X have length divisible by m.* □

The first two parts of this theorem follow from the results of the previous section. We discuss part (c), but do not give a complete proof of it, since we will not need its full strength.

Suppose A is the adjacency matrix of a connected graph X, with spectral radius ρ, and assume that θ is an eigenvalue of A such that $|\theta| = \rho$. If $\theta \neq \rho$, then $\theta = -\rho$, and so θ/ρ is a root of unity. If z_0 and z_1 are eigenvectors with eigenvalues θ and ρ, respectively, then they are linearly independent, and therefore the eigenspace of A^2 with eigenvalue ρ^2 has dimension at least two. However, it is easy to see that ρ^2 is the spectral radius of A^2. As A^2 is nonnegative, it follows from part (a) of the theorem that the underlying graph of A^2 cannot be connected, and given this, it is easy to prove that X must be bipartite.

It is not hard to see that if X is bipartite, then there is a graph isomorphic to X with adjacency matrix of the form

$$A = \begin{pmatrix} 0 & B \\ B^T & 0 \end{pmatrix},$$

for a suitable 01-matrix B. If the partitioned vector (x, y) is an eigenvector of A with eigenvalue θ, then it is easy to verify that $(x, -y)$ is an eigenvector of A with eigenvalue $-\theta$. It follows that θ and $-\theta$ are eigenvalues with the same multiplicity. Thus we have the following:

Theorem 8.8.2 *Let A be the adjacency matrix of the graph X, and let ρ be its spectral radius. Then the following are equivalent:*

(a) *X is bipartite.*

(b) *The spectrum of A is symmetric about the origin, i.e., for any θ, the multiplicities of θ and $-\theta$ as eigenvalues of A are the same.*

(c) *$-\rho$ is an eigenvalue of A.* □

There are two common applications of the Perron–Frobenius theorem to connected regular graphs. Let X be a connected k-regular graph with adjacency matrix A. Then the spectral radius of A is the valency k with corresponding eigenvector $\mathbf{1}$, which implies that every other eigenspace of A is orthogonal to $\mathbf{1}$. Secondly, the graph X is bipartite if and only if $-k$ is an eigenvalue of A.

8.9 The Rank of a Symmetric Matrix

The rank of a matrix is a fundamental algebraic concept, and so it is natural to ask what information about a graph can be deduced from the rank of its adjacency matrix. In contrast to what we obtained for the incidence matrix, there is no simple combinatorial expression for the rank of the adjacency matrix of a graph. This section develops a number of preliminary results about the rank of a symmetric matrix that will be used later.

Theorem 8.9.1 *Let A be a symmetric matrix of rank r. Then there is a permutation matrix P and a principal $r \times r$ submatrix M of A such that*

$$P^T AP = \begin{pmatrix} I \\ R \end{pmatrix} M \begin{pmatrix} I & R^T \end{pmatrix}.$$

Proof. Since A has rank r, there is a linearly independent set of r rows of A. By symmetry, the corresponding set of columns is also linearly independent. The entries of A in these rows and columns determine an $r \times r$ principal submatrix M. Therefore, there is a permutation matrix P such that

$$P^T AP = \begin{pmatrix} M & N^T \\ N & H \end{pmatrix}$$

Since the first r rows of this matrix generate the row space of $P^T AP$, we have that $N = RM$ for some matrix R, and hence $H = RN^T = RMR^T$. Therefore,

$$P^T AP = \begin{pmatrix} M & MR \\ RM & RMR^T \end{pmatrix} = \begin{pmatrix} I \\ R \end{pmatrix} M \begin{pmatrix} I & R^T \end{pmatrix}$$

as claimed. □

We note an important corollary of this result.

Corollary 8.9.2 *If A is a symmetric matrix of rank r, then it has a principal $r \times r$ submatrix of full rank.* □

If a matrix A has rank one, then it is necessarily of the form $A = xy^T$ for some nonzero vectors x and y. It is not too hard to see that if a matrix can be written as the sum of r rank-one matrices, then it has rank at most

r. However, it is less well known that a matrix A has rank r if and only if it can be written as the sum of r rank-one matrices, but no fewer. If A is symmetric, the rank-one matrices in this decomposition will not necessarily be symmetric. Instead, we have the following.

Lemma 8.9.3 *Suppose A is a symmetric matrix with rank r over some field. Then there is an integer s such that A is the sum of $r-2s$ symmetric matrices with rank one and s symmetric matrices with rank two.*

Proof. Suppose A is symmetric. First we show that if A has a nonzero diagonal entry, then it is the sum of a symmetric rank-one matrix and a symmetric matrix of rank $r-1$. Let e_i denote the ith standard basis vector, and suppose that $\alpha = e_i^T A e_i \neq 0$. Let $x = Ae_i$ and define B by

$$B := A - \alpha^{-1} x x^T.$$

Then B is symmetric, and $\alpha^{-1} x x^T$ has rank one. Clearly, $Bu = 0$ whenever $Au = 0$, and so the null space of B contains the null space of A. This inclusion is proper because e_i lies in the null space of B, but not A, and so $\text{rk}(B) \leq \text{rk}(A) - 1$. Since the column space of A is spanned by the columns of B together with the vector x, we conclude that $\text{rk}(B) = \text{rk}(A) - 1$.

Next we show that if there are two diagonal entries $A_{ii} = A_{jj} = 0$ with $A_{ij} \neq 0$, then A is the sum of a symmetric rank-two matrix and a symmetric matrix of rank $r-2$. So suppose that $e_i^T A e_i = e_j^T A e_j = 0$ but that $\beta = e_i^T A e_j \neq 0$. Let $y = Ae_i$, $z = Ae_j$ and define B by

$$B := A - \beta^{-1}(yz^T + zy^T).$$

Then B is symmetric and $\beta^{-1}(yz^T + zy^T)$ has rank two. The null space of B contains the null space of A. The independent vectors e_i and e_j lie in the null space of B but not in the null space of A, and so $\text{rk}(B) \leq \text{rk}(A) - 2$. Since the column space of A is spanned by the columns of B together with the vectors y and z, we conclude that $\text{rk}(B) = \text{rk}(A) - 2$.

Therefore, by induction on the rank of A, we may write

$$A = \sum_{i=1}^{r-2s} \alpha_i^{-1} x_i x_i^T + \sum_{j=1}^{s} \beta_j^{-1}(y_j z_j^T + z_j y_j^T), \tag{8.1}$$

and thus we have expressed A as a sum of s symmetric matrices with rank two and $r-2s$ with rank one. □

Corollary 8.9.4 *Let A be a real symmetric $n \times n$ matrix of rank r. Then there is an $n \times r$ matrix C of rank r such that*

$$A = CNC^T,$$

where N is a block-diagonal $r \times r$ matrix with $r-2s$ diagonal entries equal to ± 1, and s blocks of the form

$$\begin{pmatrix} 0 & 1 \\ 1 & 0 \end{pmatrix}.$$

Proof. We note that

$$\beta^{-1}(yz^T + zy^T) = (\beta^{-1}y \;\; z)\begin{pmatrix} 0 & 1 \\ 1 & 0 \end{pmatrix}(\beta^{-1}y \;\; z)^T.$$

Therefore, if we take C to be the $n \times r$ matrix with columns $\sqrt{|\alpha_i^{-1}|}x_i$, $\beta_j^{-1}y_j$, and z_j, then

$$A = CNC^T,$$

where N is a block-diagonal matrix, with each diagonal block one of the matrices

$$(0), \qquad (\pm 1), \qquad \begin{pmatrix} 0 & 1 \\ 1 & 0 \end{pmatrix}.$$

The column space of A is contained in the space spanned by vectors x_i, y_j and z_j in (8.1); because these two spaces have the same dimension, we conclude that these vectors are a basis for the column space of A. Therefore, $\text{rk}(C) = r$. □

The previous result is an application of Lemma 8.9.3 to real symmetric matrices. In the next section we apply it to symmetric matrices over $GF(2)$.

8.10 The Binary Rank of the Adjacency Matrix

In general, there is not a great deal that can be said about a graph given the rank of its adjacency matrix over the real numbers. However, we can say considerably more if we consider the binary rank of the adjacency matrix, that is, the rank calculated over $GF(2)$. If X is a graph, then $\text{rk}_2(X)$ denotes the rank of its adjacency matrix over $GF(2)$.

First we specialize the results of the previous section to the binary case.

Theorem 8.10.1 *Let A be a symmetric $n \times n$ matrix over $GF(2)$ with zero diagonal and binary rank m. Then m is even and there is an $m \times n$ matrix C of rank m such that*

$$A = CNC^T,$$

where N is a block diagonal matrix with $m/2$ blocks of the form

$$\begin{pmatrix} 0 & 1 \\ 1 & 0 \end{pmatrix}.$$

Proof. Over $GF(2)$, the diagonal entries of the matrix $yz^T + zy^T$ are zero. Since all diagonal entries of A are zero, it follows that the algorithm implicit in the proof of Lemma 8.9.3 will express A as a sum of symmetric matrices with rank two and zero diagonals. Therefore, Lemma 8.9.3 implies that $\text{rk}(A)$ is even. The proof of Corollary 8.9.4 now yields the rest of the theorem. □

Next we develop a graphical translation of the procedure we used to prove Lemma 8.9.3. If $u \in V(X)$, the *local complement* $\sigma_u(X)$ of X at u is defined to be the graph with the same vertex set as X such that:

(a) If v and w are distinct neighbours of u, then they are adjacent in Y if and only if they are not adjacent in X.

(b) If v and w are distinct vertices of X, and not both neighbours of u, then they are adjacent in Y if and only if they are adjacent in X.

Less formally, we get Y from X by complementing the neighbourhood $X_1(u)$ of u in X. If we view σ_u as an operator on graphs with the same vertex set as X, then σ_u^2 is the identity map. If u and v are not adjacent in X, then $\sigma_u\sigma_v(X) = \sigma_v\sigma_u(X)$. We leave the proof of this as an exercise, because our concern will be with the case where u and v are adjacent. One consequence of the following theorem is that $(\sigma_u\sigma_v)^3$ is the identity map if u and v are adjacent.

Theorem 8.10.2 *Let X be a graph and suppose u and v are neighbours in X. Then $\sigma_u\sigma_v\sigma_u(X) = \sigma_v\sigma_u\sigma_v(X)$. If Y is the graph obtained by deleting u and v from $\sigma_u\sigma_v\sigma_u(X)$, then* $\operatorname{rk}_2(X) = \operatorname{rk}_2(Y) + 2$.

Proof. Let A be the adjacency matrix of X. Define a to be the characteristic vector of the set of neighbours of u that are not adjacent to v. Define b to be the characteristic vector of the set of the neighbours of v, other than u, that are not adjacent to u. Let c denote the characteristic vector of the set of common neighbours of u and v. Finally, let e_u and e_v denote the characteristic vectors of u and v.

The characteristic vector of the neighbours of u is $a+c+e_v$, and so the off-diagonal entries of $A(\sigma_u(X))$ are equal to those of

$$A_1 = A + (a+c+e_v)(a+c+e_v)^T.$$

Similarly, as $a+b+e_u$ is the characteristic vector of the neighbours of v in $\sigma_u(X)$, the off-diagonal entries of $A(\sigma_v\sigma_u(X))$ are equal to those of

$$A_2 = A_1 + (a+b+e_u)(a+b+e_u)^T.$$

Finally, the characteristic vector of the neighbours of u in $\sigma_v\sigma_u(X)$ is $b+c+e_v$, and so the off-diagonal entries of $A(\sigma_u\sigma_v\sigma_u(X))$ are equal to those of

$$A_3 = A_2 + (b+c+e_v)(b+c+e_v)^T.$$

After straightforward manipulation we find that

$$\begin{aligned} A_3 = {} & A + ab^T + ba^T + ac^T + ca^T + bc^T + cb^T \\ & + (a+b)(e_u+e_v)^T + (e_u+e_v)(a+b)^T + e_u e_u^T. \end{aligned}$$

The only nonzero diagonal entry of this matrix is the uu-entry, and so we conclude that $A(\sigma_u\sigma_v\sigma_u(X)) = A_3 + e_u e_u^T$. The previous equation shows that $A_3 + e_u e_u^T$ is unchanged if we swap a with b and e_u with e_v. Therefore, $\sigma_u\sigma_v\sigma_u(X) = \sigma_v\sigma_u\sigma_v(X)$, as claimed.

The u-column of A is $a + c + e_v$, the v-column of A is $b + c + e_u$, and $e_u^T A e_v = 1$. Therefore, the proof of Lemma 8.9.3 shows that the rank of the matrix

$$A + (a + c + e_v)(b + c + e_u)^T + (b + c + e_u)(a + c + e_v)^T$$

is equal to $\mathrm{rk}_2(A) - 2$. The u- and v-rows and columns of this matrix are zero, and so if A' is the principal submatrix obtained by deleting the u- and v-rows and columns, then $\mathrm{rk}_2(A') = \mathrm{rk}_2(A) - 2$.

To complete the proof we note that since

$$\begin{aligned}(a + c)(b + c)^T &+ (b + c)(a + b)^T \\ &= ab^T + ba^T + ac^T + ca^T + bc^T + cb^T,\end{aligned}$$

it follows that the matrix obtained by deleting the u- and v-rows and columns from A_3 is equal to A'. This matrix is the adjacency matrix of Y, and hence the second claim follows. □

We will say that the graph Y in the theorem is obtained by *rank-two reduction* of X at the edge uv.

By way of example, if X is the cycle C_n and $n \geq 5$, then the rank-two reduction of X at an edge is C_{n-2}. When $n = 4$ it is $2K_1$, and when $n = 3$ it is K_1. It follows that $\mathrm{rk}_2(C_n)$ is $n - 2$ when n is even and $n - 1$ when n is odd. Clearly, we can use Theorem 8.10.2 to determine the binary rank of the adjacency matrix of any graph, although it will not usually be as easy as it was in this case.

8.11 The Symplectic Graphs

If a graph X has two vertices with identical neighbourhoods, then deleting one of them does not alter its rank. Conversely, we can duplicate a vertex arbitrarily often without changing the rank of a graph. Similarly, isolated vertices can be added or deleted at will without changing the rank of X. Recall that a graph is reduced if it has no isolated vertices and the neighbourhoods of distinct vertices are distinct. It is clear that every graph is a straightforward modification of a reduced graph of the same rank. We are going to show that there is a unique maximal graph with binary rank $2r$ that contains every reduced graph of binary rank at most $2r$ as an induced subgraph.

Suppose that X is a reduced graph with binary rank $2r$. Relabelling vertices if necessary, Theorem 8.9.1 shows that the adjacency matrix of X can be expressed in the form

$$A(X) = \begin{pmatrix} I \\ R \end{pmatrix} M \begin{pmatrix} I & R^T \end{pmatrix},$$

where M is a $2r \times 2r$ matrix of full rank. Therefore, using Lemma 8.9.3 we see that

$$A(X) = \begin{pmatrix} I \\ R \end{pmatrix} CNC^T \begin{pmatrix} I & R^T \end{pmatrix},$$

where N is a block diagonal matrix with r blocks of the form

$$\begin{pmatrix} 0 & 1 \\ 1 & 0 \end{pmatrix}.$$

This provides an interesting vectorial representation of the graph X. The vertices of X are the columns of the matrix $C^T \begin{pmatrix} I & R^T \end{pmatrix}$, and adjacency is given by

$$u \sim v \quad \text{if and only if} \quad u^T N v = 1.$$

Therefore, X is entirely determined by the set Ω of columns of $C^T \begin{pmatrix} I & R^T \end{pmatrix}$. Since C has full rank, Ω is a spanning set of vectors. Conversely, if Ω is a spanning set of nonzero vectors in $GF(2)^{2r}$ and X is the graph with vertex set Ω and with adjacency defined by

$$u \sim v \iff u_1v_2 + u_2v_1 + \cdots + u_{2r-1}v_{2r} + u_{2r}v_{2r-1} = 1,$$

then X is a reduced graph with binary rank $2r$.

We give an example in Figure 8.5; this graph has binary rank 4, and therefore can be represented by eight vectors from $GF(2)^4$. It is easy to check that it is represented by the set

$$\Omega = \{1000, 0100, 0010, 0001, 1110, 1101, 1011, 0111\}.$$

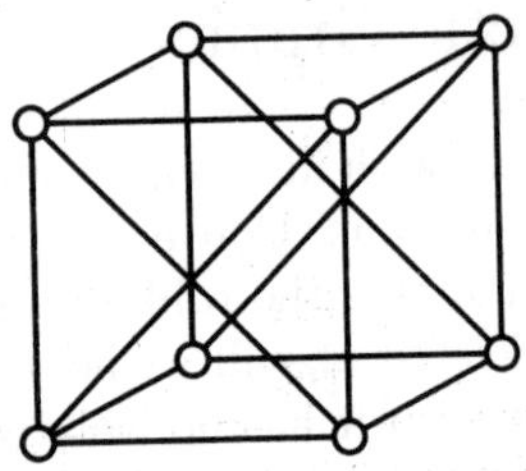

Figure 8.5. Graph with binary rank 4

Let $\text{Sp}(2r)$ be the graph obtained by taking Ω to be $GF(2)^{2r} \setminus 0$. We call it the *symplectic graph*, for reasons to be provided in Section 10.12. The next result shows that we can view $\text{Sp}(2r)$ as the *universal graph* with binary rank $2r$.

Theorem 8.11.1 *A reduced graph has binary rank at most $2r$ if and only if it is an induced subgraph of* $\text{Sp}(2r)$.

Proof. Any reduced graph X of binary rank $2r$ has a vectorial representation as a spanning set of nonzero vectors in $GF(2)^{2r}$. Therefore, the vertex set of X is a subset of the vertices of $\mathrm{Sp}(2r)$, where two vertices are adjacent in X if and only if they are adjacent in $\mathrm{Sp}(2r)$. Therefore, X is an induced subgraph of $\mathrm{Sp}(2r)$. The converse is clear. □

This implies that studying the properties of the universal graph $\mathrm{Sp}(2r)$ can yield information that applies to all graphs with binary rank $2r$. A trivial observation of this kind is that a reduced graph with binary rank $2r$ has at most $2^{2r} - 1$ vertices. A more interesting example will be given when we return to the graphs $\mathrm{Sp}(2r)$ in Section 10.12. Finally, we finish with an interesting property of the symplectic graphs.

Theorem 8.11.2 *Every graph on* $2r - 1$ *vertices occurs as an induced subgraph of* $\mathrm{Sp}(2r)$.

Proof. We prove this by induction on r. It is true when $r = 1$ because a single vertex is an induced subgraph of a triangle. So suppose that $r > 1$, and let X be an arbitrary graph on $2r-1$ vertices. If X is empty, then it is straightforward to see that it is an induced subgraph of $\mathrm{Sp}(2r)$. Otherwise, X has at least one edge uv. Let Y be the rank-two reduction of X at the edge uv. Then Y is a graph on $2r - 3$ vertices, and hence by the inductive hypothesis can be represented as a set Ω of nonzero vectors in $GF(2)^{2r-2}$. If z is a vector in Ω representing the vertex $y \in V(Y)$, then define a vector $z' \in GF(2)^{2r}$ as follows:

$$z'_i = \begin{cases} z_i, & \text{for } 1 \le i \le 2r-2; \\ 1, & \text{if } i = 2r-1 \text{ and } y \sim u \text{ in } X, \text{ or } i = 2r \text{ and } y \sim v \text{ in } X; \\ 0, & \text{otherwise.} \end{cases}$$

Then the set of vectors

$$\Omega' = \{z' : z \in S\} \cup \{e_{2r-1}, e_{2r}\}$$

is a set of $2r$ vectors in $GF(2)^{2r}$. Checking that the graph defined by Ω' is equal to X requires examining several cases, but is otherwise routine, so we leave it as Exercise 28. □

8.12 Spectral Decomposition

Let A be an $n \times n$ real symmetric matrix and let $\mathrm{ev}(A)$ denote the set of eigenvalues of A. If θ is an eigenvalue of A, let E_θ be the matrix representing orthogonal projection onto the eigenspace of θ. These are sometimes called the *principal idempotents* of A. Then

$$E_\theta^2 = E_\theta,$$

and since distinct eigenspaces of A are orthogonal, it follows that if θ and τ are distinct eigenvalues of A,

$$E_\theta E_\tau = 0.$$

Since $\mathbb{R}^n$ has a basis consisting of eigenvectors of A, we have

$$I = \sum_{\theta \in \mathrm{ev}(A)} E_\theta.$$

From this equation we see that

$$A = \sum_{\theta \in \mathrm{ev}(A)} \theta E_\theta;$$

this is known as the *spectral decomposition* of A.

More generally, if p is any polynomial, then it follows from the above that

$$p(A) = \sum_{\theta \in \mathrm{ev}(A)} p(\theta) E_\theta. \tag{8.2}$$

Since we may choose p so that it vanishes on all but one of the eigenvalues of A, it follows from (8.2) that E_θ is a polynomial in A. The matrices E_θ are linearly independent: If $\sum_\theta a_\theta E_\theta = 0$, then

$$0 = E_\tau \sum_\theta a_\theta E_\theta = a_\tau E_\tau.$$

Therefore, the principal idempotents form a basis for the vector space of all polynomials in A, and therefore this vector space has dimension equal to the number of distinct eigenvalues of A.

Lemma 8.12.1 *If X is a graph with diameter d, then $A(X)$ has at least $d+1$ distinct eigenvalues.*

Proof. We sketch the proof. Observe that the uv-entry of $(A+I)^r$ is nonzero if and only if u and v are joined by a path of length at most r. Consequently, the matrices $(A+I)^r$ for $r = 0, \dots, d$ form a linearly independent subset in the space of all polynomials in A. Therefore, $d+1$ is no greater than the dimension of this space, which is the number of primitive idempotents of A. □

A *rational function* is a function that can be expressed as the ratio q/r of two polynomials. It is not too hard to see that (8.2) still holds when p is a rational function, provided only that it is defined at each eigenvalue of A. Hence we obtain that

$$(xI - A)^{-1} = \sum_{\theta \in \mathrm{ev}(A)} (x-\theta)^{-1} E_\theta. \tag{8.3}$$

8.13 Rational Functions

In this section we explore some of the consequences of (8.3); these will be crucial to our work on interlacing in the next chapter.

Lemma 8.13.1 *Let A be a real symmetric $n \times n$ matrix and let B denote the matrix obtained by deleting the ith row and column of A. Then*

$$\frac{\phi(B,x)}{\phi(A,x)} = e_i^T (xI - A)^{-1} e_i,$$

where e_i is the ith standard basis vector.

Proof. From the standard determinantal formula for the inverse of a matrix we have

$$((xI - A)^{-1})_{ii} = \frac{\det(xI - B)}{\det(xI - A)},$$

so noting that

$$((xI - A)^{-1})_{ii} = e_i^T (xI - A)^{-1} e_i$$

suffices to complete the proof. □

Corollary 8.13.2 *For any graph X we have*

$$\phi'(X,x) = \sum_{u \in V(X)} \phi(X \setminus u, x).$$

Proof. By (8.3),

$$\operatorname{tr}(xI - A)^{-1} = \sum_\theta \frac{\operatorname{tr} E_\theta}{x - \theta}.$$

By the lemma, the left side here is equal to

$$\sum_{u \in V(X)} \frac{\phi(X \setminus u, x)}{\phi(X,x)}.$$

If m_θ denotes the multiplicity of θ as a zero of $\phi(X,x)$, then a little bit of calculus yields the partial fraction expansion

$$\frac{\phi'(X,x)}{\phi(X,x)} = \sum_\theta \frac{m_\theta}{x - \theta}.$$

Since E_θ is a symmetric matrix and $E_\theta^2 = E_\theta$, its eigenvalues are all 0 or 1, and $\operatorname{tr} E_\theta$ is equal to its rank. But the rank of E_θ is the dimension of the eigenspace associated with θ, and therefore $\operatorname{tr} E_\theta = m_\theta$. This completes the proof. □

If $f = p/q$ is a rational function, we say that f is *proper* if the degree of p is less than the degree of q. Any proper rational function has a partial fraction expansion

$$\sum_{i=1}^{r} \frac{p_i(x)}{(x-\theta_i)^{m_i}}.$$

Here m_i is a positive integer, and $p_i(x)$ is a nonzero polynomial of degree less than m_i. We call the numbers θ_i the *poles* of f; the integer m_i is the *order* of the pole at θ_i. A *simple* pole is a pole of order one. If the rational function f has a pole of order m, then f^2 has a pole of order at least $2m$. (You are invited to prove this.)

Theorem 8.13.3 *Let A be a real symmetric $n \times n$ matrix, let b be a vector of length n, and define $\psi(x)$ to be the rational function $b^T(xI - A)^{-1}b$. Then all zeros and poles of ψ are simple, and ψ' is negative everywhere it is defined. If θ and τ are consecutive poles of ψ, the closed interval $[\theta, \tau]$ contains exactly one zero of ψ.*

Proof. By (8.3),

$$b^T(xI - A)^{-1}b = \sum_{\theta \in \mathrm{ev}(A)} \frac{b^T E_\theta b}{x - \theta}. \tag{8.4}$$

This implies that the poles of ψ are simple. We differentiate both sides of (8.4) to obtain

$$\psi'(x) = -\sum_{\theta} \frac{b^T E_\theta b}{(x-\theta)^2}$$

and then observe, using (8.3), that the right side here is $-b^T(xI - A)^{-2}b$. Thus

$$\psi'(x) = -b^T(xI - A)^{-2}b.$$

Since $b^T(xI - A)^{-2}b$ is the squared length of $(xI - A)^{-1}b$, it follows that $\psi'(x) < 0$ whenever x is not a pole of ψ. This implies that each zero of ψ must be simple.

Suppose that θ and τ are consecutive poles of ψ. Since these poles are simple, it follows that ψ is a strictly decreasing function on the interval $[\theta, \tau]$ and that it is positive for values of x in this interval sufficiently close to θ, and negative when x is close enough to τ. Accordingly, this interval contains exactly one zero of ψ. □

Exercises

1. Show that $|\mathrm{Hom}(C_n, X)|$ equals the number of closed walks of length n in X, and hence that $|\mathrm{Hom}(C_n, X)|$ is the sum of the nth powers of the eigenvalues of X.

2. Let B and D be respectively the incidence and an oriented incidence matrix for the graph X. Show that X is bipartite if and only if there is a diagonal matrix M, with all diagonal entries equal to 1 or -1, such that $MD = B$. Show that X is bipartite if and only if $\Delta + A(X)$ and $\Delta - A(X)$ are similar matrices.

3. Show that cospectral graphs have the same odd girth.

4. Show that the sum of two positive semidefinite matrices is positive semidefinite, and that the sum of a positive definite and positive semidefinite matrix is positive definite.

5. Let $f_1, \ldots, f_n$ be a set of vectors in an inner product space V and let G be the $n \times n$ matrix with G_{ij} equal to the inner product of f_i and f_j. Show that G is a symmetric positive semidefinite matrix.

6. Show that any principal submatrix of a positive semidefinite matrix is positive semidefinite.

7. Let A be a symmetric positive semidefinite matrix. Show that the ith row of A is zero if and only if $A_{ii} = 0$.

8. Let X be a regular graph on n vertices with valency k and let θ be an eigenvalue of X. If u is an eigenvector for $A(X)$ with eigenvalue θ and $Ju = 0$, show that u is an eigenvector for $\overline{X}$ with eigenvalue $-1-\theta$. Use this to give an expression for $\phi(\overline{X}, x)$ in terms of $\phi(X, x)$.

9. Determine the eigenvalues of $L(P)$ in the following stages:
   (a) Determine the eigenvalues of K_n.
   (b) Find the eigenvalues of $L(K_5)$.
   (c) Find the eigenvalues of $P = \overline{L(K_5)}$.
   (d) Find the eigenvalues of $L(P)$.

10. Determine the eigenvalues of $K_{m,n}$ and their multiplicities.

11. Let P_n be the path with n vertices with vertex set $\{v_1, \ldots, v_n\}$, where $v_i \sim v_{i+1}$ for $i = 1, \ldots, n-1$. Suppose that f is an eigenvector for X with eigenvalue θ such that $f(v_1) = 1$. If we define polynomials $p_r(x)$ recursively by $p_0(x) = 1$, $p_1(x) = x$, and

$$p_{r+1}(x) = xp_r(x) - p_{r-1}(x),$$

then show that $f(v_r) = p_{r-1}(\theta)$. Deduce from this that $p_n(x)$ is the characteristic polynomial of P_n.

12. Show that when n is odd, $\phi(C_{2n}, x) = -\phi(C_n, x)\,\phi(C_n, -x)$.

13. If Y is a subgraph of X, show that $\rho(A(Y)) \le \rho(A(X))$. If X is connected, show that equality holds if and only if $Y = X$.

14. Let X be a graph with maximum valency a. Show that

$$\sqrt{a} \le \rho(A(X)) \le a$$

and characterize the cases where equality holds.

15. Let A be a symmetric matrix with distinct eigenvalues $\theta_1, \dots, \theta_n$, and for each i, let x_i be an eigenvector with length one and eigenvalue θ_i. Show that the principal idempotent E_i corresponding to θ_i equals $x_i x_i^T$.

16. A graph X is *walk regular* if for all nonnegative integers r, the diagonal entries of $A(X)^r$ are equal. (The simplest examples are the vertex-transitive graphs. Strongly regular graphs, which we study in Chapter 10, provide another less obvious class.) Show that a regular graph with at most four distinct eigenvalues is walk regular.

17. If f is a rational function with a pole of order m, show that f^2 has a pole of order at least $2m$.

18. Let B be the submatrix of the symmetric matrix A obtained by deleting the ith row and column of A. Show that if x is an eigenvector for A such that $x_i = 0$, then the vector y we get by deleting the ith coordinate from x is an eigenvector for B. We call y the *restriction* of x, and x the *extension* of y. Now, suppose that θ is a common eigenvalue of A and B, and that its multiplicity as an eigenvalue of A is m. If the multiplicity of θ as an eigenvalue of B is $m - 1$, show that each θ-eigenvector of B extends to an eigenvector for A. Using the spectral decomposition, prove that if the multiplicity of θ as an eigenvalue of B is at least m and x is a θ-eigenvector x of A, then $x_i = 0$.

19. If
$$A = \begin{pmatrix} 0 & b^T \\ b & B \end{pmatrix},$$
then show that an eigenvector of B extends to an eigenvector of A if and only if it is orthogonal to b.

20. If
$$A = \begin{pmatrix} 0 & b^T \\ b & B \end{pmatrix},$$
then show that
$$\begin{pmatrix} 1 & 0 \\ 0 & xI - B \end{pmatrix}^{-1} \begin{pmatrix} x & -b^T \\ -b & xI - B \end{pmatrix} = \begin{pmatrix} 1 & -b^T \\ 0 & I \end{pmatrix} \begin{pmatrix} x - b^T(xI - B)^{-1}b & 0 \\ -(xI - B)^{-1}b & I \end{pmatrix}.$$
Hence deduce that
$$\frac{\det(xI - A)}{\det(xI - B)} = x - b^T(xI - B)^{-1}b.$$

21. Let X be a regular graph on $2m$ vertices and suppose $S \subseteq V(X)$ such that $|S| = m$. Let X^1 be the graph we get by taking a new vertex and joining it to each vertex in S. Let X^2 be the graph we get by taking a new vertex and joining it to each vertex in $V(X)\backslash S$. Use the previous exercise to show that X^1 and X^2 are cospectral. Construct an example where X^1 and X^2 are not isomorphic.

22. Let A be an irreducible nonnegative matrix, and let L be the set of all real numbers λ such that there is a λ-subharmonic vector for A. Show directly that L is closed and bounded, and hence contains a maximum element ρ.

23. Show that an $m \times n$ matrix over a field $\mathbb{F}$ has rank r if and only if it can be written as the sum of r matrices with rank one.

24. Show that if a graph has two vertices with identical neighbourhoods, then deleting one of them does not alter its rank.

25. Show that if X is bipartite and Y is obtained from X by rank-two reduction at an edge, then Y is bipartite.

26. Suppose we consider graphs with loops, but at most one loop per vertex. Define a rank-one reduction operation, similar to local complementation at a vertex, that converts a given vertex to an isolated vertex and reduces the rank of the adjacency matrix.

27. Let A be the adjacency matrix of the Petersen graph. Compute $\mathrm{rk}_2(A)$ and $\mathrm{rk}_2(A+I)$ using rank-one and rank-two reductions.

28. Complete the details in the proof of Theorem 8.11.2 that every graph on $2r-1$ vertices occurs as an induced subgraph of $\mathrm{Sp}(2r)$.

29. A matrix A over a field of odd characteristic is *skew symmetric* if $A = -A^T$. (In even characteristic, we must add the requirement that the diagonal entries of A be zero.) An oriented graph can be represented naturally by a skew symmetric matrix over $GF(3)$. Show that there is a universal oriented graph of rank r that contains each reduced oriented graph of rank at most r as an induced subgraph.

30. Let A be the adjacency matrix of the graph X over some field $\mathbb{F}$. If $c \in \mathbb{F}$ and $c \neq 0$, show that $\alpha(X) \leq \mathrm{rk}(A+cI)$. If $c \in \mathbb{F}$ and $c \neq 1$, show that $1 + \omega(X) \leq \mathrm{rk}(A+cI)$.

31. Let A be the adjacency matrix of the graph X over some field $\mathbb{F}$. If $c \in \mathbb{F}\backslash\{0,1\}$ and $r = \mathrm{rk}(A+cI)$, show that $|V(X)| < 2^r + r$.

Notes

Detailed information and further references concerning the eigenvalues of the adjacency matrix of a graph will be found in [4, 3, 2]. Another approach, placing more emphasis on the characteristic polynomial, is presented in [5]. Most books on matrix theory include material on the Perron–Frobenius theorems (for example [8, 9]), and Minc [10] gives a detailed treatment of nonnegative matrices. We have covered some material not in the standard sources; we refer in particular to our discussions of rank and binary rank (in Section 8.9 and Section 8.10), and rational functions (in Section 8.13).

The observation that a reduced graph with binary rank $2r$ is an induced subgraph of $Sp(2r)$ is due to Rotman [11], and this is explored further in Godsil and Royle [7]. A graph is called *n-full* if it contains every graph on n vertices as an induced subgraph. Vu [12] observed that $Sp(2r)$ is $(2r-1)$-full and gave the proof presented in Theorem 8.11.2. Bollobás and Thomason [1] proved that the Paley graphs, which we will meet in Section 10.3, also contain all small graphs as induced subgraphs. More information on walk-regular graphs appears in [6].

References

[1] B. Bollobás and A. Thomason, *Graphs which contain all small graphs*, European J. Combin., 2 (1981), 13–15.

[2] D. Cvetković, P. Rowlinson, and S. Simić, *Eigenspaces of Graphs*, Cambridge University Press, Cambridge, 1997.

[3] D. M. Cvetković, M. Doob, I. Gutman, and A. Torgašev, *Recent Results in the Theory of Graph Spectra*, North-Holland, Amsterdam, 1988.

[4] D. M. Cvetković, M. Doob, and H. Sachs, *Spectra of Graphs*, Academic Press Inc., New York, 1980.

[5] C. D. Godsil, *Algebraic Combinatorics*, Chapman & Hall, New York, 1993.

[6] C. D. Godsil and B. D. McKay, *Feasibility conditions for the existence of walk-regular graphs*, Linear Algebra Appl., 30 (1980), 51–61.

[7] C. D. Godsil and G. F. Royle, *Binary rank and the chromatic number of a graph*, J. Combin. Theory Ser. B, (To appear).

[8] R. A. Horn and C. R. Johnson, *Matrix Analysis*, Cambridge University Press, Cambridge, 1990.

[9] P. Lancaster and M. Tismenetsky, *The Theory of Matrices*, Academic Press Inc., Orlando, Fla., second edition, 1985.

[10] H. Minc, *Nonnegative Matrices*, John Wiley & Sons Inc., New York, 1988.

[11] J. J. Rotman, *Projective planes, graphs, and simple algebras*, J. Algebra, 155 (1993), 267–289.

[12] V. H. Vu, *A strongly regular n-full graph of small order*, Combinatorica, 16 (1996), 295–299.

9

Interlacing

If M is a real symmetric $n \times n$ matrix, let $\theta_1(M) \geq \theta_2(M) \geq \cdots \geq \theta_n(M)$ denote its eigenvalues in nonincreasing order. Suppose A is a real symmetric $n \times n$ matrix and B is a real symmetric $m \times m$ matrix, where $m \leq n$. We say that the eigenvalues of B *interlace* the eigenvalues of A if for $i = 1, \ldots, m$,

$$\theta_{n-m+i}(A) \leq \theta_i(B) \leq \theta_i(A).$$

We will see that the eigenvalues of an induced subgraph of X interlace the eigenvalues of X. It follows that if we know enough about the spectrum of X, we can derive constraints on the subgraphs of X. We develop the theory of interlacing, equitable partitions, and generalized interlacing, and present a range of applications. These applications range from bounding the size of an independent set in a graph, and hence bounding its chromatic number, through to results related to the chemistry of the carbon molecules known as fullerenes.

9.1 Interlacing

We derive the interlacing inequalities as a consequence of our work on rational functions (in Section 8.13).

Theorem 9.1.1 *Let A be a real symmetric $n \times n$ matrix and let B be a principal submatrix of A with order $m \times m$. Then, for $i = 1, \ldots, m$,*

$$\theta_{n-m+i}(A) \leq \theta_i(B) \leq \theta_i(A).$$

Proof. We prove the result by induction on n. If $m = n$, there is nothing to prove. Assume $m = n - 1$. Then, by Lemma 8.13.1, for some i we have

$$\frac{\phi(B,x)}{\phi(A,x)} = e_i^T(xI - A)^{-1}e_i.$$

Denote this rational function by ψ. By Theorem 8.13.3, $\psi(x)$ has only simple poles and zeros, and each consecutive pair of poles is separated by a single zero. The poles of ψ are zeros of A, the zeros of ψ are zeros of B.

For a real symmetric matrix M and a real number λ, let $n(\lambda, M)$ denote the number of indices i such that $\theta_i(M) \geq \lambda$. We consider the behaviour of $n(\lambda, A) - n(\lambda, B)$ as λ decreases. If λ is greater than the largest pole of ψ, then the difference $n(\lambda, A) - n(\lambda, B)$ is initially zero. Since each pole is simple, the value of this difference increases by one each time λ passes through a pole of ψ, and since each zero is simple, its value decreases by one as it passes through a zero. As there is exactly one zero between each pair of poles, this difference alternates between 0 and 1. Therefore, it follows that $\theta_{i+1}(A) \leq \theta_i(B) \leq \theta_i(A)$ for all i.

Now, suppose that $m < n - 1$. Then B is a principal submatrix of a principal submatrix C of A with order $(n-1) \times (n-1)$. By induction we have

$$\theta_{n-1-m+i}(C) \leq \theta_i(B) \leq \theta_i(C).$$

By what we have already shown,

$$\theta_{i+1}(A) \leq \theta_i(C) \leq \theta_i(A),$$

and it follows that the eigenvalues of B interlace the eigenvalues of A. □

We will use Theorem 8.13.3 again in Chapter 13 to derive an interlacing result for the eigenvalues of the Laplacian matrix of a graph. (See Theorem 13.6.2.)

We close this section with an example. Let P be the Petersen graph and P_1 denote the subgraph obtained by deleting a single vertex. Then by Exercise 8.9, the characteristic polynomial of P is given by

$$\phi(P,x) = (x-3)(x-1)^5(x+2)^4.$$

By Corollary 8.13.2, we have

$$\phi'(P,x) = 10\,\phi(P\setminus 1, x),$$

and so

$$\phi(P_1,x) = (x^2 - 2x - 2)(x-1)^4(x+2)^3.$$

Therefore,

$$\psi(x) = \frac{(x^2-2x-2)(x-1)^4(x+2)^3}{(x-3)(x-1)^5(x+2)^4} = \frac{1/10}{(x-3)} + \frac{1/2}{(x-1)} + \frac{2/5}{(x+2)}.$$

The zeros of this are $1 \pm \sqrt{3}$, and the poles are 3, 1, and -2. Hence there is a zero between each pole, and given this it is not at all difficult to verify the the eigenvalues of $P \setminus 1$ interlace the eigenvalues of P.

9.2 Inside and Outside the Petersen Graph

We noted in Chapter 4 that the Petersen graph has no Hamilton cycle. We now give a proof of this using interlacing.

Lemma 9.2.1 *There are no Hamilton cycles in the Petersen graph P.*

Proof. First note that there is a Hamilton cycle in P if and only if there is an induced C_{10} in $L(P)$.

Now, $L(P)$ has eigenvalues 4, 2, -1, and -2 with respective multiplicities 1, 5, 4, and 5 (see Exercise 8.9). In particular, $\theta_7(L(P)) = -1$. The eigenvalues of C_{10} are

$$2, \frac{1+\sqrt{5}}{2}, \frac{-1+\sqrt{5}}{2}, \frac{1-\sqrt{5}}{2}, \frac{-1-\sqrt{5}}{2}, -2,$$

where 2 and -2 are simple eigenvalues and the others all have multiplicity two. Therefore, $\theta_7(C_{10}) \approx -0.618034$. Hence $\theta_7(C_{10}) > \theta_7(L(P))$, and so C_{10} is not an induced subgraph of $L(P)$.

It would be very interesting to find further applications of this argument. For example, there is no analogous proof that the Coxeter graph has no Hamilton cycle.

Lemma 9.2.2 *The edges of K_{10} cannot be partitioned into three copies of the Petersen graph.*

Proof. Let P and Q be two copies of Petersen's graph on the same vertex set and with no edges in common. Let R be the subgraph of K_{10} formed by the edges not in P or Q. We show that R is bipartite.

Let U_P be the eigenspace of $A(P)$ with eigenvalue 1, and let U_Q be the corresponding eigenspace for $A(Q)$. Then U_P and U_Q are 5-dimensional subspaces of $\mathbb{R}^{10}$. Since both subspaces lie in $\mathbf{1}^{\perp}$, they must have a nonzero vector u in common. Then

$$A(R)u = (J - I - A(P) - A(Q))u = (J - I)u - 2u = -3u,$$

and so -3 is an eigenvalue of $A(R)$. Since R is cubic, it follows from Theorem 8.8.2 that it must be bipartite. □

9.3 Equitable Partitions

In this section we consider partitions of the vertex set of a graph. We say that a partition π of $V(X)$ with cells $C_1, \ldots, C_r$ is *equitable* if the number of neighbours in C_j of a vertex u in C_i is a constant b_{ij}, independent of u. An equivalent definition is that the subgraph of X induced by each cell is regular, and the edges joining any two distinct cells form a semiregular bipartite graph. The directed graph with the r cells of π as its vertices and b_{ij} arcs from the ith to the jth cells of π is called the *quotient* of X over π, and denoted by X/π. Therefore, the entries of the adjacency matrix of this quotient are given by

$$A(X/\pi)_{ij} = b_{ij}.$$

One important class of equitable partitions arises from automorphisms of a graph. The orbits of any group of automorphisms of X form an equitable partition. (The proof of this is left as an exercise.) An example is given by the group of rotations of order 5 acting on the Petersen graph. The two orbits of this group, namely the 5 "inner" vertices and the 5 "outer" vertices, form an equitable partition π_1 with quotient matrix

$$A(X/\pi_1) = \begin{pmatrix} 2 & 1 \\ 1 & 2 \end{pmatrix}.$$

Another class arises from a mild generalization of the distance partitions of Section 4.5. If C is a subset of $V(X)$, let C_i denote the set of vertices in X at distance i from C. (So $C_0 = C$.) We call a subset C *completely regular* if its distance partition is equitable. Any vertex of the Petersen graph is completely regular, and the corresponding distance partition π_2 has three cells and quotient matrix

$$A(X/\pi_2) = \begin{pmatrix} 0 & 3 & 0 \\ 1 & 0 & 2 \\ 0 & 1 & 2 \end{pmatrix}.$$

If π is a partition of V with r cells, define its *characteristic matrix* P to be the $|V| \times r$ matrix with the characteristic vectors of the cells of π as its columns. Then P^TP is a diagonal matrix where $(P^TP)_{ii} = |C_i|$. Since the cells are nonempty, the matrix P^TP is invertible.

Lemma 9.3.1 *Let π be an equitable partition of the graph X, with characteristic matrix P, and let $B = A(X/\pi)$. Then $AP = PB$ and $B = (P^TP)^{-1}P^TAP$.*

Proof. We will show that for all vertices u and cells C_j we have

$$(AP)_{uj} = (PB)_{uj}.$$

The uj-entry of AP is the number of neighbours of u that lie in C_j. If $u \in C_i$, then this number is b_{ij}. Now, the uj-entry of PB is also b_{ij}, because the only nonzero entry in the u-row of P is a 1 in the i-column. Therefore, $AP = PB$, and so

$$P^TAP = P^TPB;$$

since P^TP is invertible, the second claim follows. □

We can translate the definition of an equitable partition more or less directly into linear algebra.

Lemma 9.3.2 *Let X be a graph with adjacency matrix A and let π be a partition of $V(X)$ with characteristic matrix P. Then π is equitable if and only if the column space of P is A-invariant.*

Proof. The column space of P is A-invariant if and only if there is a matrix B such that $AP = PB$. If π is equitable, then by the previous lemma we may take $B = A(X/\pi)$. Conversely, if there is such a matrix B, then every vertex in cell C_i is adjacent to b_{ij} vertices in cell C_j, and hence π is equitable. □

If $AP = PB$, then $A^rP = PB^r$ for any nonnegative integer r, and more generally, if $f(x)$ is a polynomial, then $f(A)P = Pf(B)$. If f is a polynomial such that $f(A) = 0$, then $Pf(B) = 0$. Since the columns of P are linearly independent, this implies that $f(B) = 0$. This shows that the minimal polynomial of B divides the minimal polynomial of A, and therefore every eigenvalue of B is an eigenvalue of A.

In fact, we can say more about the relationship between eigenvalues of B and eigenvalues of A. The next result implies that the multiplicity of θ as an eigenvalue of B is no greater than its multiplicity as an eigenvalue of A.

Theorem 9.3.3 *If π is an equitable partition of a graph X, then the characteristic polynomial of $A(X/\pi)$ divides the characteristic polynomial of $A(X)$.*

Proof. Let P be the characteristic matrix of π and let $B = A(X/\pi)$. If X has n vertices, then let Q be an $n \times (n - |\pi|)$ matrix whose columns, together with those of P, form a basis for $\mathbb{R}^n$. Then there are matrices C and D such that

$$AQ = PC + QD,$$

from which it follows that

$$A(\,P \quad Q\,) = (\,P \quad Q\,)\begin{pmatrix} B & C \\ 0 & D \end{pmatrix}.$$

Since $(\,P \quad Q\,)$ is invertible, it follows that $\det(xI - B)$ divides $\det(xI - A)$ as asserted. □

We can also get information about the eigenvectors of X from the eigenvectors of the quotient X/π. Suppose that $AP = PB$ and that v is an eigenvector of B with eigenvalue θ. Then $Pv \neq 0$ and

$$APv = PBv = \theta Pv;$$

hence Pv is an eigenvector of A. In this situation we say that the eigenvector v of B "lifts" to an eigenvector of A.

Alternatively, we may argue that if the column space of P is A-invariant, then it must have a basis consisting of eigenvectors of A. Each of these eigenvectors is constant on the cells of P, and hence has the form Pv, where $v \neq 0$. If $APv = \theta Pv$, then it follows that $Bv = \theta v$. □

If the column space of P is A-invariant, then so is its orthogonal complement; from this it follows that we may divide the eigenvectors of A into two classes: those that are constant on the cells of π, which have the form Pv for some eigenvector of B, and those that sum to zero on each cell of π.

For the two equitable partitions of the Petersen graph described above we have

$$\phi(X/\pi_1, x) = (x-3)(x-1)$$

and

$$\phi(X/\pi_2, x) = (x-3)(x-1)(x+2),$$

and therefore we can conclude that -2, 1, and 3 are eigenvalues of the Petersen graph.

We conclude this section with one elegant application of Theorem 9.3.3. A *perfect e-code* in a graph X is a set of vertices S such that for each vertex v of X there is a unique vertex in S at distance at most e from v.

Lemma 9.3.4 *If X is a regular graph with a perfect 1-code, then -1 is an eigenvalue of $A(X)$.*

Proof. Let S be a perfect 1-code and consider the partition π of $V(X)$ into S and its complement. If X is k-regular, then the definition of a perfect 1-code implies that π is equitable with quotient matrix

$$\begin{pmatrix} 0 & k \\ 1 & k-1 \end{pmatrix},$$

which has characteristic polynomial

$$x(x-(k-1)) - k = (x-k)(x+1).$$

Therefore, -1 is an eigenvalue of the quotient matrix, and hence an eigenvalue of $A(X)$. □

We have already seen an example of a perfect 1-code in Section 4.6: A heptad in $J(7,3,0)$ forms a perfect 1-code, because every vertex either lies in the heptad or is adjacent to a unique vertex in the heptad. In the next section we show that the eigenvalues of $J(7,3,0) = K_{7:3}$ are

$$-3, \quad -1, \quad 2, \quad 4,$$

which is reassuring.

9.4 Eigenvalues of Kneser Graphs

If X is a graph, and π an equitable partition, then in general the eigenvalues of X/π will be a proper subset of those of X. However, in certain special cases X/π retains all the eigenvalues of X, and we can get a partial converse to Theorem 9.3.3.

Theorem 9.4.1 *Let X be a vertex-transitive graph and π the orbit partition of some subgroup G of $\mathrm{Aut}(X)$. If π has a singleton cell $\{u\}$, then every eigenvalue of X is an eigenvalue of X/π.*

Proof. If f is a function on $V(X)$, and $g \in \mathrm{Aut}(X)$, then let f^g denote the function given by

$$f^g(x) = f(x^g).$$

It is routine to show that if f is an eigenvector of X with eigenvalue θ, then so is f^g.

If f is an eigenvector of X with eigenvalue θ, let $\hat{f}$ denote the average of f^g over the elements $g \in G$. Then $\hat{f}$ is constant on the cells of π, and provided that it is nonzero, it, too, is an eigenvector of X with eigenvalue θ.

Now, consider any eigenvector h of X with eigenvalue θ. Since $h \neq 0$, there is some vertex v such that $h(v) \neq 0$. Let $g \in \mathrm{Aut}(X)$ be an element such that $u^g = v$ and let $f = h^g$. Then $f(u) \neq 0$, and so

$$\hat{f}(u) = f(u) \neq 0.$$

Thus $\hat{f}$ is nonzero and constant on the cells of π. Therefore, following the discussion in Section 9.3, this implies that $\hat{f}$ is the lift of some eigenvector of X/π with the same eigenvalue. Therefore, every eigenvalue of X is an eigenvalue of X/π. □

We shall use this result to find the eigenvalues of the Kneser graphs. Assume that $v \geq 2r$, let X be the Kneser graph $K_{v:r}$, and assume that the vertices of X are the r-subsets of the set $\Omega = \{1, \ldots, v\}$. Let α be the fixed r-subset $\{1, \ldots, r\}$ and let C_i denote the r-subsets of Ω that meet α in exactly $r - i$ points. The partition π with cells $C_0, \ldots, C_r$ is the orbit partition of the subgroup of $\mathrm{Sym}(\Omega)$ that fixes α setwise, and hence π is an equitable partition satisfying the conditions of Theorem 9.4.1.

Now, we determine $A(X/\pi)$. Let β be an r-set meeting α in exactly $r-i$ points. Then the ij-entry of $A(X/\pi)$ is the number of r-subsets of Ω that are disjoint from β and meet α in exactly $r - j$ points. Hence

$$A(X/\pi)_{ij} = \binom{i}{r-j}\binom{v-r-i}{j}, \qquad 0 \leq i, j \leq r.$$

For example, if $r = 3$, then

$$A(X/\pi) = \begin{pmatrix} 0 & 0 & 0 & \binom{v-3}{3} \\ 0 & 0 & \binom{v-4}{2} & \binom{v-4}{3} \\ 0 & \binom{v-5}{1} & 2\binom{v-5}{2} & \binom{v-5}{3} \\ \binom{v-6}{0} & 3\binom{v-6}{1} & 3\binom{v-6}{2} & \binom{v-6}{3} \end{pmatrix}.$$

To determine the eigenvalues of $A(X/\pi)$, we need to carry out some computations with binomial coefficients; we note one that might not be familiar.

Lemma 9.4.2 *We have*

$$\sum_{i=0}^{h}(-1)^{h-i}\binom{h}{i}\binom{a-i}{k} = (-1)^h\binom{a-h}{k-h}.$$

Proof. Denote the sum in the statement of the lemma by $f(a,h,k)$. Since

$$\binom{a-i}{k} = \binom{a-i-1}{k} + \binom{a-i-1}{k-1},$$

we have

$$f(a,h,k) = f(a-1,h,k) + f(a-1,h,k-1).$$

We have $f(k,h,k) = (-1)^h$, while $f(a,h,0) = 0$ if $h > 0$ and $f(a,0,0) = 1$. Thus it follows by induction that

$$f(a,h,k) = (-1)^h\binom{a-h}{k-h}$$

as claimed. □

Theorem 9.4.3 *The eigenvalues of the Kneser graph $K_{v:r}$ are the integers*

$$(-1)^i\binom{v-r-i}{r-i}, \qquad i = 0,1,\ldots,r.$$

Proof. If $h(i,j)$ is a function of i and j, let $[h(i,j)]$ denote the $(r+1)\times(r+1)$ matrix with ij-entry $h(i,j)$, where $0 \le i,j \le r$. Let D be the diagonal matrix with ith diagonal entry

$$(-1)^i\binom{v-r-i}{r-i}.$$

We will prove the following identity:

$$\left[(-1)^{i-j}\binom{i}{j}\right] A(X/\pi) \left[\binom{i}{j}\right] = D\left[\binom{r-i}{r-j}\right]. \tag{9.1}$$

Here

$$\left[(-1)^{i-j}\binom{i}{j}\right]^{-1} = \left[\binom{i}{j}\right],$$

and hence this identity implies that $A(X/\pi)$ is similar to the product $D[\binom{r-i}{r-j}]$. Since $[\binom{r-i}{r-j}]$ is upper triangular with all diagonal entries equal to 1, it follows that the eigenvalues of X/π are the diagonal entries of D, and this yields the theorem.

We prove (9.1). The ik-entry of the product

$$\left[\binom{v-r-i}{j}\binom{i}{r-j}\right]\left[\binom{i}{j}\right]$$

equals

$$\sum_{j=0}^{r}\binom{v-r-i}{j}\binom{i}{r-j}\binom{j}{k}.$$

Since

$$\binom{a}{b}\binom{b}{c}=\binom{a}{c}\binom{a-c}{b-c}, \tag{9.2}$$

we can rewrite this sum as

$$\begin{aligned}\sum_{j=0}^{r}&\binom{v-r-i}{k}\binom{v-r-k-i}{j-k}\binom{i}{r-j}\\ &=\binom{v-r-i}{k}\sum_{j=0}^{r}\binom{v-r-k-i}{j-k}\binom{i}{r-j}\\ &=\binom{v-r-i}{k}\binom{v-r-k}{r-k}.\end{aligned}$$

(The last equality follows from the Vandermonde identity.)

Given this, the hk-entry of the product

$$\left[(-1)^{i-j}\binom{i}{j}\right]\left[\binom{v-r-i}{j}\binom{i}{r-j}\right]\left[\binom{i}{j}\right]$$

equals

$$\binom{v-r-k}{r-k}\sum_{i=0}^{r}(-1)^{h-i}\binom{h}{i}\binom{v-r-i}{k},$$

which by Lemma 9.4.2 is equal to

$$(-1)^h\binom{v-r-k}{r-k}\binom{v-r-h}{k-h}=(-1)^h\binom{v-r-h}{r-h}\binom{r-h}{r-k},$$

where the last equality follows from (9.2) by taking $a=v-r-h$, $b=r-h$, and $c=k-h$. This value is equal to the hk-entry of

$$D\left[\binom{r-i}{r-j}\right],$$

and so the result is proved. □

9.5 More Interlacing

We establish a somewhat more general version of interlacing. This will increase the range of application, and yields further information when some of the inequalities are tight.

We will use a tool from linear algebra known as *Rayleigh's inequalities.* Let A be a real symmetric matrix and let $u_1, \ldots, u_j$ be eigenvectors for A such that $Au_i = \theta_i(A)u_i$. Let U_j be the space spanned by the vectors $\{u_1, \ldots, u_j\}$. Then, for all u in U_j

$$\frac{u^T Au}{u^T u} \geq \theta_j(A),$$

with equality if and only if u is an eigenvector with eigenvalue $\theta_j(A)$. If $u \in U_j^{\perp}$, then

$$\frac{u^T Au}{u^T u} \leq \theta_{j+1}(A),$$

with equality if and only if u is an eigenvector with eigenvalue $\theta_{j+1}(A)$. If you prove these inequalities when A is a diagonal matrix, then it is easy to do the general case; we invite you to do so. Also, the second family of inequalities follows from the first, applied to $-A$.

Suppose that the eigenvalues of B interlace the eigenvalues of A, so that

$$\theta_{n-m+i}(A) \leq \theta_i(B) \leq \theta_i(A).$$

Then we say the interlacing is *tight* if there is some index j such that

$$\theta_i(B) = \begin{cases} \theta_i(A), & \text{for } i = 1, \ldots, j; \\ \theta_{n-m+i}(A), & \text{for } i = j+1, \ldots, m. \end{cases}$$

Informally this means that the first j eigenvalues of B are as large as possible, while the remaining $m - j$ are as small as possible.

Theorem 9.5.1 *Let A be a real symmetric $n \times n$ matrix and let R be an $n \times m$ matrix such that $R^T R = I_m$. Set B equal to $R^T AR$ and let $v_1, \ldots, v_m$ be an orthogonal set of eigenvectors for B such that $Bv_i = \theta_i(B)v_i$. Then:*

(a) *The eigenvalues of B interlace the eigenvalues of A.*

(b) *If $\theta_i(B) = \theta_i(A)$, then there is an eigenvector y of B with eigenvalue $\theta_i(B)$ such that Ry is an eigenvector of A with eigenvalue $\theta_i(A)$.*

(c) *If $\theta_i(B) = \theta_i(A)$ for $i = 1, \ldots, \ell$, then Rv_i is an eigenvector for A with eigenvalue $\theta_i(A)$ for $i = 1, \ldots, \ell$.*

(d) *If the interlacing is tight, then $AR = RB$.*

Proof. Let $u_1, \ldots, u_n$ be an orthogonal set of eigenvectors for A such that $Au_i = \theta_i(A)u_i$. Let U_j be the span of $u_1, \ldots, u_j$ and let V_j be the span of $v_1, \ldots, v_j$. For any i, the space V_i has dimension i, and the space $(R^T U_{i-1})$ has dimension at most $i - 1$. Therefore, there is a nonzero vector y in the intersection of V_i and $(R^T U_{i-1})^\perp$. Then $y^T R^T u_j = 0$ for $j = 1, \dots, i-1$, and therefore $Ry \in U_{i-1}^\perp$. By Rayleigh's inequalities this yields

$$\theta_i(A) \geq \frac{(Ry)^T ARy}{(Ry)^T Ry} = \frac{y^T By}{y^T y} \geq \theta_i(B). \tag{9.3}$$

We can now apply the same argument to the symmetric matrices $-A$ and $-B$ and conclude that $\theta_i(-B) \leq \theta_i(-A)$, and hence that $\theta_{n-m+i}(A) \leq \theta_i(B)$. Therefore, the eigenvalues of B interlace those of A, and we have proved (a).

If equality holds in (9.3), then y must be an eigenvector for B and Ry an eigenvector for A, both with eigenvalue $\theta_i(A) = \theta_i(B)$. This proves (b).

We prove (c) by induction on ℓ. If $i = 1$, we may take y in (9.3) to be v_1, and deduce that $ARv_1 = \theta_1(A)Rv_1$. So we may assume that $ARv_i = \theta_i(A)Rv_i$ for all $i < \ell$, and hence we may assume that $u_i = Rv_i$ for all $i < \ell$. But then v_ℓ lies in the intersection of V_ℓ and $(R^T U_{\ell-1})^\perp$, and thus we may choose y to be v_ℓ, which proves (c).

If the interlacing is tight, then there is some index j such that $\theta_i(B) = \theta_i(A)$ for $i \leq j$ and $\theta_i(-B) = \theta_i(-A)$ for $i \leq m - j$. Applying (c), we see that for all i,

$$RBv_i = \theta_i(B)Rv_i = ARv_i,$$

and since $v_1, \dots, v_m$ is a basis for $\mathbb{R}^m$, this implies that $RB = AR$. □

If we take R to have columns equal to the standard basis vectors e_i for i in some index set I, then $R^T AR$ is the principal submatrix of A with rows and columns indexed by I. Therefore, this result provides a considerable generalization of Theorem 9.1.1. We present an important application of this stronger version of interlacing in the next section, but before then we note the following consequence of the above theorem, which will be used in Chapter 13.

Corollary 9.5.2 *Let M be a real symmetric $n \times n$ matrix. If R is an $n \times m$ matrix such that $R^T R = I_m$, then $\operatorname{tr} R^T MR$ is less than or equal to the sum of the m largest eigenvalues of M. Equality holds if and only if the column space of R is spanned by eigenvectors belonging to these eigenvalues.* □

9.6 More Applications

Let X be a graph with adjacency matrix A, and let π be a partition, not necessarily equitable, of the vertices of X. If P is the characteristic matrix of π, then define the *quotient* of A relative to π to be the matrix $(P^T P)^{-1} P^T AP$, and denote it by A/π. We will show that the eigenvalues of A/π interlace the eigenvalues of A, and then we will give examples to show why this might be of interest.

Lemma 9.6.1 *If P is the characteristic matrix of a partition π of the vertices of the graph X, then the eigenvalues of $(P^TP)^{-1}P^TAP$ interlace the eigenvalues of A. If the interlacing is tight, then π is equitable.*

Proof. The problem with P is that its columns form an orthogonal set, not an orthonormal set, but fortunately this can easily be fixed. Recall that P^TP is a diagonal matrix with positive diagonal entries, and so there is a diagonal matrix D such that $D^2 = P^TP$. If $R = PD^{-1}$, then

$$R^TAR = D^{-1}P^TAPD^{-1} = D(D^{-2}P^TAP)D^{-1},$$

and so R^TAR is similar to $(P^TP)^{-1}P^TAP$. Furthermore,

$$R^TR = D^{-1}(P^TP)D^{-1} = D^{-1}(D^2)D^{-1} = I,$$

and therefore by Theorem 9.5.1, the eigenvalues of R^TAR interlace the eigenvalues of A. If the interlacing is tight, then the column space of R is A-invariant, and since R and P have the same column space, it follows that π is equitable. □

The ij-entry of P^TAP is the number of edges joining vertices in the ith cell of π to vertices in the jth cell. Therefore, the ij-entry of $(P^TP)^{-1}P^TAP$ is the average number of edges leading from a vertex in the ith cell of π to vertices in the jth cell.

We show how this can be used to find a bound on the size of an independent set in a regular graph. Let X be a regular graph on n vertices with valency k and let S be an independent set of vertices. Let π be the partition with two cells S and $V(X)\setminus S$ and let B be the quotient matrix A/π. There are $|S|k$ edges between S and $V(X)\setminus S$, and hence each vertex not in S has exactly $|S|k/(n-|S|)$ neighbours in S. Therefore,

$$B = \begin{pmatrix} 0 & k \\ \frac{|S|k}{n-k} & k - \frac{|S|k}{n-|S|} \end{pmatrix}.$$

Both rows of B sum to k, and thus k is one of its eigenvalues. Since

$$\operatorname{tr} B = k - \frac{|S|k}{n-k},$$

and $\operatorname{tr} B$ is the sum of the eigenvalues of B, we deduce that the second eigenvalue of B is $-k|S|/(n-|S|)$. Therefore, if τ is the smallest eigenvalue of A, we conclude by interlacing that

$$\tau \leq -\frac{k|S|}{n-|S|}. \qquad (9.4)$$

Lemma 9.6.2 *Let X be a k-regular graph on n vertices with least eigenvalue τ. Then*

$$\alpha(X) \leq \frac{n(-\tau)}{k-\tau}.$$

If equality holds, then each vertex not in an independent set of size $\alpha(X)$ has exactly $-\tau$ neighbours in it.

Proof. The inequality follows on unpacking (9.4). If S is an independent set with size meeting this bound, then the partition with S and $V(X)\backslash S$ as its cells is equitable, and so each vertex not in S has exactly $k|S|/(n-|S|) = -\tau$ neighbours in S. □

The Petersen graph P has $n = 10$, $k = 3$, and $\tau = -2$, and hence $\alpha(P) \leq 4$. The Petersen graph does have independent sets of size four, and so each vertex outside such a set has exactly two neighbours in it (and thus the complement of an independent set of size four induces a copy of $3K_2$). The Hoffman–Singleton graph has $n = 50$, $k = 7$, and as part of Exercise 10.7 we will discover that it has $\tau = -3$. Therefore, the bound on the size of a maximum independent set is 15, and every vertex not in an independent set of size 15 has exactly three neighbours in it. Thus we have another proof of Lemma 5.9.1.

We saw in Section 9.4 that the least eigenvalue of $K_{v:r}$ is

$$-\binom{v-r-1}{r-1}.$$

Since $K_{v:r}$ has valency $\binom{v-r}{r}$, we find using Lemma 9.6.2 that the size of an independent set is at most

$$\frac{\binom{v}{r}\binom{v-r-1}{r-1}}{\binom{v-r}{r}+\binom{v-r-1}{r-1}} = \frac{\binom{v}{r}}{\frac{v}{r}} = \binom{v-1}{r-1},$$

thus providing another proof of the first part of the Erdős–Ko–Rado theorem (Theorem 7.8.1). As equality holds, each vertex not in an independent set of this size has exactly $\binom{v-r-1}{r-1}$ neighbours in it.

We can use interlacing in another way to produce another bound on the size of an independent set in a graph. If A is a symmetric matrix, let $n^+(A)$ and $n^-(A)$ denote respectively the number of positive and negative eigenvalues of A.

Lemma 9.6.3 *Let X be a graph on n vertices and let A be a symmetric $n \times n$ matrix such that $A_{uv} = 0$ if the vertices u and v are not adjacent. Then*

$$\alpha(X) \leq \min\{n - n^+(A), n - n^-(A)\}.$$

Proof. Let S be the subgraph of X induced by an independent set of size s, and let B be the principal submatrix of A with rows and columns indexed by the vertices in S. (So B is the zero matrix.) By interlacing,

$$\theta_{n-s+i}(A) \leq \theta_i(B) \leq \theta_i(A).$$

But of course, $\theta_i(B) = 0$ for all i; hence we infer that

$$0 \leq \theta_s(A)$$

and that $n^-(A) \le n - s$. We can apply the same argument with $-A$ in place of A to deduce that $n^+(A) \le n - s$. □

We can always apply this result to the adjacency matrix A of X, but there are times when other matrices are more useful. One example will be offered in the next section.

9.7 Bipartite Subgraphs

We study the problem of bounding the maximum number of vertices in a bipartite induced subgraph of a Kneser graph $K_{2r+1:r}$. (The Kneser graphs with these parameters are often referred to as *odd graphs*.)

We use $\alpha_r(X)$ to denote the maximum number of vertices in an r-colourable subgraph of X. By Lemma 7.14.1, we have that

$$\alpha_r(X) = \alpha(X \square K_r),$$

from which it follows that any bound on the size of an independent set in $X \square K_r$ yields a bound on $\alpha_r(X)$. We will use the bound we derived in terms of the number of nonnegative (or nonpositive) eigenvalues. For this we need to determine the adjacency matrix of the Cartesian product. Define the *Kronecker product* $A \otimes B$ of two matrices A and B to be the matrix we get by replacing the ij-entry of A by $A_{ij}B$, for all i and j. If X and Y are graphs and $X \times Y$ is their product, as defined in Section 6.3, then you may show that

$$A(X \times Y) = A(X) \otimes A(Y).$$

For the Cartesian product $X \square Y$ we have

$$A(X \square Y) = A(X) \otimes I + I \otimes A(Y).$$

We attempt to justify this by noting that $A(X) \otimes I$ is the adjacency matrix of $|V(Y)|$ vertex-disjoint copies of X, and that $I \otimes A(Y)$ is the adjacency matrix of $|V(X)|$ vertex-disjoint copies of Y, but we omit the details.

The Kronecker product has the property that if A, B, C, and D are four matrices such that the products AC and BD exist, then

$$(A \otimes B)(C \otimes D) = AC \otimes BD.$$

Therefore, if x and y are vectors of the correct lengths, then

$$(A \otimes B)(x \otimes y) = Ax \otimes By.$$

If x and y are eigenvectors of A and B, with eigenvalues θ and τ, respectively, then

$$Ax \otimes By = \theta\tau\, x \otimes y,$$

whence $x \otimes y$ is an eigenvector of $A \otimes B$ with eigenvalue $\theta\tau$. In particular, if x and y are eigenvectors of X and Y, with respective eigenvalues θ and τ, then

$$A(X \,\square\, Y)(x \otimes y) = (\theta + \tau)\, x \otimes y.$$

This implies that if θ and τ have multiplicities a and b, respectively, then $\theta + \tau$ is an eigenvalue of $X \,\square\, Y$ with multiplicity ab. We also note that if r and s are real numbers, then $r\theta + s\tau$ is an eigenvalue of

$$rA(X) \otimes I + sI \otimes A(Y)$$

with multiplicity ab.

We are now going to apply Lemma 9.6.3 to the Kneser graphs $K_{2r+1:r}$. As we saw in Section 9.4, the eigenvalues of these graphs are the integers

$$(-1)^i(r+1-i), \qquad i = 0, \dots, r;$$

the multiplicities are known to be

$$\binom{2r+1}{i} - \binom{2r+1}{i-1}, \qquad i = 0, \dots, r,$$

with the understanding that the binomial coefficient $\binom{a}{-1}$ is zero. (We have not computed these multiplicities, and will not.)

We start with the Petersen graph $K_{5:2}$. Its eigenvalues are 3, -2 and 1 with multiplicities 1, 4, and 5. The eigenvalues of K_2 are -1 and 1, so $K_{5:2} \,\square\, K_2$ has eight negative eigenvalues, seven positive eigenvalues, and five equal to zero. This yields a bound of at most 12 vertices in a bipartite subgraph, which is certainly not wrong! There is an improvement available, though, if we work with the matrix

$$A' = A(K_{2r+1:r}) \otimes I + \frac{3}{2} I \otimes A(K_2).$$

For the Petersen graph, this matrix has seven positive eigenvalues and 13 negative eigenvalues, yielding $\alpha_2(X) \le 7$. This can be realized in two ways, shown in Figure 9.1.

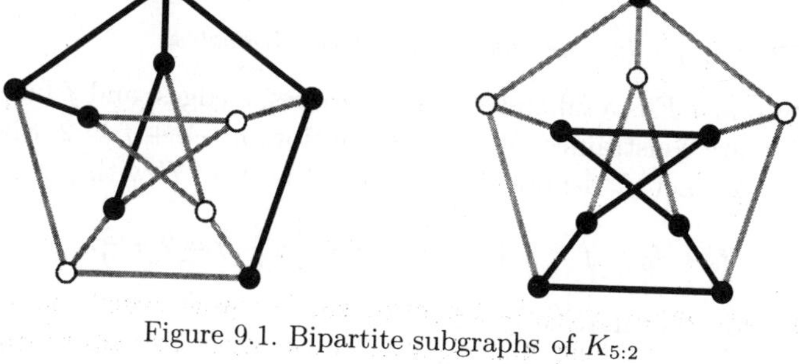

Figure 9.1. Bipartite subgraphs of $K_{5:2}$

Applying the same modification to $K_{7:3}$, we find that

$$\alpha_2(K_{7:3}) \leq 26,$$

which is again the correct bound, and again can be realized in two different ways. In general, we get the following, but we leave the proof as an exercise.

Lemma 9.7.1 *We have* $\alpha_2(K_{2r+1:r}) \leq \binom{2r}{r} + \binom{2r}{r-2}$. □

For $K_{9:4}$ this gives an upper bound of 98. The exact value is not known, but Tardif has found a bipartite subgraph of size 96.

9.8 Fullerenes

A *fullerene* is a cubic planar graph with all faces 5-cycles or 6-cycles. Fullerenes arise in chemistry as molecules consisting entirely of carbon atoms. Each carbon atom is bonded to exactly three others, thus the vertices of the graph represent the carbon atoms and the edges the bonded pairs of atoms. An example on 26 vertices is shown in Figure 9.2.

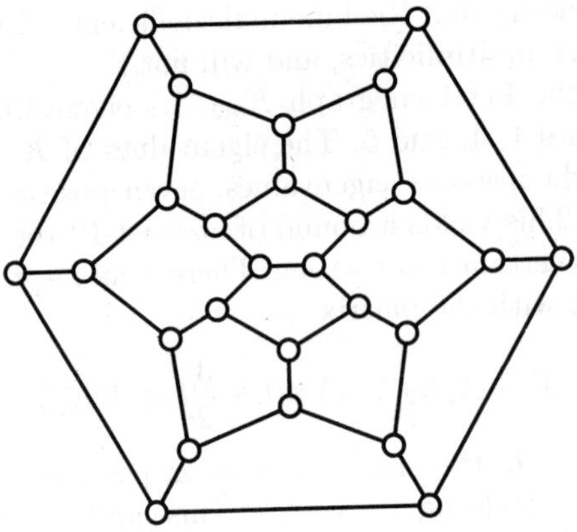

Figure 9.2. A fullerene on 26 vertices

Lemma 9.8.1 *A fullerene has exactly twelve* 5*-cycles.*

Proof. Suppose F is a fullerene with n vertices, e edges, and f faces. Then n, e, and f are constrained by Euler's relation, $n - e + f = 2$. Since F is cubic, $3n = 2e$. Let f_r denote the number of faces of F with size r. Then

$$f_5 + f_6 = f = 2 + e - n = 2 + \frac{3}{2}n - n = 2 + \frac{1}{2}n.$$

Since each edge lies in exactly two faces,

$$5f_5 + 6f_6 = 2e = 3n.$$

Solving these equations implies that $f_5 = 12$. □

It follows from the argument above that

$$n = 2f_6 + 20.$$

If $f_6 = 0$, then $n = 20$, and the dodecahedron is the unique fullerene on 20 vertices.

Most fullerene graphs do not correspond to molecules that have been observed in nature. Chemists believe that one necessary condition is that no two 5-cycles can share a common vertex—such fullerenes are called isolated pentagon fullerenes. By Lemma 9.8.1, any isolated pentagon fullerene has at least 60 vertices. There is a unique example on 60 vertices, which happens to be the Cayley graph for Alt(5) relative to the generating set

$$\{(12)(34),\ (12345),\ (15342)\}.$$

This example is shown in Figure 9.5 and is known as buckminsterfullerene.

We describe an operation on cubic planar graphs that can be used to construct fullerenes. If X is a cubic planar graph with n vertices, $m = 3n/2$ edges and f faces, then its line graph $L(X)$ is a planar 4-regular graph with m vertices and $n + f$ faces; the reason it is planar is clear from a drawing such as Figure 9.3. The $n + f$ faces consist of n triangular faces each containing a vertex of X, and f faces each completely inscribed within a face of X of the same length.

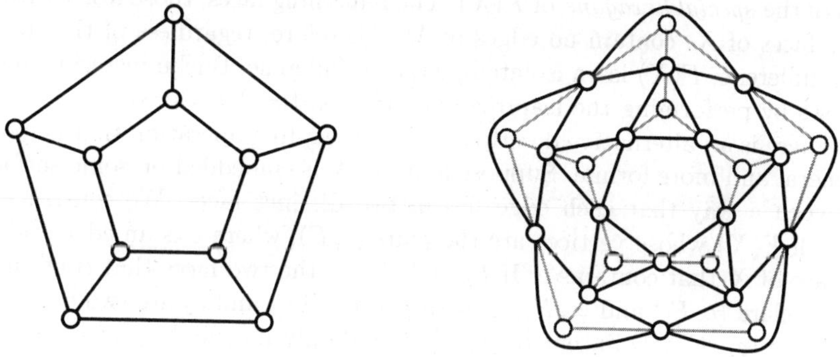

Figure 9.3. A cubic planar graph X and its line graph $L(X)$

The *leapfrog graph* $F(X)$ is formed by taking each vertex of $L(X)$ and splitting it into a pair of adjacent vertices in such a way that every triangular face around a vertex of X becomes a six-cycle; once again, a drawing such as Figure 9.4 is the easiest way to visualize this. Then $F(X)$ is a cubic planar graph on $2m$ vertices with n faces of length six and f faces of the same lengths as the faces of X. In particular, if X is a fullerene, then so is $F(X)$.

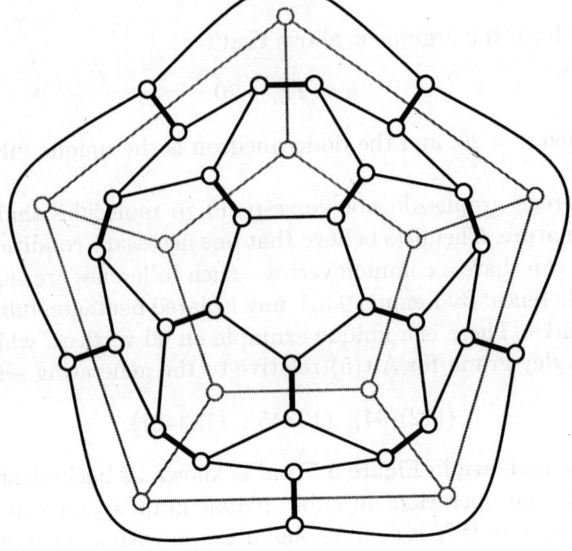

Figure 9.4. The leapfrog graph

The edges joining each pair of newly split vertices form a perfect matching M in $F(X)$. The n faces of length six around the vertices of X each contain three edges from M and three other edges, and we will call these faces the *special hexagons* of $F(X)$. The remaining faces, those arising from the faces of X, contain no edges of M. Therefore, regardless of the starting fullerene, $F(X)$ is an isolated pentagon fullerene. Buckminsterfullerene arises by performing the leapfrog operation on the dodecahedron.

There is an alternative description of the leapfrog operation that is more general and more formal. Suppose a graph X is embedded on some surface in such a way that each edge lies in two distinct faces. We construct a graph $F(X)$ whose vertices are the pairs (e, F), where e is an edge and F a face of X that contains e. If F_1 and F_2 are the two faces that contain e, we declare (e, F_1) and (e, F_2) to be adjacent. If e_1 and e_2 are two edges on F, then (e_1, F) is adjacent to (e_2, F) if and only if e_1 and e_2 have a single vertex in common. We say that $F(X)$ is obtained from X by *leapfrogging*. The edges of the first type form a perfect matching M in $F(X)$, and the edges of the second type form a disjoint collection of cycles, one for each face of X. We call M the canonical perfect matching of $F(X)$.

9.9 Stability of Fullerenes

In addition to the isolated pentagon rule, there is evidence that leads some chemists to believe that a necessary condition for the physical existence of a

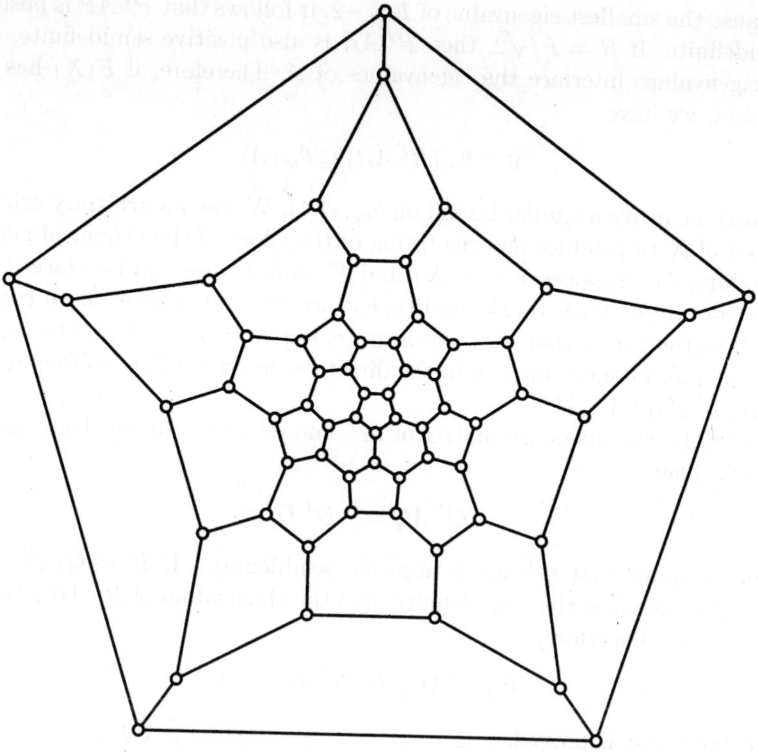

Figure 9.5. Buckminsterfullerene

particular fullerene is that the graph should have exactly half its eigenvalues positive, and half negative. In this section we use interlacing to show that a fullerene derived from the leapfrog construction has exactly half of its eigenvalues positive and half negative. Other than the leapfrog fullerenes, there are very few other fullerenes known to have this property.

Lemma 9.9.1 *If X is a cubic planar graph with leapfrog graph $F(X)$, then $F(X)$ has at most half of its eigenvalues positive and at most half of its eigenvalues negative.*

Proof. Let π be the partition whose cells are the edges of the canonical perfect matching M of $F(X)$. Since X is cubic, two distinct cells of π are joined by at most one edge. The graph defined on the cells of π where two cells are adjacent if they are joined by an edge is the line graph $L(X)$ of X.

Let P be the characteristic matrix of π, let A be the adjacency matrix of $F(X)$, and let L be the adjacency matrix of $L(X)$. Then a straightforward calculation shows that

$$P^T AP = 2I + L.$$

Because the smallest eigenvalue of L is -2, it follows that $P^T AP$ is positive semidefinite. If $R = P/\sqrt{2}$, then $R^T AR$ is also positive semidefinite, and its eigenvalues interlace the eigenvalues of A. Therefore, if $F(X)$ has $2m$ vertices, we have

$$0 \leq \theta_m(R^T AR) \leq \theta_m(A).$$

Next we prove a similar bound on $\theta_{m+1}(A)$. We use an arbitrary orientation σ of X to produce an orientation of the edges of the canonical perfect matching M. Suppose $e \in E(X)$ and F_1 and F_2 are the two faces of X that contain e. Then (e, F_1) and (e, F_2) are the end-vertices of an edge of M. We orient it so that it points from (e, F_1) to (e, F_2) if F_2 is the face on the right as we move along e in the direction determined by σ. Denote this oriented graph by M^σ.

Let Q be the incidence matrix of M^σ and let D be the incidence matrix of X^σ. Then

$$Q^T AQ = -D^T D,$$

which implies that $Q^T AQ$ is negative semidefinite. If $R = Q/\sqrt{2}$, then $R^T AR$ is also negative semidefinite, and the eigenvalues of $R^T AR$ interlace those of A. Therefore,

$$\theta_{m+1}(A) \leq \theta_1(R^T AR) \leq 0,$$

and the result is proved. □

Theorem 9.9.2 *If X is a cubic planar graph, then its leapfrog graph $F(X)$ has exactly half of its eigenvalues negative. If, in addition, X has a face of length not divisible by three, then its leapfrog graph $F(X)$ also has exactly half of its eigenvalues positive.*

Proof. By the lemma, the first conclusion follows if $\theta_{m+1}(A) \neq 0$, and the second follows if $\theta_m(A) \neq 0$.

Suppose to the contrary that $\theta_{m+1}(A) = 0$. Then by Theorem 9.5.1, there is an eigenvector f for A with eigenvalue 0 that sums to zero on each cell of π. Let $F = v_0, \dots, v_r$ be a face of $F(X)$ that is not a special hexagon. Thus each vertex v_i is adjacent to v_{i-1}, v_{i+1}, and the other vertex w_i in the same cell of π. Since f sums to zero on the cells of π, we have $f(w_i) = -f(v_i)$. Since f has eigenvalue 0, the sum of the values of f on the neighbours of v_{i+1} is 0, and similarly for v_{i+2}. Therefore (performing all subscript arithmetic modulo $r+1$), we get

$$f(v_i) - f(v_{i+1}) + f(v_{i+2}) = 0,$$
$$f(v_{i+1}) - f(v_{i+2}) + f(v_{i+3}) = 0,$$

and hence

$$f(v_{i+3}) = -f(v_i).$$

If the length of F is not divisible by six, then f is constant, and therefore zero, on the vertices of F. Any cubic planar graph has a face of length less than six, and therefore $F(X)$ has a face that is not a special hexagon on which f is zero. Every edge of M lies in two special hexagons, and if f is determined on one special hexagon, the values it takes on any "neighbouring" special hexagon are also uniquely determined. If f is zero on a special hexagon, then it is routine to confirm that it is zero on any neighbouring special hexagon, and therefore zero on every vertex of $F(X)$. Otherwise, by starting with a special hexagon H sharing an edge with F and inferring the values that f must take on the special hexagons neighbouring H and so on, it is possible to show that there is a "circuit" of special hexagons such that f takes increasingly large absolute values on every second one; we leave the details as an exercise. This, of course, is impossible, and so we conclude that there is no such eigenvector.

Next we suppose that $\theta_m(A) = 0$, in which case there is an eigenvector for A with eigenvalue 0 that is constant on each cell of π. An analogous argument to the one above yields that for a face $F = v_0, \dots, v_r$ that is not a special hexagon, we have

$$f(v_{i+3}) = f(v_i).$$

If F has length not divisible by three, then f is constant, and hence zero on every vertex of F. It is left as an exercise similar to the case above to show that this implies that $f = 0$. □

Exercises

1. What goes wrong if we apply the argument of Lemma 9.2.1 in an attempt to prove that the Petersen graph has no Hamilton path?

2. Show that the orbits of a group of automorphisms of X form an equitable partition.

3. If π is an equitable partition of the vertex set of the graph X, show that the spectral radius of $A(X/\pi)$ is equal to the spectral radius of $A(X)$.

4. Determine the graphs with $\theta_{\min} \geq -1$.

5. Let X be a vertex-transitive graph with valency k, and let θ be a simple eigenvalue of its adjacency matrix A. Show that either $\theta = k$, or $|V(X)|$ is even and $k - \theta$ is an even integer. (Hint: If P is a permutation matrix representing an automorphism of X and u is an eigenvector of A, then Pu is an eigenvector with the same eigenvalue.)

6. Let X be an arc-transitive graph with valency k. Show that if θ is a simple eigenvalue of $A(X)$, then $\theta = \pm k$.

7. Let X be a vertex-transitive graph with two simple eigenvalues, neither equal to the valency. Show that $|V(X)|$ is divisible by four.

8. Let X be a graph and π an equitable partition of $V(X)$. Show that the spectral radius of X is equal to the spectral radius of X/π.

9. Let X be a graph with largest eigenvalue θ_1 and average valency $\hat{k}$. Use interlacing to prove that $\theta_1 \leq \hat{k}$, with equality if and only if X is regular.

10. Let X be the Kneser graph $K_{v:r}$, with the r-subsets of Ω as its vertices. If $1 \in \Omega$, let π be the partition of X with two cells, one consisting of the r-subsets that contain 1, and the other of those that do not. Show that this partition is equitable, and that $-\binom{v-r-1}{r-1}$ is an eigenvalue of the quotient.

11. Let X and Y be graphs with respective equitable partitions σ and π. If $A(X/\sigma) = A(Y/\pi)$, show that there is a graph Z that covers both X and Y.

12. Let A be a symmetric $n \times n$ matrix and let R be an $n \times m$ matrix such that $R^T R = I$. Show that there is an orthogonal matrix Q whose first m columns coincide with R, and hence deduce that $R^T AR$ is a principal submatrix of a symmetric matrix similar to A.

13. Let X be a graph on n vertices and let π be an equitable partition of X. Let Q be the normalized characteristic matrix of π and assume that B is the quotient matrix, given by

$$AQ = QB.$$

If θ is an eigenvalue of A with principal idempotent E_θ, define F_θ by

$$E_\theta P = PF_\theta.$$

(Note that F_θ might be zero.) Show that F_θ is symmetric and $F_\theta^2 = F_\theta$. Show further that F_θ is one of the principal idempotents of B. If the first cell of π consists of the first vertex of X, show that

$$(E_\theta)_{11} = (F_\theta)_{11}.$$

14. Suppose X is a walk-regular graph on n vertices and π is an equitable partition of X with the first vertex of X as a cell. Assume $B = A(X/\pi)$ and θ is a simple eigenvalue of B; let x_θ be an eigenvector of B with eigenvalue θ. If m_θ is the multiplicity of θ as an eigenvalue of X, show that $m_\theta = n(x_\theta)_1^2/\|x_\theta\|^2$, where $\|x\|$ denotes the Euclidean length of a vector x. (Hint: Use the previous exercise.)

15. Suppose X is a graph on n vertices and π is an equitable partition of X with the first vertex of X as a cell. Show how to determine n from $A(X/\pi)$.

16. Let A be a real symmetric $n \times n$ matrix. A subspace U of $\mathbb{R}^n$ is *isotropic* if $x^T Ax = 0$ for all vectors x in U. Let $V(+)$ be the subspace of $\mathbb{R}^n$ spanned by the eigenvectors of A with positive eigenvalues, and let $V(-)$ be the subspace spanned by the eigenvectors with negative eigenvalues. Show that
$$V(+) \cap V(-) = \{0\},$$
and if U is isotropic, then
$$V(+) \cap U = V(-) \cap U = \{0\}.$$
Using this, deduce that $\alpha(X)$ cannot be greater than $n - n^+(A)$ or $n - n^-(A)$.

17. Compute the eigenvalues of And(k) and then use an eigenvalue bound to show that its independence number is bounded above by its valency. (See Section 6.9 for the definition of these graphs.)

18. Use a weighted adjacency matrix to prove that $\alpha(K_{7:3} \,\Box\, C_5) \leq 61$.

19. Find an expression for the eigenvalues of the lexicographic product $X\left[\overline{K_m}\right]$ in terms of the eigenvalues of X.

20. Let X be a cubic planar graph with leapfrog graph $F(X)$, and let f be an eigenvector of $F(X)$ with eigenvalue 0 that sums to zero on the cells of the canonical perfect matching. Show that $f = 0$ if and only if f is zero on the vertices of a face that is not a special hexagon.

21. Let X be a cubic planar graph with leapfrog graph $F(X)$, and let f be an eigenvector of $F(X)$ with eigenvalue 0 that is constant on the cells of the canonical perfect matching. Show that $f = 0$ if and only if f is zero on the vertices of a face that is not a special hexagon.

22. Define a generalized leapfrog operation as follows. If X is a graph, then define a graph $F'(X)$ on the vertex set $\{(e, i) : e \in E(X), i = 0, 1\}$. All the pairs of vertices $(e, 0)$ and $(e, 1)$ are adjacent, and there is a single edge between $\{(e, 0), (e, 1)\}$ and $\{(f, 0), (f, 1)\}$ if and only if e and f are incident edges in X. Show that any generalized leapfrog graph has at most half its eigenvalues positive and at most half negative.

Notes

The full power of interlacing in graph theory was most convincingly demonstrated by Haemers, in his doctoral thesis. He has exhausted his supply of copies of this, but [1] is a satisfactory substitute.

The proof that K_{10} cannot be partitioned into three copies of the Petersen graph is based on Lossers and Schwenk [3].

The bound on $\alpha(X)$ involving the least eigenvalue of X is due to Hoffman, although inspired by a bound, due to Delsarte, for strongly regular graphs. The bound on $\alpha(X)$ in terms of $n^+(A)$ and $n^-(A)$ is due to Cvetković. This bounds seems surprisingly useful, and has not received a lot of attention.

Our treatment of the stability of fullerenes follows Haemers [1], which is based in turn on Manolopoulos, Woodall, and Fowler [4].

More information related to Exercise 11 is given in Leighton [2].

The proof, in Section 9.2, that the Petersen graph has no Hamilton cycle is based on work of Mohar [5]. Some extensions to this will be treated in Section 13.6. In the notes to Chapter 3 we discussed two proofs that the Coxeter graph has no Hamilton cycle. Because we have only a very limited selection of tools for proving that a graph has no Hamilton cycle, we feel it could be very useful to have a third proof of this, using interlacing.

References

[1] W. H. Haemers, *Interlacing eigenvalues and graphs*, Linear Algebra Appl., 226/228 (1995), 593–616.

[2] F. T. Leighton, *Finite common coverings of graphs*, J. Combin. Theory Ser. B, 33 (1982), 231–238.

[3] O. Lossers and A. Schwenk, *Solutions to advanced problems, 6434*, American Math. Monthly, 94 (1987), 885–886.

[4] D. E. Manolopoulos, D. R. Woodall, and P. W. Fowler, *Electronic stability of fullerenes: eigenvalue theorems for leapfrog carbon clusters*, J. Chem. Soc. Faraday Trans., 88 (1992), 2427–2435.

[5] B. Mohar, *A domain monotonicity theorem for graphs and Hamiltonicity*, Discrete Appl. Math., 36 (1992), 169–177.

10
Strongly Regular Graphs

In this chapter we return to the theme of combinatorial regularity with the study of strongly regular graphs. In addition to being regular, a strongly regular graph has the property that the number of common neighbours of two distinct vertices depends only on whether they are adjacent or nonadjacent. A connected strongly regular graph with connected complement is just a distance-regular graph of diameter two. Any vertex-transitive graph with a rank-three automorphism group is strongly regular, and we have already met several such graphs, including the Petersen graph, the Hoffman–Singleton graph, and the symplectic graphs of Section 8.11.

We present the basic theory of strongly regular graphs, primarily using the algebraic methods of earlier chapters. We show that the adjacency matrix of a strongly regular graph has just three eigenvalues, and develop a number of conditions that these eigenvalues satisfy, culminating in an elementary proof of the Krein bounds. Each of these conditions restricts the structure of a strongly regular graph, and most of them yield some additional information about the possible subgraphs of a strongly regular graph.

Although many strongly regular graphs have large and interesting groups, this is not at all typical, and it is probably true that "almost all" strongly regular graphs are asymmetric. We show how strongly regular graphs arise from Latin squares and designs, which supply numerous examples of strongly regular graphs with no reason to have large automorphism groups.

10.1 Parameters

Let X be a regular graph that is neither complete nor empty. Then X is said to be *strongly regular* with *parameters*

$$(n, k, a, c)$$

if it is k-regular, every pair of adjacent vertices has a common neighbours, and every pair of distinct nonadjacent vertices has c common neighbours. One simple example is the 5-cycle C_5, which is a 2-regular graph such that adjacent vertices have no common neighbours and distinct nonadjacent vertices have precisely one common neighbour. Thus it is a $(5, 2, 0, 1)$ strongly regular graph.

It is straightforward to show that if X is strongly regular with parameters (n, k, a, c), then its complement $\overline{X}$ is also strongly regular with parameters $(n, \bar{k}, \bar{a}, \bar{c})$, where

$$\begin{aligned} \bar{k} &= n - k - 1, \\ \bar{a} &= n - 2 - 2k + c, \\ \bar{c} &= n - 2k + a. \end{aligned}$$

A strongly regular graph X is called *primitive* if both X and its complement are connected, otherwise *imprimitive*. The next lemma shows that there is only one class of imprimitive strongly regular graphs.

Lemma 10.1.1 *Let X be an (n, k, a, c) strongly regular graph. Then the following are equivalent:*

(a) *X is not connected,*

(b) $c = 0$,

(c) $a = k - 1$,

(d) *X is isomorphic to mK_{k+1} for some $m > 1$.*

Proof. Suppose that X is not connected and let X_1 be a component of X. A vertex in X_1 has no common neighbours with a vertex not in X_1, and so $c = 0$. If $c = 0$, then any two neighbours of a vertex $u \in V(X)$ must be adjacent, and so $a = k - 1$. Finally, if $a = k - 1$, then the component containing any vertex must be a complete graph K_{k+1}, and hence X is a disjoint union of complete graphs. □

Two simple families of examples of strongly regular graphs are provided by the line graphs of K_n and $K_{n,n}$. The graph $L(K_n)$ has parameters

$$(n(n-1)/2,\ 2n-4,\ n-2,\ 4),$$

while $L(K_{n,n})$ has parameters

$$(n^2,\ 2n-2,\ n-2,\ 2).$$

These graphs are sometimes referred to as the *triangular graphs* and the *square lattice graphs*, respectively.

The parameters of a strongly regular graph are not independent. We can find some relationships between them by simple counting. Every vertex u has k neighbours, and hence $n-k-1$ non-neighbours. We will count the total number of edges between the neighbours and non-neighbours of u in two ways. Each of the k neighbours of u is adjacent to u itself, to a neighbours of u, and thus to $k-a-1$ non-neighbours of u, for a total of $k(k-a-1)$ edges. On the other hand, each of the $n-k-1$ non-neighbours of u is adjacent to c neighbours of u for a total of $(n-k-1)c$ edges. Therefore,

$$k(k-a-1) = (n-k-1)c. \tag{10.1}$$

The study of strongly regular graphs often proceeds by constructing a list of possible parameter sets, and then trying to find the actual graphs with those parameter sets. We can view the above equation as a very simple example of a *feasibility condition* that must be satisfied by the parameters of any strongly regular graph.

10.2 Eigenvalues

Suppose A is the adjacency matrix of the (n, k, a, c) strongly regular graph X. We can determine the eigenvalues of the matrix A from the parameters of X and thereby obtain some strong feasibility conditions.

The uv-entry of the matrix A^2 is the number of walks of length two from the vertex u to the vertex v. In a strongly regular graph this number is determined only by whether u and v are equal, adjacent, or distinct and nonadjacent. Therefore, we get the equation

$$A^2 = kI + aA + c(J - I - A),$$

which can be rewritten as

$$A^2 - (a-c)A - (k-c)I = cJ.$$

We can use this equation to determine the eigenvalues of A. Since X is regular with valency k, it follows that k is an eigenvalue of A with eigenvector $\mathbf{1}$. By Lemma 8.4.1 any other eigenvector of A is orthogonal to $\mathbf{1}$. Let z be an eigenvector for A with eigenvalue $\theta \neq k$. Then

$$A^2z - (a-c)Az - (k-c)Iz = cJz = 0,$$

so

$$\theta^2 - (a-c)\theta - (k-c) = 0.$$

Therefore, the eigenvalues of A different from k must be zeros of the quadratic $x^2 - (a-c)x - (k-c)$. If we set $\Delta = (a-c)^2 + 4(k-c)$ (the discriminant of the quadratic) and denote the two zeros of this polynomial by θ and τ, we get

$$\theta = \frac{(a-c)+\sqrt{\Delta}}{2},$$

$$\tau = \frac{(a-c)-\sqrt{\Delta}}{2}.$$

Now, $\theta\tau = (c-k)$, and so, provided that $c < k$, we get that θ and τ are nonzero with opposite signs. We shall usually assume that $\theta > 0$. We see that the eigenvalues of a strongly regular graph are determined by its parameters (although strongly regular graphs with the same parameters need not be isomorphic).

The multiplicities of the eigenvalues are also determined by the parameters. To see this, let m_θ and m_τ be the multiplicities of θ and τ, respectively. Since k has multiplicity equal to one and the sum of all the eigenvalues is the trace of A (which is 0), we have

$$m_\theta + m_\tau = n-1, \quad m_\theta\theta + m_\tau\tau = -k.$$

Hence

$$m_\theta = -\frac{(n-1)\tau + k}{\theta - \tau}, \quad m_\tau = \frac{(n-1)\theta + k}{\theta - \tau}. \tag{10.2}$$

Now,

$$(\theta-\tau)^2 = (\theta+\tau)^2 - 4\theta\tau = (a-c)^2 + 4(k-c) = \Delta.$$

Substituting the values for θ and τ into the expressions for the multiplicities, we get

$$m_\theta = \frac{1}{2}\left((n-1) - \frac{2k+(n-1)(a-c)}{\sqrt{\Delta}}\right)$$

and

$$m_\tau = \frac{1}{2}\left((n-1) + \frac{2k+(n-1)(a-c)}{\sqrt{\Delta}}\right).$$

This argument yields a powerful feasibility condition. Given a parameter set we can compute m_θ and m_τ using these equations. If the results are not integers, then there cannot be a strongly regular graph with these parameters. In practice this is a very useful condition, as we shall see in Section 10.5. The classical application of this idea is to determine the possible valencies for a Moore graph with diameter two. We leave this as Exercise 7.

Lemma 10.2.1 *A connected regular graph with exactly three distinct eigenvalues is strongly regular.*

Proof. Suppose that X is connected and regular with eigenvalues k, θ, and τ, where k is the valency. If $A = A(X)$, then the matrix polynomial

$$M := \frac{1}{(k-\theta)(k-\tau)}(A-\theta I)(A-\tau I)$$

has all its eigenvalues equal to 0 or 1. Any eigenvector of A with eigenvalue θ or τ lies in the kernel of M, whence we see that the rank of M is equal to the multiplicity of k as an eigenvalue. Since X is connected, this multiplicity is one, and as $M\mathbf{1} = \mathbf{1}$, it follows that $M = \frac{1}{n}J$.

We have shown that J is a quadratic polynomial in A, and thus A^2 is a linear combination of I, J, and A. Accordingly, X is strongly regular. □

10.3 Some Characterizations

We begin this section with an important class of strongly regular graphs. Let q be a prime power such that $q \equiv 1 \bmod 4$. Then the *Paley graph* $P(q)$ has as vertex set the elements of the finite field $GF(q)$, with two vertices being adjacent if and only if their difference is a nonzero square in $GF(q)$. The congruence condition on q implies that -1 is a square in $GF(q)$, and hence the graph is undirected. The Paley graph $P(q)$ is strongly regular with parameters

$$(q,\ (q-1)/2,\ (q-5)/4,\ (q-1)/4).$$

By using the equations above we see that the eigenvalues θ and τ are $(-1 \pm \sqrt{q})/2$ and that they have the same multiplicity $(q-1)/2$.

Lemma 10.3.1 *Let X be strongly regular with parameters (n,k,a,c) and distinct eigenvalues k, θ, and τ. Then*

$$m_\theta m_\tau = \frac{nk\bar{k}}{(\theta-\tau)^2}.$$

Proof. The proof of this lemma is left as an exercise. □

Lemma 10.3.2 *Let X be strongly regular with parameters (n,k,a,c) and eigenvalues k, θ, and τ. If $m_\theta = m_\tau$, then $k = (n-1)/2$, $a = (n-5)/4$, and $c = (n-1)/4$.*

Proof. If $m_\theta = m_\tau$, then they both equal $(n-1)/2$, which we denote by m. Then m is coprime to n, and therefore it follows from the previous lemma that m^2 divides $k\bar{k}$. Since $k + \bar{k} = n-1$, it must be the case that $k\bar{k} \le (n-1)^2/4 = m^2$, with equality if and only if $k = \bar{k}$. Therefore, we must have equality, and so $k = \bar{k} = m$. Since $m(\theta+\tau) = -k$, we see that $\theta+\tau = a-c = -1$, and so $a = c-1$. Finally, because $k(k-a-1) = \bar{k}c$ we see that $c = k-a-1$, and hence $c = k/2$. Putting this all together shows that X has the stated parameters. □

A graph with $m_\theta = m_\tau$ is called a *conference graph*. Therefore, all Paley graphs are conference graphs (but the converse is not true; there are conference graphs that are not Paley graphs).

The difference between m_θ and m_τ is given by

$$m_\tau - m_\theta = \frac{2k + (n-1)(a-c)}{\sqrt{\Delta}}.$$

If the numerator of this expression is nonzero, then Δ must be a perfect square, and the eigenvalues θ and τ are rational. Because they are the roots of a monic quadratic with integer coefficients, they are integers. This gives us the following lemma.

Lemma 10.3.3 *Let X be strongly regular with parameters (n, k, a, c) and eigenvalues k, θ, and τ. Then either*

(a) *X is a conference graph, or*

(b) *$(\theta - \tau)^2$ is a perfect square and θ and τ are integers.* □

The graph $L(K_{3,3})$ satisfies both (a) and (b), and more generally, all Paley graphs $P(q)$ where q is a square satisfy both these conditions.

We can use our results to severely restrict the parameter sets of the strongly regular graphs on p or $2p$ vertices when p is prime.

Lemma 10.3.4 *Let X be a strongly regular graph with p vertices, where p is prime. Then X is a conference graph.*

Proof. By Lemma 10.3.1 we have

$$(\theta - \tau)^2 = \frac{pk\bar{k}}{m_\theta m_\tau}. \qquad (10.3)$$

If X is not a conference graph, then $(\theta - \tau)^2$ is a perfect square. But since k, $\bar{k}$, m_θ, and m_τ are all nonzero values smaller than p, the right-hand side of (10.3) is not divisible by p^2, which is a contradiction. □

Lemma 10.3.5 *Let X be a primitive strongly regular graph with an eigenvalue θ of multiplicity $n/2$. If $k < n/2$, then the parameters of X are*

$$\left((2\theta+1)^2 + 1,\ \theta(2\theta+1),\ \theta^2 - 1,\ \theta^2\right).$$

Proof. Since $m_\theta = n - 1 - m_\tau$, we see that $m_\theta \neq m_\tau$, and hence that θ and τ are integers.

First we will show by contradiction that θ must be the eigenvalue of multiplicity $n/2$. Suppose instead that $m_\tau = n/2$. From above we know that $m_\tau = (n\theta + k - \theta)/(\theta - \tau)$, and because m_τ divides n, it must also divide $k - \theta$. But since X is primitive, $0 < \theta < k$, and so $0 < k - \theta < n/2$, and hence m_τ cannot divide $k - \theta$. Therefore, we conclude that $m_\theta = n/2$.

Now, $m_\theta = (n\tau + k - \tau)/(\tau - \theta)$, and since m_θ divides n, it must also divide $k - \tau$. But since $-k \leq \tau$, we see that $k - \tau < n$, and hence it must be the case that $m_\theta = k - \tau$.

Set $m = m_\theta = n/2$. Then $m_\tau = m-1$, and since $\operatorname{tr} A = 0$ and $\operatorname{tr} A^2 = nk$, we have

$$k + m\theta + (m-1)\tau = 0, \qquad k^2 + m\theta^2 + (m-1)\tau^2 = 2mk.$$

By expanding the terms $(m-1)\tau$ and $(m-1)\tau^2$ in the above expressions and then substituting $k - m$ for the second occurrence of τ in each case we get

$$1 + \theta + \tau = 0, \qquad \theta^2 + \tau^2 = m,$$

respectively. Combining these we get that $m = \theta^2 + (\theta+1)^2$, and hence that $k = \theta(2\theta+1)$. Finally, we know that $c - k = \theta\tau = -(\theta^2+\theta)$, and so $c = \theta^2$; we also know that $a - c = \theta + \tau = -1$, so $a = \theta^2 - 1$. Hence the result is proved. □

Corollary 10.3.6 *Let X be a primitive strongly regular graph with $2p$ vertices where p is prime. Then the parameters of X or its complement are*

$$\left((2\theta+1)^2+1,\ \theta(2\theta+1),\ \theta^2-1,\ \theta^2\right).$$

Proof. By taking the complement if necessary, we may assume that $k \leq (n-1)/2$. The graph X cannot be a conference graph (because for a conference graph $n = 2m_\tau + 1$ is odd), and hence θ and τ are integers. Since $(\theta-\tau)^2 m_\theta m_\tau = 2pk\bar{k}$, we see that either p divides $(\theta-\tau)^2$ or p divides $m_\theta m_\tau$. If p divides $(\theta-\tau)^2$, then so must p^2, and hence p must divide $k\bar{k}$. Since $k \leq (2p-1)/2$, this implies that p must divide $\bar{k}$, and hence $\bar{k} = p$ and $k = p - 1$. It is left as Exercise 1 to show that there are no primitive strongly regular graphs with k and $\bar{k}$ coprime. On the other hand, if p divides $m_\theta m_\tau$, then either $m_\theta = p$ or $m_\tau = p$, and the result follows from Lemma 10.3.5. □

Examples of such graphs can be obtained from certain Steiner systems. In particular, if D is a $2-(2\theta(\theta+1)+1, \theta+1, 1)$ design, then the complement of the block graph of D is a graph with these parameters. Such designs are known only for $\theta \leq 4$. However, there are many further examples of strongly regular graphs with these parameters.

10.4 Latin Square Graphs

In this section we consider an extended example, namely the graphs arising from Latin squares. Recall from Section 4.5 that a Latin square of order n is an $n \times n$ matrix with entries from a set of size n such that each row and column contains each symbol precisely once. Here are three examples:

$$L_1 = \begin{pmatrix} 1 & 2 & 3 & 4 \\ 2 & 1 & 4 & 3 \\ 3 & 4 & 1 & 2 \\ 4 & 3 & 2 & 1 \end{pmatrix}, \quad L_2 = \begin{pmatrix} 1 & 3 & 4 & 2 \\ 4 & 2 & 1 & 3 \\ 2 & 4 & 3 & 1 \\ 3 & 1 & 2 & 4 \end{pmatrix}, \quad L_3 = \begin{pmatrix} 1 & 2 & 3 & 4 \\ 4 & 1 & 2 & 3 \\ 3 & 4 & 1 & 2 \\ 2 & 3 & 4 & 1 \end{pmatrix}.$$

Two Latin squares $L = (l_{ij})$ and $M = (m_{ij})$ are said to be *orthogonal* if the n^2 pairs

$$(l_{ij}, m_{ij}), \qquad \text{where} \qquad 1 \le i, j \le n,$$

are all distinct. The Latin squares L_1 and L_2 shown above are orthogonal, as can easily be checked by "superimposing" the two squares and then checking that every ordered pair of elements from $\{1, 2, 3, 4\}$ occurs twice in the resulting array:

$$\begin{matrix} 1^1 & 2^3 & 3^4 & 4^2 \\ 2^4 & 1^2 & 4^1 & 3^3 \\ 3^2 & 4^4 & 1^3 & 2^1 \\ 4^3 & 3^1 & 2^2 & 1^4 \end{matrix}$$

A set S of Latin squares is called *mutually orthogonal* if every pair of squares in S is orthogonal.

An *orthogonal array* with parameters k and n is a $k \times n^2$ array with entries from a set N of size n such that the n^2 ordered pairs defined by any two rows of the matrix are all distinct. We denote an orthogonal array with these parameters by $OA(k, n)$. A Latin square immediately gives rise to an $OA(3, n)$ by taking the three rows of the orthogonal array to be "row number," "column number," and "symbol":

$$\begin{matrix} 1 & 1 & 1 & 1 & 2 & 2 & 2 & 2 & 3 & 3 & 3 & 3 & 4 & 4 & 4 & 4 \\ 1 & 2 & 3 & 4 & 1 & 2 & 3 & 4 & 1 & 2 & 3 & 4 & 1 & 2 & 3 & 4 \\ 1 & 2 & 3 & 4 & 2 & 1 & 4 & 3 & 3 & 4 & 1 & 2 & 4 & 3 & 2 & 1 \end{matrix}$$

Conversely, any $OA(3, n)$ gives rise to a Latin square (or in fact several Latin squares) by reversing this procedure and regarding two of the rows of the orthogonal array as the row and column indices of a Latin square.

In an analogous fashion an $OA(k, n)$ gives rise to a set of $k - 2$ Latin squares. It is easy to see that these Latin squares are mutually orthogonal, and hence we have one half of the following result. (The other half is easy. and is left to the reader to ponder.)

Theorem 10.4.1 *An $OA(k, n)$ is equivalent to a set of $k - 2$ mutually orthogonal Latin squares.* □

Given an $OA(k, n)$, we can define a graph X as follows: The vertices of X are the n^2 columns of the orthogonal array (viewed as column vectors of length k), and two vertices are adjacent if they have the same entries in one coordinate position.

Theorem 10.4.2 *The graph defined by an $OA(k, n)$ is strongly regular with parameters*

$$(n^2,\ (n-1)k,\ n-2+(k-1)(k-2),\ k(k-1)). \qquad \square$$

Notice that the graph defined by an $OA(2, n)$ is isomorphic to $L(K_{n,n})$. We can say more about the structure of strongly regular graphs arising from orthogonal arrays.

Lemma 10.4.3 *Let L be a Latin square of order n and let X be the graph of the corresponding $OA(3, n)$. Then the maximum number of vertices $\alpha(X)$ in an independent set of X is n, and the chromatic number $\chi(X)$ of X is at least n. If L is the multiplication table of a group, then $\chi(X) = n$ if and only if $\alpha(X) = n$.*

Proof. If we identify the n^2 vertices of X with the n^2 cells of the Latin square L, then it is clear that an independent set of X can contain at most one cell from each row of L. Therefore, $\alpha(X) \leq n$, which immediately implies that $\chi(X) \geq n$.

Assume now that L is the multiplication table of a group G and denote the ij-entry of L by $i \circ j$. An independent set of size n contains precisely one cell from each row of L, and hence we can describe such a set by giving for each row the column number of that cell. Therefore, an independent set of size n is determined by giving a permutation π of $N := \{1, 2, \ldots, n\}$ such that the map $i \mapsto i \circ i^\pi$ is also a permutation of N. But if $k \in N$, then the permutation π_k which maps i to $k \circ i^\pi$ will also provide an independent set either equal to or disjoint from the one determined by π. Thus we obtain n independent sets of size n, and so X has chromatic number n. $\square$

Lemma 10.4.4 *Let L be a Latin square arising from the multiplication table of the cyclic group G of order $2n$ and let X be the graph of the corresponding $OA(3, 2n)$. Then X has no independent sets of size $2n$.*

Proof. Suppose on the contrary that X does have an independent set of $2n$ vertices, described by the permutation π. There is a unique element τ of order two in G, and so all the remaining nonidentity elements can be paired with their inverses. It follows that the product of all the entries in G is equal to τ. Hence

$$\tau = \sum_{i \in G} i \circ i^\pi = \sum_{i \in G} i \sum_{i \in G} i^\pi = \tau^2 = 1.$$

This contradiction shows that such a permutation π cannot exist. $\square$

An orthogonal array $OA(k, n)$ is called *extendible* if it occurs as the first k rows of an $OA(k+1, n)$.

Theorem 10.4.5 *An $OA(k, n)$ is extendible if and only if its graph has chromatic number n.*

Proof. Let X be the graph of an $OA(k, n)$. Suppose first that $\chi(X) = n$. Then the n^2 vertices of X fall into n colour classes $V(X) = V_1 \cup \cdots \cup V_n$. Define the $(k+1)$st row of the orthogonal array by setting the entry to i if the column corresponds to a vertex in V_i. Conversely, if the orthogonal array is extendible, then the sets of columns on which the $(k+1)$st row is constant form n independent sets of size n in X. □

Up to permutations of rows, columns, and symbols there are only two Latin squares of order 4, one coming from the group table of the cyclic group of order 4 and the other from the group table of the noncyclic group of order 4. Of the three Latin squares above, L_1 is the multiplication table of the noncyclic group, L_2 a permuted version of the same thing, and L_3 the multiplication table of the cyclic group. From Lemma 10.4.4 the orthogonal array corresponding to L_3 is not extendible (and hence we cannot find a Latin square orthogonal to L_3). However, L_1 and L_2 are orthogonal, and hence give us an $OA(4, 4)$. The graph corresponding to this orthogonal array is the complement of $4K_4$, and hence is 4-colourable. Thus the $OA(4, 4)$ can be extended to an $OA(5, 4)$ whose graph is the complete graph K_{16}. This shows that the graph of L_1 is the complement of the graph defined by just two rows of the orthogonal array, which is simply $L(K_{4,4})$.

10.5 Small Strongly Regular Graphs

In this section we describe the small primitive strongly regular graphs, namely those on up to 25 vertices. First we construct a list of the parameter sets that satisfy equation (10.1) and for which the multiplicities of the eigenvalues are integral. This is not hard to do by hand; indeed, we recommend it.

The parameter sets in Table 10.1 pass this test. (We list only those where $k \leq n/2$.)

Graphs that we have already met dominate this list. Paley graphs occur on 5, 9, 13, 17, and 25 vertices. The Petersen graph is a strongly regular cubic graph on 10 vertices, isomorphic to $\overline{L(K_5)}$. The complement of $L(K_6)$ is a 6-valent graph on 15 vertices, while $L(K_7)$ and its complement are strongly regular graphs with parameters $(21, 10, 5, 4)$ and $(21, 10, 3, 6)$, respectively. There are two nonisomorphic Latin squares of order four, and the complements of their associated graphs provide two $(16, 6, 2, 2)$ strongly regular graphs. The line graph $L(K_{5,5})$ is a $(25, 8, 3, 2)$ strongly regular graph. It is possible to show that the parameter vector $(16, 6, 2, 2)$ is realized only by the two Latin square graphs, and that each of the othe parameter vectors just mentioned is realized by a unique graph.

There is a $(16, 5, 0, 2)$ strongly regular graph called the *Clebsch graph*, shown in Figure 10.1. We will discuss this graph, and prove that it is unique, in Section 10.6. There are two Latin squares of order 5, yielding

n	k	a	c	θ	τ	m_θ	m_τ
5	2	0	1	$(-1+\sqrt{5})/2$	$(-1-\sqrt{5})/2$	2	2
9	4	1	2	1	-2	4	4
10	3	0	1	1	-2	5	4
13	6	2	3	$(-1+\sqrt{13})/2$	$(-1-\sqrt{13})/2$	6	6
15	6	1	3	1	-3	9	5
16	5	0	2	1	-3	10	5
16	6	2	2	2	-2	6	9
17	8	3	4	$(-1+\sqrt{17})/2$	$(-1-\sqrt{17})/2$	8	8
21	10	3	6	1	-4	14	6
21	10	4	5	$(-1+\sqrt{21})/2$	$(-1-\sqrt{21})/2$	10	10
21	10	5	4	3	-2	6	14
25	8	3	2	3	-2	8	16
25	12	5	6	2	-3	12	12

Table 10.1. Parameters of small strongly regular graphs

two $(25, 12, 5, 6)$ strongly regular graphs. An exhaustive computer search has shown that there are 10 strongly regular graphs with these parameters. Such exhaustive computer searches have also been performed for a very limited number of other parameter sets. The smallest parameter set for which the exact number of strongly regular graphs is unknown is $(37, 18, 8, 9)$.

The only parameter set remaining from Table 10.1 is $(21, 10, 4, 5)$. A graph with these parameters would be a conference graph. However, it is known that a conference graph on n vertices exists if and only if n is the sum of two squares. Therefore, there is no strongly regular graph with these parameters.

10.6 Local Eigenvalues

Let X be a strongly regular graph and choose a vertex $u \in V(X)$. We can write the adjacency matrix A of X in partitioned form:

$$A = \begin{pmatrix} 0 & \mathbf{1}^T & 0 \\ \mathbf{1} & A_1 & B^T \\ 0 & B & A_2 \end{pmatrix}$$

Here A_1 is the adjacency matrix of the subgraph of X induced by the neighbours of u, and A_2 is the adjacency matrix of the subgraph induced by the vertices at distance two from u. We call these two subgraphs respectively the first and second *subconstituents* of X relative to u. Our goal in this

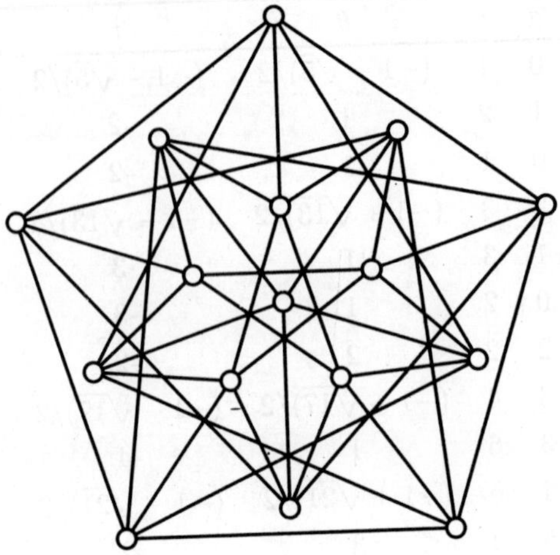

Figure 10.1. The Clebsch graph

section is to describe the relation between the eigenvalues of X and the eigenvalues of its subconstituents.

Suppose that X has parameters (n, k, a, c). Then

$$A^2 - (a-c)A - (k-c)I = cJ.$$

On the other hand, we also have

$$A^2 = \begin{pmatrix} k & \mathbf{1}^T A_1 & \mathbf{1}^T B^T \\ A_1\mathbf{1} & J + A_1^2 + B^T B & A_1 B^T + B^T A_2 \\ B\mathbf{1} & BA_1 + A_2 B & A_2^2 + BB^T \end{pmatrix},$$

and so, comparing these expressions, we obtain the following three conditions:

$$\begin{aligned} A_1^2 - (a-c)A_1 - (k-c)I + B^T B &= (c-1)J, \\ A_2^2 - (a-c)A_2 - (k-c)I + BB^T &= cJ, \\ BA_1 + A_2 B &= (a-c)B + cJ. \end{aligned}$$

We say that an eigenvalue of A_i is *local* if it is not equal to an eigenvalue of A and has an eigenvector orthogonal to $\mathbf{1}$. We have the following characterization.

Lemma 10.6.1 *Let X be strongly regular with eigenvalues $k > \theta > \tau$. Suppose that x is an eigenvector of A_1 with eigenvalue σ_1 such that $\mathbf{1}^T x = 0$. If $Bx = 0$, then $\sigma_1 \in \{\theta, \tau\}$, and if $Bx \neq 0$, then $\tau < \sigma_1 < \theta$.*

Proof. Since $\mathbf{1}^T x = 0$, we have

$$(A_1^2 - (a-c)A_1 - (k-c)I)x = -B^T Bx,$$

and since X is strongly regular with eigenvalues k, θ, and τ, we have

$$(A_1^2 - (a-c)A_1 - (k-c)I)x = (A_1 - \theta I)(A_1 - \tau I)x.$$

Therefore, if x is an eigenvector of A_1 with eigenvalue σ_1,

$$(\sigma_1 - \theta)(\sigma_1 - \tau)x = -B^T Bx.$$

If $Bx = 0$, then $(\sigma_1 - \theta)(\sigma_1 - \tau) = 0$ and $\sigma_1 \in \{\theta, \tau\}$. If $Bx \neq 0$, then $B^T Bx \neq 0$, and so x is an eigenvector for the positive semidefinite matrix $B^T B$ with eigenvalue $-(\sigma_1 - \theta)(\sigma_1 - \tau)$. It follows that $(\sigma_1 - \theta)(\sigma_1 - \tau) < 0$, whence $\tau < \sigma_1 < \theta$. □

Either using similar arguments to those above or taking complements we obtain the following result.

Lemma 10.6.2 *Let X be a strongly regular graph with eigenvalues $k > \theta > \tau$. Suppose that y is an eigenvector of A_2 with eigenvalue σ_2 such that $\mathbf{1}^T y = 0$. If $B^T y = 0$, then $\sigma_2 \in \{\theta, \tau\}$, and if $B^T y \neq 0$, then $\tau < \sigma_2 < \theta$.* □

Theorem 10.6.3 *Let X be an (n, k, a, c) strongly regular graph. Then σ is a local eigenvalue of one subconstituent of X if and only if $a - c - \sigma$ is a local eigenvalue of the other, with equal multiplicities.*

Proof. Suppose that σ_1 is a local eigenvalue of A_1 with eigenvector x. Then, since $\mathbf{1}^T x = 0$,

$$BA_1 + A_2 B = (a-c)B + cJ$$

implies that

$$A_2 Bx = (a-c)Bx - \sigma_1 Bx = (a - c - \sigma_1)Bx.$$

Therefore, since $Bx \neq 0$, it is an eigenvector of A_2 with eigenvalue $a - c - o_1$. Since $\mathbf{1}^T B = (k-1-a)\mathbf{1}^T$, we also have $\mathbf{1}^T Bx = 0$, and so $a - c - \sigma_1$ is a local eigenvalue for A_2.

A similar argument shows that if σ_2 is a local eigenvalue of A_2 with eigenvector y, then $a - c - \sigma_2$ is a local eigenvalue of A_1 with eigenvector $B^T y$.

Finally, note that the mapping B from the σ_1-eigenspace of A_1 into the $(a - c - \sigma_1)$-eigenspace of A_2 is injective and the mapping B^T from the $(a - c - \sigma_1)$-eigenspace of A_2 into the σ_1-eigenspace of A_1 is also injective. Therefore, the dimension of these two subspaces is equal. □

These results also give us some information about the eigenvectors of A. Since the distance partition is equitable, the three eigenvectors of the quotient matrix yield three eigenvectors of A that are constant on u, $V(X_1)$, and $V(X_2)$. The remaining eigenvectors may all be taken to sum to zero on u, $V(X_1)$, and $V(X_2)$. If x is an eigenvector of A_1 with eigenvector σ_1

that sums to zero on $V(X_1)$, then define a vector z by

$$z = \begin{pmatrix} 0 \\ x \\ \alpha Bx \end{pmatrix}.$$

We will show that for a suitable choice for α, the vector z is an eigenvector of A. If $Bx = 0$, then it is easy to see that z is an eigenvector of A with eigenvalue σ_1, which must therefore be equal to either θ or τ.

If $Bx \neq 0$, then

$$Az = \begin{pmatrix} 0 \\ \sigma_1 x + \alpha B^T Bx \\ Bx + \alpha A_2 Bx \end{pmatrix} = \begin{pmatrix} 0 \\ (\sigma_1 - \alpha(\sigma_1 - \theta)(\sigma_1 - \tau))x \\ Bx + \alpha A_2 Bx \end{pmatrix}.$$

Now, $A_2 Bx = (a-c-\sigma_1)Bx$, and so by taking $\alpha = (\sigma_1 - \tau)^{-1}$ and recalling that $\theta = a - c - \tau$, we deduce that

$$Az = \begin{pmatrix} 0 \\ \theta x \\ \theta\alpha Bx \end{pmatrix}.$$

Therefore, z is an eigenvector of A with eigenvalue θ. Taking $\alpha = (\sigma_1 - \theta)^{-1}$ yields an eigenvector of A with eigenvalue τ.

We finish with a result that uses local eigenvalues to show that the Clebsch graph is unique.

Theorem 10.6.4 *The Clebsch graph is the unique strongly regular graph with parameters* $(16, 5, 0, 2)$.

Proof. Suppose that X is a $(16, 5, 0, 2)$ strongly regular graph, which therefore has eigenvalues 5, 1, and -3. Let X_2 denote the second subconstituent of X. This is a cubic graph on 10 vertices, and so has an eigenvalue 3 with eigenvector **1**. All its other eigenvectors are orthogonal to **1**. Since 0 is the only eigenvalue of the first subconstituent, the only other eigenvalues that X_2 can have are 1, -3, and the local eigenvalue -2. Since -1 is not in this set, X_2 can not have K_4 as a component, and so X_2 is connected. This implies that its diameter is at least two; therefore, X_2 has at least three eigenvalues. Hence the spectrum of X_2 is not symmetric about zero, and so X_2 is not bipartite. Consequently, -3 is not an eigenvalue of X_2. Therefore, X_2 is a connected cubic graph with exactly the three eigenvalues 3, 1, and -2. By Lemma 10.2.1 it is strongly regular, and hence isomorphic to the Petersen graph. The neighbours in X_2 of any fixed vertex of the first subconstituent form an independent set of size four in X_2. Because the Petersen graph has exactly five independent sets of size four, each vertex of the first subconstituent is adjacent to precisely one of these independent sets. Therefore, we conclude that X is uniquely determined by its parameters. □

10.7 The Krein Bounds

This section is devoted to proving the following result, which gives inequalities between the parameters of a strongly regular graph. The bounds implied by these inequalities are known as the *Krein bounds*, as they apply to strongly regular graphs. (There are related inequalities for distance-regular graphs and, more generally, for association schemes.) The usual proof of these inequalities is much less elementary, and does not provide information about the cases where equality holds.

Theorem 10.7.1 *Let X be a primitive (n, k, a, c) strongly regular graph, with eigenvalues k, θ, and τ. Let m_θ and m_τ denote the multiplicities of θ and τ, respectively. Then*

$$\theta\tau^2 - 2\theta^2\tau - \theta^2 - k\theta + k\tau^2 + 2k\tau \geq 0,$$
$$\theta^2\tau - 2\theta\tau^2 - \tau^2 - k\tau + k\theta^2 + 2k\theta \geq 0.$$

If the first inequality is tight, then $k \geq m_\theta$, and if the second is tight, then $k \geq m_\tau$. If either of the inequalities is tight, then one of the following is true:

(a) *X is the 5-cycle C_5.*

(b) *Either X or its complement $\overline{X}$ has all its first subconstituents empty, and all its second subconstituents strongly regular.*

(c) *All subconstituents of X are strongly regular.* □

Our proof is long, and somewhat indirect, but involves nothing deeper than an application of the Cauchy–Schwarz inequality. We break the argument into a number of lemmas. First, however, we introduce some notation which is used throughout this section. Let X be a primitive (n, k, a, c) strongly regular graph with eigenvalues k, θ, and τ, where we make no assumption concerning the signs of θ and τ (that is, either θ or τ may be the positive eigenvalue). Let u be an arbitrary vertex of X and let X_1 and X_2 be the first and second subconstituents relative to u. The adjacency matrix of X_1 is denoted by A_1. We use m for m_θ where needed in the proofs, but not the statements, of the series of lemmas.

Lemma 10.7.2 *If $k \geq m_\theta$, then τ is an eigenvalue of the first subconstituent of X with multiplicity at least $k - m_\theta$.*

Proof. Let U denote the space of functions on $V(X)$ that sum to zero on each subconstituent of X relative to u. This space has dimension $n - 3$. Let T be the space spanned by the eigenvectors of X with eigenvalue τ that sum to zero on $V(X_1)$; this has dimension $n - m - 2$ and is contained in U. Let N denote the space of functions on $V(X)$ that sum to zero and have their support contained in $V(X_1)$; this has dimension $k - 1$ and is also contained in U. If $k > m$, then $\dim N + \dim T > \dim U$, and therefore

$\dim N \cap T \geq k - m$. Each function in $N \cap T$ is an eigenvector of X with eigenvalue τ, and its restriction to $V(X_1)$ is an eigenvector of X_1 with the same eigenvalue. □

The next result is the heart of the theorem. It is essentially the first of the two Krein inequalities.

Lemma 10.7.3 *If $k \geq m_\theta$, then*

$$(m_\theta - 1)(ka - a^2 - (k - m_\theta)\tau^2) - (a + (k - m_\theta)\tau)^2 \geq 0.$$

Proof. We know that a is an eigenvalue of A_1 with multiplicity at least one, and that τ is an eigenvalue with multiplicity at least $k - m$. This leaves $m - 1$ eigenvalues as yet unaccounted for; we denote them by $\sigma_1, \ldots, \sigma_{m-1}$. Then

$$0 = \operatorname{tr}(A_1) = a + (k - m)\tau + \sum_i \sigma_i$$

and

$$ka = \operatorname{tr}(A_1^2) = a^2 + (k - m)\tau^2 + \sum_i \sigma_i^2.$$

By the Cauchy–Schwarz inequality,

$$(m - 1) \sum_i \sigma_i^2 \geq \left(\sum_i \sigma_i \right)^2,$$

with equality if and only if the $m - 1$ eigenvalues σ_i are all equal. Using the two equations above, we obtain the inequality in the statement of the lemma. □

Lemma 10.7.4 *If $k < m_\theta$, then*

$$(m_\theta - 1)(ka - a^2 - (k - m_\theta)\tau^2) - (a + (k - m_\theta)\tau)^2 > 0.$$

Proof. Define the polynomial $p(x)$ by

$$p(x) := (m - 1)(ka - a^2 - (k - m)x^2) - (a + (k - m)x)^2.$$

Then

$$p(x) = (m - 1)ka - ma^2 + 2a(m - k)x + (k - 1)(m - k)x^2,$$

and after some computation, we find that its discriminant is

$$-4a(m - k)(m - 1)k(k - 1 - a).$$

Since $k < m$ and $1 < m$, we see that this is negative unless $a = 0$. If $a = 0$, then

$$p(x) = (k - 1)(m - k)x^2,$$

and consequently $p(\tau) \neq 0$, unless $\tau = 0$.

If $a = 0$ and $\tau = 0$, then X is the complete bipartite graph $K_{k,k}$ with eigenvalues k, 0, and $-k$. However, if $\tau = 0$, then $\theta = -k$ and $m = 1$, which contradicts the condition that $k < m$. □

Note that the proof of Lemma 10.7.4 shows that if $k < m_\theta$, then $p(x) \geq 0$ for any choice of x, while the proof of Lemma 10.7.3 shows that if $k \geq m_\theta$, then the eigenvalue τ must satisfy $p(\tau) \geq 0$. Therefore, only Lemma 10.7.3 provides an actual constraint on the parameters of a strongly regular graph.

We have now shown that whether or not $k \geq m_\theta$,

$$(m_\theta - 1)(ka - a^2 - (k - m_\theta)\tau^2) - (a + (k - m_\theta)\tau)^2 \geq 0.$$

Using Exercise 5, we can write this in terms of k, θ, and τ, obtaining

$$-\frac{k\tau(\tau+1)(\theta+1)}{(k+\theta\tau)(\theta-\tau)}(2\theta^2\tau + \theta^2 - \theta\tau^2 + k\theta - k\tau^2 - 2k\tau) \geq 0. \qquad (10.4)$$

(Don't try this at home on your own! Use Maple or some approximation thereto.) Also from Exercise 5 we find that

$$\ell := -\frac{k(\theta+1)(\tau+1)}{k+\theta\tau}$$

is the number of vertices of X_2, and since X is primitive, this is strictly positive. Therefore,

$$-\frac{k\tau(\tau+1)(\theta+1)}{(k+\theta\tau)(\theta-\tau)} = \frac{\ell\tau}{\theta-\tau}.$$

Since X is primitive, $\tau \neq 0$, and so $\tau(\theta-\tau)^{-1} < 0$. Therefore, (10.4) implies that

$$2\theta^2\tau + \theta^2 - \theta\tau^2 + k\theta - k\tau^2 - 2k\tau \leq 0.$$

Thus we have proved the first inequality in the statement of our theorem. Because our proofs made no assumption about which eigenvalue was the positive one, the second inequality follows immediately from the first by exchanging θ and τ.

Next we consider the case where one of the inequalities is tight.

Lemma 10.7.5 *If*

$$(m_\theta - 1)(ka - a^2 - (k - m_\theta)\tau^2) - (a + (k - m_\theta)\tau)^2 = 0,$$

then $k \geq m_\theta$. In addition, either each first subconstituent of X is strongly regular, or $k = m_\theta$ and $a = 0$.

Proof. By Lemma 10.7.4, equality cannot occur if $k < m_\theta$, and so $k \geq m_\theta$. If equality holds in the Cauchy–Schwarz bound in the proof of Lemma 10.7.3, then the eigenvalues σ_i must all be equal; we denote their common value by σ. Therefore, X_1 has at most three distinct eigenvalues a, σ, and τ.

If $k = m$, then

$$0 = (k-1)(ka - a^2) - a^2 = k^2a - ka^2 - ka = ka(k - a - 1).$$

Since X is neither empty nor complete, $k \neq 0$ and $k \neq a+1$, which implies that $a = 0$.

Therefore, we assume that $k > m$ and consider separately the cases where X_1 has one, two, or three distinct eigenvalues. If X_1 has just one eigenvalue, then it is empty, and so $a = \sigma = \tau = 0$. Since $\theta\tau = c - k$, this implies that $c = k$ and X is a complete bipartite graph, which is not primitive.

If X_1 has exactly two distinct eigenvalues, then by Lemma 8.12.1, each component of X_1 has diameter at most 1, and so X_1 is a union of cliques. Since X is not complete, there are at least two cliques, and since X_1 has two eigenvalues, these cliques have size at least two. Therefore, X_1 is strongly regular.

If X_1 has three distinct eigenvalues, then X_1 is a regular graph whose largest eigenvalue is simple. By the Perron–Frobenius theorem the multiplicity of the largest eigenvalue of a regular graph equals the number of components, and so it follows that X_1 is connected. Therefore, by Lemma 10.2.1, X_1 is strongly regular. □

To complete our proof we consider the complement of X, which is strongly regular with parameters

$$(n,\ n-1-k,\ n-2-2k+c,\ n-2k+a)$$

and has eigenvalues $n-k-1$, $-1-\tau$, and $-1-\theta$ with multiplicities 1, m_τ, and m_θ, respectively. If we set ℓ equal to $n-1-k$ and b equal to $n-2-2k+c$, then

$$(m_\theta - 1)(\ell b - b^2 - (\ell - m_\theta)(\tau+1)^2) - (b + (\ell - m_\theta)(-\tau-1))^2 \geq 0.$$

If we write the left-hand side of this in terms of k, θ, and τ (Maple again!), then the surprising conclusion is that it is also equal to

$$-\frac{k\tau(\tau+1)(\theta+1)}{(k+\theta\tau)(\theta-\tau)}(2\theta^2\tau + \theta^2 - \theta\tau^2 + k\theta - k\tau^2 - 2k\tau).$$

Therefore, if equality holds for X, it also holds for $\overline{X}$, and we conclude that either $\ell = m_\theta$ and $b = 0$ or the first subconstituent of $\overline{X}$ is strongly regular. Consequently, the second subconstituent of X is either complete or strongly regular.

Therefore, when equality holds, X_1 is empty or strongly regular, and X_2 is complete or strongly regular. It is straightforward to see that C_5 is the only primitive strongly regular graph with X_1 empty and X_2 complete. The Clebsch graph provides an example where X_1 is empty and X_2 is strongly regular, and we will discuss a graph with both subconstituents strongly regular in the next section.

We give one example of using the Krein bound to show that certain feasible parameter sets cannot be realized.

Corollary 10.7.6 *There is no strongly regular graph with parameter set* $(28, 9, 0, 4)$.

Proof. The parameter set $(28, 9, 0, 4)$ is feasible, and a strongly regular graph with these parameters would have spectrum

$$\{9,\ 1^{(21)},\ -5^{(6)}\}.$$

But if $k = 9$, $\theta = 1$, and $\tau = -5$, then

$$\theta^2\tau - 2\theta\tau^2 - \tau^2 - k\tau + k\theta^2 + 2k\theta = -8,$$

and hence there is no such graph. □

Although we cannot go into the matter in any depth here, the case where $a = 0$ and $k = m_\theta$ is extremely interesting. If $k = m_\theta$ (or $k = m_\tau$), then X is said to be formally self-dual. Mesner has shown that other than the conference graphs, there are just two classes of such graphs. The first are the strongly regular graphs constructed from orthogonal arrays, and these are not triangle-free. The second are known as negative Latin square graphs, and the Clebsch graph is an example.

10.8 Generalized Quadrangles

We recall from Section 5.4 that a generalized quadrangle is an incidence structure such that:

(a) any two points are on at most one line (and hence any two lines meet in at most one point,

(b) if P is a point not on a line ℓ, then there is a unique point on ℓ collinear with P.

If every line contains $s+1$ points, and every point lies on $t+1$ lines, then the generalized quadrangle has order (s, t). As we saw in Section 5.4, the edges and one-factors of K_6 form a generalized quadrangle of order $(2, 2)$.

The *point graph* of a generalized quadrangle is the graph with the points of the quadrangle as its vertices, with two points adjacent if and only if they are collinear. The point graph of the generalized quadrangle on the edges and one-factors of K_6 is $\overline{L(K_6)}$, which is strongly regular. This is no accident, as the next result shows that the point graph of any nontrivial generalized quadrangle is strongly regular.

Lemma 10.8.1 *Let X be the point graph of a generalized quadrangle of order (s, t). Then X is strongly regular with parameters*

$$((s+1)(st+1),\ s(t+1),\ s-1,\ t+1).$$

Proof. Each point P of the generalized quadrangle lies on $t+1$ lines of size $s+1$, any two of which have exactly P in common. Hence X has valency

$s(t+1)$. The graph induced by the points collinear with P consists of $t+1$ vertex-disjoint cliques of size s, whence $a = s - 1$. Let Q be a point not collinear with P. Then Q is collinear with exactly one point on each of the lines through P. This shows that $c = t+1$.

Finally, we determine the number of vertices in the graph. Let ℓ be a line of the quadrangle. Each point not on ℓ is collinear with a unique point on ℓ; consequently, there are st points collinear with a given point of ℓ and not on ℓ. This gives us exactly $st(s+1)$ points not on ℓ, and $(s+1)(st+1)$ points in total. □

Lemma 10.8.2 *The eigenvalues of the point graph of a generalized quadrangle of order* (s,t) *are* $s(t+1)$, $s-1$, *and* $-t-1$, *with respective multiplicities*

$$1, \quad \frac{st(s+1)(t+1)}{s+t}, \quad \frac{s^2(st+1)}{s+t}.$$

Proof. Let X be the point graph of a generalized quadrangle of order (s,t). From Section 10.2, the eigenvalues of X are its valency $s(t+1)$ and the two zeros of the polynomial

$$x^2-(a-c)x-(k-c) = x^2-(s-t-2)x-(s-1)(t+1) = (x-s+1)(x+t+1).$$

Thus the nontrivial eigenvalues are $s-1$ and $-t-1$. Their multiplicities now follow from (10.2). □

The fact that these expressions for the multiplicities must be integers provides a nontrivial constraint on the possible values of s and t. A further constraint comes from the Krein inequalities.

Lemma 10.8.3 *If* $\mathcal{G}$ *is a generalized quadrangle of order* (s,t) *with* $s>1$ *and* $t>1$, *then* $s \le t^2$ *and* $t \le s^2$.

Proof. Let X be the point graph of $\mathcal{G}$. Substituting $k = s(t+1)$, $\theta = s-1$, and $\tau = -t-1$ into the second Krein inequality

$$\theta^2\tau - 2\theta\tau^2 - \tau^2 - k\tau + k\theta^2 + 2k\theta \ge 0$$

and factoring yields

$$(s^2-t)(t+1)(s-1) \ge 0.$$

Since $s>1$, this implies that $t \le s^2$. Since we may apply the same argument to the point graph of the dual quadrangle, we also find that $s \le t^2$. □

Generalized quadrangles with lines of size three will arise in the next chapter, where we study graphs with smallest eigenvalue at least -2. There we will need the following result.

Lemma 10.8.4 *If a generalized quadrangle of order* $(2,t)$ *exists, then* $t \in \{1,2,4\}$.

Proof. If $s = 2$, then $-t-1$ is an eigenvalue of the point graph with multiplicity

$$\frac{8t+4}{t+2} = 8 - \frac{12}{t+2}.$$

Therefore, $t+2$ divides 12, which yields that $t \in \{1, 2, 4, 10\}$. The case $t = 10$ is excluded by the Krein bound. □

Note that a generalized quadrangle of order $(2, t)$ has $6t+3$ points; thus the possible number of points is 9, 15, or 27.

10.9 Lines of Size Three

At the end of the previous section we saw that if there is a generalized quadrangle of order $(2, t)$, then $t \in \{1, 2, 4\}$. In this section we will show that there is a unique example for each of these three values of t.

Lemma 10.9.1 *Let X be a strongly regular graph with parameters*

$$(6t+3, 2t+2, 1, t+1).$$

The spectrum of the second subconstituent of X is

$$\{(t+1)^{(1)},\ 1^{(x)},\ (1-t)^{(t+1)},\ (-t-1)^{(y)}\}$$

where

$$x = \frac{4(t^2-1)}{t+2}, \quad y = \frac{t(4-t)}{t+2}.$$

Proof. The first subconstituent of X has valency one, and hence consists of $t+1$ vertex-disjoint edges. Its eigenvalues are 1 and -1, each with multiplicity $t+1$, and so -1 is the unique local eigenvalue of the first subconstituent. Therefore, the nonlocal eigenvalues of the second subconstituent of X are $t+1$ (its valency) and a subset of $\{1, -1-t\}$. The only local eigenvalue of the second subconstituent is $1-(t+1)-(-1) = 1-t$ with multiplicity $t+1$. We also see that $t+1$ is a simple eigenvalue, for if it had multiplicity greater than one, then it would a local eigenvalue. Therefore, letting x denote the multiplicity of 1 and y the multiplicity of $-1-t$ we get the spectrum as stated.

Then, since the second subconstituent has $4t$ vertices and as its eigenvalues sum to zero, we have

$$\begin{aligned} 1 + (t+1) + x + y &= 4t, \\ t+1+(t+1)(1-t) + x - y(t+1) &= 0. \end{aligned}$$

Solving this pair of equations yields the stated expression for the multiplicities. □

We will use these results to show that there is a unique strongly regular graph with these parameters for $t \in \{1, 2, 4\}$. Throughout we will let X be the point graph of the generalized quadrangle, and X_1 and X_2 will denote the first and second subconstituents of X relative to an arbitrary vertex.

The three proofs below all follow a common strategy. First we show that the second subconstituent X_2 is uniquely determined from its spectrum. Then it remains only to determine the edges between X_1 and X_2. Because the first subconstituent of any vertex consists of $t+1$ disjoint edges, each vertex of X_1 is adjacent to t vertex-disjoint edges in X_2. If xy is an edge of X_1, then the t edges adjacent to x together with the t edges adjacent to y form a one-factor of X_2. Since $a = 1$, the graph X has no induced 4-cycles. This implies that the endpoints of the t edges adjacent to x have vertex-disjoint neighbourhoods. We say that a one-factor of size $2t$ has a *proper partition* if it can be divided into two sets of t edges such that the endpoints of the edges in each set have vertex-disjoint neighbourhoods. If X_2 is connected, then a one-factor has at most one proper partition.

In each of the following proofs we show that every edge of X_2 lies in a unique one-factor with a proper partition, and thus find exactly $t+1$ one-factors of X_2 with proper partitions. Every way of assigning the $t+1$ one-factors to the $t+1$ edges of X_1 is equivalent. The one-factor assigned to the edge xy has a unique proper partition. It is clear that both ways of assigning the two parts of the one-factor to x and y yield isomorphic graphs.

Lemma 10.9.2 *The graph $L(K_{3,3})$ is the unique strongly regular graph with parameters* $(9, 4, 1, 2)$.

Proof. Let X be a strongly regular graph with parameters $(9, 4, 1, 2)$. Every second subconstituent X_2 is a connected graph with valency two on four vertices, and so is C_4. Every edge of C_4 lies in a unique one-factor, and so in a unique one-factor with a proper partition. □

Lemma 10.9.3 *The graph $\overline{L(K_6)}$ is the unique strongly regular graph with parameters* $(15, 6, 1, 3)$.

Proof. The second subconstituent X_2 is a connected cubic graph on 8 vertices. By Lemma 10.9.1 we find that its spectrum is symmetric, and therefore X_2 is bipartite. From this we can see that X_2 cannot have diameter two, and therefore it has diameter at least three. By considering two vertices at distance three and their neighbours, it follows quickly that X_2 is the cube.

Consider any edge $e = uv$ of the cube. There is only one edge that does not contain u or v or any of their neighbours. This pair of edges can be completed uniquely to a one-factor, and so every edge lies in a unique one-factor with a proper partition. □

In Section 11.7 we will see a strongly regular graph with parameters $(27, 16, 10, 8)$ known as the Schläfli graph.

Lemma 10.9.4 *The complement of the Schläfli graph is the unique strongly regular graph with parameters* $(27, 10, 1, 5)$.

Proof. The second subconstituent X_2 is a connected graph on 16 vertices with valency 5. Using Lemma 10.9.1 we find that X_2 has exactly three eigenvalues, and so is strongly regular with parameters $(16, 5, 0, 2)$. We showed in Section 10.6 that the Clebsch graph is the only strongly regular graph with these parameters, and so X_2 is the Clebsch graph.

Let uv be an edge of the Clebsch graph. The non-neighbours of u form a copy P of the Petersen graph, and the neighbours of v form an independent set S of size four in P. This leaves three edges, all in P, that together with uv form a set of four vertex-disjoint edges. This set of four edges can be uniquely completed to a one-factor of the Clebsch graph by taking the four edges each containing a vertex of S, but not lying in P. Therefore, every edge lies in a unique one-factor with a proper partition. □

Corollary 10.9.5 *There is a unique generalized quadrangle of each order* $(2, 1)$, $(2, 2)$, *and* $(2, 4)$.

Proof. We have shown that the point graph of a generalized quadrangle of these orders is uniquely determined. Therefore, it will suffice to show that the generalized quadrangle can be recovered from its point graph. If X is a strongly regular graph with parameters $(6t+3, 2t+2, 1, t+1)$, then define an incidence structure whose points are the vertices of X and whose lines are the triangles of X. It is routine to confirm that the properties of the strongly regular graph imply that the incidence structure satisfies the axioms for a generalized quadrangle. Therefore, the previous three results imply that there is a unique generalized quadrangle of each order $(2, 1)$, $(2, 2)$, and $(2, 4)$. □

We will use the results of the section in Chapter 12 to characterize the graphs with smallest eigenvalue at least -2.

10.10 Quasi-Symmetric Designs

A 2-design $\mathcal{D}$ is *quasi-symmetric* if there are constants ℓ_1 and ℓ_2 such that any two distinct blocks of $\mathcal{D}$ have exactly ℓ_1 or ℓ_2 points in common. For example, a Steiner triple system is a quasi-symmetric design with (ℓ_1, ℓ_2) equal to $(0, 1)$. We call the integers ℓ_i the *intersection numbers* of the design. Our next result provides the reason we wish to consider this class of designs.

Lemma 10.10.1 *Let $\mathcal{D}$ be a quasi-symmetric* 2-(v, k, λ) *design with intersection numbers ℓ_1 and ℓ_2. Let X be the graph with the blocks of $\mathcal{D}$ as its* *vertices, and with two blocks adjacent if and only if they have exactly ℓ_1 points in common. If X is connected, then it is strongly regular.*

Proof. Suppose that $\mathcal{D}$ has b blocks and that each point lies in r blocks. If N is the $v \times b$ incidence matrix of $\mathcal{D}$, then from the results in Section 5.10 we have

$$NN^T = (r - \lambda)I + \lambda J$$

and

$$NJ = rJ, \quad N^T J = kJ.$$

Let A be the adjacency matrix of X. Since $\mathcal{D}$ is quasi-symmetric, we have

$$N^T N = kI + \ell_1 A + \ell_2(J - I - A) = (k - \ell_2)I + (\ell_1 - \ell_2)A + \ell_2 J.$$

Since $N^T N$ commutes with J, it follows that A commutes with J, and therefore X is a regular graph.

We now determine the eigenvalues of $N^T N$. The vector $\mathbf{1}$ is an eigenvector of $N^T N$ with eigenvalue rk, and so

$$rk\mathbf{1} = (k - \ell_2 + b\ell_2)\mathbf{1} + (\ell_1 - \ell_2)A\mathbf{1},$$

from which we see that $\mathbf{1}$ is an eigenvector for A, and hence the valency of X is

$$\frac{rk - k + \ell_2 - b\ell_2}{\ell_1 - \ell_2}.$$

Because $N^T N$ is symmetric, we can assume that the remaining eigenvectors are orthogonal to $\mathbf{1}$. Suppose that x is such an eigenvector with eigenvalue θ. Then

$$\theta x = (k - \ell_2)x + (\ell_1 - \ell_2)Ax,$$

and so x is also an eigenvector for A with eigenvalue

$$\frac{\theta - k + \ell_2}{\ell_1 - \ell_2}.$$

By Lemma 8.2.4, the matrices NN^T and $N^T N$ have the same nonzero eigenvalues with the same multiplicities. Since $NN^T = (r-\lambda)I + \lambda J$, we see that it has eigenvalues rk with multiplicity one, and $r - \lambda$ with multiplicity $v - 1$. Therefore, $N^T N$ has eigenvalues rk, $r - \lambda$, and 0 with respective multiplicities 1, $v - 1$, and $b - v$.

Hence the remaining eigenvalues of A are

$$\frac{r - \lambda - k + \ell_2}{\ell_1 - \ell_2}, \qquad \frac{\ell_2 - k}{\ell_1 - \ell_2},$$

with respective multiplicities $v - 1$ and $b - v$.

We have shown that X is a regular graph with at most three eigenvalues. If X is connected, then it has exactly three eigenvalues, and so it is strongly regular by Lemma 10.2.1. □

It is possible that a graph obtained from a quasi-symmetric design is not connected. This occurs when the valency of X coincides with one of the other eigenvalues, and so is not simple. If $\mathcal{D}$ is the design obtained by taking two or more copies of the same symmetric design, then X is complete multipartite or a disjoint union of cliques.

10.11 The Witt Design on 23 Points

The Witt design on 23 points is a 4-(23, 7, 1) design. It is one of the most remarkable structures in all of combinatorics.

The design can be described as follows. Over $GF(2)$, the polynomial $x^{23} - 1$ factors as

$$(x-1)g(x)h(x),$$

where

$$g(x) = x^{11} + x^9 + x^7 + x^6 + x^5 + x + 1$$

and

$$h(x) = x^{11}g(x^{-1}) = x^{11} + x^{10} + x^6 + x^5 + x^4 + x^2.$$

Both $g(x)$ and $h(x)$ are irreducible polynomials of degree 11. Let R denote the ring of polynomials over $GF(2)$, modulo $x^{23} - 1$, and let C be the ideal in this ring generated by $g(x)$, that is, all the polynomials in R divisible by $g(x)$. (In this case, though not in general, it suffices to take just the powers of $g(x)$ modulo $x^{23} - 1$.) Each element of R is represented by a polynomial over $GF(2)$ with degree at most 22. In turn, any such polynomial f can be represented by a vector of length 23 over $GF(2)$ with the ith coordinate position (counting from 0) being the coefficient of x^i in f. Though we will not prove it, the ideal C contains 2048 polynomials, whose associated vectors form the binary Golay code. Exactly 253 of these vectors have support of size 7, and the supports of these vectors form a 4-(23, 7, 1) design known as the *Witt design* on 23 points.

It is known that the Witt design is the only design with these parameters, but we will not need this fact and shall not prove it. However, we will prove that this design is quasi-symmetric with intersection numbers 1 and 3.

First, as in Section 5.10, we compute that a 4-(23, 7, 1) design has 253 blocks, 77 blocks on each point and 23 blocks on each pair of points; that is,

$$b = 253, \quad r = 77, \quad \lambda_2 = 21.$$

Let B be an arbitrarily chosen block of the 4-(23, 7, 1) design $\mathcal{D}$. By relabelling the points if necessary, we may assume that $B = \{1, 2, 3, 4, 5, 6, 7\}$. For $i + j \leq 7$, define the parameter $\lambda_{i,j}$ to be the number of blocks of $\mathcal{D}$

that contain the first $i-j$ points of B, but none of the next j points. Thus we have

$$\lambda_{0,0}=253, \quad \lambda_{1,0}=77, \quad \lambda_{2,0}=21, \quad \lambda_{3,0}=5,$$

and $\lambda_{i,0}=1$ if $4 \le i \le 7$, since no two distinct blocks have four points in common.

Given these values, we can compute the remaining ones, because if $i, j \ge 1$ and $i+j \le 7$, then

$$\lambda_{i,j}=\lambda_{i-1,j-1}-\lambda_{i,j-1},$$

the proof of which we leave as an exercise. For any chosen initial block B, the values of $\lambda_{i,j}$ are given by the following lower triangular matrix:

$$\begin{pmatrix} 253 \\ 77 & 176 \\ 21 & 56 & 120 \\ 5 & 16 & 40 & 80 \\ 1 & 4 & 12 & 28 & 52 \\ 1 & 0 & 4 & 8 & 20 & 32 \\ 1 & 0 & 0 & 4 & 4 & 16 & 16 \\ 1 & 0 & 0 & 0 & 4 & 0 & 16 & 0 \end{pmatrix}.$$

We note that for a general t-design, the values $\lambda_{i,j}$ are guaranteed to be independent of the initial block only when $i+j \le t$. However, the Witt design has $\lambda_4=1$, and this is sufficient for the result to hold as stated.

From the last row we infer that any two blocks meet in either one or three points, and therefore $\mathcal{D}$ is quasi-symmetric. By the results of the previous section, the graph on the blocks of $\mathcal{D}$ where two blocks are adjacent if they intersect in one point is strongly regular with parameters

$$(253, 112, 36, 60)$$

and eigenvalues 112, 2, and -26.

In the next chapter we show how to use the Witt design to construct a strongly regular graph on 275 vertices with strongly regular subconstituents on 112 and 162 vertices.

10.12 The Symplectic Graphs

In this section we examine an interesting family $\mathrm{Sp}(2r)$ of strongly regular graphs known as the symplectic graphs; we first encountered these graphs in Section 8.10.

For any $r>0$, let N be the $2r \times 2r$ block diagonal matrix with r blocks of the form

$$\begin{pmatrix} 0 & 1 \\ 1 & 0 \end{pmatrix}.$$

The symplectic graph Sp$(2r)$ is the graph whose vertex set is the set of all nonzero vectors in $GF(2)^{2r}$, with adjacency defined by

$$x \sim y \quad \text{if and only if} \quad x^T N y = 1$$

with all calculations over $GF(2)$. The name "symplectic graph" arises because the function

$$f(x, y) = x^T N y$$

is known as a *symplectic form*. Two vectors x and y are *orthogonal* with respect to f if $f(x, y) = 0$. Therefore, Sp$(2r)$ is the nonorthogonality graph of $GF(2)^{2r} \setminus 0$ with respect to the symplectic form f.

Lemma 10.12.1 *The graph* Sp$(2r)$ *is strongly regular with parameters*

$$\left(2^{2r} - 1,\ 2^{2r-1},\ 2^{2r-2},\ 2^{2r-2}\right),$$

and it has eigenvalues

$$2^{2r-1},\ 2^{r-1},\ \text{and}\ -2^{r-1}.$$

Proof. In $GF(2)^{2r}$, the number of vectors orthogonal to a given vector y is the number of solutions to the equation $y^T N x = 0$. Now, if y is nonzero, the rank of $y^T N$ is one, and hence its null space has dimension $2r - 1$. Therefore, the number of nonzero vectors that are not orthogonal to y is $2^{2r} - 2^{2r-1} = 2^{2r-1}$. If x and z are distinct nonzero vectors in $GF(2)^{2r}$, then the vectors mutually orthogonal to both form a subspace of dimension $2r - 2$, which leaves 2^{2r-2} vectors mutually nonorthogonal to both x and y. Therefore, Sp$(2r)$ has parameters as given. The eigenvalues of Sp$(2r)$ follow directly from the results of Section 10.2. □

By Theorem 8.11.1, every reduced graph X of binary rank $2r$ is an induced subgraph of Sp$(2r)$. One implication of this is that the chromatic number of X is no more than the chromatic number of Sp$(2r)$. We will use the algebraic techniques of Section 9.6 to bound the chromatic number of Sp$(2r)$. We start by bounding the size of an independent set in Sp$(2r)$. The graph Sp$(2r)$ is k-regular with $k = 2^{2r-1}$ and has minimum eigenvalue $\tau = -2^{r-1}$. By Lemma 9.6.2 we have

$$\alpha(\text{Sp}(2r)) \le \frac{n(-\tau)}{k - \tau} = \frac{(2^{2r} - 1)2^{r-1}}{2^{2r-1} + 2^{r-1}} = 2^r - 1.$$

We can easily find independent sets of this size. Any set of r linearly independent mutually orthogonal vectors generates a subspace that contains $2^r - 1$ mutually orthogonal nonzero vectors (check this!). One such set is given by the standard basis vectors $e_2, e_4, \ldots, e_{2r}$.

Therefore, the chromatic number of Sp$(2r)$ satisfies the inequality

$$\chi(\text{Sp}(2r)) \ge n/\alpha(\text{Sp}(2r)) = 2^r + 1.$$

The chromatic number of Sp$(2r)$ is equal to $2^r + 1$ if and only if the vertex set can be partitioned into independent sets of this maximum size.

The graph $\mathrm{Sp}(4)$ is a $(15, 8, 4, 4)$ strongly regular graph, and therefore isomorphic to $L(K_6)$. An independent set of size three in $L(K_6)$ is a 1-factor of K_6, and a partition into independent sets of size three is a 1-factorization of K_6. The results of Section 4.7 show that 1-factorizations of K_6 exist, and hence $\chi(\mathrm{Sp}(4)) = 5$. Although we shall not prove it, there is always such a partition, and so we get the following result.

Theorem 10.12.2 *The chromatic number of* $\mathrm{Sp}(2r)$ *is* $2^r + 1$. □

Corollary 10.12.3 *Let X be a graph with binary rank $2r$. Then* $\chi(X) \leq 2^r + 1$.

Proof. Duplicating vertices or adding isolated vertices does not alter the chromatic number of a graph. Therefore, we can assume without loss of generality that X is a reduced graph. Thus it is an induced subgraph of $\mathrm{Sp}(2r)$ and can be coloured with at most $2^r + 1$ colours. □

Exercises

1. If X is a strongly regular graph with k and $\bar{k} = n - k - 1$ coprime, show that X is imprimitive. Deduce that if p is a prime, all strongly regular graphs on $p + 1$ vertices are imprimitive.
2. What are the parameters of mK_n and $\overline{mK_n}$?
3. Determine the strongly regular graphs with $c = k$.
4. Prove that $m_\theta m_\tau(\theta - \tau)^2 = nk\bar{k}$.
5. Let X be an (n, k, a, c) strongly regular graph with eigenvalues k, θ, and τ. Let ℓ be $n - k - 1$, which is the number of vertices in the second subconstituent of X. The following identities express various parameters of X as functions of the eigenvalues. Verify them.
$$\begin{aligned} a &= k + \theta + \tau + \theta\tau, \\ c &= k + \theta\tau, \\ n &= \frac{(k-\theta)(k-\tau)}{k+\theta\tau}, \\ \ell &= -\frac{k(\theta+1)(\tau+1)}{k+\theta\tau}, \\ m_\theta &= \frac{k(k-\tau)(\tau+1)}{(k+\theta\tau)(\tau-\theta)}. \end{aligned}$$
6. Let X be a k-regular graph on n vertices and, if $u \in V(X)$, let t_u denote the number of triangles that contain u. Let v be a fixed vertex in X and let π be the partition of $V(X)$ with three cells: $\{v\}$, the

neighbours of v, the vertices distinct from and not adjacent to v. If $A = A(X)$, show that the quotient A/π is given by

$$A/\pi = \begin{pmatrix} 0 & k & 0 \\ 1 & \frac{2t_v}{k} & \frac{k^2-k-2t_v}{k} \\ 0 & \frac{k^2-k-2t_v}{n-1-k} & \frac{nk-2k^2+2t_v}{n-1-k} \end{pmatrix}.$$

If $\theta_1, \dots, \theta_n$ are the eigenvalues of X in nonincreasing order, show that

$$k\frac{nk - 2k^2 + 2t_v}{n-1-k} = -\det(A/\pi) \leq k\theta_2\theta_n.$$

Deduce from this that

$$2t_v \leq 2k^2 - nk - \theta_2\theta_n(n-1-k)$$

and that if equality holds for each vertex, then X is strongly regular.

7. Let X be a Moore graph of diameter two and valency k. Compute the multiplicities of the eigenvalues of X, and hence show that $k \in \{2, 3, 7, 57\}$.

8. Prove that an orthogonal array $OA(k, n)$ is equivalent to a set of $k-2$ mutually orthogonal Latin squares.

9. Prove that the graph of an $OA(k, n)$ is strongly regular with parameters

$$\left(n^2,\ (n-1)k,\ n-2+(k-1)(k-2),\ k(k-1)\right).$$

10. Let X be the graph of an $OA(k, n)$. Show that the size of a maximum clique in X is n.

11. Let X be the graph of an $OA(k, n)$. If X has no independent sets of size n, then show that $\chi(X) \geq n + 2$.

12. Call two triples from a fixed set of v points adjacent if they have exactly one common point, and let X denote the resulting graph on the triples of a set of size v. Show that X is strongly regular if $v = 5$, $v = 7$, or $v = 10$, and determine its parameters.

13. Let Ω denote the set of all partitions of a set of nine elements into three triples. If π and σ are two of these partitions, define their product to be the partition whose cells are all possible nonempty intersections of the cells of π with those of σ. Define two elements of Ω to be adjacent if their product contains exactly five cells. Show that the graph on Ω with this adjacency relation is strongly regular, and determine its parameters.

14. Show that C_5 is the only primitive strongly regular graph with all first subconstituents empty and all second subconstituents complete.

15. Show that if there is a thick generalized quadrangle of order $(3, t)$, then $t \in \{3, 5, 6, 9\}$.

16. Show that if there is a thick generalized quadrangle of order $(4, t)$, then $t \in \{2, 4, 6, 8, 11, 12, 16\}$.

17. If X is the point graph of a generalized quadrangle of order (s, t), then show that the second subconstituent X_2 has spectrum

$$\{(s-1)(t+1)^{(1)},\ (s-1)^{(x)},\ (s-t-1)^{(s-1)(t+1)},\ (-t-1)^{(y)}\}$$

where

$$x = \frac{s^2(t^2-1)}{s+t}, \quad y = \frac{t(s-1)(s^2-t)}{s+t}.$$

Under what conditions is the second subconstituent strongly regular?

18. Suppose we have a finite group of people such that any two have exactly one friend in common. Show that there must be a politician. (A politician is a person who is everyone's friend. Friendship is understood to be a symmetric, irreflexive relation.)

19. Show that if there exists a quasi-symmetric 2-(v, k, λ) design with intersection numbers ℓ_1 and ℓ_2, then $\ell_1 - \ell_2$ divides $r - \lambda$.

20. Determine the parameters of the strongly regular graphs obtained from the blocks of the Witt design on 23 points.

21. There are 77 blocks in the Witt design on 23 points that contain a given point. Show that the set of blocks we get by deleting the common point from each of these 77 blocks is a 3-$(22, 6, 1)$ design. Show that this is a quasi-symmetric design with intersection numbers 0 and 2, and determine the parameters and eigenvalues of the associated strongly regular graphs.

Notes

Further information on strongly regular graphs can be found in [2, 3, 5].

Our treatment of the Krein bounds in Section 10.7 is equivalent to that offered by Thas in [8].

Quasi-symmetric designs are studied at length in [7].

The solution to Exercise 13 is in [6]. Exercise 6 is based on Theorem 7.1 in [4]. The result of Exercise 18 is sometimes called the "friendship theorem." It is easily seen to be equivalent to the fact, due to Baer [1], that a polarity of a finite projective plane must have absolute points. (The adjacency matrix of the friendship relation is the incidence matrix of a projective plane, where each line is the set of friends of some person.)

We do not know whether the number of primitive triangle-free strongly regular graphs is finite. The largest known such graph is the Higman–Sims graph, with parameters $(100, 22, 0, 6)$.

References

[1] R. BAER, *Polarities in finite projective planes*, Bull. Amer. Math. Soc., 52 (1946), 77–93.

[2] A. E. BROUWER AND J. H. VAN LINT, *Strongly regular graphs and partial geometries*, in Enumeration and design (Waterloo, Ont., 1982), Academic Press, Toronto, Ont., 1984, 85–122.

[3] P. J. CAMERON AND J. H. VAN LINT, *Designs, Graphs, Codes and their Links*, Cambridge University Press, Cambridge, 1991.

[4] W. H. HAEMERS, *Interlacing eigenvalues and graphs*, Linear Algebra Appl., 226/228 (1995), 593–616.

[5] X. L. HUBAUT, *Strongly regular graphs*, Discrete Math., 13 (1975), 357–381.

[6] R. MATHON AND A. ROSA, *A new strongly regular graph*, J. Combin. Theory Ser. A, 38 (1985), 84–86.

[7] M. S. SHRIKHANDE AND S. S. SANE, *Quasi-Symmetric Designs*, Cambridge University Press, Cambridge, 1991.

[8] J. A. THAS, *Interesting pointsets in generalized quadrangles and partial geometries*, Linear Algebra Appl., 114(115) (1989), 103–131.

11
Two-Graphs

The problem that we are about to discuss is one of the founding problems of algebraic graph theory, despite the fact that at first sight it has little connection to graphs. A *simplex* in a metric space with distance function d is a subset S such that the distance $d(x, y)$ between any two distinct points of S is the same. In $\mathbb{R}^d$, for example, a simplex contains at most $d+1$ elements. However, if we consider the problem in real projective space then finding the maximum number of points in a simplex is not so easy. The points of this space are the lines through the origin of $\mathbb{R}^d$, and the distance between two lines is determined by the angle between them. Therefore, a simplex is a set of lines in $\mathbb{R}^d$ such that the angle between any two distinct lines is the same. We call this a set of *equiangular lines*. In this chapter we show how the problem of determining the maximum number of equiangular lines in $\mathbb{R}^d$ can be expressed in graph-theoretic terms.

11.1 Equiangular Lines

We can represent a line in $\mathbb{R}^d$ by giving a unit vector x that spans it, and so a set of lines can be represented by a set $\Omega = \{x_1, \dots, x_n\}$ of unit vectors. Of course, $-x$ represents the same line as x, so Ω is not unique. If Ω represents an equiangular set of lines where the angle between any two distinct lines is θ, then for $i \neq j$,

$$x_i^T x_j = \pm \cos\theta.$$

We can get an example by taking the 28 unit vectors in $\mathbb{R}^8$ of the form

$$x_i^T = \sqrt{1/24}\,(3, 3, -1, -1, -1, -1, -1, -1)$$

with two entries equal to 3 and the remaining six equal to -1. For $i \neq j$, we have $x_i^T x_j = \pm\frac{1}{3}$, where the positive sign is taken if and only if x_i and x_j have an entry of 3 in the same coordinate. Therefore, we have a set of 28 equiangular lines in $\mathbb{R}^8$. Since all of the vectors are orthogonal to $\mathbf{1}$, they lie in the 7-dimensional subspace $\mathbf{1}^\perp$, so in fact there are 28 equiangular lines in $\mathbb{R}^7$.

Given a set $\Omega = \{x_1, \ldots, x_n\}$ of vectors in $\mathbb{R}^d$, let U be the $d \times n$ matrix with the elements of Ω as its columns. Then

$$G = U^T U$$

is the Gram matrix of the vectors in Ω. Thus G is a symmetric positive semidefinite matrix with the same rank as U. Conversely, given a symmetric positive semidefinite $n \times n$ matrix G of rank d, it is possible to find a $d \times n$ matrix U such that $G = U^T U$ (see the proof of Lemma 8.6.1). Therefore, for our purposes it suffices to represent Ω by its Gram matrix G.

If Ω is a set of unit vectors representing a set of equiangular lines with $x_i^T x_j = \pm\alpha$, then its Gram matrix G has the form

$$G = I + \alpha S,$$

where S is a symmetric $(0, \pm 1)$-matrix with all diagonal entries zero and all off-diagonal entries nonzero. By taking -1 to represent adjacent and 1 to represent nonadjacent we can view this as a kind of nonstandard "adjacency matrix" of a graph X. This matrix is called the *Seidel matrix* of X and is related to the usual adjacency matrix $A(X)$ by

$$S(X) = J - I - 2A(X).$$

Suppose now that we start with a graph X on n vertices and form its Seidel matrix S. Then since $\operatorname{tr} S = 0$ and $S \neq 0$, the least eigenvalue of S is negative. If this eigenvalue is $-\alpha$, then

$$I + \frac{1}{\alpha} S$$

is a positive semidefinite matrix. If its rank is d, then it is the Gram matrix of a set of n equiangular lines in $\mathbb{R}^d$ with mutual cosine $\pm 1/\alpha$. Therefore, the geometric problem of finding the least integer d such that there are n equiangular lines in $\mathbb{R}^d$ is equivalent to the graph-theoretic problem of finding graphs X on n vertices such that the multiplicity of the least eigenvalue of $S(X)$ is as large as possible.

We describe one example. If $X = \overline{L(K_8)}$, then by Lemma 8.2.5, the eigenvalues of $A(X)$ are 15, 1, and -5 with multiplicities 1, 7, and 20, respectively. Hence the eigenvalues of $S(X)$ are 9 and -3 with multiplicities

7 and 21, respectively. Therefore,

$$I + \frac{1}{3}S(X)$$

is the Gram matrix of a set of 28 equiangular lines in $\mathbb{R}^7$. It is easy to see that the 28 lines given at the start of this section yield the graph $\overline{L(K_8)}$, and we will see later that this is the maximum number possible in $\mathbb{R}^7$.

Choosing different sets of vectors to represent the set of equiangular lines (that is, replacing some of the x_i with $-x_i$) will yield different graphs. This is explored in more depth in Section 11.5.

11.2 The Absolute Bound

In this section we will derive an upper bound on the number of equiangular lines in $\mathbb{R}^d$. It is called the absolute bound because the expression is independent of the angle between the lines.

Let x be a unit vector in $\mathbb{R}^d$ and let $X = xx^T$. Then X is a symmetric $d \times d$ matrix and $X^2 = X$. Therefore, X represents orthogonal projection onto its column space, which is the line spanned by x. Furthermore, replacing x by $-x$ does not change the matrix X (in general, the form of a projection onto a subspace does not depend on the basis chosen for the subspace).

If y is a second unit vector in $\mathbb{R}^d$ and $Y = yy^T$, then

$$XY = xx^T yy^T = (x^T y)xy^T,$$

and so

$$\operatorname{tr}(XY) = (x^T y)^2.$$

Therefore, the trace of XY is the square of the cosine of the angle between the lines spanned by x and y. Also,

$$\operatorname{tr}(X) = \operatorname{tr}(xx^T) = \operatorname{tr}(x^T x) = 1.$$

Theorem 11.2.1 (The Absolute Bound) *Let $X_1, \ldots, X_n$ be the projections onto a set of n equiangular lines in $\mathbb{R}^d$. Then these matrices form a linearly independent set in the space of symmetric matrices, and consequently $n \leq \binom{d+1}{2}$.*

Proof. Let α be the cosine of the angle between the lines. If $Y = \sum_i c_i X_i$, then

$$\begin{aligned}\operatorname{tr}(Y^2) &= \sum_{i,j} c_i c_j \operatorname{tr}(X_i X_j) \\ &= \sum_i c_i^2 + \sum_{i,j:i \neq j} c_i c_j \alpha^2\end{aligned}$$

$$= \alpha^2 \left(\sum_i c_i\right)^2 + (1-\alpha^2)\sum_i c_i^2.$$

It follows that $\operatorname{tr}(Y^2) = 0$ if and only if $c_i = 0$ for all i, so the X_i are linearly independent. The space of symmetric $d \times d$ matrices has dimension $\binom{d+1}{2}$, so the result follows. □

11.3 Tightness

We have just seen that if there is a set of n equiangular lines in $\mathbb{R}^d$, then $n \le \binom{d+1}{2}$. If equality holds, then the projections $X_1, \dots, X_n$ onto the lines form a basis for the space of symmetric $d \times d$ matrices. In particular, this means that there are scalars $c_1, \dots, c_n$ such that

$$I = \sum_i c_i X_i.$$

The fact that I is in the span of the projections $X_1, \dots, X_n$ has significant consequences whether or not the absolute bound is met.

Lemma 11.3.1 *Suppose that $X_1, \dots, X_n$ are the projections onto a set of equiangular lines in $\mathbb{R}^d$ and that the cosine of the angle between the lines is α. If $I = \sum_i c_i X_i$, then $c_i = d/n$ for all i and*

$$n = \frac{d - d\alpha^2}{1 - d\alpha^2}.$$

The Seidel matrix determined by any set of n unit vectors spanning these lines has eigenvalues

$$-\frac{1}{\alpha}, \quad \frac{n-d}{d\alpha}$$

with multiplicities $n-d$ and d, respectively. If $n \neq 2d$, then α is an integer.

Proof. For any j we have

$$X_j = \sum_i c_i X_i X_j,$$

and so by taking the trace we get

$$1 = \operatorname{tr}(X_j) = \sum_i c_i \operatorname{tr}(X_i X_j) = (1-\alpha^2)c_j + \alpha^2 \sum_i c_i. \tag{11.1}$$

The first consequence of this is that all the c_i's are equal. Since $d = \operatorname{tr} I = \sum_i c_i$, we see that $c_i = d/n$ for all i. Substituting this back into (11.1) gives the stated expression for d.

Now, let $x_1, \dots, x_n$ be a set of unit vectors representing the equiangular lines, so $X_i = x_i x_i^T$. Let U be the $d \times n$ matrix with $x_1, \dots, x_n$ as its columns. Then

$$UU^T = \sum_i x_i x_i^T = \sum_i X_i = \frac{n}{d} I$$

and

$$U^T U = I + \alpha S.$$

By Lemma 8.2.4, UU^T and U^TU have the same nonzero eigenvalues with the same multiplicities. We deduce that $I + \alpha S$ has eigenvalues 0 with multiplicity $n-d$ and n/d with multiplicity d. Therefore, the eigenvalues of S are as claimed. Since the entries of S are integers, the eigenvalues of S are algebraic integers. Therefore, either they are integers or are algebraically conjugate, and so have the same multiplicity. If $n \neq 2d$, the multiplicities are different, so $1/\alpha$ is an integer. □

Now, suppose the absolute bound is tight and that there is a set of $n = \binom{d+1}{2}$ equiangular lines in $\mathbb{R}^d$. Then the previous result shows that $d+2 = 1/\alpha^2$. If $d \neq 3$, then $n \neq 2d$, and so $d+2$ must be a perfect square. So we get the following table:

d	n	$1/\alpha$
3	6	$\sqrt{5}$
7	28	3
14	105	4
23	276	5

Six equiangular lines in $\mathbb{R}^3$ can be constructed by taking the six diagonals of the icosahedron. We have already seen a collection of 28 equiangular lines in $\mathbb{R}^7$, and in Section 11.8 we will see a collection of 276 equiangular lines in $\mathbb{R}^{23}$. Later we will see that $1/\alpha$ must be an odd integer, so there cannot be a set of 105 equiangular lines in $\mathbb{R}^{14}$. Are there further examples where the absolute bound is tight? We do not know.

11.4 The Relative Bound

In this section we consider the relative bound, which bounds the maximum number of equiangular lines in $\mathbb{R}^d$ as a function of both d and the cosine of the angle between the lines.

Lemma 11.4.1 *Suppose that there are n equiangular lines in $\mathbb{R}^d$ and that α is the cosine of the angle between them. If $\alpha^{-2} > d$, then*

$$n \leq \frac{d - d\alpha^2}{1 - d\alpha^2}.$$

If $X_1, \ldots, X_n$ are the projections onto these lines, then equality holds if and only if $\sum_i X_i = (n/d)I$.

Proof. Put

$$Y := I - \frac{d}{n}\sum_i X_i.$$

Because Y is symmetric, we have $\operatorname{tr}(Y^2) \geq 0$, with equality if and only if $Y = 0$. Now,

$$Y^2 = I - \frac{2d}{n}\sum_i X_i + \frac{d^2}{n^2}\left(\sum_i X_i\right)^2,$$

so

$$\operatorname{tr}(Y^2) = d - 2d + \frac{d^2}{n^2}(n + \alpha^2 n(n-1)) \geq 0.$$

This reduces to

$$d - d\alpha^2 \geq n(1 - d\alpha^2),$$

which, provided that $1 - d\alpha^2$ is positive, yields the result. Equality holds if and only if $\operatorname{tr}(Y^2) = 0$, in which case $Y = 0$ and $\sum_i X_i = (n/d)I$. □

The Petersen graph provides an example where the relative bound is tight. The eigenvalues of the Seidel matrix of the Petersen graph are ± 3 with equal multiplicity 5. Thus we can find 10 equiangular lines in $\mathbb{R}^5$ where the cosine of the angle between the lines is equal to $-\frac{1}{3}$. This meets the relative bound, but not the absolute bound.

If equality holds in the relative bound, then I is in the span of the projections X_i, and so the results of Lemma 11.3.1 hold. In particular, the Seidel matrix determined by the set of lines has only two eigenvalues. Conversely, if S is a Seidel matrix with two eigenvalues, then the set of equiangular lines that it determines meets the relative bound.

11.5 Switching

Given a set of n equiangular lines, there are 2^n possible choices for the set $\Omega = \{x_1, \ldots, x_n\}$ of unit vectors representing those lines. Different choices for Ω may have different Gram matrices, hence yield different graphs. In this section we consider the relationship between these graphs. Let σ be a subset of $\{1, \ldots, n\}$ and suppose that we form Ω' from Ω by replacing each vector x_i by $-x_i$ if $i \in \sigma$. The Gram matrix for Ω' is obtained from the Gram matrix for Ω by multiplying the rows and columns corresponding to σ by -1. If X is the graph obtained from Ω, and X' is the graph obtained from Ω', then X' arises from X by changing all the edges between σ and

$V(X) \setminus \sigma$ to nonedges, and all the nonedges between σ and $V(X) \setminus \sigma$ to edges. This operation is called *switching* on the subset σ.

If X is a graph and $\sigma \subseteq V(X)$, then X^σ denotes the graph obtained from X by switching on σ. If $\sigma, \tau \subseteq V(X)$, then

$$X^\sigma = X^{V(X)\setminus\sigma}$$

and

$$(X^\sigma)^\tau = (X^\tau)^\sigma = X^{\sigma\Delta\tau},$$

where Δ is the symmetric difference operator.

The collection of graphs that can be obtained from X by switching on every possible subset of $V(X)$ is called the *switching class* of X. A switching class of graphs is also known as a *two-graph*. It may seem unnecessary to have two names for the same thing, but "two-graph" is also used to refer to other combinatorial objects that are equivalent to a switching class of graphs. Thus a set of equiangular lines in $\mathbb{R}^d$ determines a two-graph.

Certain graph-theoretical parameters are constant across all the graphs in a two-graph, and hence can usefully be regarded as parameters of the two-graph itself. For example, the next result shows that the Seidel matrices of all the graphs in a two-graph have the same eigenvalues; we call these the eigenvalues of the two-graph.

Lemma 11.5.1 *If X is a graph and σ is a subset of $V(X)$, then $S(X)$ and $S(X^\sigma)$ have the same eigenvalues.*

Proof. Let D be the diagonal matrix with $D_{uu} = -1$ if $u \in \sigma$ and 1 otherwise. Then $D^2 = I$, so D is its own inverse. Then

$$S(X^\sigma) = DS(X)D,$$

so $S(X)$ and $S(X^\sigma)$ are similar and have the same eigenvalues. □

If Y is isomorphic to X^σ for some σ, we say that X and Y are *switching equivalent*. If $N(v)$ denotes the neighbourhood of the vertex v, then $X^{N(v)}$ is a graph with the vertex v isolated. We denote by X_v the graph on $n-1$ vertices obtained from $X^{N(v)}$ by deleting the isolated vertex v, and say that X_v is obtained by *switching off* v. For any vertex v, there is a unique graph in the switching class with v isolated, so the collection of graphs $\{X_v : v \in V(X)\}$ is independent of the choice of X. Therefore, this collection of graphs is determined only by the two-graph; we call these graphs the *neighbourhoods* of the two-graph. This shows that determining switching equivalence is polynomially reducible to the graph isomorphism problem.

Given a graph X, we define the *switching graph* $\mathrm{Sw}(X)$ of X as follows: The vertex set of $\mathrm{Sw}(X)$ is $V(X) \times \{0,1\}$. If $u \sim v$ in X, then join $(u,0)$ to $(v,0)$ and $(u,1)$ to $(v,1)$, and if $u \not\sim v$, then join $(u,0)$ to $(v,1)$ and $(u,1)$ to $(v,0)$. The neighbourhoods of the vertices $(v,0)$ and $(v,1)$ are both isomorphic to X_v. Since $\mathrm{Sw}(X)$ is determined completely by any one of its neighbourhoods, we see that the switching graph is determined only by the two-graph, rather than the particular choice of X used to construct it. Therefore, X and Y are switching equivalent if and only if $\mathrm{Sw}(X)$ is isomorphic to $\mathrm{Sw}(Y)$.

If we consider the complement $\overline{X}$ of a graph X, it is straightforward to see that $S(\overline{X}) = -S(X)$. The neighbourhoods of the two-graph containing $\overline{X}$ are the complements of the neighbourhoods of the two-graph containing X.

11.6 Regular Two-Graphs

If a set of equiangular lines meets the absolute bound or the relative bound, then the associated two-graph has only two eigenvalues. This is a very strong condition: A symmetric matrix with only one eigenvalue must be a multiple of the identity. Motivated by this, we define a *regular two-graph* to be a two-graph with only two eigenvalues. The switching classes of the complete graph and the empty graph are regular two-graphs; the neighbourhoods of these two-graphs are all complete or empty, respectively. We refer to these two-graphs as trivial, and usually exclude them from discussion.

Theorem 11.6.1 *Let Φ be a nontrivial two-graph on $n+1$ vertices. Then the following are equivalent:*

(a) *Φ is a regular two-graph.*
(b) *All the neighbourhoods of Φ are regular graphs.*
(c) *All the neighbourhoods of Φ are (n, k, a, c) strongly regular graphs with $k = 2c$.*
(d) *One neighbourhood of Φ is an (n, k, a, c) strongly regular graph with $k = 2c$.*

Proof. (a) $\Rightarrow$ (b) Let S be the Seidel matrix of any neighbourhood of Φ. Then the matrix

$$T = \begin{pmatrix} 0 & \mathbf{1}^T \\ \mathbf{1} & S \end{pmatrix}$$

has two eigenvalues, and so it satisfies an equation of the form

$$T^2 + aT + bI = 0$$

for some constants a and b. Since

$$T^2 + aT + bI = \begin{pmatrix} n+b & \mathbf{1}^T S + a\mathbf{1}^T \\ S\mathbf{1} + a\mathbf{1} & J + S^2 + aS + bI \end{pmatrix},$$

we see that $S\mathbf{1} = -a\mathbf{1}$, which implies that S has constant row sum $-a$. Therefore, S is the Seidel matrix of a regular graph.

(b) $\Rightarrow$ (c) Let X be a neighbourhood of Φ and let

$$V(X) = \{v\} \cup N(v) \cup \overline{N}(v),$$

where $N(v)$ and $\overline{N}(v)$ are nonempty. Then $(X \cup K_1)^{N(v)} = Y \cup K_1$, and both X and Y are k-regular.

Let w be the isolated vertex in $X \cup K_1$. Then in $Y \cup K_1$, the vertex v is now isolated, w is adjacent to $N(v)$, and the edges between $N(v)$ and $\overline{N}(v)$ have been complemented.

Consider a vertex in $N(v)$, and suppose that it is adjacent to r vertices of $\overline{N}(v)$ in X. Then its valency in Y is

$$k + |\overline{N}(v)| - 2r,$$

and so $r = |\overline{N}(v)|/2$. Therefore, every vertex in $N(v)$ is adjacent to the same number of vertices in $\overline{N}(v)$, and hence to the same number of vertices in $N(v)$.

Now, consider a vertex in $\overline{N}(v)$, and suppose that it is adjacent to s vertices of $N(v)$ in X. Then its valency in Y is

$$k + |N(v)| - 2s,$$

and so $s = |N(v)|/2 = k/2$. Therefore, every vertex in $\overline{N}(v)$ is adjacent to $k/2$ vertices in $N(v)$.

As v was an arbitrarily chosen vertex of X, this shows that X is a strongly regular graph with $c = k/2$, and the claim follows.

(c) $\Rightarrow$ (d) This is obvious.

(d) $\Rightarrow$ (a) Let X be an (n, k, a, c) strongly regular graph with $k = 2c$. Let A be the adjacency matrix of X with eigenvalues k, θ and τ, and let $S = J - I - 2A$ be the Seidel matrix of X. Now, we wish to consider the eigenvalues of

$$T = \begin{pmatrix} 0 & \mathbf{1}^T \\ \mathbf{1} & S \end{pmatrix}.$$

If z is an eigenvector of S orthogonal to $\mathbf{1}$, then $\binom{0}{z}$ is an eigenvector of T with the same eigenvalue. Therefore, T has $n-1$ eigenvectors with eigenvalues $-2\theta - 1$ or $-2\tau - 1$.

The above partition of the matrix T is equitable, with quotient matrix

$$Q = \begin{pmatrix} 0 & n \\ 1 & n-1-2k \end{pmatrix}.$$

Therefore, any eigenvector of Q yields an eigenvector of T that is constant on the two cells of the partition, and so in particular is not among the $n-1$ eigenvectors that we have already found. Therefore, the remaining eigenvalues of T are precisely the two eigenvalues of Q.

Using $k - c = -\theta\tau$, $k = 2c$, and $a - c = \theta + \tau$ we can express all the parameters of X in terms of θ and τ, yielding

$$\begin{aligned} n &= -(2\theta + 1)(2\tau + 1), \\ k &= -2\theta\tau, \\ a &= \theta + \tau - \theta\tau, \\ c &= -\theta\tau. \end{aligned}$$

Therefore,

$$Q = \begin{pmatrix} 0 & -2(\theta + 1)(2\tau + 1) \\ 1 & -2(\theta + \tau + 1) \end{pmatrix},$$

which has eigenvalues $-2\theta - 1$ and $-2\tau - 1$, and so we can conclude that T has precisely two eigenvalues. □

Corollary 11.6.2 *A nontrivial regular two-graph has an even number of vertices.*

Proof. From the above proof, it follows that $n = -(4\theta\tau + 2(\theta + \tau) + 1)$. Because both $\theta\tau$ and $\theta + \tau$ are integers, this shows that n is odd; hence $n + 1$ is even. □

The Paley graphs are strongly regular graphs with $k = 2c$, so they provide a family of examples of regular two-graphs. Although the corresponding set of equiangular lines meets the relative bound, this yields only $4k+2$ lines in $\mathbb{R}^{2k+1}$. More generally, any conference graph will yield a regular two-graph.

Finally, we note that Theorem 11.6.1 can be used to find strongly regular graphs. Suppose we start with a strongly regular graph X with $k = 2c$. Then by forming the graph $X \cup K_1$ and switching off every vertex in turn, we construct other strongly regular graphs with $k = 2c$, which are sometimes nonisomorphic to X.

11.7 Switching and Strongly Regular Graphs

There is another connection between regular two-graphs and strongly regular graphs. This arises by considering when a regular two-graph contains a regular graph.

Theorem 11.7.1 *Let X be a k-regular graph on n vertices not switching equivalent to the complete or empty graph. Then $S(X)$ has two eigenvalues if and only if X is strongly regular and $k - n/2$ is an eigenvalue of $A(X)$.*

Proof. Any eigenvector of $A(X)$ orthogonal to $\mathbf{1}$ with eigenvalue θ is an eigenvector of $S(X)$ with eigenvalue $-2\theta - 1$, while $\mathbf{1}$ itself is an eigenvector of $S(X)$ with eigenvalue $n - 1 - 2k$. Therefore, if X is strongly regular with $k - n/2$ equal to θ or τ, then $S(X)$ has just two eigenvalues.

For the converse, suppose that X is a graph such that $S(X)$ has two eigenvalues. First we consider the case where X is connected. Since X is not complete, $A(X)$ has at least three distinct eigenvalues (Lemma 8.12.1). Since $S(X)$ has only two eigenvalues, this implies that $A(X)$ must have precisely three eigenvalues k, θ, and τ and also that $n-1-2k$ must equal either $-2\theta-1$ or $-2\tau-1$. Therefore, by Lemma 10.2.1, X is strongly regular, and $k-n/2$ is either θ or τ.

Now, suppose that X is not connected. Then there is an eigenvector of $A(X)$ with eigenvalue k orthogonal to $\mathbf{1}$. Hence $n-1-2k$ and $-1-2k$ are the two eigenvalues of $S(X)$. Therefore, every component of X has at most one eigenvalue θ other than k, and this eigenvalue must satisfy $n-1-2k=-1-2\theta$. Since X is nonempty, every component of X has exactly one further eigenvalue θ, and so is complete. Thus $\theta=-1$, $k=(n/2)-1$ and $X=2K_{(n/2)-1}$, which is easily seen to be switching equivalent to the complete graph. □

If $X=L(K_8)$, then X is a $(28,12,6,4)$ strongly regular graph with eigenvalues $\theta=4$ and $\tau=-2$, so the two-graph Φ containing X is regular. Switching off a vertex yields the *Schläfli graph*, which is a $(27,16,10,8)$ strongly regular graph whose uniqueness was demonstrated in Lemma 10.9.4.

Does Φ contain any regular graphs other than $L(K_8)$? To answer this we need to find all the proper subsets σ of $V(X)$ such that X^σ is regular. If a vertex in σ is adjacent to a vertices in σ, then it is adjacent to $k-a$ vertices in $V(X)\setminus\sigma$. After switching, its valency increases by $n-|\sigma|-2(k-a)$. Since this must be the same for every vertex in σ, we conclude that a is independent of the choice of vertex. Arguing similarly for $V(X)\setminus\sigma$ we conclude that the partition $\{\sigma, V(X)\setminus\sigma\}$ is equitable.

The partition of the vertices of $L(K_8)$ into σ and its complement is equivalent to a partition of the edges of K_8 into two graphs X_1 and X_2. The partition is equitable if and only if both $L(X_1)$ and $L(X_2)$ are regular graphs. From Lemma 1.7.5 we see that this implies that X_1 and X_2 are regular or bipartite and semiregular. However, if X_1 is bipartite and semiregular, then $L(X_2)$ consists of two cliques of different sizes. Hence both X_1 and X_2 are regular. Conversely, if X_1 is an r-regular graph on 8 vertices, then $|\sigma|=4r$, $a=2(r-1)$, and routine calculations show that X^σ is 12-regular.

We can assume that $|\sigma|\leq 14$, and hence $r\leq 3$. If $r=1$, then we can take $X=4K_2$; if $r=2$, then X_1 is one of C_8, $C_5\cup C_3$, and $C_4\cup C_4$; and if $r=3$, there are six possible cubic graphs on eight vertices. Some of these choices give isomorphic graphs. It is relatively straightforward to show that $L(K_8)$ and the three graphs obtained by taking $X_1=C_8$, $C_5\cup C_3$, and $C_4\cup C_4$ are pairwise nonisomorphic. These latter three graphs are called the *Chang graphs*. It is a little harder to show that every other choice for X_1 leads to a graph isomorphic to one of the three Chang graphs.

11.8 The Two-Graph on 276 Vertices

The Witt design on 23 points is a 4-$(23, 7, 1)$ design, which we studied in Section 10.11. There we found that if N is the incidence matrix of this design, then

$$NN^T = 56I + 21J$$

and

$$NJ = 77J, \qquad N^T J = 7J, \qquad N^T NJ = 539J.$$

Further, this design is quasi-symmetric, with intersection numbers 1 and 3. Hence there is a 01-matrix A such that

$$N^T N = 7I + A + 3(J - I - A) = 4I - 2A + 3J.$$

Define the matrix S by

$$S = \begin{pmatrix} J - I & J - 2N \\ J - 2N^T & N^T N - 5I - 2J \end{pmatrix}.$$

Since $N^T N - 5I - 2J = -2A - I + J$ is a Seidel matrix, it follows that S is a Seidel matrix. To prove that S determines a regular two-graph, we aim to show that it has only two eigenvalues. The obvious approach, which is to show that S^2 can be expressed as a linear combination of S and I, rapidly leads to unpleasant algebraic manipulations. Rather than this, we use a method similar to that used in Section 10.6 to find local eigenvectors, and determine a complete collection of eigenvectors of S.

Theorem 11.8.1 *The matrix S defined above has two eigenvalues, which are -5 with multiplicity 253 and 55 with multiplicity 23.*

Proof. We work with partitioned vectors of the form

$$\begin{pmatrix} x \\ y \end{pmatrix},$$

where x has length 23 and y length 253. First we compute

$$S\begin{pmatrix} \alpha\mathbf{1} \\ \mathbf{1} \end{pmatrix} = \begin{pmatrix} 22\alpha\mathbf{1} + 99\,\mathbf{1} \\ 9\alpha\mathbf{1} + 28\,\mathbf{1} \end{pmatrix}.$$

Hence we get an eigenvector for S with eigenvalue θ if and only if

$$\begin{pmatrix} 22 & 99 \\ 9 & 28 \end{pmatrix}\begin{pmatrix} \alpha \\ 1 \end{pmatrix} = \theta\begin{pmatrix} \alpha \\ 1 \end{pmatrix}.$$

The eigenvalues of this 2×2 matrix are -5 and 55, and a simple calculation shows that

$$\begin{pmatrix} -\frac{11}{3}\mathbf{1} \\ \mathbf{1} \end{pmatrix}, \qquad \begin{pmatrix} \mathbf{1} \\ \frac{1}{3}\mathbf{1} \end{pmatrix}$$

are eigenvectors for S with eigenvalues -5 and 55, respectively.

Next we look for eigenvectors of S orthogonal to the pair we have just found. Suppose $y \in \mathbb{R}^{253}$ and $\mathbf{1}^T y = 0$. Then $\mathbf{1}^T N y = 0$, and hence

$$S \begin{pmatrix} Ny \\ \beta y \end{pmatrix} = \begin{pmatrix} -Ny - 2\beta Ny \\ -2N^T Ny + \beta N^T Ny - 5\beta y \end{pmatrix}$$
$$= \begin{pmatrix} -(1+2\beta)Ny \\ (\beta - 2)N^T Ny - 5\beta y \end{pmatrix}.$$

If we take y to be an eigenvector of $N^T N$ with eigenvalue θ, then

$$S \begin{pmatrix} Ny \\ \beta y \end{pmatrix} = \begin{pmatrix} (-1-2\beta)Ny \\ ((\beta-2)\theta - 5\beta)y \end{pmatrix},$$

and therefore, if we select β such that $\beta(-1-2\beta) = (\beta-2)\theta - 5\beta$, we will find an eigenvector of S. Solving this last equation implies that we must select $\beta = 2$ or $\beta = -\theta/2$.

Now, we need to know the eigenvalues of $N^T N$ in order to find the eigenvectors that are orthogonal to $\mathbf{1}$. By Lemma 8.2.4, the nonzero eigenvalues of $N^T N$ equal the nonzero eigenvalues of NN^T, and have the same multiplicities. Since $NN^T = 56I + 21J$, its eigenvalues are 539 and 56, with multiplicities 1 and 22, respectively. Hence $N^T N$ has eigenspaces of dimension 1, 22, and 230, with the latter two consisting of eigenvectors orthogonal to $\mathbf{1}$.

Now, if we take $\beta = 2$, then we get 252 linearly independent eigenvectors of S of the form

$$\begin{pmatrix} Ny \\ 2y \end{pmatrix},$$

all of which have eigenvalue $-1 - 2\beta = -5$.

In addition, $N^T N$ has 22 linearly independent eigenvectors with eigenvalue $\theta = 56$, and so taking $\beta = -\theta/2$ yields 22 eigenvectors of the form

$$\begin{pmatrix} Ny \\ -28y \end{pmatrix},$$

all with eigenvalue $-1 - 2(-28) = 55$. These are necessarily independent of the 252 eigenvectors with eigenvalue -5. The 230 eigenvectors of $N^T N$ with eigenvalue 0 do not produce any further eigenvectors of S because both Ny and βy are zero for all of these. This is just as well, because the 274 eigenvectors that we have just found, together with the two initial eigenvectors, form a set of 276 linearly independent eigenvectors. Therefore, there are no further eigenvectors of S, and it has just two eigenvalues -5 and 55. □

By Theorem 11.6.1, the neighbourhoods of a regular two-graph on 276 vertices are strongly regular graphs. If X is such a neighbourhood with eigenvalues k, θ, and τ, then we must have

$$-1-2\theta=-5,$$
$$-1-2\tau=55,$$

and so $\theta = 2$ and $\tau = -28$. Using the expressions given in the proof of Theorem 11.6.1, we see that X has parameters $(275, 112, 30, 56)$. These values give equality in the second Krein bound, whence we deduce from Theorem 10.7.1 that both subconstituents of X are strongly regular. It can also be shown that the subconstituents of these subconstituents are strongly regular too, but we leave this an exercise.

Exercises

1. If n is odd, show that a switching class of graphs on n vertices contains a unique graph in which all vertices have even valency.

2. Show that the strongly regular graph that arises by switching a vertex off the Petersen graph is $L(K_{3,3})$.

3. Let S be the Seidel matrix for $\overline{C_5} \cup K_1$. Show that S has only two eigenvalues and that there is no regular graph in its switching class.

4. Let X be the block graph of a Steiner triple system on v points. Show that the switching class of X is a regular two-graph if and only if $v = 13$. Show that the switching class of $X \cup K_1$ is a regular two-graph if and only if $v = 15$.

5. Let X be a strongly regular graph constructed from a Latin square of order n. Show that the switching class of $X \cup K_1$ is a regular two-graph if and only if $n = 5$. Show that the switching class of X is a regular two-graph if and only if $n = 6$.

6. Let X be a k-regular graph on n vertices. If there is a nontrivial switching σ such that X^σ is k-regular, show that $k-\frac{n}{2}$ is an eigenvalue of $A(X)$.

7. Show that the Petersen graph can be switched into its complement.

8. Let X be the neighbourhood of a vertex in the regular two-graph on 276 vertices. Determine the eigenvalues and their multiplicities for each of the subconstituents of X, and show that the subconstituents of the subconstituents are strongly regular.

Notes

Seidel's selected works [3] contains a number of papers on two-graphs, including two surveys.

The regular two-graph on 276 vertices is a remarkable object. Its automorphism group is Conway's simple group $\cdot 3$. Goethals and Seidel [2] provide a short proof that there is a unique regular two-graph on 276 vertices; their proof reduces to the uniqueness of the ternary Golay code. The switching class of this two-graph contains a strongly regular graph with parameters $(276, 135, 78, 54)$.

One of the oldest open problems concerning two-graphs is the question of whether there exist regular two-graphs on 76 and 96 vertices. The corresponding strongly regular graphs have parameters $(75, 32, 10, 16)$ and $(95, 40, 12, 20)$. The switching classes of the two-graphs could contain strongly regular graphs on 76 or 96 vertices.

It would be interesting to have better information about the maximum number of equiangular lines in $\mathbb{R}^n$. The absolute bound gives an upper bound of order $n^2/2$. Dom de Caen [1] has a class of examples that provides a set of size $2q^2$ in $\mathbb{R}^{3q-1}$, where $q = 2^{2t-1}$. These examples do not form regular two-graphs, which makes them even more interesting. It would be very surprising if there was a further set of lines realizing the absolute bound.

References

[1] D. de Caen, *Large equiangular sets of lines in Euclidean space*, Electron. J. Combin., 7 (2000), Research Paper 55, 3 pp. (electronic).

[2] J.-M. Goethals and J. J. Seidel, *The regular two-graph on 276 vertices*, Discrete Math., 12 (1975), 143–158.

[3] J. J. Seidel, *Geometry and Combinatorics*, Academic Press Inc., Boston, MA, 1991.

12

Line Graphs and Eigenvalues

If X is a graph with incidence matrix B, then the adjacency matrix of its line graph $L(X)$ is equal to $B^T B - 2I$. Because $B^T B$ is positive semidefinite, it follows that the minimum eigenvalue of $L(X)$ is at least -2. This chapter is devoted to showing how close this property comes to characterizing line graphs. The main result is a beautiful characterization of all graphs with minimum eigenvalue at least -2. One surprise is that the proof uses several seemingly unrelated combinatorial objects. These include generalized quadrangles with lines of size three and root systems, which arise in connection with a number of important problems, including the classification of Lie algebras.

12.1 Generalized Line Graphs

Suppose that A is a symmetric matrix with zero diagonal such that $A+2I$ is positive semidefinite. Then $A+2I = UU^T$ for some matrix U, and thus $A+2I$ is the Gram matrix of a set of vectors, each with length $\sqrt{2}$. Conversely, given a set of vectors with length $\sqrt{2}$ and pairwise inner products 0 and 1, we can construct a graph with minimum eigenvalue at least -2. Our eventual aim is to characterize all such sets of vectors, and hence the graphs with minimum eigenvalue at least -2. However, it is necessary to approach this task by first considering sets of vectors of length $\sqrt{2}$ whose pairwise inner products are allowed to be 0, 1, or -1. We begin by defining an important set of vectors with this property.

Let $e_1, \ldots, e_n$ be the standard basis for $\mathbb{R}^n$ and let D_n be the set of all vectors of the form

$$\pm e_i \pm e_j, \quad i \neq j.$$

All vectors in D_n have squared length 2, and the inner product of two distinct vectors from D_n is 0, 1, or -1. Hence if S is a subset of D_n such that the inner product of any two elements of S is nonnegative, then the Gram matrix of S determines a graph with minimum eigenvalue at least -2.

Since D_n contains all the vectors $e_i + e_j$, it follows that the columns of the incidence matrix of any graph on n vertices lie in D_n, and so all line graphs can be obtained in this way. This motivates us to call a graph X a *generalized line graph* if $2I + A(X)$ is the Gram matrix of a subset of D_n for some n.

All line graphs are generalized line graphs, but there are many generalized line graphs that are not line graphs. A simple example is given by the graph $\overline{4K_2}$ which has adjacency matrix A satisfying $A + 2I = D^T D$, where D is the matrix

$$\begin{pmatrix} 1 & 1 & 1 & 1 & 1 & 1 & 1 & 1 \\ 1 & -1 & 0 & 0 & 0 & 0 & 0 & 0 \\ 0 & 0 & 1 & -1 & 0 & 0 & 0 & 0 \\ 0 & 0 & 0 & 0 & 1 & -1 & 0 & 0 \\ 0 & 0 & 0 & 0 & 0 & 0 & 1 & -1 \end{pmatrix}.$$

Since $\overline{4K_2}$ contains a copy of K_5 with an edge removed as an induced subgraph, it is not a line graph. This example is easily generalized (sorry!) to yield that $\overline{rK_2}$ is a generalized line graph for all $r \geq 4$. (The graph $\overline{3K_2}$ is the line graph of K_4.)

12.2 Star-Closed Sets of Lines

If x is a vector in $\mathbb{R}^n$, let $\langle x \rangle$ denote the line spanned by x. Suppose that S is a set of vectors in $\mathbb{R}^m$ of length $\sqrt{2}$ such that their pairwise inner products lie in $\{-1, 0, 1\}$. Then the set of lines $\mathcal{L} = \{\langle x \rangle : x \in S\}$ has the property that any two distinct lines are at an angle of 60° or 90°. Our aim is to classify such sets of lines.

We call such a set *maximal* if there is no way to add a new line at 60° or 90° to those already given. All maximal sets are finite. To see this, consider the points formed by the intersection of the lines with the unit sphere in $\mathbb{R}^n$. Since the angle formed at the origin by any two points is at least 60°, the distance in $\mathbb{R}^n$ between any two points is at least 1. Since the unit sphere has finite area, this means we can have only finitely many points. Thus it suffices to classify the maximal sets of lines at 60° and 90°.

A *star* is a set of three coplanar lines, with any two at an angle of 60°. A set of lines $\mathcal{L}$ is *star-closed* if for any star ℓ, m, and n such that ℓ and m lie in $\mathcal{L}$, the line n also lies in $\mathcal{L}$. The *star-closure* of a set of lines $\mathcal{L}$ is the intersection of all the star-closed sets of lines that contain $\mathcal{L}$. It is immediate that the star-closure of a set of lines is itself star-closed.

Theorem 12.2.1 *A maximal set of lines at 60° and 90° in $\mathbb{R}^n$ is star-closed.*

Proof. Let $\mathcal{L}$ be a set of lines at 60° and 90°, and suppose that $\langle a\rangle$, $\langle b\rangle \in \mathcal{L}$ are two lines at 60°. We can assume that a and b have length $\sqrt{2}$ and choose b such that $\langle a, b\rangle = -1$. Then $a+b$ has length $\sqrt{2}$, and $\langle a+b\rangle$ forms a star with $\langle a\rangle$ and $\langle b\rangle$. We show that $\langle a+b\rangle$ is either in $\mathcal{L}$ or is at 60° or 90° to every line in $\mathcal{L}$. Let x be a vector spanning a line of $\mathcal{L}$. Then $\langle x, a+b\rangle = \langle x, a\rangle + \langle x, b\rangle$, and so $\langle x, a+b\rangle \in \{-2, -1, 0, 1, 2\}$. If $\langle x, a+b\rangle = \pm 2$, then $x = \pm(a+b)$, and so $\langle a+b\rangle \in \mathcal{L}$. Otherwise, it is at 60° or 90° to every line of $\mathcal{L}$, and so can be added to $\mathcal{L}$ to form a larger set of lines. □

We note that the converse of Theorem 12.2.1 is false: For example, the four vectors $e_1 \pm e_2$ and $e_3 \pm e_4$ in D_4 are pairwise orthogonal, and therefore span a star-closed set of lines at 60° and 90°.

We record a result that we will need later; the proof is left as an exercise.

Lemma 12.2.2 *The set of lines spanned by the vectors of D_n is star-closed.* □

12.3 Reflections

We can characterize star-closed sets of lines at 60° and 90° in terms of their symmetries. If h is a vector in $\mathbb{R}^n$, then there is a unique hyperplane through the origin perpendicular to h. Let ρ_h denote the operation of reflection in this hyperplane. Simple calculations reveal that for all x,

$$\rho_h(x) = x - 2\frac{\langle x, h\rangle}{\langle h, h\rangle}h.$$

We make a few simple observations. It is easy to check that $\rho_h(h) = -h$. Also, $\rho_h(x) = x$ if and only if $\langle h, x\rangle = 0$. The product $\rho_a\rho_b$ of two reflections is not in general a reflection. It can be shown that $\rho_a\rho_b = \rho_b\rho_a$ if and only if either $\langle a\rangle = \langle b\rangle$ or $\langle a, b\rangle = 0$.

Lemma 12.3.1 *Let $\mathcal{L}$ be a set of lines at 60° and 90° in $\mathbb{R}^n$. Then $\mathcal{L}$ is star-closed if and only if for every vector h that spans a line in $\mathcal{L}$, the reflection ρ_h fixes $\mathcal{L}$.*

Proof. Let h be a vector of length $\sqrt{2}$ spanning a line in $\mathcal{L}$. From our comments above, ρ_h fixes $\langle h\rangle$ and all the lines orthogonal to $\langle h\rangle$. So suppose that $\langle a\rangle$ is a line of $\mathcal{L}$ at 60° to $\langle h\rangle$. Without loss of generality we can assume that a has length $\sqrt{2}$ and that $\langle h, a\rangle = -1$. Now,

$$\rho_h(a) = a - 2\frac{(-1)}{2}h = a + h,$$

and $\langle a+h\rangle$ forms a star with $\langle a\rangle$ and $\langle h\rangle$. This implies that ρ_h fixes $\mathcal{L}$ if and only if $\mathcal{L}$ is star-closed. □

A *root system* is a set S of vectors in $\mathbb{R}^n$ such that

(a) if $h \in S$, then $\langle h\rangle \cap S = \{h, -h\}$;

(b) if $h \in S$, then $\rho_h(S) = S$.

Lemma 12.3.1 shows that if $\mathcal{L}$ is a star-closed set of lines at 60° and 90° in $\mathbb{R}^m$, then the vectors of length $\sqrt{2}$ that span these lines form a root system. For example, the set D_n is a root system.

The group generated by the reflections ρ_h, for h in S, is the *reflection group* of the root system. The *symmetry group* of a set of lines or vectors in $\mathbb{R}^n$ is the group of all orthogonal transformations that take the set to itself. The symmetry group of a root system or of a set of lines always contains multiplication by -1.

12.4 Indecomposable Star-Closed Sets

A set $\mathcal{L}$ of lines at 60° and 90° is called *decomposable* if it can be partitioned into two subsets $\mathcal{L}_1$ and $\mathcal{L}_2$ such that every line in $\mathcal{L}_1$ is orthogonal to every line in $\mathcal{L}_2$. If there is no such partition, then it is called *indecomposable*.

Lemma 12.4.1 *For $n \geq 2$, the set of lines $\mathcal{L}$ spanned by the vectors in D_n is indecomposable.*

Proof. The lines $\langle e_1 + e_i\rangle$ for $i \geq 2$ have pairwise inner products equal to 1, and hence must be in the same part of any decomposition of $\mathcal{L}$. It is clear, however, that any other vector in D_n has nonzero inner product with at least one of these vectors, and so there are no lines orthogonal to all of this set. □

Theorem 12.4.2 *Let $\mathcal{L}$ be a star-closed indecomposable set of lines at 60° and 90°. Then the reflection group of $\mathcal{L}$ acts transitively on ordered pairs of nonorthogonal lines.*

Proof. First we observe that the reflection group acts transitively on the lines of $\mathcal{L}$. Suppose that $\langle a\rangle$ and $\langle b\rangle$ are two lines that are not orthogonal, and that $\langle a, b\rangle = -1$. Then $c = -a - b$ spans the third line in the star with $\langle a\rangle$ and $\langle b\rangle$, and the reflection ρ_c swaps $\langle a\rangle$ and $\langle b\rangle$. Therefore, $\langle a\rangle$ can be mapped on to any line not orthogonal to it. Let $\mathcal{L}'$ be the orbit of $\langle a \rangle$ under the reflection group of $\mathcal{L}$. Then every line in $\mathcal{L} \setminus \mathcal{L}'$ is orthogonal to every line of $\mathcal{L}'$. Since $\mathcal{L}$ is indecomposable, this shows that $\mathcal{L}' = \mathcal{L}$.

Now, suppose that $(\langle a \rangle, \langle b \rangle)$ and $(\langle a \rangle, \langle c \rangle)$ are two ordered pairs of nonorthogonal lines. We will show that there is a reflection that fixes $\langle a \rangle$ and exchanges $\langle b \rangle$ and $\langle c \rangle$. Assume that a, b, and c have length $\sqrt{2}$ and that $\langle a, b \rangle = \langle a, c \rangle = -1$. Then the vector $-a - b$ has length $\sqrt{2}$ and spans a line in $\mathcal{L}$. Now,

$$1 = \langle c, -a \rangle = \langle c, b \rangle + \langle c, -a - b \rangle.$$

If $c = b$ or $c = -a - b$, then $\langle c \rangle$ and $\langle b \rangle$ are exchanged by the identity reflection or ρ_a, respectively. Otherwise, c has inner product 1 with precisely one of the vectors in $\{b, -a-b\}$, and is orthogonal to the other. Exchanging the roles of b and $-a - b$ if necessary, we can assume that $\langle c, b \rangle = 1$. Then $\langle b - c \rangle \in \mathcal{L}$, and the reflection ρ_{b-c} fixes $\langle a \rangle$ and exchanges $\langle b \rangle$ and $\langle c \rangle$. □

Now, suppose that X is a graph with minimum eigenvalue at least -2. Then $A(X) + 2I$ is the Gram matrix of a set of vectors of length $\sqrt{2}$ that span a set of lines at 60° and 90°. Let $\mathcal{L}(X)$ denote the star-closure of this set of lines. Notice that the Gram matrix determines the set of vectors up to orthogonal transformations of the underlying vector space, and therefore $\mathcal{L}(X)$ is uniquely determined up to orthogonal transformations.

Lemma 12.4.3 *If X is a graph with minimum eigenvalue at least -2, then the star-closed set of lines $\mathcal{L}(X)$ is indecomposable if and only if X is connected.*

Proof. First suppose that X is connected. Let $\mathcal{L}'$ be the lines spanned by the vectors whose Gram matrix is $A(X) + 2I$. Lines corresponding to adjacent vertices of X are not orthogonal, and hence must be in the same part of any decomposition of $\mathcal{L}(X)$. Therefore, all the lines in $\mathcal{L}'$ are in the same part. Any line lying in a star with two other lines is not orthogonal to either of them, and therefore lies in the same part of any decomposition of $\mathcal{L}(X)$. Hence the star-closure of $\mathcal{L}'$ is all in the same part of any decomposition, which shows that $\mathcal{L}(X)$ is indecomposable.

If X is not connected, then $\mathcal{L}'$ has a decomposition into two parts. Any line orthogonal to two lines in a star is orthogonal to all three lines of the star, and so any line added to $\mathcal{L}'$ to complete a star can be assigned to one of the two parts of the decomposition, eventually yielding a decomposition of $\mathcal{L}$.

□

Therefore, we see that any connected graph with minimum eigenvalue at least -2 is associated with a star-closed indecomposable set of lines. Our strategy will be to classify all such sets, and thereby classify all the graphs with minimum eigenvalue at least -2.

12.5 A Generating Set

We now show that any indecomposable star-closed set of lines $\mathcal{L}$ at 60° and 90° is the star-closure of a special subset of those lines. Eventually, we will see that the structure of this subset is very restricted.

Lemma 12.5.1 *Let $\mathcal{L}$ be an indecomposable star-closed set of lines at 60° and 90°, and let $\langle a \rangle$, $\langle b \rangle$, and $\langle c \rangle$ form a star in $\mathcal{L}$. Every other line of $\mathcal{L}$ is orthogonal to either one or three lines in the star.*

Proof. Without loss of generality we may assume that a, b, and c all have length $\sqrt{2}$ and that

$$\langle a, b \rangle = \langle b, c \rangle = \langle c, a \rangle = -1.$$

It follows then that $c = -a - b$, and so for any other line $\langle x \rangle$ of $\mathcal{L}$ we have

$$\langle x, a \rangle + \langle x, b \rangle + \langle x, c \rangle = 0.$$

Because each of the terms is in $\{0, \pm 1\}$, we see that either all three terms are zero or the three terms are 1, 0, and -1 in some order. □

Now, fix a star $\langle a \rangle$, $\langle b \rangle$, and $\langle c \rangle$, and as above choose a, b, and c to be vectors of length $\sqrt{2}$ with pairwise inner products -1. Let D be the set of lines of $\mathcal{L}$ that are orthogonal to all three lines in the star. The remaining lines of $\mathcal{L}$ are orthogonal to just one line of the star, so can be partitioned into three sets A, B, and C, consisting of those lines orthogonal to $\langle a \rangle$, $\langle b \rangle$, and $\langle c \rangle$, respectively.

Lemma 12.5.2 *The set $\mathcal{L}$ is the star-closure of $\langle a \rangle$, $\langle b \rangle$, and C.*

Proof. Let $\mathcal{M}$ denote the set of lines $\{\langle a \rangle, \langle b \rangle\} \cup C$. Clearly, $\langle c \rangle$ lies in the star-closure of $\mathcal{M}$, and so it suffices to show that every line in A, B, and D lies in a star with two lines chosen from $\mathcal{M}$. Suppose that $\langle x \rangle \in A$, and without loss of generality, select x such that $\langle b, x \rangle = -1$. Then $-b - x \in \mathcal{L}$ and straightforward calculation shows that $\langle -b - x \rangle \in C$. Thus $\langle x \rangle$ is in the star containing $\langle b \rangle$ and $\langle -b - x \rangle$. An analogous argument deals with the case where $\langle x \rangle \in B$, leaving only the lines in D. Let x be a line in D, and first suppose that there is some line in C, call it z, not orthogonal to x. Then we can assume that $\langle x, z \rangle = -1$, and hence that $-x - z \in \mathcal{L}$. Once again, straightforward calculations show that $-x - z \in C$, and so $\langle x \rangle$ lies in a star with two lines from C. Therefore, the only lines remaining are those in D that are orthogonal to every line of C. Let D' denote this set of lines. Every line in D' is orthogonal to every line in $\mathcal{M}$, and hence to every line in the star-closure of $\mathcal{M}$, which we have just seen contains $\mathcal{L} \setminus D'$. Since $\mathcal{L}$ is indecomposable, this implies that D' is empty. □

Clearly, an analogous proof could be used to show that $\mathcal{L}$ is the star-closure of $\langle a \rangle$, B, and $\langle c \rangle$ and that $\mathcal{L}$ is the star-closure of A, $\langle b \rangle$, and $\langle c \rangle$.

However, it is not necessary to do so because Theorem 12.4.2 implies that all choices of a pair of lines are equivalent under the reflection group of $\mathcal{L}$.

12.6 The Classification

We now need to select a spanning vector for each of the lines in A, B, and C. We assume that every vector chosen has length $\sqrt{2}$, but this still leaves us two choices for each line. We fix the choice of one vector from each line by defining A^*, B^*, and C^* as follows:

$$\begin{aligned}
A^* &= \{x : \langle x \rangle \in \mathcal{L}, \langle a, x\rangle = 0, \langle b, x\rangle = 1, \langle c, x\rangle = -1\},\\
B^* &= \{x : \langle x \rangle \in \mathcal{L}, \langle a, x\rangle = -1, \langle b, x\rangle = 0, \langle c, x\rangle = 1\},\\
C^* &= \{x : \langle x \rangle \in \mathcal{L}, \langle a, x\rangle = 1, \langle b, x\rangle = -1, \langle c, x\rangle = 0\}.
\end{aligned}$$

With these definitions, it is routine to confirm that

$$\begin{aligned}
A^* &= \rho_b(C^*) = C^* + b,\\
B^* &= \rho_c(A^*) = A^* + c,\\
C^* &= \rho_a(B^*) = B^* + a.
\end{aligned}$$

Lemma 12.6.1 *If x and y are orthogonal vectors in C^*, then there is a unique vector in C^* orthogonal to both of them.*

Proof. Suppose that vectors x, $y \in C^*$ are orthogonal. Then by our comments above, we see that $x + b \in A^*$ and that $y - a \in B^*$ and that $\langle x + b, y - a\rangle = -1$. Therefore, $\langle a - b - x - y\rangle \in \mathcal{L}$, and calculation shows that $a - b - x - y \in C^*$. Now,

$$\langle a - b - x - y, x\rangle = \langle a - b - x - y, y\rangle = 0,$$

and so $a - b - x - y$ is orthogonal to both x and y. If z is any other vector in C^* orthogonal to x and y, then

$$\langle z, a - b - x - y\rangle = \langle z, a\rangle - \langle z, b\rangle = 2,$$

and hence $z = a - b - x - y$. □

Using this lemma we see that C^* has a very special structure. Define the incidence structure $\mathcal{Q}$ to have the vectors of C^* as its points, and the orthogonal triples of vectors as its lines. The previous lemma shows that any two points lie in at most one line, and hence $\mathcal{Q}$ is a partial linear space. However, we can say more about $\mathcal{Q}$.

Theorem 12.6.2 *Let $\mathcal{Q}$ be the incidence structure whose points are the vectors of C^*, and whose lines are triples of mutually orthogonal vectors. Then either $\mathcal{Q}$ has no lines, or $\mathcal{Q}$ is a generalized quadrangle, possibly degenerate, with lines of size three.*

Proof. A generalized quadrangle has the property that given any line ℓ and a point P off that line, there is a unique point on ℓ collinear with P. We show that $\mathcal{Q}$ satisfies this axiom.

Suppose that x, y, and $a - b - x - y$ are the three points of a line of $\mathcal{Q}$, and let z be an arbitrary vector in C^*, not equal to any of these three. Then

$$\langle z, x\rangle + \langle z, y\rangle + \langle z, a - b - x - y\rangle = \langle z, a - b\rangle = 2.$$

Since each of the three terms is either 0 or 1, it follows that there is a unique term equal to 0, and hence z is collinear with exactly one of the three points of the line.

Therefore, $\mathcal{Q}$ is a generalized quadrangle with lines of size three. □

From our earlier work on generalized quadrangles with lines of size three, we get the following result.

Corollary 12.6.3 *If $\mathcal{Q}$ is the incidence structure arising from a star-closed indecomposable set of lines at $60°$ and $90°$, then one of the following holds:*

(a) *$\mathcal{Q}$ has no lines;*
(b) *$\mathcal{Q}$ is a set of concurrent lines of size three;*
(c) *$\mathcal{Q}$ is the unique generalized quadrangle of order $(2,1)$, $(2,2)$, or $(2,4)$.* □

In the next section we will describe families of lines that realize each of the five cases enumerated above.

12.7 Root Systems

In this section we present five root systems, known as D_n, A_n, E_8, E_7, and E_6. We will show that the corresponding sets of lines are indecomposable and star-closed, and that they realize the five possibilities of Corollary 12.6.3.

We have already defined D_n, and shown that the corresponding set of lines is star-closed and indecomposable. We leave it as an exercise to confirm that the corresponding incidence structure $\mathcal{Q}$ is a set of concurrent lines of size three.

The next root system is A_n, which consists of all nonzero vectors of the form $e_i - e_j$, where e_i and e_j run over the standard basis of $\mathbb{R}^{n+1}$. This is a subset of D_{n+1}, and in fact is the set of all vectors orthogonal to $\mathbf{1}$. We leave the proof of the next result as an exercise.

Lemma 12.7.1 *The set of lines corresponding to the root system A_n is star-closed and indecomposable. The incidence structure $\mathcal{Q}$ has no lines.* □

Our next root system is called E_8, and lives in $\mathbb{R}^8$. It contains the vectors of D_8, together with the 128 vectors x such that $x_i \in \{-\frac{1}{2}, \frac{1}{2}\}$ for $i = 1, \dots, 8$ and the number of positive entries is even.

Theorem 12.7.2 *The root system E_8 contains exactly 240 vectors. The lines spanned by these vectors form an indecomposable star-closed set of lines at 60° and 90° in $\mathbb{R}^8$. The generalized quadrangle $\mathcal{Q}$ associated with this set of lines is the unique generalized quadrangle of order* $(2, 4)$.

Proof. This is immediate, since D_8 contains 112 vectors, and there are 128 further vectors.

First we show that the set of lines spanned by E_8 is indecomposable. Since the set of lines spanned by D_8 is indecomposable, any decomposition will have all the lines spanned by D_8 in one part. Any vector in $E_8 \setminus D_8$ that is orthogonal to $e_1 + e_2$ has its first two entries of opposite sign, while any vector orthogonal to $e_1 - e_2$ has its first two entries of the same sign. Therefore, there are no vectors in $E_8 \setminus D_8$ orthogonal to all the vectors in D_8.

To show that E_8 is star-closed, we consider pairs of vectors x, y that have inner product -1, and show that in all cases $-x - y \in E_8$. Observe that permuting coordinates and reversing the sign of an even number of entries are operations that fix E_8 and preserve inner products, and so we can freely use these to simplify our calculations. Suppose firstly that x and y are both in $E_8 \setminus D_8$. Then we can assume that $x = \frac{1}{2}\mathbf{1}$, and so y has six entries equal to $-\frac{1}{2}$ and two equal to $\frac{1}{2}$. Therefore, $-x - y \in D_8$, and so this star can be closed. Secondly, suppose that x is in $E_8 \setminus D_8$ and that $y \in D_8$. Once again we can assume that $x = \frac{1}{2}\mathbf{1}$, and therefore y has two entries equal to -1. Then $-x-y$ has two entries of $\frac{1}{2}$ and six equal to $-\frac{1}{2}$, and so lies in $E_8 \setminus D_8$. Finally, if x and y are both in D_8, then we appeal to the fact that D_8 is star-closed.

By Theorem 12.4.2 we can select any pair of nonorthogonal lines as $\langle a \rangle$ and $\langle b \rangle$, so choose $a = e_1 + e_2$ and $b = -e_1 + e_3$, which implies that $c = -e_2 - e_3$. We count the number of lines orthogonal to this star. A vector x is orthogonal to this star if and only if its first three coordinates are $x_1 = \alpha$, $x_2 = -\alpha$, $x_3 = \alpha$. Therefore, if $x \in D_8$, we have $\alpha = 0$, and there are thus $4\binom{5}{2} = 40$ such vectors. If $x \in E_8 \setminus D_8$, then the remaining five coordinates have either 1, 3, or 5 negative entries, and so there are $\binom{5}{1} + \binom{5}{3} + \binom{5}{5} = 16$ such vectors. Because α can be $\pm\frac{1}{2}$, this yields 32 vectors. Therefore, there are 72 vectors in E_8 orthogonal to the star, or 36 lines orthogonal to the star. Since there are 120 lines altogether, this means that the star together with A, B, and C contain 84 lines. Since A, B, and C have the same size, this shows that they each contain 27 lines. Thus $\mathcal{Q}$ is a generalized quadrangle with 27 points, and so is the unique generalized quadrangle of order $(2, 4)$. □

We define two further root systems. First E_7 is the set of vectors in E_8 orthogonal to a fixed vector, while E_6 is the subset of E_8 formed by the set of vectors orthogonal to a fixed pair of vectors with inner product ± 1. The next result outlines the properties of these root systems; the proofs are similar to those for E_8, and so are left as exercises.

Lemma 12.7.3 *The root systems E_6 and E_7 contain 72 and 126 vectors respectively. The sets of lines spanned by the vectors of E_6 and E_7 are star-closed and indecomposable. The associated generalized quadrangles are the unique generalized quadrangles of order $(2,1)$ and $(2,2)$, respectively.*

Theorem 12.7.4 *An indecomposable star-closed set of lines at 60° and 90° is the set of lines spanned by the vectors in one of the root systems E_6, E_7, E_8, A_n, or D_n (for some n).*

Proof. The Gram matrix of the vectors in C^* determines the Gram matrix of the entire collection of lines in $\mathcal{L}$, which in turn determines $\mathcal{L}$ up to an orthogonal transformation. Since these five root systems give the only five possible Gram matrices for the vectors in C^*, there are no further indecomposable star-closed sets of lines at 60° and 90°. □

We summarize some of the properties of our five root systems in the following table.

Name	Size	$\|C^*\|$
D_n	$n(2n-2)$	$2n-5$
A_n	$n(n+1)$	$n-2$
E_8	240	27
E_7	126	15
E_6	72	9

12.8 Consequences

We begin by translating Theorem 12.7.4 into graph theory, then determine some of its consequences.

Corollary 12.8.1 *Let X be a connected graph with smallest eigenvalue at least -2, and let A be its adjacency matrix. Then either X is a generalized line graph, or $A+2I$ is the Gram matrix of a set of vectors in E_8.*

Proof. Let S be a set of vectors with Gram matrix $2I+A$. Then the star-closure of S is contained in the set of lines spanned by the vectors in E_8 or D_n. □

This implies that a connected graph with minimum eigenvalue at least -2 and more than 120 vertices must be a generalized line graph. We can be more precise than this, at the cost of some effort.

Theorem 12.8.2 *Let X be a graph with least eigenvalue at least -2. If X has more than 36 vertices or maximum valency greater than 28, it is a generalized line graph.*

Proof. If X is not a generalized line graph, then $A(X) + 2I$ is the Gram matrix of a set of vectors in E_8. So let S be a set of vectors from E_8 with nonnegative pairwise inner products. First we will show that $|S| \leq 36$. For any vector $x \in \mathbb{R}^8$, let P_x be the 8×8 matrix xx^T. The matrices P_x span a subspace of the real vector space formed by the 8×8 symmetric matrices, which has dimension $\binom{9}{2} = 36$. To prove the first part of the lemma, it will suffice to show that the matrices P_x for $x \in S$ are linearly independent.

Suppose that there are real numbers a_x such that

$$\sum_{x \in S} a_x P_x = 0.$$

Then we have

$$\begin{aligned} 0 = \operatorname{tr}\left(\left(\sum_x a_x P_x\right)^2\right) &= \sum_{x,y} a_x a_y \operatorname{tr}(P_x P_y) \\ &= \sum_{x,y} a_x a_y \operatorname{tr}(xx^T yy^T) \\ &= \sum_{x,y} a_x a_y (x^T y)^2. \end{aligned}$$

The last sum can be written in the form $a^T Da$, where D is the matrix obtained by replacing each entry of the Gram matrix of S by its square. This Gram matrix is equal to $2I + A(X)$, and since $A(X)$ is a 01-matrix, it follows that $D = 4I + A(X)$. But as $2I + A(X)$ is positive semidefinite, D is positive definite, and so $a = 0$. Therefore, the matrices P_x are linearly independent, and so $|S| \leq 36$.

It remains for us to prove the claim about the maximum valency. Suppose that $a \in S$, and let $\langle a \rangle$, $\langle b \rangle$, and $\langle c \rangle$ be a star containing $\langle a \rangle$. The vectors whose inner product with a is 1 are the vectors $-b$, $-c$, and the vectors in C^* and $-B^*$. If $x \in C^*$, then $x - a \in B^*$, and so $a - x \in -B^*$. Then

$$\langle x, a - x \rangle = 1 - 2 = -1,$$

and so S cannot contain both x and $a - x$. Similarly, S cannot contain both $-b$ and $-c$, because their inner product is -1. Thus S can contain at most one vector from each of the pairs $\{x, a - x\}$ for $x \in C^*$, together with at most one vector from $\{-b, -c\}$. Thus S contains at most 28 vectors with positive inner product with a, and so any vertex of X has valency at most 28. □

12.9 A Strongly Regular Graph

Finally, we show how to construct a strongly regular graph from the root system E_8.

Let $\mathcal{L}$ be the set of lines spanned by the vectors in E_8. Let X be the graph with the lines of $\mathcal{L}$ as its vertices, with two lines adjacent if they are distinct and are not perpendicular. We will prove that X is a strongly regular graph with parameters $(120, 56, 28, 24)$.

Choose one vector from each pair of vectors $\{x, -x\}$ in E_8, and let $\mathcal{P}$ be the collection of matrices P_x. The map that takes a pair of symmetric matrices A and B to $\operatorname{tr}(AB)$ is an inner product on the space of symmetric matrices. Using this inner product, we can form the Gram matrix D of the elements of $\mathcal{P}$. Then

$$D = 4I + A,$$

where A is the adjacency matrix of X. To show that X is strongly regular we will determine the eigenvalues of D.

For each vector x in E_8 there are $56 = 2(27 + 1)$ vectors y in E_8 such that $\langle x, y \rangle = 1$; hence X has valency 56, and the row sum of D is 60. This provides one eigenvalue of D.

The span of $\mathcal{P}$ has dimension at most 36, and so the rank of the Gram matrix of $\mathcal{P}$ is at most 36. Therefore, D has 0 as an eigenvalue with multiplicity at least $120 - 36 = 84$.

Now, we have found at least 85 eigenvalues of D. Let $\theta_1, \ldots, \theta_{35}$ denote the remaining ones, which may also be equal to 0 or 60. Then

$$480 = \operatorname{tr} D = 60 + \sum_i \theta_i$$

and

$$\operatorname{tr} D^2 = (60)^2 + \sum_i \theta_i^2.$$

Since $\operatorname{tr} D^2$ is the sum of the squares of the entries of D, we have

$$\operatorname{tr} D^2 = 120(16 + 56) = 8640.$$

Thus our two equations yield that

$$\sum_i \theta_i = 420, \qquad \sum_i \theta_i^2 = 5040.$$

Now,

$$35 \sum_{i=1}^{35} \theta_i^2 \geq \left(\sum_{i=1}^{35} \theta_i \right)^2,$$

with equality if and only if the eigenvalues θ_i are all equal. But in this case both sides equal $(420)^2$, whence $\theta_i = 12$ for all i.

Therefore, the eigenvalues of D are 60, 12, and 0 with respective multiplicities 1, 35, and 84. The eigenvalues of $A = D - 4I$ are 56, 8, and -4, which shows that X is a regular graph with exactly three eigenvalues. Since the largest eigenvalue is simple, X is connected, and so by Lemma 10.2.1, we find that X is strongly regular. Given the eigenvalues of X, we can compute that the parameter vector for X is $(120, 56, 28, 24)$. Note that

$$k - \frac{n}{2} = 56 - 60 = -4;$$

hence the switching class of X is a regular two-graph, by Theorem 11.7.1.

Each second subconstituent of X is a strongly regular graph on 63 vertices, namely the graph Sp(6).

Exercises

1. Let a and b be two nonzero vectors in $\mathbb{R}^m$. Show that $\rho_a\rho_b = \rho_b\rho_a$ if and only if either $\langle a\rangle = \langle b\rangle$ or $a^Tb = 0$.
2. Show that the set of lines spanned by the vectors of D_n is star-closed and indecomposable.
3. If $\mathcal{L}$ is a set of lines, then select a pair $\{x, -x\}$ of vectors of length $\sqrt{2}$ spanning each line. Define a graph Y on this set of vectors where adjacency is given by having inner product 1. If $\mathcal{L}$ is an indecomposable star-closed set of lines 60° and 90°, then find the diameter of Y.
4. Let Y be the graph defined on an indecomposable set of lines at 60° and 90° as in Exercise 3. Let a be a vertex of Y and let N be the set of neighbours of a. Show that each vector in N is adjacent to exactly one vector in $-N$.
5. Let S be a set of vectors in A_n. If the inner product of any pair of vectors in S is nonnegative, then show that the Gram matrix of S is equal to $2I + A$, where A is the adjacency matrix of the line graph of a bipartite graph.
6. Let X be the orthogonality graph of C^*. Let u and v be two nonadjacent vertices in X. Show that there is an automorphism of X that swaps u and v, but fixes all vertices adjacent to both u and v and all vertices adjacent to neither.
7. Let X be a graph with minimum eigenvalue at least -2. Show that if $\alpha(X) \geq 9$, then X is a generalized line graph.
8. If X is a line graph and u and v are vertices in it with the same set of neighbours, show that $u = v$.
9. If X is a generalized line graph, but not a line graph, show that there must be a pair of vertices in X with the same neighbours.

10. Show that Petersen's graph is neither a line graph nor a generalized line graph.
11. In which of the root systems E_6, E_7, and E_8 is Petersen's graph contained, and in which is it not? (It is contained in at least one, and it is not necessary to construct an embedding.)
12. Show that the set of lines spanned by the vectors in E_7 is star-closed and indecomposable.
13. Show that the set of lines spanned by the vectors in E_6 is star-closed and indecomposable.
14. Let X be a graph with least eigenvalue greater than -2. Show that if X is not a generalized line graph, then $|V(X)| \le 8$.
15. Determine the graphs with largest eigenvalue at most 2.

Notes

The results in this chapter are based on the fundamental paper by Cameron et al [2]. Alternative expositions are given in [1, 3]. These place less emphasis on the role of the generalized quadrangles with lines of size three. There is still scope for improvement in the results: In [4] Hoffman shows that if the minimum valency of X is large and $\theta_{\min}(X) > -1-\sqrt{2}$, then $\theta_{\min}(X) \ge -2$ and X is a generalized line graph. Neumaier and Woo [5] completely determine the structure of the graphs with $\theta_{\min}(X) \ge -1-\sqrt{2}$ and with sufficiently large minimum valency.

References

[1] A. E. Brouwer, A. M. Cohen, and A. Neumaier, *Distance-Regular Graphs*, Springer-Verlag, Berlin, 1989.

[2] P. J. Cameron, J.-M. Goethals, J. J. Seidel, and E. E. Shult, *Line graphs, root systems, and elliptic geometry*, J. Algebra, 43 (1976), 305–327.

[3] P. J. Cameron and J. H. van Lint, *Designs, Graphs, Codes and their Links*, Cambridge University Press, Cambridge, 1991.

[4] A. J. Hoffman, *On graphs whose least eigenvalue exceeds $-1-\sqrt{2}$*, Linear Algebra and Appl., 16 (1977), 153–165.

[5] R. Woo and A. Neumaier, *On graphs whose smallest eigenvalue is at least $-1-\sqrt{2}$*, Linear Algebra Appl., 226/228 (1995), 577–591.

13

The Laplacian of a Graph

The Laplacian is another important matrix associated with a graph, and the Laplacian spectrum is the spectrum of this matrix. We will consider the relationship between structural properties of a graph and the Laplacian spectrum, in a similar fashion to the spectral graph theory of previous chapters. We will meet Kirchhoff's expression for the number of spanning trees of a graph as the determinant of the matrix we get by deleting a row and column from the Laplacian. This is one of the oldest results in algebraic graph theory. We will also see how the Laplacian can be used in a number of ways to provide interesting geometric representations of a graph. This is related to work on the Colin de Verdière number of a graph, which is one of the most important recent developments in graph theory.

13.1 The Laplacian Matrix

Let σ be an arbitrary orientation of a graph X, and let D be the incidence matrix of X^σ. Then the *Laplacian* of X is the matrix $Q(X) = DD^T$. It is a consequence of Lemma 8.3.2 that the Laplacian does not depend on the orientation σ, and hence is well-defined.

Lemma 13.1.1 *Let X be a graph with n vertices and c connected components. If Q is the Laplacian of X, then* $\operatorname{rk} Q = n - c$.

Proof. Let D be the incidence matrix of an arbitrary orientation of X. We shall show that $\operatorname{rk} D = \operatorname{rk} D^T = \operatorname{rk} DD^T$, and the result then follows from Theorem 8.3.1. If $z \in \mathbb{R}^n$ is a vector such that $DD^T z = 0$, then

$z^T DD^T z = 0$. But this is the squared length of the vector $D^T z$, and hence we must have $D^T z = 0$. Thus any vector in the null space of DD^T is in the null space of D^T, which implies that $\operatorname{rk} DD^T = \operatorname{rk} D$. □

Let X be a graph on n vertices with Laplacian Q. Since Q is symmetric, its eigenvalues are real, and by Theorem 8.4.5, $\mathbb{R}^n$ has an orthogonal basis consisting of eigenvectors of Q. Since $Q = DD^T$, it is positive semidefinite, and therefore its eigenvalues are all nonnegative. We denote them by $\lambda_1(Q)$, ..., $\lambda_n(Q)$ with the assumption that

$$\lambda_1(Q) \leq \lambda_2(Q) \leq \cdots \leq \lambda_n(Q).$$

We use $\lambda_i(X)$ as shorthand for $\lambda_i(Q(X))$, or simply λ_i when Q is clear from the context or unimportant. We will also use λ_∞ to denote λ_n. For any graph, $\lambda_1 = 0$, because $Q\mathbf{1} = 0$. By Lemma 13.1.1, the multiplicity of zero as an eigenvalue of Q is equal to the number of components of X, and so for connected graphs, λ_2 is the smallest nonzero eigenvalue. Much of what follows will concentrate on the information determined by this particular eigenvalue.

If X is a regular graph, then the eigenvalues of the Laplacian are determined by the eigenvalues of the adjacency matrix.

Lemma 13.1.2 *Let X be a regular graph with valency k. If the adjacency matrix A has eigenvalues $\theta_1, \ldots, \theta_n$, then the Laplacian Q has eigenvalues $k - \theta_1, \ldots, k - \theta_n$.*

Proof. If X is k-regular, then $Q = \Delta(X) - A = kI - A$. Thus every eigenvector of A with eigenvalue θ is an eigenvector of Q with eigenvalue $k - \theta$. □

This shows that if two regular graphs are cospectral, then they also have the same Laplacian spectrum. However, this is not true in general; the two graphs of Figure 8.1 have different Laplacian spectra.

The next result describes the relation between the Laplacian spectrum of X and the Laplacian spectrum of its complement $\overline{X}$.

Lemma 13.1.3 *If X is a graph on n vertices and $2 \leq i \leq n$, then $\lambda_i(\overline{X}) = n - \lambda_{n-i+2}(X)$.*

Proof. We start by observing that

$$Q(X) + Q(\overline{X}) = nI - J. \tag{13.1}$$

The vector $\mathbf{1}$ is an eigenvector of $Q(X)$ and $Q(\overline{X})$ with eigenvalue 0. Let x be another eigenvector of $Q(X)$ with eigenvalue λ; we may assume that x is orthogonal to $\mathbf{1}$. Then $Jx = 0$, so

$$nx = (nI - J)x = Q(X)x + Q(\overline{X})x = \lambda x + Q(\overline{X})x.$$

Therefore, $Q(\overline{X})x = (n - \lambda)x$, and the lemma follows. □

Note that $nI - J = Q(K_n)$; thus (13.1) can be rewritten as

$$Q(X) + Q(\overline{X}) = Q(K_n).$$

From the proof of Lemma 13.1.3 it follows that the eigenvalues of $Q(K_n)$ are n, with multiplicity $n-1$, and 0, with multiplicity 1. Since $K_{m,n}$ is the complement of $K_m \cup K_n$, we can use this fact, along with Lemma 13.1.3, to determine the eigenvalues of the complete bipartite graph. We leave the pleasure of this computation to the reader, noting only the result that the characteristic polynomial of $Q(K_{m,n})$ is

$$t(t-m)^{n-1}(t-n)^{m-1}(t-m-n).$$

We note another useful consequence of Lemma 13.1.3.

Corollary 13.1.4 *If X is a graph on n vertices, then $\lambda_n(X) \leq n$. If $\overline{X}$ has $\bar{c}$ connected components, then the multiplicity of n as an eigenvalue of $Q(X)$ is $\bar{c} - 1$.* □

Our last result in this section is a property of the Laplacian that will provide us with a lot of information about its eigenvalues.

Lemma 13.1.5 *Let X be a graph on n vertices with Laplacian Q. Then for any vector x,*

$$x^T Q x = \sum_{uv \in E(X)} (x_u - x_v)^2.$$

Proof. This follows from the observations that

$$x^T Q x = x^T D D^T x = (D^T x)^T (D^T x)$$

and that if $uv \in E(X)$, then the entry of $D^T x$ corresponding to uv is $\pm(x_u - x_v)$. □

13.2 Trees

In this section we consider a classical result of algebraic graph theory, which shows that the number of spanning trees in a graph is determined by the Laplacian.

First we need some preparatory definitions. Let X be a graph, and let $e = uv$ be an edge of X. The graph $X \setminus e$ with vertex set $V(X)$ and edge set $E(X) \setminus e$ is said to be obtained by *deleting* the edge e. The graph X/e constructed by identifying the vertices u and v and then deleting e is said to be obtained by *contracting* e. Deletion and contraction are illustrated in Figure 13.1. If a vertex x is adjacent to both u and v, then there will be multiple edges between x and the newly identified vertex in X/e. Furthermore, if X itself has multiple edges, then any edges between

u and v other than e itself become loops on the newly identified vertex in X/e. Depending on the situation, it is sometimes possible to ignore loops, multiple edges, or both.

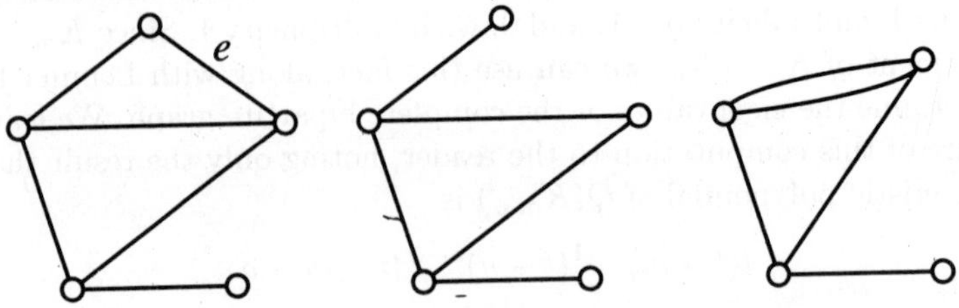

Figure 13.1. Graph Y, deletion $Y \setminus e$, and contraction Y/e

If M is a symmetric matrix with rows and columns indexed by the set V and if $S \subseteq V$, then let $M[S]$ denote the submatrix of M obtained by deleting the rows and columns indexed by elements of S.

Theorem 13.2.1 *Let X be a graph with Laplacian matrix Q. If u is an arbitrary vertex of X, then* $\det Q[u]$ *is equal to the number of spanning trees of X.*

Proof. We prove the theorem by induction on the number of edges of X.

Let $\tau(X)$ denote the number of spanning trees of X. If e is an edge of X, then every spanning tree either contains e or does not contain e, so we can count them according to this distinction. There is a one-to-one correspondence between spanning trees of X that contain e and spanning trees of X/e, so there are $\tau(X/e)$ such trees. Any spanning tree of X that does not contain e is a spanning tree of $X \setminus e$, and so there are $\tau(X \setminus e)$ of these. Therefore,

$$\tau(X) = \tau(X/e) + \tau(X \setminus e). \tag{13.2}$$

In this situation, multiple edges are retained during contraction, but we may ignore loops, because they cannot occur in a spanning tree.

Now, assume that $e = uv$, and let E be the $n \times n$ diagonal matrix with E_{vv} equal to 1, and all other entries equal to 0. Then

$$Q[u] = Q(X \setminus e)[u] + E,$$

from which we deduce that

$$\det Q[u] = \det Q(X \setminus e)[u] + \det Q(X \setminus e)[u, v]. \tag{13.3}$$

Note that $Q(X \setminus e)[u, v] = Q[u, v]$.

Assume that in forming X/e we contract u onto v, so that $V(X/e) = V(X) \setminus u$. Then $Q(X/e)[v]$ has rows and columns indexed by $V(X) \setminus \{u, v\}$ with the xy-entry being equal to Q_{xy}, and so we also have that $Q(X/e)[v] = Q[u, v]$.

Thus we can rewrite (13.3) as

$$\det Q[u] = \det Q(X \setminus e)[u] + \det Q(X/e)[v].$$

By induction, $\det Q(X\setminus e)[u] = \tau(X\setminus e)$ and $\det Q(X/e)[v] = \tau(X/e)$; hence (13.2) implies the theorem. □

It follows from Theorem 13.2.1 that $\det Q[u]$ is independent of the choice of the vertex u.

Corollary 13.2.2 *The number of spanning trees of K_n is n^{n-2}.*

Proof. This follows directly from the fact that $Q[u] = nI_{n-1} - J$ for any vertex u. □

If M is a square matrix, then denote by $M(i,j)$ the matrix obtained by deleting row i and column j from M. The *ij-cofactor* of M is the value

$$(-1)^{i+j} \det M(i,j).$$

The transposed matrix of cofactors of M is called the *adjugate* of M and denoted by $\operatorname{adj} M$. The ij-entry of $\operatorname{adj} M$ is the ji-cofactor of M. The most important property of the adjugate is that

$$M \operatorname{adj}(M) = (\det M)I.$$

If M is invertible, it implies that $M^{-1} = (\det M)^{-1} \operatorname{adj}(M)$. Theorem 13.2.1 implies that if Q is the Laplacian of a graph, then the diagonal entries of $\operatorname{adj}(Q)$ are all equal. The full truth is somewhat surprising: All of the entries of $\operatorname{adj}(Q)$ are equal.

Lemma 13.2.3 *Let $\tau(X)$ denote the number of spanning trees in the graph X and let Q be its Laplacian. Then* $\operatorname{adj}(Q) = \tau(X)J$.

Proof. Suppose that X has n vertices. Assume first that X is not connected, so that $\tau(X) = 0$. Then Q has rank at most $n-2$, so any submatrix of Q of order $(n-1) \times (n-1)$ is singular and $\operatorname{adj}(Q) = 0$.

Thus we may assume that X is connected. Then $\operatorname{adj}(Q) \neq 0$, but nonetheless $Q \operatorname{adj}(Q) = 0$. Because X is connected, $\ker Q$ is spanned by $\mathbf{1}$, and therefore each column of $\operatorname{adj}(Q)$ must be a constant vector. Since $\operatorname{adj}(Q)$ is symmetric, it follows that it is a nonzero multiple of J; now the result follows at once from Theorem 13.2.1. □

To prove the next result we need some information about the characteristic polynomial of a matrix. If A and B are square $n \times n$ matrices, then $\det(A+B)$ may be computed as follows. For each subset S of $\{1,\ldots,n\}$, let A_S be the matrix obtained by replacing the rows of A indexed by elements of S with the corresponding rows of B. Then

$$\det(A+B) = \sum_S \det A_S.$$

Applying this to $tI + (-A)$, we deduce that the coefficient of t^{n-k} in $\det(tI - A)$ is $(-1)^k$ times the sum of the determinants of the principal $k \times k$ submatrices of A. (This is a classical result, due to Laplace.)

Lemma 13.2.4 *Let X be a graph on n vertices, and let $\lambda_1, \ldots, \lambda_n$ be the eigenvalues of the Laplacian of X. Then the number of spanning trees in X is $\frac{1}{n}\prod_{i=2}^{n} \lambda_i$.*

Proof. The result clearly holds if X is not connected, so we may assume without loss that X is connected. Let $\phi(t)$ denote the characteristic polynomial $\det(tI - Q)$ of the Laplacian Q of X. The zeros of $\phi(t)$ are the eigenvalues of Q. Since $\lambda_1 = 0$, its constant term is zero and the coefficient of t is

$$(-1)^{n-1} \prod_{i=2}^{n} \lambda_i.$$

On the other hand, by our remarks just above, the coefficient of the linear term in $\phi(t)$ is

$$(-1)^{n-1} \sum_{u \in V(X)} \det Q[u].$$

This yields the lemma immediately. □

13.3 Representations

Define a *representation* ρ of a graph X in $\mathbb{R}^m$ to be a map ρ from $V(X)$ into $\mathbb{R}^m$. Informally, we think of a representation as the positions of the vertices in an m-dimensional drawing of a graph. Figure 13.2 shows a representation of the cube in $\mathbb{R}^3$.

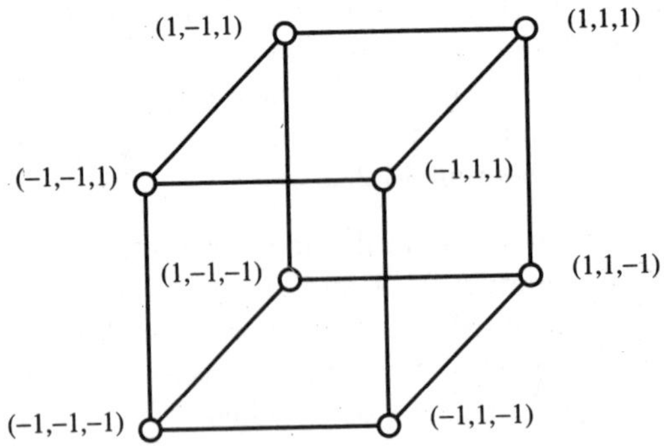

Figure 13.2. The cube in $\mathbb{R}^3$

We regard the vectors $\rho(u)$ as row vectors, and thus we may represent ρ by the $|V(X)| \times m$ matrix R with the images of the vertices of X as its rows.

Suppose then that ρ maps $V(X)$ into $\mathbb{R}^m$. We say ρ is *balanced* if

$$\sum_{u \in V(X)} \rho(u) = 0.$$

Thus ρ is balanced if and only if $\mathbf{1}^T R = 0$. The representation of Figure 13.2 is balanced. A balanced representation has its "centre of gravity" at the origin, and clearly we can translate any representation so that it is balanced without losing any information. Henceforth we shall assume that a representation is balanced.

If the columns of the matrix R are not linearly independent, then the image of X is contained in a proper subspace of $\mathbb{R}^m$, and ρ is just a lower-dimensional representation embedded in $\mathbb{R}^m$. Any maximal linearly independent subset of the columns of R would suffice to determine all the properties of the representation. Therefore, we will furthermore assume that the columns of R are linearly independent.

We can imagine building a physical model of X by placing the vertices in the positions specified by ρ and connecting adjacent vertices by identical springs. It is natural to consider a representation to be better if it requires the springs to be less extended. Letting $\|x\|$ denote the Euclidean length of a vector x, we define the *energy* of a representation ρ to be the value

$$\mathcal{E}(\rho) = \sum_{uv \in E(X)} \|\rho(u) - \rho(v)\|^2,$$

and hope that natural or good drawings of graphs correspond to representations with low energy. (Of course, the representation with least energy is the one where each vertex is mapped to the zero vector. Thus we need to add further constraints, to exclude this.)

We can go further by dropping the assumption that the springs are identical. To model this, let ω be a function from the edges of X to the positive real numbers, and define the energy $\mathcal{E}(\rho)$ of a representation ρ of X by

$$\mathcal{E}(\rho) = \sum_{uv \in E(X)} \omega_{uv} \|\rho(u) - \rho(v)\|^2,$$

where ω_{uv} denotes the value of ω on the edge uv. Let W be the diagonal matrix with rows and columns indexed by the edges of X and with the diagonal entry corresponding to the edge uv equal to ω_{uv}.

The next result can be viewed as a considerable generalization of Lemma 13.1.5.

Lemma 13.3.1 *Let ρ be a representation of the edge-weighted graph X, given by the $|V(X)| \times m$ matrix R. If D is an oriented incidence matrix* *for X, then*

$$\mathcal{E}(\rho) = \operatorname{tr} R^T DWD^T R.$$

Proof. The rows of $D^T R$ are indexed by the edges of X, and if $uv \in E(X)$, then the uv-row of $D^T R$ is $\pm(\rho(u) - \rho(v))$. Consequently, the diagonal entries of $D^T RR^T D$ have the form $\|\rho(u) - \rho(v)\|^2$, where uv ranges over the edges of X. Hence

$$\mathcal{E}(\rho) = \operatorname{tr} WD^T RR^T D = \operatorname{tr} R^T DWD^T R$$

as required. □

We may view $Q = DWD^T$ as a *weighted Laplacian.* If $uv \in E(X)$, then $Q_{uv} = -\omega_{uv}$, and for each vertex u in X,

$$Q_{uu} = \sum_{v \sim u} \omega_{uv}.$$

Thus $Q\mathbf{1} = 0$. Conversely, any symmetric matrix Q with nonpositive off-diagonal entries such that $Q\mathbf{1} = 0$ is a weighted Laplacian.

Note that $R^T DWD^T R$ is an $m \times m$ symmetric matrix; hence its eigenvalues are real. The sum of the eigenvalues is the trace of the matrix, and hence the energy of the representation is given by the sum of the eigenvalues of $R^T DWD^T R$.

For the normalized representation of the cube we have (with $W = I$)

$$R^T = \frac{1}{\sqrt{8}} \begin{pmatrix} 1 & 1 & -1 & -1 & 1 & 1 & -1 & -1 \\ 1 & -1 & -1 & 1 & 1 & -1 & -1 & 1 \\ 1 & 1 & 1 & 1 & -1 & -1 & -1 & -1 \end{pmatrix},$$

$$Q = \begin{pmatrix} 3 & -1 & 0 & -1 & -1 & 0 & 0 & 0 \\ -1 & 3 & -1 & 0 & 0 & -1 & 0 & 0 \\ 0 & -1 & 3 & -1 & 0 & 0 & -1 & 0 \\ -1 & 0 & -1 & 3 & 0 & 0 & 0 & -1 \\ -1 & 0 & 0 & 0 & 3 & -1 & 0 & -1 \\ 0 & -1 & 0 & 0 & -1 & 3 & -1 & 0 \\ 0 & 0 & -1 & 0 & 0 & -1 & 3 & -1 \\ 0 & 0 & 0 & -1 & -1 & 0 & -1 & 3 \end{pmatrix},$$

which implies that

$$R^T QR = \begin{pmatrix} 2 & 0 & 0 \\ 0 & 2 & 0 \\ 0 & 0 & 2 \end{pmatrix}$$

and $\mathcal{E}(\rho) = 6$. This can be confirmed directly by noting that each of the 12 edges of the cube has length $1/\sqrt{2}$.

13.4 Energy and Eigenvalues

We now show that the energy of certain representations of a graph X are determined by the eigenvalues of the Laplacian of X. If M is an invertible $m \times m$ matrix, then the map that sends u to $\rho(u)M$ is another representation of X. This representation is given by the matrix RM and provides as much information about X as does ρ. From this point of view the representation is determined by its column space. Therefore, we may assume that the columns of R are orthogonal to each other, and as above that each column has norm 1. In this situation the matrix R satisfies $R^T R = I_m$, and the representation is called an *orthogonal representation.*

Theorem 13.4.1 *Let X be a graph on n vertices with weighted Laplacian Q. Assume that the eigenvalues of Q are $\lambda_1 \leq \cdots \leq \lambda_n$ and that $\lambda_2 > 0$. The minimum energy of a balanced orthogonal representation of X in $\mathbb{R}^m$ equals $\sum_{i=2}^{m+1} \lambda_i$.*

Proof. By Lemma 13.3.1 the energy of a representation is tr $R^T QR$. From Corollary 9.5.2, the energy of an orthogonal representation in $\mathbb{R}^\ell$ is bounded below by the sum of the ℓ smallest eigenvalues of Q. We can realize this lower bound by taking the columns of R to be vectors $x_1, \ldots, x_\ell$ such that $Qx_i = \lambda_i x_i$.

Since $\lambda_2 > 0$, we must have $x_1 = \mathbf{1}$, and therefore by deleting x_1 we obtain a balanced orthogonal representation in $\mathbb{R}^{\ell-1}$, with the same energy. Conversely, we can reverse this process to obtain an orthogonal representation in $\mathbb{R}^\ell$ from a balanced orthogonal representation in $\mathbb{R}^{\ell-1}$ such that these two representations have the same energy. Therefore, the minimum energy of a balanced orthogonal representation of X in $\mathbb{R}^m$ equals the minimum energy of an orthogonal representation in $\mathbb{R}^{m+1}$, and this minimum equals $\lambda_2 + \cdots + \lambda_{m+1}$. □

This result provides an intriguing automatic method for drawing a graph in any number of dimensions. Compute an orthonormal basis of eigenvectors $x_1, \ldots, x_n$ for the Laplacian Q and let the columns of R be $x_2, \ldots, x_{m+1}$. Theorem 13.4.1 implies that this yields an orthogonal balanced representation of minimum energy. The representation is not necessarily unique, because it may be the case that $\lambda_{m+1} = \lambda_{m+2}$, in which case there is no reason to choose between x_{m+1} and x_{m+2}.

Figure 13.3 shows that such a representation (in $\mathbb{R}^2$) can look quite appealing, while Figure 13.4 shows that it may be less appealing.

Both of these graphs are planar graphs on 10 vertices, but in both cases the drawing is not planar. Worse still, in general there is no guarantee that the images of the vertices are even distinct. The representation of the cube in $\mathbb{R}^3$ given above can be obtained by this method.

More generally, any pairwise orthogonal triple of eigenvectors of Q provides an orthogonal representation in $\mathbb{R}^3$, and this representation may have

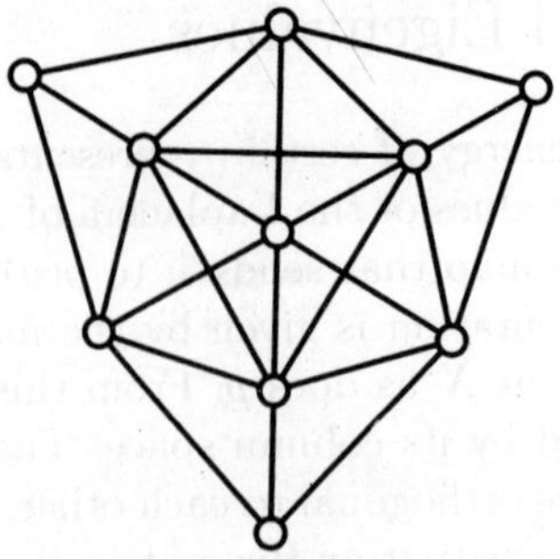

Figure 13.3. A planar triangulation represented in $\mathbb{R}^2$

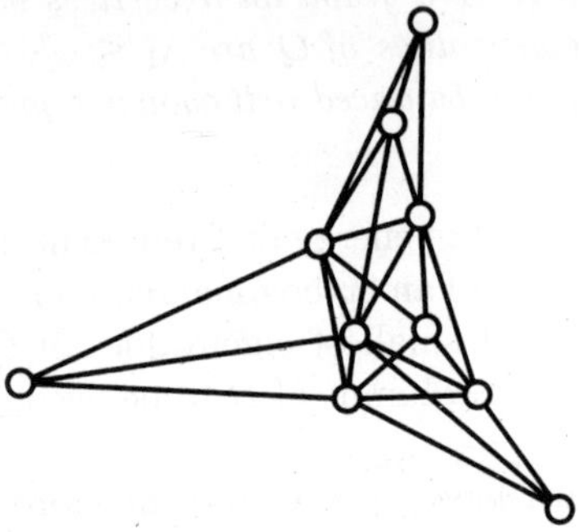

Figure 13.4. A planar triangulation represented in $\mathbb{R}^2$

pleasing properties, even if we do not choose the eigenvectors that minimize the energy.

We finish this section with a corollary to Theorem 13.4.1.

Corollary 13.4.2 *Let X be a graph on n vertices. Then the minimum value of*

$$\frac{\sum_{uv \in E(X)} (x_u - x_v)^2}{\sum_u x_u^2},$$

as x ranges over all nonzero vectors orthogonal to $\mathbf{1}$, *is $\lambda_2(X)$. The maximum value is $\lambda_\infty(X)$.* □

13.5 Connectivity

Our main result in this section is a consequence of the following bound.

Theorem 13.5.1 *Suppose that S is a subset of the vertices of the graph X. Then $\lambda_2(X) \leq \lambda_2(X \setminus S) + |S|$.*

Proof. Let z be a unit vector of length n such that (when viewed as a function on $V(X)$) its restriction to S is zero, and its restriction to $V(X)\backslash S$ is an eigenvector of $Q(X \setminus S)$ orthogonal to $\mathbf{1}$ and with eigenvalue θ. Then by Corollary 13.4.2

$$\lambda_2(X) \le \sum_{uv \in E(X)} (z_u - z_v)^2.$$

Hence by dividing the edges into those with none, one, or two endpoints in $X \setminus S$ we get

$$\lambda_2(X) \le \sum_{u \in S} \sum_{v \sim u} z_v^2 + \sum_{uv \in E(X \setminus S)} (z_u - z_v)^2 \le |S| + \theta.$$

We may take $\theta = \lambda_2(X \setminus S)$, and hence the result follows. □

If S is a vertex-cutset, then $X \setminus S$ is disconnected, so $\lambda_2(X \setminus S) = 0$, and we have the following bound on the vertex connectivity of a graph.

Corollary 13.5.2 *For any graph X we have $\lambda_2(X) \le \kappa_0(X)$.* □

It follows from our observation in Section 13.1 or from Exercise 4 that the characteristic polynomial of $Q(K_{1,n})$ is $t(t-1)^{n-1}(t-n-1)$. This provides one family of examples where λ_2 equals the vertex connectivity.

Provided that X is not complete, the vertex connectivity of X is bounded above by the edge connectivity, which, in turn, is bounded above by the minimum valency $\delta(X)$ of a vertex in X. We thus have the following useful inequalities for noncomplete graphs:

$$\lambda_2(X) \le \kappa_0(X) \le \kappa_1(X) \le \delta(X).$$

Note that deleting a vertex may increase λ_2. For example, suppose $X = K_n$, where $n \ge 3$, and Y is constructed by adding a new vertex adjacent to two distinct vertices in X. Then $\lambda_2(Y) \le 2$, since $\delta(Y) = 2$, but $\lambda_2(X) = n$.

Recall that a bridge is an edge whose removal disconnects a graph, and thus a graph has edge-connectivity one if and only if it has a bridge. In this case, the above result shows that $\lambda_2(X) \le 1$ unless $X = K_2$. It has been noted empirically that λ_2 seems to give a fairly natural measure of the "shape" of a graph. Graphs with small values of λ_2 tend to be elongated graphs of large diameter with bridges, whereas graphs with larger values of λ_2 tend to be rounder with smaller diameter, and larger girth and connectivity.

For cubic graphs, this observation can be made precise, at least as regards the minimum values for λ_2. If $n \ge 10$ and $n \equiv 2 \bmod 4$, the graphs shown in Figure 13.5 have the smallest value of λ_2 among all cubic graphs on n vertices. If $n \ge 12$ and $n \equiv 0 \bmod 4$, the graphs shown in Figure 13.6 have the smallest value of λ_2 among all cubic graphs on n vertices. In both cases these graphs have the maximum diameter among all cubic graphs on n vertices.

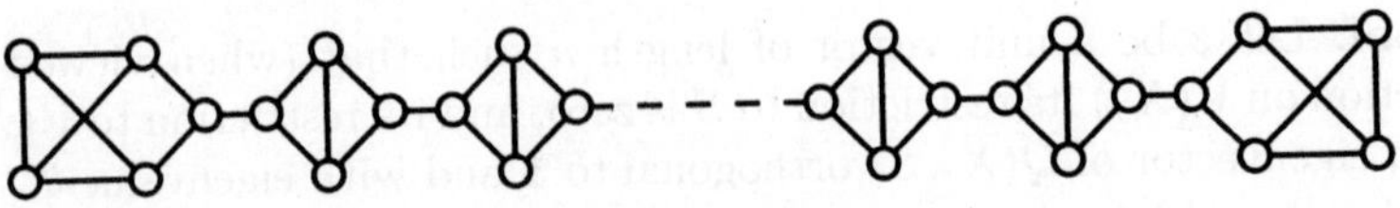

Figure 13.5. Cubic graph with minimum λ_2 on $n \equiv 2 \bmod 4$ vertices

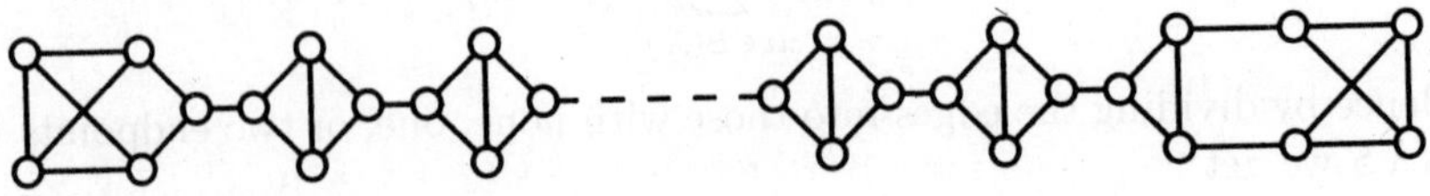

Figure 13.6. Cubic graph with minimum λ_2 on $n \equiv 0 \bmod 4$ vertices

13.6 Interlacing

We now consider what happens to the eigenvalues of $Q(X)$ when we add an edge to X.

Lemma 13.6.1 *Let X be a graph and let Y be obtained from X by adding an edge joining two distinct vertices of X. Then*

$$\lambda_2(X) \leq \lambda_2(Y) \leq \lambda_2(X) + 2.$$

Proof. Suppose we get Y by joining vertices r and s of X. For any vector z we have

$$z^T Q(Y) z = \sum_{uv \in E(Y)} (z_u - z_v)^2 = (z_r - z_s)^2 + \sum_{uv \in E(X)} (z_u - z_v)^2.$$

If we choose z to be a unit eigenvector of $Q(Y)$, orthogonal to **1**, and with eigenvalue $\lambda_2(Y)$, then by Corollary 13.4.2 we get

$$\lambda_2(Y) \geq \lambda_2(X) + (z_r - z_s)^2. \tag{13.4}$$

On the other hand, if we take z to be a unit eigenvector of $Q(X)$, orthogonal to **1**, with eigenvalue $\lambda_2(X)$, then by Corollary 13.4.2 we get

$$\lambda_2(Y) \leq \lambda_2(X) + (z_r - z_s)^2. \tag{13.5}$$

It follows from (13.4) that $\lambda_2(X) \leq \lambda_2(Y)$. We can complete the proof by appealing to (13.5). Since $z_r^2 + z_s^2 \leq 1$, it is straightforward to see that $(z_r - z_s)^2 \leq 2$, and the result is proved. □

A few comments on the above proof. If we add an edge joining the two vertices in $2K_1$ (to get K_2), then λ_2 increases from 0 to 2. Although this example might not be impressive, it does show that the upper bound can be tight. The full story is indicated in Exercise 8.

Next, the reader may well have thought that we forgot to insist that the edge added to X in the lemma has to join two distinct and nonadjacent vertices. In fact, the proof works without alteration even if the two vertices chosen are adjacent. We say no more, because here we are not really interested in graphs with multiple edges.

Theorem 13.6.2 *Let X be a graph with n vertices and let Y be obtained from X by adding an edge joining two distinct vertices of X. Then $\lambda_i(X) \le \lambda_i(Y)$, for all i, and $\lambda_i(Y) \le \lambda_{i+1}(X)$ if $i < n$.*

Proof. Suppose we add the edge uv to X to get Y. Let z be the vector of length n with u-entry and v-entry 1 and -1, respectively, and all other entries equal to 0. Then $Q(Y) = Q(X) + zz^T$, and if we use Q to denote $Q(X)$, we have

$$tI - Q(Y) = tI - Q - zz^T = (tI - Q)(I - (tI - Q)^{-1}zz^T).$$

By Lemma 8.2.4,

$$\det(I - (tI - Q)^{-1}zz^T) = 1 - z^T(tI - Q)^{-1}z,$$

and therefore

$$\frac{\det(tI - Q(Y))}{\det(tI - Q(X))} = 1 - z^T(tI - Q)^{-1}z.$$

The result now follows from Theorem 8.13.3, applied to the rational function $\psi(t) = 1 - z^T(tI - Q)^{-1}z$, and the proof of Theorem 9.1.1. □

One corollary of this and Theorem 13.4.1 is that if X is a spanning subgraph of Y, then the energy of any balanced orthogonal representation of Y can never be less than the energy of the induced representation of X.

As another corollary of the theorem, we prove again that the Petersen graph does not have a Hamilton cycle. The eigenvalues of the adjacency matrix of the Petersen graph are 3, 1, and -2, with multiplicities 1, 5, and 4, respectively. Therefore, the eigenvalues of the Laplacian matrix for the Petersen graph are 0, 2, and 5, with multiplicities 1, 5, and 4, respectively. The eigenvalues of the adjacency matrix of C_{10} are $2\cos(\pi r/5)$, for $r = 0, 1, \ldots, 9$. It follows that

$$\lambda_6(C_{10}) = (3 + \sqrt{5})/2 > \lambda_6(P) = 2.$$

Consequently, the eigenvalues of the Laplacian matrix of C_{10} do not interlace the eigenvalues of the Laplacian matrix of the Petersen graph, and therefore the Petersen graph does not have a Hamilton cycle.

We present two further examples in Figure 13.7; we can prove that these graphs are not hamiltonian by considering their Laplacians in this fashion. These two graphs are of some independent interest. They are cubic hypohamiltonian graphs, which are somewhat rare. The first graph, on 18 vertices, is one of the two smallest cubic hypohamiltonian graphs after the Petersen graph. Like the Petersen graph it cannot be 3-edge coloured (it is one of the *Blanuša snarks*). The second graph, on 22 vertices, belongs to an infinite family of hypohamiltonian graphs.

It is interesting to note that the technique described in Section 9.2 using the adjacency matrix is not strong enough to prove that these two graphs are not hamiltonian. However, there are cases where the adjacency matrix technique works, but the Laplacian technique does not.

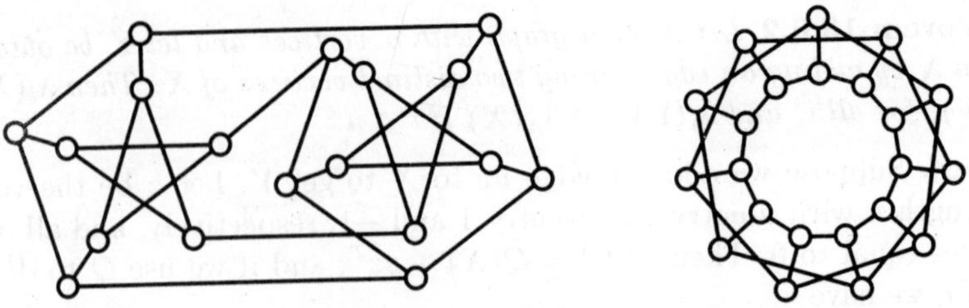

Figure 13.7. Two nonhamiltonian graphs

13.7 Conductance and Cutsets

We now come to some of the most important applications of λ_2. If X is a graph and $S \subseteq V(X)$, let ∂S denote the set of edges with one end in S and the other in $V(X) \setminus S$.

Lemma 13.7.1 *Let X be a graph on n vertices and let S be a subset of $V(X)$. Then*

$$\lambda_2(X) \leq \frac{n|\partial S|}{|S|(n-|S|)}.$$

Proof. Suppose $|S| = a$. Let z be the vector (viewed as a function on $V(X)$) whose value is $n-a$ on the vertices in S and $-a$ on the vertices not in S. Then z is orthogonal to $\mathbf{1}$, so by Corollary 13.4.2

$$\lambda_2(X) \leq \frac{\sum_{uv \in E(X)}(z_u - z_v)^2}{\sum_u z_u^2} = \frac{|\partial S|n^2}{a(n-a)^2 + (n-a)a^2}.$$

The lemma follows immediately from this. □

By way of a simple example, if S is a single vertex with valency k, then the lemma implies that $\lambda_2(X) \leq kn/(n-1)$. This is weaker than Fiedler's result that λ_2 is no greater than the minimum valency of X (Theorem 13.5.1), although not by much.

Our next application is much more important. Define the *conductance* $\Phi(X)$ of a graph X to be the minimum value of

$$\frac{|\partial S|}{|S|},$$

where S ranges over all subsets of $V(X)$ of size at most $|V(X)|/2$. (Many authors refer to this quantity as the *isoperimetric number* of a graph. We follow Lovász, which seems safe.) From Lemma 13.7.1 we have at once the following:

Corollary 13.7.2 *For any graph X we have $\Phi(X) \geq \lambda_2(X)/2$.* □

The real significance of this bound is that λ_2 can be computed to a given number of digits in polynomial time, whereas determining the conductance of a graph is an NP-hard problem. A family of graphs with constant valency and conductance bounded from below by a positive constant is called a family of *expanders*. These are important in theoretical computer science, if not in practice.

The *bisection width* of a graph on n vertices is the minimum value of $|\partial S|$, for any subset S of size $\lfloor n/2 \rfloor$. Again, this is NP-hard to compute, but we do have the following:

Corollary 13.7.3 *The bisection width of a graph X on $2m$ vertices is at least $m\lambda_2(X)/2$.* □

We apply this to the k-cube Q_k. In Exercise 13 it is established that $\lambda_2(Q_k) = 2$, from which it follows that the bisection width of the k-cube is at least 2^{k-1}. Since this value is easily realized, we have thus found the exact value.

Let $\operatorname{bip}(X)$ denote the maximum number of edges in a spanning bipartite subgraph of X. This equals the maximum value of $|\partial S|$, where S ranges over all subsets of $V(X)$ with size at most $|V(X)|/2$.

Lemma 13.7.4 *If X is a graph with n vertices, then* $\operatorname{bip}(X) \leq n\lambda_\infty(X)/4$.

Proof. By applying Lemma 13.7.1 to the complement of X we get

$$|\partial S| \leq |S|(n - |S|)\lambda_\infty(X)/n \leq n\lambda_\infty(X)/4,$$

which is the desired inequality. □

13.8 How to Draw a Graph

We will describe a remarkable method, due to Tutte, for determining whether a 3-connected graph is planar.

Lemma 13.8.1 *Let S be a set of points in $\mathbb{R}^m$. Then the vector x in $\mathbb{R}^m$ minimizes $\sum_{y \in S} \|x - y\|^2$ if and only if*

$$x = \frac{1}{|S|} \sum_{y \in S} y.$$

Proof. Let $\hat{y}$ be the centroid of the set S, i.e.,

$$\hat{y} = \frac{1}{|S|} \sum_{y \in S} y.$$

Then

$$\sum_{y \in S} \|x - y\|^2 = \sum_{y \in S} \|(x - \hat{y}) + (\hat{y} - y)\|^2$$

$$= |S|\,\|x-\hat{y}\|^2 + \sum_{y\in S}\|\hat{y}-y\|^2 + 2\sum_{y\in S}\langle x-\hat{y}, \hat{y}-y\rangle$$

$$= |S|\,\|x-\hat{y}\|^2 + \sum_{y\in S}\|\hat{y}-y\|^2.$$

Therefore, this is a minimum if and only if $x=\hat{y}$. □

We say that a representation ρ of X is *barycentric* relative to a subset F of $V(X)$ if for each vertex u not in F, the vector $\rho(u)$ is the centroid of the images of the neighbours of u. A barycentric representation can easily be made balanced, but will normally not be orthogonal. If the images of the vertices in F are specified, then a barycentric embedding has minimum energy. Our next result formalizes the connection with the Laplacian.

Lemma 13.8.2 *Let F be a subset of the vertices of X, let ρ be a representation of X, and let R be the matrix whose rows are the images of the vertices of X. Let Q be the Laplacian of X. Then ρ is barycentric relative to F if and only if the rows of QR corresponding to the vertices in $X\setminus F$ are all zero.*

Proof. The vector x is the centroid of the vectors in S if and only if

$$\sum_{y\in S}(x-y)=0.$$

If u has valency d, the u-row of QR is equal to

$$d\rho(u)-\sum_{v\sim u}\rho(v)=\sum_{v\sim u}\rho(u)-\rho(v).$$

The lemma follows. □

Lemma 13.8.3 *Let X be a connected graph, let F be a subset of the vertices of X, and let σ be a map from F into $\mathbb{R}^m$. If $X\setminus F$ is connected, there is a unique m-dimensional representation ρ of X that extends σ and is barycentric relative to F.*

Proof. Let Q be the Laplacian of X. Assume that we have

$$Q=\begin{pmatrix} Q_1 & B^T \\ B & Q_2 \end{pmatrix},$$

where the rows and columns of Q_1 are indexed by the vertices of F. Let R be the matrix describing the representation ρ. We may assume

$$R=\begin{pmatrix} R_1 \\ R_2 \end{pmatrix},$$

where R_1 gives the values of σ on F. Then ρ extends σ and is barycentric (relative to F) if and only if

$$\begin{pmatrix} Q_1 & B^T \\ B & Q_2 \end{pmatrix}\begin{pmatrix} R_1 \\ R_2 \end{pmatrix}=\begin{pmatrix} Y_1 \\ 0 \end{pmatrix}.$$

Then $BR_1 + Q_2R_2 = 0$, and so if Q_2 is invertible, this yields that

$$R_2 = -Q_2^{-1}BR_1, \qquad Y_1 = (Q_1 - B^TQ_2B)R_1.$$

We complete the proof by showing that since $X \setminus F$ is connected, Q_2 is invertible. Let $Y = X \setminus F$. Then there is a nonnegative diagonal matrix Δ_2 such that

$$Q_2 = Q(Y) + \Delta_2.$$

Since X is connected, $\Delta_2 \neq 0$. We prove that Q_2 is positive definite. We have

$$x^TQ_2x = x^TQ(Y)x + x^T\Delta_2x.$$

Because $x^TQ(Y)x = \sum_{ij\in E(Y)}(x_i - x_j)^2$, we see that $x^TQ(Y)x \geq 0$ and that $x^TQ(Y)x = 0$ if and only if $x = c\mathbf{1}$ for some c. But now $x^T\Delta_2x = c^2\mathbf{1}^T\Delta_2\mathbf{1}$, and this is positive unless $c = 0$. Therefore, $x^TQ_2x > 0$ unless $x = 0$; in other words, Q_2 is positive definite, and consequently it is invertible. □

Tutte showed that each edge in a 3-connected graph lies in a cycle C such that no edge not in C joins two vertices of C and $X \setminus C$ is connected. He called these *peripheral cycles*. For example, any face of a 3-connected planar graph can be shown to be a peripheral cycle.

Suppose that C is a peripheral cycle of size r in a 3-connected graph X and suppose that we are given a mapping σ from $V(C)$ to the vertices to a convex r-gon in $\mathbb{R}^2$, such that adjacent vertices in C are adjacent in the polygon. It follows from Lemma 13.8.3 that there is a unique barycentric representation ρ of X relative to F. This determines a drawing of X in the plane, with all vertices of $X \setminus C$ inside the image of C. Tutte proved the truly remarkable result that this drawing has no crossings if and only if X is planar.

Peripheral cycles can be found in polynomial time, and given this, Lemma 13.8.3 provides an automatic method for drawing 3-connected planar graphs. Unfortunately, from an aesthetic viewpoint, the quality of the output is variable. Sometimes there is a good choice of outside face, maybe a large face as in Figure 13.8 or one that is preserved by an automorphism as in Figure 13.9.

However, particularly if there are a lot of triangular faces, the algorithm tends to produce a large number of faces-within-faces, many of which are minuscule.

13.9 The Generalized Laplacian

The rest of this chapter is devoted to a generalization of the Laplacian matrix of a graph. There are many generalized Laplacians associated with

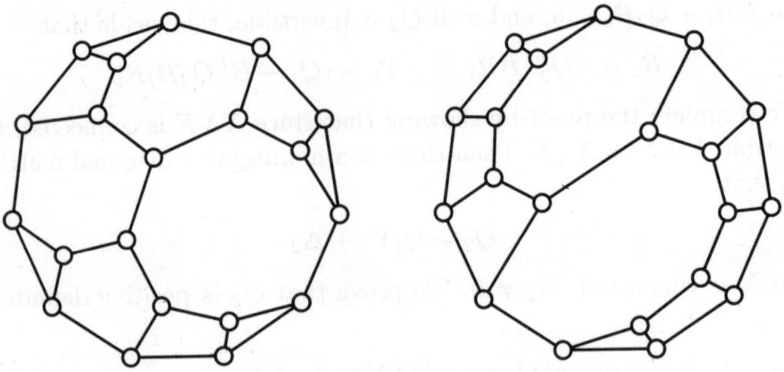

Figure 13.8. Tutte embeddings of cubic planar graphs

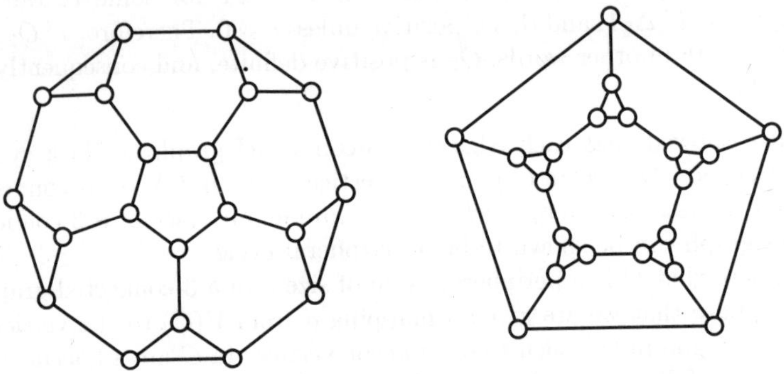

Figure 13.9. Different Tutte embeddings of the same graph

each graph, which at first sight seem only tenuously related. Nevertheless, graph-theoretical properties of a graph constrain the algebraic properties of the entire class of generalized Laplacians associated with it. The next few sections provide an introduction to this important and recent development.

Let X be a graph with n vertices. We call a symmetric $n \times n$ matrix Q a *generalized Laplacian* of X if $Q_{uv} < 0$ when u and v are adjacent vertices of X and $Q_{uv} = 0$ when u and v are distinct and not adjacent. There are no constraints on the diagonal entries of Q; in particular, we do not require that $Q\mathbf{1} = 0$. The ordinary Laplacian is a generalized Laplacian, and if A is the adjacency matrix of X, then $-A$ is a generalized Laplacian.

As with the usual Laplacian, we will denote the eigenvalues of a generalized Laplacian Q by

$$\lambda_1(Q) \leq \lambda_2(Q) \leq \cdots \leq \lambda_n(Q).$$

We will be concerned with the eigenvectors in the λ_2-eigenspace of Q. If Q is a generalized Laplacian of X, then for any c, the matrix $Q - cI$ is a generalized Laplacian with the same eigenvectors as Q. Therefore, we can freely assume that $\lambda_2(Q) = 0$, whenever it is convenient to do so.

Lemma 13.9.1 *Let X be a graph with a generalized Laplacian Q. If X is connected, then $\lambda_1(Q)$ is simple and the corresponding eigenvector can be taken to have all its entries positive.*

Proof. Choose a constant c such that all diagonal entries of $Q - cI$ are nonpositive. By the Perron–Frobenius theorem (Theorem 8.8.1), the largest eigenvalue of $-Q+cI$ is simple and the associated eigenvector may be taken to have only positive entries. □

If x is a vector with entries indexed by the vertices of X, then the *positive support* $\mathrm{supp}_+(x)$ consists of the vertices u such that $x_u > 0$, and the *negative support* $\mathrm{supp}_-(x)$ of the vertices u such that $x_u < 0$. A *nodal domain* of x is a component of one of the subgraphs induced by $\mathrm{supp}_+(x)$ or $\mathrm{supp}_-(x)$. A nodal domain is *positive* if it is a component of $\mathrm{supp}_+(x)$; otherwise, it is *negative*.

If Y is a nodal domain of x, then x_Y is the vector given by

$$(x_Y)_u = \begin{cases} |x_u|, & u \in Y; \\ 0, & \text{otherwise.} \end{cases}$$

If Y and Z are distinct nodal domains with the same sign, then since no edges of X join vertices in Y to vertices in Z,

$$x_Y^T Q x_Z = 0. \tag{13.6}$$

Lemma 13.9.2 *Let x be an eigenvector of Q with eigenvalue λ and let Y be a positive nodal domain of x. Then $(Q - \lambda I)x_Y \leq 0$.*

Proof. Let y denote the restriction of x to $V(Y)$ and let z be the restriction of x to $V(X) \setminus \mathrm{supp}_+(x)$. Let Q_Y be the submatrix of Q with rows and columns indexed by $V(Y)$, and let B_Y be the submatrix of Q with rows indexed by $V(Y)$ and with columns indexed by $V(X) \setminus \mathrm{supp}_+(x)$. Since $Qx = \lambda x$, we have

$$Q_Y y + B_Y z = \lambda y. \tag{13.7}$$

Since B_Y and z are nonpositive, $B_Y z$ is nonnegative, and therefore

$$Q_Y y \leq \lambda y.$$

□

It is not necessary for x to be an eigenvector for the conclusion of this lemma to hold; it is sufficient that $(Q - \lambda I)x \leq 0$. Given our discussion in Section 8.7, we might say that it suffices that x be λ-*superharmonic*.

Corollary 13.9.3 *Let x be an eigenvector of Q with eigenvalue λ, and let U be the subspace spanned by the vectors x_Y, where Y ranges over the positive nodal domains of x. If $u \in U$, then $u^T(Q - \lambda I)u \leq 0$.*

Proof. If $u = \sum_Y a_Y x_Y$, then using (13.6), we find that

$$u^T(Q - \lambda I)u = \sum_Y a_Y^2 x_Y^T(Q - \lambda I)x_Y,$$

and so the claim follows from the previous lemma. □

Theorem 13.9.4 *Let X be a connected graph, let Q be a generalized Laplacian of X, and let x be an eigenvector for Q with eigenvalue $\lambda_2(Q)$. If x has minimal support, then $\operatorname{supp}_+(x)$ and $\operatorname{supp}_-(x)$ induce connected subgraphs of X.*

Proof. Suppose that v is a λ_2-eigenvector with distinct positive nodal domains Y and Z. Because X is connected, λ_1 is simple and the span of v_Y and v_Z contains a vector, u say, orthogonal to the λ_1-eigenspace.

Now, u can be expressed as a linear combination of eigenvectors of Q with eigenvalues at least λ_2; consequently, $u^T(Q - \lambda_2 I)u \geq 0$ with equality if and only if u is a linear combination of eigenvectors with eigenvalue λ_2.

On the other hand, by Corollary 13.9.3, we have $u^T(Q - \lambda_2 I)u \leq 0$, and so $u^T(Q - \lambda_2 I)u = 0$. Therefore, u is an eigenvector of Q with eigenvalue λ_2 and support equal to $V(Y) \cup V(Z)$.

Any λ_2-eigenvector has both positive and negative nodal domains, because it is orthogonal to the λ_1-eigenspace. Therefore, the preceding argument shows that an eigenvector with distinct nodal domains of the same sign does not have minimal support. Therefore, since x has minimal support, it must have precisely one positive and one negative nodal domain. □

Lemma 13.9.5 *Let Q be a generalized Laplacian of a graph X and let x be an eigenvector of Q. Then any vertex not in $\operatorname{supp}(x)$ either has no neighbours in $\operatorname{supp}(x)$, or has neighbours in both $\operatorname{supp}_+(x)$ and $\operatorname{supp}_-(x)$.*

Proof. Suppose that $u \notin \operatorname{supp}(x)$, so $x_u = 0$. Then

$$0 = (Qx)_u = Q_{uu}x_u + \sum_{v \sim u} Q_{uv}x_v = \sum_{v \sim u} Q_{uv}x_v.$$

Since $Q_{uv} < 0$ when v is adjacent to u, either $x_v = 0$ for all vertices adjacent to u, or the sum has both positive and negative terms. In the former case u is not adjacent to any vertex in $\operatorname{supp}(x)$; in the latter it is adjacent to vertices in both $\operatorname{supp}_+(x)$ and $\operatorname{supp}_-(x)$. □

13.10 Multiplicities

In this section we show that if X is 2-connected and outerplanar, then λ_2 has multiplicity at most two, and that if X is 3-connected and planar, then λ_2 has multiplicity at most three. In the next section we show that if equality holds in the latter case, then the representation provided by the λ_2-eigenspace yields a planar embedding of X.

Lemma 13.10.1 *Let Q be a generalized Laplacian for the graph X. If X is 3-connected and planar, then no eigenvector of Q with eigenvalue $\lambda_2(Q)$ vanishes on three vertices in the same face of any embedding of X.*

Proof. Let x be an eigenvector of Q with eigenvalue λ_2, and suppose that u, v, and w are three vertices not in $\text{supp}(x)$ lying in the same face. We may assume that x has minimal support, and hence $\text{supp}_+(x)$ and $\text{supp}_-(x)$ induce connected subgraphs of X. Let p be a vertex in $\text{supp}_+(x)$. Since X is 3-connected, Menger's theorem implies that there are three paths in X joining p to u, v, and w such that any two of these paths have only the vertex p in common. It follows that there are three vertex-disjoint paths P_u, P_v, and P_w joining u, v, and w, respectively, to some triple of vertices in $N(\text{supp}_+(x))$. Each of these three vertices is also adjacent to a vertex in $\text{supp}_-(x)$. Since both the positive and negative support induce connected graphs, we may now contract all vertices in $\text{supp}_+(x)$ to a single vertex, all vertices in $\text{supp}_-(x)$ to another vertex, and each of the paths P_u, P_v, and P_w to u, v, and w, respectively. The result is a planar graph which contains a copy of $K_{2,3}$ with its three vertices of valency two all lying on the same face. This is impossible. □

Corollary 13.10.2 *Let Q be a generalized Laplacian for the graph X. If X is 3-connected and planar, then $\lambda_2(Q)$ has multiplicity at most three.*

Proof. If λ_2 has multiplicity at least four, then there is an eigenvector in the associated eigenspace whose support is disjoint from any three given vertices. Thus we conclude that λ_2 has multiplicity at most three. □

The graph $K_{2,n}$ is 2-connected and planar. Its adjacency matrix A has eigenvalues $\pm\sqrt{2n}$, both simple, and 0 with multiplicity $n-2$. Taking $Q = -A$, we see that we cannot drop the assumption that X is 3-connected in the last result.

Lemma 13.10.3 *Let X be a 2-connected plane graph with a generalized Laplacian Q, and let x be an eigenvector of Q with eigenvalue $\lambda_2(Q)$ and with minimal support. If u and v are adjacent vertices of a face F such that $x_u = x_v = 0$, then F does not contain vertices from both the positive and negative support of x.*

Proof. Since X is 2-connected, the face F is a cycle. Suppose that F contains vertices p and q such that $x_p > 0$ and $x_q < 0$. Without loss of generality we can assume that they occur in the order u, v, q, and p clockwise around the face F, and that the portion of F from q to p contains only vertices not in $\text{supp}(x)$. Let v' be the first vertex not in $\text{supp}(x)$ encountered moving anticlockwise around F from q, and let u' be the first vertex not in $\text{supp}(x)$ encountered moving clockwise around F

from p. Then u', v', q, and p are distinct vertices of F and occur in that order around F. Let P be a path from v' to p all of whose vertices other than v' are in $\mathrm{supp}_+(x)$, and let N be a path from u' to q all of whose vertices other than u' are in $\mathrm{supp}_-(x)$. The existence of the paths P and N is a consequence of Corollary 13.9.4 and Lemma 13.9.5. Because F is a face, the paths P and N must both lie outside F, and since their endpoints are interleaved around F, they must cross. This is impossible, since P and N are vertex-disjoint, and so we have the necessary contradiction. □

We call a graph *outerplanar* if it has a planar embedding with a face that contains all the vertices. Neither $K_{2,3}$ nor K_4 is outerplanar, and it is known that a graph is outerplanar if and only if it has no minor isomorphic to one of these two graphs. (A *minor* of a graph X is a graph obtained by contracting edges in a subgraph of X.)

Corollary 13.10.4 *Let X be a graph on n vertices with a generalized Laplacian Q. If X is 2-connected and outerplanar, then $\lambda_2(Q)$ has multiplicity at most two.*

Proof. If λ_2 had multiplicity greater than two, then we could find an eigenvector x with eigenvalue λ_2 such that x vanished on two adjacent vertices in the sole face of X. However, since x must be orthogonal to the eigenvector with eigenvalue λ_1, both $\mathrm{supp}_+(x)$ and $\mathrm{supp}_-(x)$ must be nonempty. □

The tree $K_{1,n}$ is outerplanar, but if A is its adjacency matrix, then $-A$ is a generalized Laplacian for it with λ_2 having multiplicity greater than two. Hence we cannot drop the assumption in the corollary that X be 2-connected.

13.11 Embeddings

We have seen that if X is a 3-connected planar graph and Q is a generalized Laplacian for X, then $\lambda_2(Q)$ has multiplicity at most three. The main result of this section is that if $\lambda_2(Q)$ has multiplicity exactly three, then the representation ρ provided by the λ_2-eigenspace of Q provides a planar embedding of X on the unit sphere.

As a first step we need to verify that in the case just described, no vertex is mapped to zero by ρ. This, and more, follows from the next result.

Lemma 13.11.1 *Let X be a 3-connected planar graph with a generalized Laplacian Q such that $\lambda_2(Q)$ has multiplicity three. Let ρ be a representation given by a matrix U whose columns form a basis for the λ_2-eigenspace of Q. If F is a face in some planar embedding of X, then the images under ρ of any two vertices in F are linearly independent.*

Proof. Assume by way of contradiction that u and v are two vertices in a face of X such that $\rho(u) = \alpha\rho(v)$ for some real number α, and let w be a third vertex in the same face. Then we can find a linear combination of the columns of U that vanishes on the vertices u, v, and w, thus contradicting Lemma 13.10.1. □

If ρ is a representation of X that maps no vertex to zero, then we define the *normalized representation* $\hat{\rho}$ by

$$\hat{\rho} = \|\rho(u)\|^{-1}\rho(u).$$

Suppose that X is a 3-connected planar graph with a generalized Laplacian Q such that $\lambda_2(Q)$ has multiplicity three, and let ρ be the representation given by the λ_2-eigenspace. By the previous lemma, the corresponding normalized representation $\hat{\rho}$ is well-defined and maps every vertex to a point of the unit sphere. If u and v are adjacent in X, then $\hat{\rho}(u) \neq \pm\hat{\rho}(v)$, so there is a unique geodesic on the sphere joining the images of u and v. Thus we have a well-defined embedding of the graph X on the unit sphere, and our task is to show that this embedding is planar, i.e., distinct edges can meet only at a vertex.

If $C \subseteq \mathbb{R}^n$, then the *convex cone* generated by C is the set of all nonnegative linear combinations of the elements of C. A subset of the unit sphere is *spherically convex* if whenever it contains points u and v, it contains all points on any geodesic joining u to v. The intersection of the unit sphere with a convex cone is spherically convex. Suppose that F is a face in some planar drawing of X, and consider the convex cone C generated by the images under $\hat{\rho}$ of the vertices of F. This meets the unit sphere in a convex spherical polygon, and by Lemma 13.10.3, each edge of F determines an edge of this polygon.

This does not yet imply that our embedding of X on the sphere has no crossings; for example, the images of distinct faces of X could overlap. Our next result removes some of the difficulty.

Lemma 13.11.2 *Let X be a 2-connected planar graph. Suppose it has a planar embedding where the neighbours of the vertex u are, in cyclic order, $v_1, \dots, v_k$. Let Q be a generalized Laplacian for X such that $\lambda_2(Q)$ has multiplicity three. Then the planes spanned by the pairs $\{\rho(u), \rho(v_i)\}$ are arranged in the same cyclic order around the line spanned by $\rho(u)$ as the vertices v_i are arranged around u.*

Proof. Let x be an eigenvector with eigenvalue λ_2 with minimal support such that $x(u) = x(v_1) = 0$. (Here we are viewing x as a function on $V(X)$.) By Lemma 13.10.1, we see that neither $x(v_2)$ nor $x(v_k)$ can be zero, and replacing x by $-x$ if needed, we may suppose that $x(v_2) > 0$. Given this, we prove that $x(v_k) < 0$.

Suppose that there are some values h, i, and j such that $2 \leq h < i < j \leq k$ and $x(v_h) > 0$, $x(v_j) > 0$, and $x(v_i) \leq 0$. Since $\text{supp}_+(x)$ is connected, the vertices v_h and v_j are joined in X by a path with all vertices in $\mathrm{supp}_+(x)$. Taken with u, this path forms a cycle in X that separates v_1 from v_i. Since X is 2-connected, there are two vertex-disjoint paths P_1 and P_i joining v_1 and v_i respectively to vertices in $N(\mathrm{supp}_+(x))$. The end-vertices of these paths other than v_1 and v_i are adjacent to vertices in $\mathrm{supp}_-(x)$, and thus we have found two vertices in $\mathrm{supp}_-(x)$ that are separated by vertices in $\mathrm{supp}_+(x)$. This contradicts the fact that $\mathrm{supp}_-(x)$ is connected.

It follows that there is exactly one index i such that $x(v_i) > 0$ and $x(v_{i+1}) \leq 0$. Since $x(u) = 0$ and $x(v_2) > 0$, it follows from Lemma 13.9.5 that u has a neighbour in $\mathrm{supp}_-(x)$, and therefore $x(v_k)$ must be negative.

From this we see that if we choose x such that $x(u) = x(v_i) = 0$ and $x(v_{i+1}) > 0$, then $x(v_{i-1}) < 0$ (where the subscripts are computed modulo k). The lemma follows at once from this. □

We now prove that the embedding provided by $\hat{\rho}$ has no crossings. The argument is topological.

Suppose that X is drawn on a sphere S_a without crossings. Let S_b be a unit sphere, with X embedded on it using $\hat{\rho}$, as described above. The normalized representation $\hat{\rho}$ provides an injective map from the vertices of X in S_a to the vertices of X in S_b. By our remarks preceding the previous lemma, this map extends to a continuous map ψ from S_a to S_b, which injectively maps each face on S_a to a spherically convex region on S_b. From Lemma 13.11.2, it even follows that ψ is injective on the union of the faces of X that contain a given vertex. Hence ψ is a continuous locally injective map from S_a to S_b.

It is a standard result that such a map must be injective; we outline a proof. First, since ψ is continuous and locally injective, there is an integer k such that $|\psi^{-1}(x)| = k$ for each point x on S_b. Let Y be any graph embedded on S_b, with v vertices, e edges, and f faces. Then $\psi^{-1}(Y)$ is a plane graph on S_a with kv vertices, ke edges, and kf faces. By Euler's formula,

$$2 = kv - ke + kf = k(v - e + f) = 2k,$$

and therefore $k = 1$.

Thus we have shown that ψ is injective, and therefore it is a homeomorphism. We conclude that $\hat{\rho}$ embeds X without crossings.

Exercises

1. If D is the incidence matrix of an oriented graph, then show that any square submatrix of D has determinant 0, 1, or -1.

2. Show that the determinant of a square submatrix of $B(X)$ is equal to 0 or $\pm 2^r$, for some integer r.

3. If M is a matrix, let $M(i|j)$ denote the submatrix we get by deleting row i and column j. Define a 2-forest in a graph to be a spanning forest with exactly two components. Let Q be the Laplacian of X. If u, p, and q are vertices of X and $p \neq q$, show that $\det Q[u](p|q)$ is equal to the number of 2-forests with u in one component and p and q in the other.
4. Determine the characteristic polynomial of $Q(K_{m,n})$.
5. An *arborescence* is an acyclic directed graph with a root vertex u such that u has in-valency 0 and each vertex other than u has in-valency 1 and is joined to u by a directed path. (In other words, it is a tree oriented so that all arcs point away from the root.) Let Y be a directed graph with adjacency matrix A and let D be the diagonal matrix with ith diagonal entry equal to the in-valency of the ith vertex of Y. Show that the number of spanning arborescences in Y rooted at a given vertex u is equal to $\det((D - A)[u])$.
6. Show that if X is connected and has n vertices, then

$$\lambda_2(X) = \min_x \frac{n \sum_{ij \in E(X)} (x_i - x_j)^2}{\sum_{i<j} (x_i - x_j)^2},$$

where the minimum is taken over all nonconstant vectors x.

7. Show that if T is a tree with at least three vertices, then $\lambda_2(T) \leq 1$, with equality if and only if T is a star (i.e., is isomorphic to $K_{1,n}$).
8. Let r and s be distinct nonadjacent vertices in the graph X. If $e \in E(X)$, show that $\lambda_2(X \setminus e) = \lambda_2(X) - 2$ if and only if X is complete.
9. Let D be an oriented incidence matrix for the graph X. Let d_i denote the valency of the vertex i in X. Show that the largest eigenvalue of $D^T D$ is bounded above by the maximum value of $d_i + d_j$, for any two adjacent vertices i and j in X. Deduce that this is also an upper bound on λ_∞. (And for even more work, show that this bound is tight if and only if X is bipartite and semiregular.)
10. Let X be a connected graph on n vertices. Show that there is a subset of $V(X)$ such that $2|S| \leq n$,

$$\frac{|\partial S|}{|S|} = \Phi(X),$$

and the subgraphs induced by S and $V \setminus S$ are both connected.

11. Let X be a graph on n vertices with diameter d. Show that $\lambda_2 \geq 1/nd$.
12. If X is the Cartesian product of two graphs Y and Z, show that $\lambda_2(X)$ is the minimum of $\lambda_2(Y)$ and $\lambda_2(Z)$. (Hint: Find eigenvectors for X, and hence determine all eigenvalues of X in terms of those of its factors.)

13. Use Exercise 12 to show that $\lambda_2(Q_k) = 2$.

14. If X is an arc-transitive graph with diameter d and valency r, show that $\Phi(X) \geq r/2d$.

15. Show that a cycle in a 3-connected planar graph is a peripheral cycle if and only if it is a face in every planar embedding of the graph.

16. Let X be a connected graph and let z be an eigenvector of $Q(X)$ with eigenvalue λ_2. Call a path $u_1, \ldots, u_r$ strictly decreasing if the values of z on the vertices of the path form a strictly decreasing sequence. Show that if $u \in V(X)$ and $z_u > 0$, then u is joined by a strictly decreasing path to some vertex v such that $z_v \leq 0$.

17. Let X be a connected graph. Show that if $Q(X)$ has exactly three distinct eigenvalues, then there is a constant μ such that any pair of distinct nonadjacent vertices in X have exactly μ common neighbours. Show further that there is a constant $\bar{\mu}$ such that any pair of distinct nonadjacent vertices in $\overline{X}$ have exactly $\bar{\mu}$ common neighbours. Find a graph X with this property that is not regular. (A regular graph would be strongly regular.)

18. Let Q be a generalized Laplacian for a connected graph X. If x is an eigenvector for Q with eigenvalue λ_2 and u is a vertex in X such that x_u is maximal, prove that

$$Q_{uu} + \sum_{v \sim u} Q_{uv} \leq \lambda_2.$$

19. Let Q be a generalized Laplacian for a connected graph X and consider the representation ρ provided by the λ_2-eigenspace. Show that if $\rho(u)$ does not lie in the convex hull of the set

$$N := \{\rho(v) : v \sim u\} \cup \{0\},$$

then there is a vector a such that $a^T\rho(u) > a^T\rho(v)$, for any neighbour v of u. (Do not struggle with this; quote a result from optimization.) Deduce that if $\rho(u)$ does not lie in the convex hull of N, then

$$Q_{uu} + \sum_{v \sim u} Q_{uv} < \lambda_2.$$

20. Let Q be a generalized Laplacian for a path. Show that all the eigenvalues of Q are simple.

21. Let Q be a generalized Laplacian for a connected graph X and let x be an eigenvector for Q with eigenvalue λ_2. Show that if no entries of x are zero, then both $\text{supp}_+(x)$ and $\text{supp}_-(x)$ are connected.

22. Let Q be a generalized Laplacian for a connected graph X, let x be an eigenvector for Q with eigenvalue λ_2 and let C be the vertex set of some component of $\text{supp}(x)$. Show that $N(C) = N(\text{supp}(x))$.

Notes

Theorem 13.4.1 comes from Pisanski and Shawe-Taylor [8], and our discussion in Section 13.3 and Section 13.4 follows their treatment. Fiedler [2] introduced the study of λ_2. He called it the *algebraic connectivity* of a graph. In [3], Fiedler proves that if z is an eigenvector for the connected graph X with eigenvalue λ_2 and $c \leq 0$, then the graph induced by the set

$$\{u \in V(X) : z_u \geq c\}$$

is connected. Exercise 16 shows that it suffices to prove this when $c = 0$.

Our work in Section 13.6 is a modest extension of an idea due to Mohar, which we treated in Section 9.2. Van den Heuvel [11] offers further applications of this type.

Alon uses Lemma 13.7.4 to show that there is a positive constant c such that for every e there is a triangle-free graph with e edges and

$$\mathrm{bip}(X) \leq \frac{e}{2} + ce^{4/5}.$$

Lovász devotes a number in exercises in Chapter 11 of [4] to conductance.

Section 13.8 is, of course, based on [9], one of Tutte's many masterpieces.

The final sections are based on work of van der Holst, Schrijver, and Lovász [12], [13], [5]. These papers are motivated by the study of the *Colin de Verdière number* of a graph. This is defined to be the maximum corank of a generalized Laplacian Q that also satisfies the additional technical condition that there is no nonzero matrix B such that $QB = 0$ and $B_{uv} = 0$ when u is equal or adjacent to v. For an introduction to this important subject we recommend [13].

For a solution to Exercise 5, see [4]. The result in Exercise 9 comes from [1]. Exercise 10 comes from Mohar [6]. B. D. McKay proved that if X has n vertices and diameter d, then $d \geq 4/n\lambda_2$. This is stronger than the result of Exercise 11, and is close to optimal for trees. For a proof of the stronger result, see Mohar [7]; for the weaker bound, try [4]. It might appear that we do not need lower bounds on the diameter, as after all, it can be computed in polynomial time. The problem is that this is polynomial time in the number of vertices of a graph. However, we may wish to bound the diameter of a Cayley graph given by its connection set; in this case we need to compute the diameter in time polynomial in the size of the connection set, i.e., the valency of the graph. Exercise 17 is based on van Dam and Haemers [10].

The Colin de Verdière number of a graph is less than or equal to three if and only if the graph is planar, and in this case, we can find a generalized Laplacian of maximum corank. The null space of this generalized Laplacian then yields a planar representation of the graph, using the method described in Section 13.11. However, for a general graph X, we do not know how to find a suitable generalized Laplacian with corank equal to the Colin de

Verdière number of X, nor do we know any indirect method to determine it.

References

[1] W. N. ANDERSON JR. AND T. D. MORLEY, *Eigenvalues of the Laplacian of a graph*, Linear and Multilinear Algebra, 18 (1985), 141–145.

[2] M. FIEDLER, *Algebraic connectivity of graphs*, Czechoslovak Math. J., 23(98) (1973), 298–305.

[3] ——, *A property of eigenvectors of nonnegative symmetric matrices and its application to graph theory*, Czechoslovak Math. J., 25(100) (1975), 619–633.

[4] L. LOVÁSZ, *Combinatorial Problems and Exercises*, North-Holland Publishing Co., Amsterdam, second edition, 1993.

[5] L. LOVÁSZ AND A. SCHRIJVER, *On the null space of a Colin de Verdière matrix*, Ann. Inst. Fourier (Grenoble), 49 (1999), 1017–1026.

[6] B. MOHAR, *Isoperimetric numbers of graphs*, J. Combin. Theory Ser. B, 47 (1989), 274–291.

[7] ——, *Eigenvalues, diameter, and mean distance in graphs*, Graphs Combin., 7 (1991), 53–64.

[8] T. PISANSKI AND J. SHAWE-TAYLOR, *Characterising graph drawing with eigenvectors*, J. Chem. Inf. Comput. Sci, 40 (2000), 567–571.

[9] W. T. TUTTE, *How to draw a graph*, Proc. London Math. Soc. (3), 13 (1963), 743–767.

[10] E. R. VAN DAM AND W. H. HAEMERS, *Graphs with constant μ and $\overline{\mu}$*, Discrete Math., 182 (1998), 293–307.

[11] J. VAN DEN HEUVEL, *Hamilton cycles and eigenvalues of graphs*, Linear Algebra Appl., 226/228 (1995), 723–730.

[12] H. VAN DER HOLST, *A short proof of the planarity characterization of Colin de Verdière*, J. Combin. Theory Ser. B, 65 (1995), 269–272.

[13] H. VAN DER HOLST, L. LOVÁSZ, AND A. SCHRIJVER, *The Colin de Verdière graph parameter*, in Graph theory and combinatorial biology (Balatonlelle, 1996), János Bolyai Math. Soc., Budapest, 1999, 29–85.

14
Cuts and Flows

Let X be a graph with an orientation σ and let D be the incidence matrix of X^σ. In this chapter we continue the study of how graph-theoretic properties of X are reflected in the algebraic properties of D. As previously, the orientation is merely a device used to prove the results, and the results themselves are independent of which particular orientation is chosen. Let $\mathbb{R}^E$ and $\mathbb{R}^V$ denote the real vector spaces with coordinates indexed by the edges and vertices of X, respectively. Then the column space of D^T is a subspace of $\mathbb{R}^E$, called the *cut space* of X. The orthogonal complement of this vector space is called the *flow space* of X.

One of the aims of this chapter is to study the relationship between the properties of X and its cut space and flow space. In particular, we observe that there are significant parallels between the properties of the cut space and the flow space of X, and that if X is planar, this extends to a formal duality.

We also study the set of integral vectors in the cut space and the flow space; these form a *lattice*, and the geometric properties of this lattice yield further information about the structure of X.

We may also choose to view D as a matrix over a field other than the real numbers, in particular the finite field $GF(2)$. Since the cut space and the flow space are orthogonal, they intersect trivially when considered as vector spaces over the real numbers. Over $GF(2)$, however, their intersection may be nontrivial; it is known as the *bicycle space* of X, and will play a role in our work on knots in Chapter 17.

14.1 The Cut Space

Let X be a graph with an orientation σ, and let D be the incidence matrix of X^σ. The column space of D^T is known as the cut space of the oriented graph. Clearly, the cut space depends not only on X, but also on the orientation assigned to X. However, the results in this chapter are independent of the particular orientation, and so we abuse notation by referring simply to the *cut space* of X with the understanding that σ is some fixed, but arbitrarily chosen, orientation.

The first result is an immediate consequence of Theorem 8.3.1:

Theorem 14.1.1 *If X is a graph with n vertices and c connected components, then its cut space has dimension $n - c$.* □

Next we shall examine the elements of the cut space of X that justify its name. If (U, V) is a partition of $V(X)$ into two nonempty subsets, the set of edges uv with u in U and v in V is a *cut*. We call U and V the *shores* of the cut. A nonempty cut that is minimal (with respect to inclusion) is called a *bond*. If X is connected, then the shores of a bond are connected.

An *oriented cut* is a cut with one shore declared to be the positive shore $V(+)$, and the other the negative shore $V(-)$. Using the orientation of X, an oriented cut determines a vector $z \in \mathbb{R}^E$ as follows:

$$z_e = \begin{cases} 0, & \text{if } e \text{ is not in } C; \\ 1, & \text{if the head of } e \text{ is in } V(+); \\ -1, & \text{if the head of } e \text{ is in } V(-). \end{cases}$$

We refer to z as the *signed characteristic vector* of the cut C. The sign on an edge in the cut is $+1$ if the edge is oriented in the same sense as the cut, and -1 otherwise.

Now, each vertex u determines an oriented cut $C(u)$ with positive shore $\{u\}$ and negative shore $V(X) \setminus u$. The u-column of D^T is the signed characteristic vector of the cut $C(u)$, and so these vectors lie in the cut space of X. In fact, we can say much more.

Lemma 14.1.2 *If X is a graph, then the signed characteristic vector of each cut lies in the cut space of X. The nonzero elements of the cut space with minimal support are scalar multiples of the signed characteristic vectors of the bonds of X.*

Proof. First let C be a cut in X and suppose that $V(+)$ and $V(-)$ are its shores. Let y be the characteristic vector in $\mathbb{R}^V$ of $V(+)$ and consider the vector $D^T y$. It takes the value 0 on any edge with both ends in the same shore of C and is equal to ± 1 on the edges C; its value is positive on e if and only if the head of e lies in $V(+)$. So $D^T y$ is the signed characteristic vector of C.

Now, let x be a nonzero element of the cut space of X. Then $x = D^T y$ for some nonzero vector $y \in \mathbb{R}^V$. The vector y determines a partition of

$V(X)$ with cells

$$S(\alpha) = \{u \in V(X) \mid y_u = \alpha\}.$$

An edge is in $\text{supp}(x)$ if and only if its endpoints lie in distinct cells of this partition, so the support of x is determined only by this partition. If there are edges between more than one pair of cells, x does not have minimal support, because a partition created by merging two of the cells would determine an element x' with $\text{supp}(x') \subset \text{supp}(x)$.

Therefore, if x has minimal support, the only edges between distinct cells all lie between two cells $S(\alpha)$ and $S(\beta)$. This implies that x is a scalar multiple of the signed characteristic vector of the cut with shores $S(\alpha)$ and $V(X)\backslash S(\alpha)$. Finally, we observe that if x is the signed characteristic vector of a cut, then it has minimal support if and only if that cut is a bond. □

There are a number of natural bases for the cut space of X. If X has c connected components, then we can form a basis for the cut space by taking the signed characteristic vectors of the cuts $C(u)$, as u ranges over every vertex but one in each component. This yields $n - c$ independent vectors, which therefore form a basis.

There is a second class of bases that will also prove useful. First we consider the case where X is connected. Let T be a spanning tree of X and let e be an edge of T. Then $T \setminus e$ has exactly two connected components, so we can define an oriented cut $C(T, e)$ whose positive shore is the component containing the head of e and whose negative shore is the component containing the tail of e.

Lemma 14.1.3 *Let X be a connected graph and let T be a spanning tree of X. Then the signed characteristic vectors of the $n-1$ cuts $C(T,e)$, for $e \in E(T)$, form a basis for the cut space of X.*

Proof. An edge $e \in E(T)$ is in the cut $C(T, e)$ but not in any of the cuts $C(T, f)$ for $f \neq e$. Therefore, the signed characteristic vectors of the cuts are linearly independent, and since there are $n - 1$ vectors, they form a basis. □

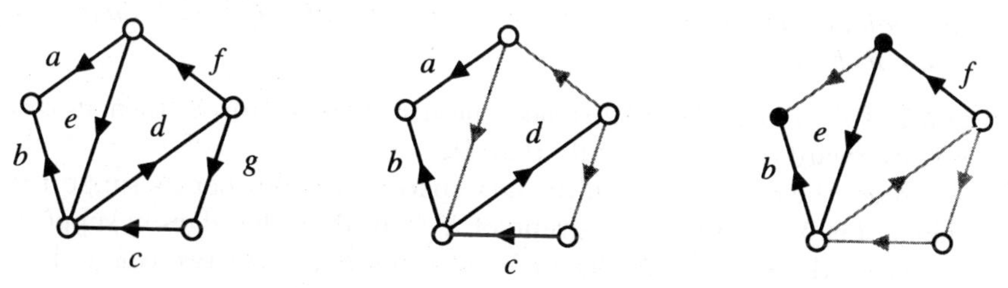

Figure 14.1. An oriented graph X, spanning tree T, and cut $C(T, b)$

Figure 14.1 shows a graph, together with a spanning tree T and a cut defined in this fashion.

This result is easily extended to the case where X is not connected. Recall that in a graph with c connected components, a maximal spanning forest is a spanning forest of c trees, each spanning one connected component. If F is a maximal spanning forest and e is an edge in F, then $F \setminus e$ has exactly $c+1$ components. Define a cut whose positive shore contains the component containing the head of e and whose negative shore contains the component containing the tail of e. Denote the resulting cut (which is independent of which shore the other components are assigned to) by $C(F, e)$. Then the signed characteristic vectors of the $n - c$ cuts $C(F, e)$ form a basis for the cut space of X.

14.2 The Flow Space

The *flow space* of a graph X is the orthogonal complement of its cut space, so therefore consists of all the vectors x such that $Dx = 0$. The dimension of the flow space of X follows directly from elementary linear algebra.

Theorem 14.2.1 *If X is a graph with n vertices, m edges, and c connected components, then its flow space has dimension $m - n + c$.* □

Let C be a cycle in a graph X. An *oriented cycle* is obtained by choosing a particular cyclic order $u_1, \ldots, u_r$ for the vertices of C. Using the orientation of X, each oriented cycle C determines an element $z \in \mathbb{R}^E$ as follows:

$$z_e = \begin{cases} 0, & \text{if } e \text{ is not in } C; \\ 1, & \text{if } e = u_i u_{i+1} \text{ and } u_{i+1} \text{ is the head of } e; \\ -1, & \text{if } e = u_i u_{i+1} \text{ and } u_{i+1} \text{ is the tail of } e. \end{cases}$$

We will refer to z as the signed characteristic vector of the oriented cycle C. The sign on an edge in the cycle is $+1$ if the edge is oriented in the same sense as the cycle, and -1 otherwise.

Theorem 14.2.2 *If X is a graph, then the signed characteristic vector of each cycle lies in the flow space of X. The nonzero elements of the flow space with minimal support are scalar multiples of the signed characteristic vectors of the cycles of X.*

Proof. If C is a cycle with signed characteristic vector z, then it is a straightforward exercise to verify that $Dz = 0$.

Suppose, then, that y lies in the flow space of X and that its support is minimal. Let Y denote the subgraph of X formed by the edges in $\operatorname{supp}(y)$. Any vertex that lies in an edge of Y must lie in at least two edges of Y. Hence Y has minimum valency at least two, and therefore it contains a cycle.

Suppose that C is a cycle formed from edges in Y. Then, for any real number α, the vector

$$y' = y + \alpha z$$

is in the flow space of X and has support contained in Y.

By choosing α appropriately, we can guarantee that there is an edge of C not in $\mathrm{supp}(y')$. But since y has minimal support, this implies that $y' = 0$, and hence y is a scalar multiple of z. □

If x and y are vectors with the same support S, then either x is a scalar multiple of y, or there is a vector in the span of x and y with support a proper subset of S. This implies that the set of minimal supports of vectors in a finite-dimensional vector space is finite, and that the vectors with minimal support in a subspace must span it. This has the following consequence:

Corollary 14.2.3 *The flow space of X is spanned by the signed characteristic vectors of its cycles.* □

There are also a number of natural bases for the flow space of a graph. If F is a maximal spanning forest of X, then any edge not in F is called a *chord* of F. If e is a chord of F, then e together with the path in F from the head of e to the tail of e is a cycle in X. If X has n vertices, m edges, and c connected components, then this provides us with $m - n + c$ cycles. Since each chord is in precisely one of these cycles, the signed characteristic vectors of these cycles are linearly independent in $\mathbb{R}^E$, and hence they form a basis for the flow space of X.

The spanning tree depicted in Figure 14.1 has three chords; Figure 14.2 shows the cycles corresponding to these chords.

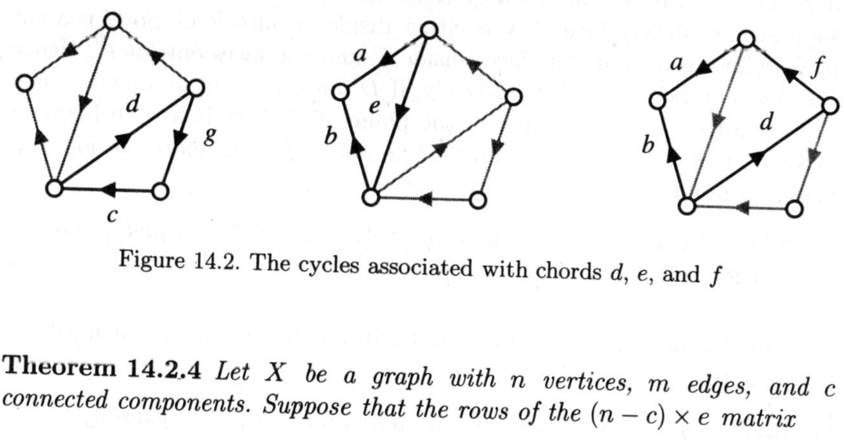

Figure 14.2. The cycles associated with chords d, e, and f

Theorem 14.2.4 *Let X be a graph with n vertices, m edges, and c connected components. Suppose that the rows of the $(n-c) \times e$ matrix*

$$M = (I \quad R)$$

form a basis for the cut space of X. Then the rows of the $(m-n+c)\times e$ matrix

$$N = (\,-R^T \quad I\,)$$

form a basis for the flow space of X.

Proof. It is obvious that $MN^T = 0$. Therefore, the rows of N are in the flow space of X, and since they are linearly independent, they form a basis for the flow space of X. □

Let F be a maximal spanning forest of X, and label the edges of X so that the edges of F come first. Then the matrix whose rows are the signed characteristic vectors of the cuts $C(F,e)$ has the form $M = (\,I \quad R\,)$. By Theorem 14.2.4, the rows of the matrix $N = (\,-R^T \quad I\,)$ are a basis for the flow space. In fact, they are the signed characteristic vectors of the cycles corresponding to the chords of F.

14.3 Planar Graphs

The results in the previous sections indicate that there are a number of similarities between properties of the cut space and properties of the flow space of a graph. For planar graphs we can show that these similarities are not accidental. Our next result provides the basic reason.

Theorem 14.3.1 *If X is a plane graph, then a set of edges is a cycle in X if and only if it is a bond in the dual graph X^*.*

Proof. We shall show that a set of edges $D \subseteq E(X)$ contains a cycle of X if and only if it contains a bond of X^* (here we are identifying the edges of X and X^*). If D contains a cycle C, then this forms a closed curve in the plane, and every face of X is either inside or outside C. So C is a cut in X^* whose shores are the faces inside C and the faces outside C. Hence D contains a bond of X^*. Conversely, if D does not contain a cycle, then D does not enclose any region of the plane, and there is a path between any two faces of X^* that uses only edges not in D. Therefore, D does not contain a bond of X^*. □

Corollary 14.3.2 *The flow space of a plane graph X is equal to the cut space of its dual graph X^*.* □

The full significance of the next result will not become apparent until the next chapter.

Lemma 14.3.3 *If X is a connected plane graph and T a spanning tree of X, then $E(X)\setminus E(T)$ is a spanning tree of X^*.*

Proof. The tree T contains no cycle of X, and therefore T contains no bond of X^*. Therefore, the graph with vertex set $V(X^*)$ and edge set $E(X) \setminus E(T)$ is connected. Euler's formula shows that $|E(X) \setminus E(T)| = |V(X^*)| - 1$, and the result follows. □

14.4 Bases and Ear Decompositions

Let $\mathcal{C}$ be a set of cycles in the graph X and let H be the matrix with the signed characteristic vectors of these cycles as its rows. We say that $\mathcal{C}$ is *triangular* if, possibly after permuting some of the rows and columns of H, some set of columns of H forms an upper triangular matrix. Clearly, the set of signed characteristic vectors of a triangular set of cycles is linearly independent. An example is provided by the set of cycles determined by the chords of a spanning tree. This is a triangular set because the identity matrix is upper triangular. In this section we present a large class of graphs with triangular bases for the flow space.

Let Y be a graph and let P be a path with at least one edge. If the graph X is produced by identifying the end-vertices of P with two vertices in Y, we say that X comes from *adding an ear* to Y. The path P may have length one, in which case we have just offered a complicated description of how to add an edge to Y. The end-vertices of P may be identified with the same vertex of Y, but if the two vertices from Y are distinct, we say that the ear is *open*. Any graph that can be constructed from a single vertex by successively adding ears is said to have an *ear decomposition*. An ear decomposition is *open* if all of its ears after the initial one are open. Figure 14.3 shows an open ear decomposition of the cube.

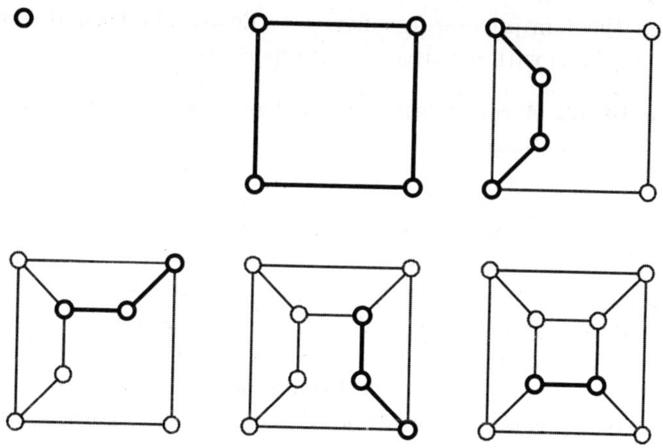

Figure 14.3. An open ear decomposition of the cube

If Y' is obtained by adding an ear to the connected graph Y, then

$$|E(Y')| - |V(Y')| = |E(Y)| - |V(Y)| + 1.$$

Therefore, there are exactly $|E(X)| - |V(X)| + 1$ ears in any ear decomposition of a connected graph X.

Ear decompositions can be used to construct triangular bases, but before we can see this, we need some information about graphs with ear decompositions. Recall that a cut-edge in a connected graph X is an edge whose removal disconnects X. An edge is a cut-edge if and only if it is not contained in a cycle.

Lemma 14.4.1 *Let Y be a graph with no cut-edges and let X be a graph obtained by adding an ear to Y. Then X has no cut-edges.*

Proof. Assume that X is obtained from Y by adding a path P and suppose $e \in E(X)$. If $e \in E(P)$, then $X \setminus e$ is connected. If $e \in Y$, then $Y \setminus e$ is connected (because Y has no cut-edges), and hence $X \setminus e$ is connected. This shows that X has no cut-edges. □

Suppose that X is a graph with an ear decomposition, and that the paths $P_0, \ldots, P_r$ are the successive ears. Let Y_i be the graph obtained after the addition of ear P_i. (So Y_0 is a cycle and $Y_r = X$.) Choose an edge e_0 in Y_0, and for $i = 1, \ldots, r$, let e_i be an edge in P_i that is not in $E(Y_{i-1})$. For each i, the graph Y_i has no cut-edges, so it has a cycle C_i containing the edge e_i. Clearly, if $j > i$, then $e_j \notin C_i$. If we form a matrix with the signed characteristic vectors of these cycles as the rows, then the columns corresponding to the edges $e_0, \ldots, e_r$ form a lower triangular matrix. Therefore, the set of cycles $C_0, \ldots, C_r$ is a triangular set of cycles. Since an ear decomposition of X contains exactly $|E(X)| - |V(X)| + 1$ ears, it follows that the signed characteristic vectors of these cycles form a basis for the flow space.

The next theorem provides a converse to Lemma 14.4.1 and shows that the class of graphs with ear decompositions is very large.

Theorem 14.4.2 *A connected graph X has an ear decomposition if and only if it has no cut-edges.*

Proof. It remains only to prove that X has an ear decomposition if it has no cut-edges. In fact, we will prove something slightly stronger, which is that X has an ear decomposition starting with any cycle. Let Y_0 be any cycle of X and form a sequence of graphs Y_0, Y_1, ... as follows. If $Y_i \neq X$, then there is an edge $e = uv \in E(X) \setminus E(Y_i)$, and because X is connected, we may assume that $u \in V(Y_i)$. Since e is not a cut-edge, it lies in a cycle of X, and the portion of this cycle from u along e until it first returns to $V(Y_i)$ is an ear; this may be only one edge if $v \in V(Y_i)$. Form Y_{i+1} from Y_i by the addition of this ear. Because X is finite, this process must terminate, yielding an ear decomposition of X. □

14.5 Lattices

Up till now we have worked with the row space of the incidence matrix over the rationals. From the combinatorialist's point of view, the most interesting vectors in this row space are likely to be integer vectors. The set of integer vectors in a subspace forms an abelian group under addition, and is a prominent example of a lattice. We turn to the study of these objects.

Let V be a finite-dimensional vector space over $\mathbb{R}$. Then V is an abelian group under addition. A subgroup $\mathcal{L}$ of V is *discrete* if there is a positive constant ϵ such that if $x \in \mathcal{L}$ and $x \neq 0$, then $\langle x, x \rangle \geq \epsilon$. We define a *lattice* to be a discrete subgroup of V. The *rank* of a lattice in V is the dimension of its span. A *sublattice* $\mathcal{M}$ of a lattice $\mathcal{L}$ is simply a subgroup of $\mathcal{L}$; we call it a *full* sublattice if it has the same rank as $\mathcal{L}$. The set $\mathbb{Z}^n$ of all vectors in $\mathbb{R}^n$ with integer coordinates provides a simple example of a lattice, and the subset $(2\mathbb{Z})^n$ of all vectors with even integer coordinates is a full sublattice of $\mathbb{Z}^n$.

Lemma 14.5.1 *The set of all integer linear combinations of a set of linearly independent vectors in V is a lattice.*

Proof. Suppose that M is a matrix with linearly independent columns. It will be enough to show that there is a positive constant ϵ such that if y is a nonzero integer vector, then $y^T M^T M y \geq \epsilon$. But if y is an integer vector, then $y^T y \geq 1$. Since $M^T M$ is positive definite, its least eigenvalue is positive; and since it equals

$$\min_{\|y\| \geq 1} y^T M^T M y,$$

we may take ϵ to be least eigenvalue of $M^T M$. □

We state without proof two important results about lattices. An *integral basis* for a lattice $\mathcal{L}$ of rank r is a set of vectors $x_1, \ldots, x_r$ from $\mathcal{L}$ such that each vector in $\mathcal{L}$ is an integral linear combination of $x_1, \ldots, x_r$. The standard basis vectors $e_1, \ldots, e_r$ form an integral basis for $\mathbb{Z}^r$. It is by no means clear in advance that a lattice should have an integral basis, but in fact, this is always the case.

Theorem 14.5.2 *Every lattice has an integral basis.* □

One consequence of this result is that every lattice in V is the set of all integer linear combinations of a set of linearly independent vectors in V. This is often used as the definition of a lattice.

Theorem 14.5.3 *Suppose A is a matrix whose columns form an integral basis for $\mathcal{L}$, and B is a matrix whose columns form an integral basis for a full sublattice $\mathcal{M}$ of $\mathcal{L}$. Then there is an integer matrix G such that $AG = B$. The absolute value of $\det G$ is equal to the number of cosets of $\mathcal{M}$ in $\mathcal{L}$.* □

This result implies that the index of $\mathcal{M}$ in $\mathcal{L}$ must be finite; hence the quotient group $\mathcal{L}/\mathcal{M}$ is a finite abelian group of order $\det G$.

14.6 Duality

If $\mathcal{L}$ is a lattice, then the *dual* $\mathcal{L}^*$ of $\mathcal{L}$ is defined by

$$\mathcal{L}^* = \{x \in \operatorname{span}(\mathcal{L}) : \langle x, y\rangle \in \mathbb{Z}, \forall y \in \mathcal{L}\}.$$

We say that $\mathcal{L}$ is an *integral lattice* if it is contained in its dual. Therefore, a lattice is integral if and only if $\langle x, y\rangle \in \mathbb{Z}$ for all $x, y \in \mathcal{L}$. Any sublattice of $\mathbb{Z}^n$ is certainly integral, but in general, the vectors in an integral lattice need not be integer vectors. The dual of a lattice is again a lattice. To see this, suppose that M is a matrix whose columns form an integral basis for $\mathcal{L}$. Then $\mathcal{L}^*$ consists of the vectors x in the column space of M such that $x^T M$ is an integer vector. It is easy to write down an integral basis for $\mathcal{L}^*$.

Lemma 14.6.1 *If the columns of the matrix M form an integral basis for the lattice $\mathcal{L}$, then the columns of $M(M^TM)^{-1}$ are an integral basis for its dual, $\mathcal{L}^*$.*

Proof. Let $a_1, \dots, a_r$ denote the columns of M and $b_1, \dots, b_r$ the columns of $M(M^TM)^{-1}$. Clearly, the vectors $b_1, \dots, b_r$ lie in the column space of M, and because $M^TM(M^TM)^{-1} = I$ we have

$$\langle a_i, b_j\rangle = \begin{cases} 1, & i = j; \\ 0, & \text{otherwise}. \end{cases}$$

Therefore, the vectors $b_1, \dots, b_r$ lie in $\mathcal{L}^*$.

Now, consider any vector $x \in \mathcal{L}^*$, and define

$$\bar{x} := \sum_i \langle x, a_i\rangle b_i.$$

Then $\bar{x}$ is an integer linear combination of the vectors b_i. Since $\langle x - \bar{x}, a_i\rangle = 0$, we have that $(x - \bar{x})^T M = 0$. Therefore, $x - \bar{x}$ belongs to both the column space of M and its orthogonal complement, and so $x - \bar{x} = 0$. Therefore, x is an integer linear combination of the basis vectors $x_1, \dots, x_r$. □

As an example of this result, consider the lattice with integral basis given by the columns of the matrix

$$M = \begin{pmatrix} 1 & 0 \\ -1 & 1 \\ 0 & -1 \end{pmatrix}.$$

Then we have

$$M^TM = \begin{pmatrix} 2 & -1 \\ -1 & 2 \end{pmatrix}, \qquad M(M^TM)^{-1} = \begin{pmatrix} 2/3 & 1/3 \\ -1/3 & 1/3 \\ -1/3 & -2/3 \end{pmatrix}.$$

If $\mathcal{L}$ is an integral lattice, then by Theorem 14.5.3, with $A = M(M^TM)^{-1}$ and $G = M^TM$, the index of $\mathcal{L}$ in $\mathcal{L}^*$ is equal to $\det M^TM$. This may explain why this index is called the *determinant* of $\mathcal{L}$. A lattice is *unimodular* if its determinant is 1, in which case it is equal to its dual. Note in addition that an integral lattice is a full sublattice of its dual.

For the next results, we need a preliminary result from linear algebra.

Theorem 14.6.2 *If M is a matrix with linearly independent columns, then projection onto the column space of M is given by the matrix*

$$P = M(M^TM)^{-1}M^T.$$

This matrix has the properties that $P = P^T$ and $P^2 = P$. □

Suppose that $\mathcal{L}$ is a lattice in $V = \mathbb{R}^n$ and that M is an integer matrix whose columns form an integral basis for $\mathcal{L}$. If P is the matrix representing orthogonal projection onto the column space of M, then for any vector x in V and any vector y in $\mathcal{L}$ we have

$$\langle Px, y\rangle = (Px)^Ty = x^TP^Ty = x^T(Py) = \langle x, Py\rangle = \langle x, y\rangle.$$

If x is an integer vector in V, then $\langle x, y\rangle \in \mathbb{Z}$, and it follows that $Px \in \mathcal{L}^*$. Thus P maps $\mathbb{Z}^n$ into $\mathcal{L}^*$. We will show that in many important cases this map is onto.

Theorem 14.6.3 *Suppose that M is an $n \times r$ matrix whose columns form an integral basis for the lattice $\mathcal{L}$. Let P be the matrix representing orthogonal projection from $\mathbb{R}^n$ onto the column space of M. If the greatest common divisor of the $r \times r$ minors of M is 1, then $\mathcal{L}^*$ is generated by the columns of P.*

Proof. From Theorem 14.6.2 we have that $P = M(M^TM)^{-1}M^T$, and from Lemma 14.6.1 we know that the columns of $M(M^TM)^{-1}$ form an integral basis for $\mathcal{L}^*$. Therefore, it is sufficient to show that if $y \in \mathbb{Z}^r$, then $y = M^Tx$ for some integer vector x. Equivalently, we must show that the lattice generated by the rows of M is $\mathbb{Z}^r$.

Let $\mathcal{M}$ denote the lattice generated by the rows of M. There is an $r \times r$ matrix N whose rows form an integral basis for $\mathcal{M}$. The rows of N are an integral basis, so there is some integer matrix X such that $XN = M$. Because N is invertible, we have $X = MN^{-1}$, so we conclude that MN^{-1} has integer entries. If M' is any $r \times r$ submatrix of M, then $M'N^{-1}$ is also an integer matrix; hence $\det N$ must divide $\det M'$. So our hypothesis implies that $\det N = 1$, and therefore $\mathcal{M} = \mathbb{Z}^r$. □

14.7 Integer Cuts and Flows

The set of all integer vectors in the cut space of a graph is a lattice; we call it the lattice of integer cuts or the *cut lattice*. If x is an integer cut and

D is the incidence matrix of X, then $x = D^T y$ for some vector y. Hence, if $uv \in E(X)$, then $y_u - y_v$ is an integer, and it follows that there is an integer vector, y' say, such that $x = D^T y'$. Thus the lattice of integer cuts of X is the set of all integer linear combinations of the columns of D^T. If X is connected with n vertices, then any set of $n-1$ columns of D^T forms an integral basis for this lattice.

The set of all integer vectors in the flow space of a graph is the lattice of integer flows, or the *flow lattice*. The bases for the flow space of X that we described in Section 14.2 and Section 14.4 are integral bases for this lattice.

The *norm* of a vector x in a lattice is $\langle x, x\rangle$ (which, we admit, would be called the square of the norm in most other contexts). A vector is *even* if it has even norm. An integral lattice $\mathcal{L}$ is even if all its elements are even; it is *doubly even* if the norm of each element is divisible by four. A graph is called *even* if each vertex has even valency.

Lemma 14.7.1 *The flow lattice of a graph X is even if and only if X is bipartite. The cut lattice of X is even if and only if X is even.*

Proof. If x and y are even vectors, then

$$\langle x+y, x+y\rangle = \langle x, x\rangle + 2\langle x, y\rangle + \langle y, y\rangle,$$

and so $x+y$ is also even. If X is bipartite, then all cycles in it have even length. It follows that the flow lattice of integer flows is spanned by a set of even vectors; therefore, it is even. If X is even, then each column of D^T is even, so the cut lattice is also spanned by a set of even vectors.

The converse is trivial in both cases. □

Theorem 14.7.2 *The determinant of the cut lattice of a connected graph X is equal to the number of spanning trees of X.*

Proof. Let D be the oriented incidence matrix of X, let u be a vertex of X, and let D_u be the matrix obtained by deleting the row corresponding to u from D. Then the columns of D_u^T form an integral basis for the lattice, and so its determinant is $\det(D_u D_u^T)$. By Theorem 13.2.1, this equals the number of spanning trees of X. □

Theorem 14.7.3 *The determinant of the flow lattice of a connected graph X is equal to the number of spanning trees of X.*

Proof. Suppose the rows of the matrix

$$M = (I \quad R)$$

form a basis for the cut space of X (the existence of such a basis is guaranteed by the spanning-tree construction of Section 14.1). Then the rows of the matrix

$$N = (-R^T \quad I)$$

form a basis for the flow space of X.

The columns of M^T and N^T form integral bases for the cut lattice and the flow lattice, respectively. Therefore, the determinant of the cut lattice is $\det MM^T$, and the determinant of the flow lattice is $\det NN^T$.

But

$$MM^T = I + RR^T, \qquad NN^T = I + R^T R,$$

and so by Lemma 8.2.4,

$$\det MM^T = \det NN^T,$$

and the result follows from Theorem 14.7.2. □

We leave to the reader the task of showing that for a disconnected graph X, the determinant of the cut lattice and the flow lattice is equal to the number of maximal spanning forests of X.

14.8 Projections and Duals

In the previous section we saw that the cut lattice $\mathcal{C}$ and flow lattice $\mathcal{F}$ of a graph have the same determinant. This means that the quotient groups $\mathcal{C}^*/\mathcal{C}$ and $\mathcal{F}^*/\mathcal{F}$ have the same order. In fact, they are isomorphic, and this can be proved using the theory we have already developed. But we take an alternative approach.

Assume that X is connected and has n vertices and m edges. Let u be a vertex in X and put $M = D_u^T$. Every $(n-1) \times (n-1)$ minor of M is either 0, 1, or -1, and since the columns of M are linearly independent, at least one of them is nonzero. Therefore, the greatest common divisor of the minors is 1. By Theorem 14.6.3 we conclude that if P represents orthogonal projection onto the column space of M, then the columns of P generate $\mathcal{C}^*$.

We note that the kernel of P is the flow space of X; hence P maps $\mathbb{Z}^m/\mathcal{F}$ onto $\mathcal{C}^*$. Since $Px = x$ for each x in the cut space of X, it follows that P maps $\mathbb{Z}^m/(\mathcal{C} \oplus \mathcal{F})$ onto $\mathcal{C}^*/\mathcal{C}$. It is not hard to see that this mapping is injective; hence we conclude that

$$\frac{\mathbb{Z}^m}{\mathcal{C} \oplus \mathcal{F}} \cong \frac{\mathcal{C}^*}{\mathcal{C}}.$$

Turning to flows, we note that $I - P$ represents orthogonal projection onto the flow space of X, and a straightforward modification of the previous argument yields that

$$\frac{\mathbb{Z}^m}{\mathcal{C} \oplus \mathcal{F}} \cong \frac{\mathcal{F}^*}{\mathcal{F}}.$$

Therefore, the groups $\mathcal{C}^*/\mathcal{C}$ and $\mathcal{F}^*/\mathcal{F}$ are isomorphic.

Now, we derive an expression for the matrix P that represents orthogonal projection onto the cut space of a graph X. First we need some notation. If T is a spanning tree of X, then define the matrix N_T with rows and columns indexed by $E(X)$ as follows: The e-column of N_T is zero if $e \notin E(T)$ and is the signed characteristic vector of the cut $C(T,e)$ if $e \in E(T)$.

Theorem 14.8.1 *If X is a connected graph, then the matrix*

$$P = \frac{1}{\tau(X)} \sum_T N_T$$

represents orthogonal projection onto the cut space of X.

Proof. To prove the result we will show that P is symmetric, that $Px = x$ for any vector in the cut space of X, and that $Px = 0$ for any vector in the flow space of X.

Now,

$$(N_T)^2_{eg} = \sum_{f \in E(X)} (N_T)_{ef} (N_T)_{fg},$$

and $(N_T)_{fg}$ can be nonzero only when $f \in E(T)$, but then $(N_T)_{fg} = 0$ unless $g = f$. Hence the sum reduces to $(N_T)_{eg}(N_T)_{gg} = (N_T)_{eg}$, and so $N_T^2 = N_T$. Because the column space of N_T is the cut space of X, we deduce from this that $N_T x = x$ for each vector in the cut space of X, and thus that $Px = x$.

Next we prove that the sum of the matrices N_T over all spanning trees of X is symmetric. Suppose e and f are edges of X. Then $(N_T)_{ef}$ is zero unless $f \in E(T)$ and $e \in C(T,f)$. Let $\mathcal{T}_f$ denote the set of trees T such that $f \in E(T)$ and $e \in C(T,f)$; let $\mathcal{T}_e$ denote the set of trees T such that $e \in E(T)$ and $f \in C(T,e)$. Let $\mathcal{T}_f(+)$ and $\mathcal{T}_f(-)$ respectively denote the set of trees in $\mathcal{T}_f$ such that the head of e lies in the positive or negative shore of $C(T,f)$, and define $\mathcal{T}_e(+)$ and $\mathcal{T}_e(-)$ similarly. Note next that if $T \in \mathcal{T}_f$, then $(T \setminus f) \cup e$ is a tree in $\mathcal{T}_e$. This establishes a bijection from $\mathcal{T}_e$ to $\mathcal{T}_f$, and this bijection maps $\mathcal{T}_e(+)$ to $\mathcal{T}_f(+)$.

Since $(N_T)_{ef}$ equals 1 or -1 according as the head of e is in the positive or negative shore of $C(T,f)$, it follows that

$$\sum_T (N_T)_{ef} = \sum_T (N_T)_{fe},$$

and therefore that P is symmetric.

Finally, if x lies in the flow space, then $x^T N_T = 0$, for any tree T, and so $x^T P = 0$, and taking the transpose we conclude that $Px = 0$. $\square$

14.9 Chip Firing

We are going to discuss some games on graphs; the connection to lattices will only slowly become apparent. We start with a connected graph X and a set of N chips, which are placed on the vertices of the graph. One step of the game is taken by choosing a vertex that has at least as many chips as its valency, then moving one chip from it to each of its neighbours. We call this *firing* a vertex. The game can continue as long as there is a vertex with as many chips on it as its valency. The basic question is, given an initial configuration, whether the game must terminate in a finite number of steps. We start with a simple observation.

Lemma 14.9.1 *In an infinite chip-firing game, every vertex is fired infinitely often.*

Proof. Since there are only a finite number of ways to place N chips on X, some configuration, say s, must reappear infinitely often. Let σ be the sequence of vertices fired between two occurrences of s. If there is a vertex v that is not fired in this sequence, then the neighbours of v are also not fired, for otherwise the number of chips on v would increase. Since X is connected, either σ is empty or every vertex occurs in σ. □

Before we prove the next result we need some additional information about orientations of a graph. An orientation of a graph X is called *acyclic* if the oriented graph contains no directed cycles. It is easy to see that every graph has an acyclic orientation.

Theorem 14.9.2 *Let X be a graph with n vertices and m edges and consider the chip-firing games on X with N chips. Then*

(a) *If $N > 2m - n$, the game is infinite.*
(b) *If $m \leq N \leq 2m - n$, the game may be finite or infinite.*
(c) *If $N < m$, the game is finite.*

Proof. Let $d(v)$ be the valency of the vertex v. If each vertex has at most $d(v) - 1$ chips on it, then

$$N \leq \sum_v (d(v) - 1) = 2m - n.$$

So, if $N > 2m - n$, there is always a vertex with as least as many chips on it as its valency. We also see that for $N \leq 2m - n$ there are initial configurations where no vertex can be fired.

Next we show that if $N \geq m$, there is an initial configuration that leads to an infinite game. It will be enough to prove that there are infinite games when $N = m$. Suppose we are given an acyclic orientation of X, and let $d^+(v)$ denote the out-valency of the vertex v with respect to this orientation. Every acyclic orientation determines a configuration with $N = e$ obtained by placing $d^+(v)$ chips on each vertex v. If an orientation is acyclic, there is a vertex u such that $d^+(u) = d(u)$; this vertex can be fired. After u has been fired it has no chips, and every neighbour of u has one additional chip. However, this is simply the configuration determined by another acyclic orientation, the one obtained by reversing the orientation of every edge on u. Therefore, the game can be continued indefinitely.

Finally, we prove that the game is finite if $N < m$. Assume that the chips are distinguishable and let an edge capture the first chip that uses it. Once a chip is captured by an edge, assume that each time a vertex is fired, it either stays fixed, or moves to the other end of the edge it belongs to. Since $N < m$, it follows that there is an edge that never has a chip assigned to it. Hence neither of the vertices in this edge is ever fired, and therefore the game is finite. □

Now, consider a chip-firing game with N chips where $m \leq N \leq 2m - n$. We will show that whether the game is finite or infinite depends only on the initial configuration, and not on the particular sequence of moves that are made.

First we need some additional terminology. At any stage in a chip-firing game, the *state* is the vector indexed by $V(X)$ giving the number of chips on each vertex; we will also use this term as a synonym for configuration. The *score* of a sequence of firings is the vector indexed by $V(X)$ that gives the number of times each vertex has been fired. If a chip-firing game is in state s, then after applying a firing sequence σ with score x, the resulting state is $t = s - Qx$, where Q is the Laplacian of X. In this situation we say that σ is a firing sequence leading from s to t.

If x and y are two scores, then we define the score $x \vee y$ by

$$(x \vee y)_u = \max\{x_u, y_u\}.$$

If σ and τ are two sequences, then we construct a new sequence $\sigma \setminus \tau$ as follows: For each vertex u, if u is fired i times in τ, then delete the first i occurrences of u from σ (deleting all of them if there are fewer than i).

Theorem 14.9.3 *Let X be a connected graph and let σ and τ be two firing sequences starting from the same state s with respective scores x and y. Then τ followed by $\sigma \setminus \tau$ is a firing sequence starting from s having score $x \vee y$.*

Proof. We leave the proof of this result as a useful exercise. □

Corollary 14.9.4 *Let X be a connected graph, and s a given initial state. Then either every chip-firing game starting from s is infinite, or all such games terminate in the same state.*

Proof. Let τ be the firing sequence of a terminating game starting from s, and let σ be the firing sequence of another game starting from s. Then by Theorem 14.9.3, $\sigma \setminus \tau$ is necessarily empty, and hence σ is finite.

Now, suppose that σ is the firing sequence of another terminating game starting from s. Then both $\sigma \setminus \tau$ and $\tau \setminus \sigma$ are empty, and hence σ and τ

have the same score. Since the state of a chip-firing game depends only on the initial state and the score of the firing sequence, all terminating games must end in the same state. □

14.10 Two Bounds

We consider two bounds on the length of a terminating chip-firing game.

Lemma 14.10.1 *Suppose u and v are adjacent vertices in the graph X. At any stage of a chip-firing game on X with N chips, the difference between the number of times that u has been fired and the number of times that v has been fired is at most N.*

Proof. Suppose that u has been fired a times and v has been fired b times, and assume without loss of generality that $a < b$. Let H be the subgraph of X induced by the vertices that have been fired at most a times. Consider the number of chips currently on the subgraph H. Along every edge between H and $V(X)\setminus H$ there has been a net movement of chips from $V(X)\setminus H$ to H, and in particular, the edge uv has contributed $b-a$ chips to this total. Since H cannot have more than N chips on it, we have $b - a \leq N$. □

Theorem 14.10.2 *If X is a connected graph with n vertices, e edges, and diameter D, then a terminating chip-firing game on X ends within $2neD$ moves.*

Proof. If every vertex is fired during a game, then the game is infinite, and so in a terminating game there is at least one vertex v that is never fired. By Lemma 14.10.1, a vertex at distance d from v has fired at most dN times, and so the total number of moves is at most nDN. By Theorem 14.9.2, $N < 2e$, and so the game terminates within $2neD$ moves. □

Next we derive a bound on the length of a terminating chip-firing game, involving the eigenvalue λ_2. This requires some preparation.

Lemma 14.10.3 *Let M be a positive semidefinite matrix, with largest eigenvalue ρ. Then, for all vectors y and z,*

$$|y^T Mz| \leq \rho\|y\|\,\|z\|.$$

Proof. Since M is positive semidefinite, for any real number t we have

$$(y+tz)^T M(y+tz) \geq 0.$$

The left side here is a quadratic polynomial in t, and the inequality implies that its discriminant is less than or equal to zero. This yields the following extension of the Cauchy–Schwarz inequality:

$$(y^T Mz)^2 \leq y^T My\, z^T Mz.$$

Since $\rho I - M$ is also positive semidefinite, for any vector x,

$$0 \le x^T(\rho I - M)x,$$

and therefore

$$x^T Mx \le \rho x^T x.$$

The lemma now follows easily. □

Theorem 14.10.4 *Let X be a connected graph with n vertices and let Q be the Laplacian of X. If $Qx = y$ and $x_n = 0$, then*

$$|\mathbf{1}^T x| \le \frac{n}{\lambda_2}\|y\|.$$

Proof. Since Q is a symmetric matrix, the results of Section 8.12 show that Q has spectral decomposition

$$Q = \sum_{\theta \in \mathrm{ev}(Q)} \theta E_\theta.$$

Since X is connected, $\ker Q$ is spanned by $\mathbf{1}$, and therefore

$$E_0 = \frac{1}{n}J.$$

Define the matrix $Q^\dagger$ by

$$Q^\dagger := \sum_{\theta \ne 0} \theta^{-1} E_\theta.$$

The eigenvalues of $Q^\dagger$ are 0, together with the reciprocals of the nonzero eigenvalues of Q. Therefore, it is positive semidefinite, and its largest eigenvalue is λ_2^{-1}. Since the idempotents E_θ are pairwise orthogonal, we have

$$Q^\dagger Q = \sum_{\theta \ne 0} E_\theta = I - E_0 = I - \frac{1}{n}J.$$

Therefore, if $Qx = y$, then

$$\left(I - \frac{1}{n}J\right)x = Q^\dagger y.$$

Multiplying both sides of this equality on the left by e_n^T, and recalling that $e_n^T x = 0$, we get

$$-\frac{1}{n}\mathbf{1}^T x = e_n^T Q^\dagger y.$$

By Lemma 14.10.3,

$$|e_n^T Q^\dagger y| \le \lambda_2^{-1}\|e_n\|\,\|y\|,$$

from which the theorem follows. □

Corollary 14.10.5 *Let X be a connected graph with n vertices. A terminating chip-firing game on X with N chips has length at most $\sqrt{2}nN/\lambda_2$.*

Proof. Let s be the initial state and t the final state of a chip-firing game with score x. Since the game uses N chips, we have $\|s\| \leq N$ and $\|t\| \leq N$, and since both s and t are nonnegative, $\|s-t\| \leq \sqrt{2}N$. Because the game is finite, some vertex is not fired, and we may assume without loss that it is the last vertex. Since $Qx = s - t$, the result follows directly by applying the theorem with $s - t$ in place of y. □

14.11 Recurrent States

We say that a state s is *recurrent* if there is some firing sequence leading from state s back to s. Clearly, a chip-firing game is infinite if and only if there is some firing sequence leading from the initial state to a recurrent state. A state s is *diffuse* if and only if for every induced subgraph $Y \subseteq X$ there is some vertex of Y with at least as many chips as its valency in Y.

Theorem 14.11.1 *A state is recurrent if and only if it is diffuse.*

Proof. Suppose that s is a recurrent state, and that σ is a firing sequence leading from s back to itself. Let $Y \subseteq X$ be an induced subgraph of X, and let v be the vertex of Y that first finishes firing. Then every neighbour of v in Y is fired at least once in the remainder of σ, and so by the time s recurs, v has at least as many chips as its valency in Y.

For the converse, suppose that the state s is diffuse. Then we will show that some permutation of the vertices of X is a firing sequence from s. Since X is an induced subgraph of itself, some vertex is ready to fire. Now, consider the situation after some set W of vertices has been fired exactly once each. Let U be the subgraph induced by the unfired vertices. In the initial state s, some vertex $u \in U$ has at least as many chips on it as its valency in U. After the vertices of W have been fired, u has gained one chip from each of its neighbours in W. Therefore, u now has at least as many chips as its valency in X, and hence is ready to fire. By induction on $|W|$, some permutation of the vertices is a firing sequence from s. □

One consequence of the proof of this result is that it is easy to identify diffuse states. Given a state, simply fire vertices at most once each in any order. This process terminates in at most n steps, with the state being diffuse if and only if every vertex has been fired.

Theorem 14.11.2 *Let X be a connected graph with m edges. Then there is a one-to-one correspondence between diffuse states with m chips and acyclic orientations of X.*

Proof. Let s be a state given, as in the proof of Theorem 14.9.2, by an acyclic orientation of X. If Y is an induced subgraph of X, then the restriction of the acyclic orientation of X to Y is an acyclic orientation of Y. Hence there is some vertex whose out-valency in Y is equal to its valency in Y, and so this vertex has at least as many chips as its valency in Y. Therefore, s is diffuse.

Conversely, let s be a diffuse state with m chips, and let the permutation σ of $V(X)$ be a firing sequence leading from s to itself. Define an acyclic orientation of X by orienting the edge ij from i to j if i precedes j in σ. For any vertex v, let U and W denote the vertices that occur before and after v in σ, respectively, and let n_U and n_W denote the number of neighbours of v in U and W, respectively. When v fires it has at least $d(v) = n_U + n_W$ chips on it, of which n_U were accumulated while the vertices of U were fired. Therefore, in the initial state, v has at least $n_W = d^+(v)$ chips on it. This is true for every vertex, and since $N = m = \sum_u d^+(u)$, every vertex v has exactly $d^+(v)$ chips on it. □

14.12 Critical States

We call a state *q-stable* if q is the only vertex ready to be fired, and *q-critical* if it is both q-stable and recurrent. By Theorem 14.9.2, the number of chips in a q-critical state is at least the number of edges. Applying Theorem 14.11.2 and the fact that q is the only vertex ready to be fired, we get the following characterization of the q-critical states with the minimum number of chips.

Lemma 14.12.1 *Let X be a connected graph with m edges. There is a one-to-one correspondence between q-critical states with m chips and acyclic orientations of X with q as the unique source.* □

Lemma 14.12.2 *Let X be a connected graph with m edges, and let t be a q-critical state. Then there is a q-critical state s with m chips such that $s_v \leq t_v$ for every vertex v.*

Proof. The state t is recurrent if and only if there is a permutation σ of $V(X)$ that is a legal firing sequence from t. Suppose that during this firing sequence, v is the first vertex with more than $d(v)$ chips on it when fired. Then the state obtained from t by reducing the number of chips on v by the amount of this excess is also q-critical. If every vertex has precisely $d(v)$ chips on it when fired, then there are m chips in total. □

This result shows that the q-critical states with m chips are the "minimal" q-critical states. Every q-critical state with more than m chips is obtained from a minimal q-critical state by increasing the number of chips on some of the vertices.

Although the number of chips on q in q-critical state can be arbitrarily large, the number of chips on any other vertex v is at most $d(v) - 1$. This implies that the number of q-critical states with $d(q)$ chips on q is finite, and also yields the following bounds.

Lemma 14.12.3 *Let X be a connected graph on n vertices with m edges. The number of chips N in a q-critical state with $d(q)$ chips on q satisfies*

$$m \leq N \leq 2m - n + 1.$$

□

Suppose that we call two states *equivalent* if they differ only in the number of chips on q. Then we have proved that there are only finitely many equivalence classes of q-critical states. In the next section we prove that the number of equivalence classes of q-critical states equals the number of spanning trees of X.

14.13 The Critical Group

In this section we will investigate the number of equivalence classes of q-critical states in the chip-firing game. One problem is that two equivalent states may lead to inequivalent games, one terminating and the other infinite. However, this problem can be overcome by introducing a slightly modified game known as the *dollar game*. We arrive at the dollar game by introducing a separate rule for firing the special vertex q:

(a) Vertex q can be fired if and only if no other vertex can be fired.

An immediate consequence of this rule change is that every game is infinite. If q has fewer than $d(q)$ chips on it when no other vertex can be fired, then it is fired anyway and just ends up with a negative number of chips. The second important consequence of this is that the possible games leading from one state to another are now independent of the initial number of chips on q, and so are determined by the equivalence classes of the two states.

It is straightforward to verify that a state is q-critical in the dollar game if and only if it is equivalent to a q-critical state in the chip-firing game. We will select one representative state from each equivalence class of states in the dollar game. Define a state s to be *balanced* if its entries sum to zero, that is, $s^T \mathbf{1} = 0$, and it is nonnegative on $V(X) \setminus q$. It is immediate that each equivalence class of states in the chip-firing game contains a unique balanced state. If the dollar game starts from a state s, then after a sequence of firings the new state is represented by $t = s - Qx$, and so t is balanced if and only s is balanced. Therefore, we can view the dollar game as a game on balanced states only, with the property that q-critical states in the dollar game correspond precisely to equivalence classes of q-critical states in the chip-firing game. Henceforth we adopt this view, and so any state in the dollar game is balanced.

Now, let X be a connected graph on n vertices with Laplacian Q, and let $\mathcal{L}(Q)$ denote the lattice generated by the columns of Q. Every column of Q is orthogonal to $\mathbf{1}$, so this is a sublattice of $\mathbb{Z}^n \cap \mathbf{1}^\perp$. Hence we deduce that s and all the states reachable from s lie in the same coset of $\mathcal{L}(Q)$ in $\mathbb{Z}^n \cap \mathbf{1}^\perp$. We will show that the q-critical states form a complete set of coset representatives for $\mathcal{L}(Q)$.

Suppose we play the dollar game on X. From a q-stable state, vertex q is fired a number of times and then the game is played under the rules of the chip-firing game until another q-stable state is reached. We can view one of these portions of a dollar game as a terminating chip-firing game, which by Corollary 14.9.4 ends in the same q-stable state regardless of the moves chosen. Therefore, every dollar game from a given initial state passes through the same infinite sequence of q-stable states. Since there are finitely many q-stable states, some of them recur, and so every game from s passes through the same infinite sequence of q-critical states. Our next result implies that there is only one q-critical state in this sequence.

Lemma 14.13.1 *In the dollar game, after q has been fired, no other vertex can be fired twice before q is fired again.*

Proof. Suppose that no vertex has yet been fired twice after q and consider the number of chips on any vertex u that has been fired exactly once since q. Immediately before q was last fired, u had at most $d(u) - 1$ chips on it. Since then, u has gained at most $d(u)$ chips, because no vertex has been fired twice, and has lost $d(u)$ chips when it was fired. Therefore, u is not ready to fire. □

Now, suppose that s is a q-critical state and that σ is a nonempty firing sequence with score x that starts and finishes at s. Then we have

$$s - Qx = s,$$

which implies that $Qx = 0$. Since X is connected, the kernel of Q is spanned by $\mathbf{1}$, and hence $x = m\mathbf{1}$, for some positive integer m. By the previous lemma, each vertex must be fired exactly once between each firing of q, so $\mathbf{1}$ is a legal firing sequence from s. Since all games starting at s pass through the same sequence of q-stable states, no q-stable states other than s can be reached.

Lemma 14.13.2 *If s and t are q-critical states such that $s - t = Qx$ for some integer vector x, then $s = t$.*

Proof. We shall show that x is necessarily a constant vector, so $Qx = 0$, and hence $s = t$. Assume for a contradiction that x is not constant. Then, exchanging s and t if necessary, we may assume that x_q is not a maximum coordinate of x. Let the permutation τ be a legal firing sequence starting and ending at t, and let $v \neq q$ be the first vertex in τ such that x_v is one of the maximum coordinates of x. Let W be the neighbours of v that occur before v in τ. Then

$$\begin{aligned} s_v &= t_v + x_v d(v) - \sum_{w \sim v} x_w \\ &= t_v + \sum_{w \sim v} (x_v - x_w) \\ &\geq t_v + \sum_{w \in W} 1 \\ &\geq d(v). \end{aligned}$$

This contradicts the fact that s is a q-critical configuration, because v is ready to be fired. □

Theorem 14.13.3 *Let X be a connected graph on n vertices. Each coset of $\mathcal{L}(Q)$ in $\mathbb{Z}^n \cap \mathbf{1}^\perp$ contains a unique q-critical state for the dollar game.*

Proof. Given a coset of $\mathcal{L}(Q)$, choose an element s in the coset that represents a valid initial state for the dollar game. By the discussion above, every game with initial state s eventually falls into a loop containing a unique q-critical state. Therefore, each coset of $\mathcal{L}(Q)$ contains a q-critical state, and by Lemma 14.13.2 no coset contains more than one q-critical state. □

Thus we have shown that the q-critical states form a complete set of coset representatives for $\mathcal{L}(Q)$. The determinant of $\mathcal{L}(Q)$ is equal to the number of spanning trees of X. Therefore, the number of q-critical states is equal to the number of spanning trees of X, regardless of the choice of vertex q.

We can now consider the quotient group

$$(\mathbb{Z}^n \cap \mathbf{1}^T)/\mathcal{L}(Q)$$

to be an abelian group defined on the q-critical states. If s and t are q-critical states, then their sum is the unique q-critical state in the same coset as $s+t$. It seems necessary to actually play the dollar game from $s+t$ in order to identify this q-critical state in any particular instance!

We leave as an exercise the task of showing that this quotient group is isomorphic to $\mathcal{C}^*/\mathcal{C}$ and $\mathcal{F}^*/\mathcal{F}$, and therefore independent of q. It is known as the *critical group* of X.

14.14 Voronoi Polyhedra

Let $\mathcal{L}$ be a lattice in a vector space V and let a be an element of $\mathcal{L}$. The *Voronoi cell* of a is the set of all vectors in V that are as close to a as they are to any other point of $\mathcal{L}$. The Voronoi cells of distinct elements of

$\mathcal{L}$ differ only by a translation, so we will usually consider only the Voronoi cell of the origin, which we denote by $\mathcal{V}$. This consists of the vectors x in V such that for all vectors a in $\mathcal{L}$,

$$\langle x, x\rangle \leq \langle x-a, x-a\rangle. \tag{14.1}$$

Let $H(a)$ denote the half-space of V defined by the inequality

$$\langle x, a\rangle \leq \frac{1}{2}\langle a, a\rangle.$$

Since $\langle x-a, x-a\rangle = \langle x,x\rangle - 2\langle x,a\rangle + \langle a,a\rangle$, (14.1) implies that $\mathcal{V}$ is the intersection of the half-spaces $H(a)$, where a ranges over the nonzero elements of $\mathcal{L}$. One consequence of this is that the Voronoi cells of a lattice are closed convex sets; one difficulty with this description is that it uses an infinite set of half-spaces.

Our next result will eliminate this difficulty.

Lemma 14.14.1 *Let a and b be elements of the lattice $\mathcal{L}$ with $\langle a, b\rangle \geq 0$. Then $H(a)\cap H(b) \subseteq H(a+b)$.*

Proof. Suppose $x \in H(a)\cap H(b)$. Then

$$\langle x, a+b\rangle = \langle x,a\rangle + \langle x,b\rangle \leq \frac{1}{2}\langle a,a\rangle + \frac{1}{2}\langle b,b\rangle.$$

Since $\langle a,b\rangle \geq 0$, we have that

$$\langle a+b, a+b\rangle \geq \langle a,a\rangle + \langle b,b\rangle.$$

It follows that $x \in H(a+b)$. □

Define an element a of $\mathcal{L}$ to be *indecomposable* if it is nonzero and cannot be written as $b+c$, where b and c are nonzero elements of $\mathcal{L}$ and $\langle b,c\rangle \geq 0$. Then our last result implies that $\mathcal{V}$ is the intersection of the half-spaces $H(a)$, where a runs over the indecomposable elements of $\mathcal{L}$.

We first show that indecomposable elements exist, then that there are only finitely many.

Any element of minimum norm is indecomposable: If $a = b+c$, then

$$\langle a,a\rangle = \langle b+c, b+c\rangle = \langle b,b\rangle + \langle c,c\rangle + 2\langle b,c\rangle;$$

if $\langle b,c\rangle \geq 0$ and $b, c \neq 0$, this implies that the norm of a is not minimal.

Lemma 14.14.2 *An element a of $\mathcal{L}$ is indecomposable if and only if a and $-a$ are the two elements of minimum norm in the coset $a + 2\mathcal{L}$.*

Proof. Suppose $a \in \mathcal{L}$. If $x \in \mathcal{L}$, then $a = a - x + x$, whence we see that a is indecomposable if and only if

$$\langle a-x, x\rangle < 0$$

for all elements of $\mathcal{L}\setminus\{0,a\}$. Since

$$\langle a-2x, a-2x\rangle = \langle a,a\rangle - 4(\langle a,x\rangle - \langle x,x\rangle),$$

this condition holds if and only if for any x in $\mathcal{L}\setminus\{0,a\}$,

$$\langle a-2x, a-2x\rangle > \langle a,a\rangle,$$

which implies that any other elements of the coset have greater norm than a. □

If $\mathcal{L}$ has rank n, then $2\mathcal{L}$ has index 2^n in $\mathcal{L}$. In particular, there are only finitely many distinct cosets of $2\mathcal{L}$ in $\mathcal{L}$. It follows that $\mathcal{V}$ is the intersection of a finite number of closed half-spaces, and therefore it is a convex polytope.

A *face* of a convex polytope is the intersection of the polytope with a *supporting hyperplane*. Informally, a hyperplane is supporting if it contains points of the polytope, but does not divide it into two parts. Formally, we may define a hyperplane H to be supporting if whenever p and q are two points in the polytope and the line segment joining p to q meets H, then p or q (or both) lie in H. A facet of a polytope is a face with dimension one less than the dimension of the polytope. It is not hard to show that if a polytope is presented as the intersection of a finite number of closed half-spaces, then each face must be contained in a hyperplane bounding one of the half-spaces.

Theorem 14.14.3 *Let $\mathcal{V}$ be the Voronoi cell of the origin in the lattice $\mathcal{L}$. Then $\mathcal{V}$ is the intersection of the closed half-spaces $H(a)$, where a ranges over the indecomposable elements of $\mathcal{L}$. For each such a, the intersection $\mathcal{V}\cap H(a)$ is a facet.*

Proof. We must show that $\mathcal{V}\cap H(a)$ has dimension one less than the dimension of the polytope. So let a be a fixed indecomposable element of $\mathcal{L}$ and let u be any vector orthogonal to a. If b is a second indecomposable element of $\mathcal{L}$, then

$$\left\langle b, \frac{1}{2}(a+\epsilon u)\right\rangle = \frac{1}{2}(\langle b,a\rangle + \epsilon\langle b,u\rangle).$$

If $b\neq \pm a$, then $\langle a,b\rangle < \langle b,b\rangle$; hence for all sufficiently small values of ϵ we have

$$\left\langle b, \frac{1}{2}(a+\epsilon u)\right\rangle \leq \frac{1}{2}\langle b,b\rangle.$$

This shows that for all vectors u orthogonal to a and all sufficiently small values of ϵ, the vector $\frac{1}{2}(a+\epsilon u)$ lies in the face $H(a)\cap\mathcal{V}$. Therefore, this face is a facet. □

The next theorem identifies the indecomposable vectors in the flow lattice of a connected graph.

Theorem 14.14.4 *The indecomposable vectors in the flow lattice of a connected graph X are the signed characteristic vectors of the cycles.*

Proof. Let $\mathcal{L}$ be a lattice contained in $\mathbb{Z}^n$, and let x be an element of $\mathcal{L}$ of minimal support which has all its entries in $\{-1, 0, 1\}$. For any element $y \in \mathcal{L}$ we have $(x+2y)_i \neq 0$ whenever $x_i \neq 0$, and so $\langle x+2y, x+2y \rangle \geq \langle x, x \rangle$. Equality holds only if $\operatorname{supp}(y) = \operatorname{supp}(x)$, and then y is a multiple of x. Therefore, by Lemma 14.14.2, x is indecomposable. Since the signed characteristic vectors of the cycles of X have minimal support, they are indecomposable elements of the flow lattice.

Conversely, suppose that x is a vector in the flow lattice $\mathcal{F}$ of X. Since x is a flow, there is a cycle C that supports a flow c such that $c_e x_e \geq 0$ for all $e \in E(C)$. (Prove this.) Then $\langle x, c \rangle \geq 0$, and so either x is decomposable, or $x = \pm c$. □

14.15 Bicycles

We consider some properties of the cut and flow spaces of a graph X over $GF(2)$. This topic is interesting in its own right, and the information we present will be needed in our work on knots in Chapter 17. Some aspects of our work are simplified by working over $GF(2)$. Firstly, there is no need to introduce orientations, since the oriented incidence matrix D is equal to the ordinary incidence matrix B. Secondly, every vector in $GF(2)^E$ is the characteristic vector of a subset of the edges of X. Therefore, we can easily visualize any vector as a subgraph of X. The binary addition of two vectors in $GF(2)^E$ corresponds to taking the symmetric difference of the edge-sets of two subgraphs of X.

Lemma 14.15.1 *Let X be a graph with n vertices and c components, with incidence matrix B. Then the 2-rank of B is $n - c$.*

Proof. The argument given in Theorem 8.3.1 remains valid over $GF(2)$. (The argument in Theorem 8.2.1 implicitly uses the fact that $-1 \neq 1$, and hence fails over $GF(2)$.) □

Let X be a graph with incidence matrix B and let S be a subset of $E(X)$ with characteristic vector x. Then $Bx = 0$ if and only if each vertex of X lies in an even number of edges from S. Equivalently, each vertex in the subgraph of X formed by the edges in S has even valency. Therefore, the flow space of X consists of the characteristic vectors of the even subgraphs.

We say that S is a *bicycle* if the characteristic vector x of S lies in both the flow space and cut space of X. Thus S is an edge cutset that is also an even subgraph. We admit the empty subgraph as a bicycle, and provide two more interesting examples in Figure 14.4, where the bicycles are shown with thick edges.

Let C denote the cut space of X and F its flow space. Then each element of $C \cap F$ is the characteristic vector of a bicycle; conversely, the characteristic vector of a bicycle lies in $C \cap F$. Therefore, we call $C \cap F$ the *bicycle*

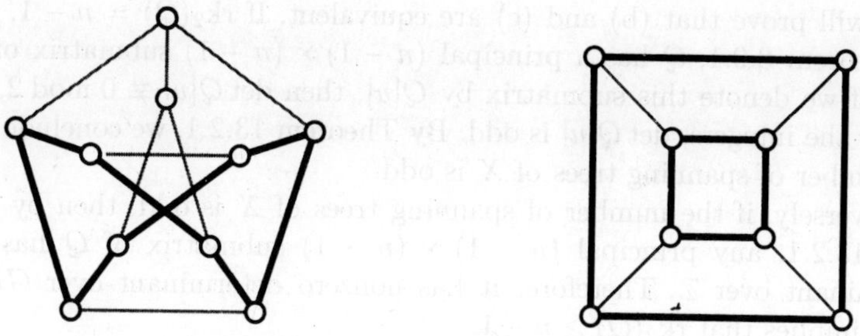

Figure 14.4. Bicycles in the Petersen graph and the cube

space of X. If X has no nonempty bicycles, then it is called *pedestrian*. Trees provide simple examples of pedestrian graphs, as do the complete graphs on an odd number of vertices.

Now,

$$\dim(C+F)=\dim(C)+\dim(F)-\dim(C\cap F),$$

whence we see that $C+F$ is the space of all binary vectors of length $|E(X)|$ if and only if $C\cap F$ is the empty subspace.

Lemma 14.15.2 *A graph X is pedestrian if and only if each subgraph of X is the symmetric difference of an even subgraph and an edge cutset.*

Proof. A subgraph of X is the symmetric difference of an even subgraph and an edge cutset if and only if its characteristic vector lies in $C+F$. □

Lemma 14.15.3 *Let X be a graph with c components on n vertices, and let Q be its Laplacian. Then the dimension of the bicycle space of X is $n-c-\mathrm{rk}_2\,Q$.*

Proof. Since we are working over $GF(2)$, we have $Q=BB^T$. The bicycle space of X may be identified with the set of vectors B^Tx such that $BB^Tx=0$. In other words, it is the image of the null space of BB^T under B^T. Hence the dimension of the bicycle space is

$$\dim\ker BB^T-\dim\ker B^T.$$

Since $\dim\ker BB^T=n-\mathrm{rk}_2(Q)$ and $\dim\ker B^T=c$, the result follows. □

Theorem 14.15.4 *For a connected graph X on n vertices, the following assertions are equivalent:*

(a) *X is pedestrian.*

(b) *The Laplacian Q of X has binary rank $n-1$.*

(c) *The number of spanning trees of X is odd.*

Proof. It follows at once from the previous lemma that (a) and (b) are equivalent.

We will prove that (b) and (c) are equivalent. If $\text{rk}_2(Q) = n - 1$, then by Theorem 8.9.1, Q has a principal $(n-1) \times (n-1)$ submatrix of full rank. If we denote this submatrix by $Q[u]$, then $\det Q[u] \not\equiv 0 \bmod 2$, and so over the integers, $\det Q[u]$ is odd. By Theorem 13.2.1, we conclude that the number of spanning trees of X is odd.

Conversely, if the number of spanning trees of X is odd, then by Theorem 13.2.1, any principal $(n-1) \times (n-1)$ submatrix of Q has odd determinant over $\mathbb{Z}$. Therefore, it has nonzero determinant over $GF(2)$, which implies that $\text{rk}_2(Q) \geq n - 1$. □

14.16 The Principal Tripartition

We can use the bicycles of a graph to determine a natural partition of its edges into three classes, called the *principal tripartition.*

Suppose $e \in E(X)$, and identify e with its characteristic vector in $GF(2)^E$. If e lies in a bicycle with characteristic vector b, then $b^T e \neq 0$, and therefore

$$e \notin (C \cap F)^\perp = (C^\perp + F^\perp) = F + C.$$

If e does not lie in a bicycle, then e is orthogonal to all vectors in $C \cap F$, and so $e \in F + C$. In other words, either e is contained in a bicycle, or e is the symmetric difference of a cut and an even subgraph.

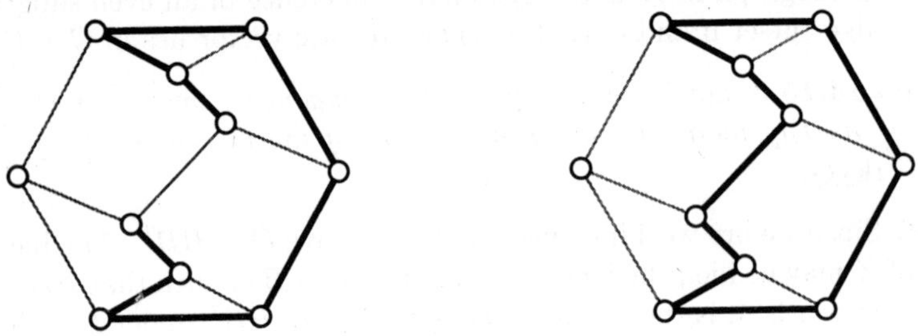

Figure 14.5. A cut and an even subgraph

Figure 14.5 shows a cut and an even subgraph whose symmetric difference is a single edge. In this figure we see that the edge is in the even subgraph, but not in the cut. There may be more than one way of representing a particular edge as the symmetric difference of a cut and an even subgraph, but we claim that in any such representation the edge will either always lie in the cut, or always lie in the even subgraph. For suppose that $e = c + f = c' + f'$, where c and c' lie in C and f and f' lie in F. Then $c + c' \in C$ and $f + f' \in F$, and so $c + c' = f + f'$ is a bicycle. Since e does not lie in a bicycle, we see that it must lie in both or neither of c and c', and similarly for f and f'. This can be summarized in the following result:

Theorem 14.16.1 *Let e be an edge in the graph X. Then precisely one of the following holds:*

(a) *e is contained in a bicycle,*
(b) *e lies in a cut S such that $S \setminus e$ is an even subgraph, or*
(c) *e lies in an even subgraph T such that $T \setminus e$ is a cut.* □

This provides us with a partition of the edges of X into three classes: *bicycle-type*, *cut-type*, and *flow-type* according as case (a), (b), or (c), respectively, holds. This is known as the principal tripartition of the edges of X. Figure 14.6 shows the principal tripartition of the graph of Figure 14.5.

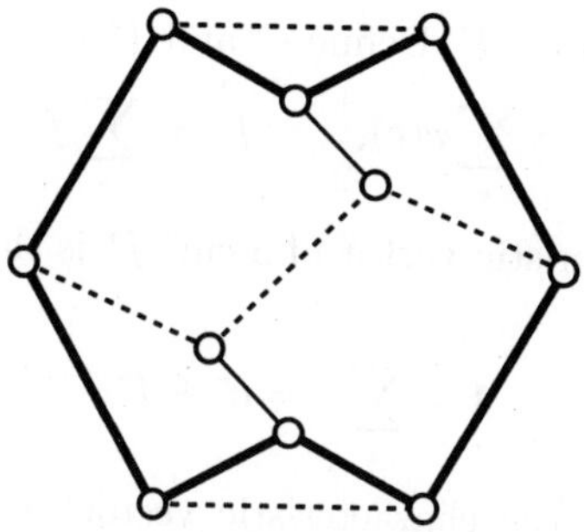

Figure 14.6. Bicycle-type (thick), cut-type (light), and flow-type (dashed) edges

If X is a plane graph, then clearly the bicycles of X are bicycles of its dual graph, and it is not hard to see that edges of cut-type in X correspond to edges of flow-type in X^*.

It is not difficult to find the principal tripartition of the edge set of a graph. The next result shows that it can be determined by solving a system of linear equations over $GF(2)$ with augmented matrix $(Q \quad B)$, and therefore can be found using Gaussian elimination.

Theorem 14.16.2 *Let y be the column of the incidence matrix B corresponding to the edge e, and consider the system of linear equations $Qx = y$. Then:*

(a) *If $Qx = y$ has no solution, then e is of bicycle-type.*
(b) *If $Qx = y$ has a solution x, where $x^TQx \neq 0$, then e is of cut-type.*
(c) *If $Qx = y$ has a solution x, where $x^TQx = 0$, then e is of flow-type.*

Proof. Identify e with its characteristic vector in $GF(2)^E$, and suppose that there is a vector x such that $Qx = y$. Then $BB^Tx = y = Be$, and as $\ker B = F$, we have

$$B^Tx \in F + e.$$

Now, $B^T x$ is a cut, and so this provides us with a representation of e as the symmetric difference of a cut and an even subgraph. Therefore, if e is of bicycle-type, the system of linear equations has no solution. If there is a solution, then the additional condition merely checks whether e is in the cut or not. □

Theorem 14.16.3 *In a pedestrian graph X, the union of the cut-type edges is a cut, and the union of the flow-type edges is a flow.*

Proof. Since X is pedestrian, any edge e has a unique expression of the form

$$e = c(e) + f(e),$$

where $c(e) \in C$ and $f(e) \in F$. Define c^* and f^* by

$$c^* := \sum_e c(e), \qquad f^* := \sum_e f(e).$$

Then c^* is the characteristic vector of a cut, f^* is the characteristic vector of a flow, and

$$\mathbf{1} = \sum_e e = c^* + f^*.$$

Using the fact that the characteristic vector of any cut is orthogonal to the characteristic vector of any flow, we obtain the following string of equalities:

$$\begin{aligned} c(e)^T e = c(e)^T (c(e) + f(e)) &= c(e)^T c(e) \\ &= c(e)^T \mathbf{1} \\ &= c(e)^T (c^* + f^*) \\ &= c(e)^T c^* \\ &= (e + f(e))^T c^* \\ &= e^T c^*. \end{aligned}$$

Now, $c(e)^T e = 1$ if and only if e is of cut-type, and therefore $e^T c^* = 1$ if and only if e is of cut-type. □

Exercises

1. If z is the signed characteristic vector of a cycle of X, show that $Dz = 0$.

2. Show that a graph has an open ear decomposition if and only if it contains no cut-vertex.

3. Show that if X is connected and has no cut-vertices and $e \in E(X)$, then there is a cycle basis consisting of cycles that contain e. (You

may need to use the fact that in a graph without cut-vertices any two edges lie in a cycle.)

4. A *strong orientation* is an orientation of a graph such that the resulting directed graph is strongly connected. Show that a connected graph with no cut-edge has a strong orientation.

5. Determine the rank over $GF(p)$ of the incidence matrix B of a graph X.

6. Let X be a connected plane graph with n vertices and e edges, and with faces $F_1, \dots, F_{e-n+2}$. Each face F_i is a cycle in X, and hence determines an element x_i of its flow space. Show that the cycles can be oriented so that $\sum_i x_i = 0$, and that there is no other linear relation satisfied by these vectors. (Hence, if we drop any one of them, the remaining $e - n + 1$ form a basis for the flow space.)

7. Show that if x belongs to the flow lattice $\mathcal{F}$ of X, then there is a signed circuit with characteristic vector c such that $x_e c_e \geq 0$ for each edge e of X. Prove similarly that if $x \in \mathcal{C}$, then there is a signed bond with characteristic vector b such that $x_e b_e \geq 0$ for all edges e.

8. A *source* in an oriented graph is a vertex with $d^+(u) = d(u)$, and a *sink* is a vertex with $d^-(u) = d(u)$. Show that a graph with an acyclic orientation has at least one sink and at least one source.

9. In the chip-firing game, show that a game is infinite if and only if every vertex is fired at least once.

10. Let X be a pedestrian graph. Then the number $\tau(X)$ of spanning trees of X is odd, and so exactly one of $\tau(X \setminus e)$ and $\tau(X/e)$ is odd. Show that if e is of cut-type, then $\tau(X/e)$ is odd, and if e is of flow-type, then $\tau(X \setminus e)$ is odd.

11. Show that if X is an even graph on n vertices, then the dimension of the bicycle space of X is congruent to $(n - 1) \bmod 2$. (So an even graph on an even number of vertices must have a nonempty bicycle.)

12. If X is an even graph, show that it has no edges of cut-type. If X is bipartite, show that it has no edges of flow-type.

13. Show that each subgraph obtained from the Petersen graph by deleting a vertex has two cycles of length nine in it (i.e., two Hamilton cycles). Show that the Petersen graph contains at least 12 pentagons, 10 hexagons, 15 octagons, 20 nonagons, and 6 subgraphs formed from two vertex-disjoint pentagons. Verify that the even subgraphs just listed are all the nonempty even subgraphs of the Petersen graph. Which even subgraphs form the bicycle space? [You may use any standard facts about automorphisms of the Petersen graph to shorten your work.]

14. Let X be a pedestrian graph with $\tau(X)$ spanning trees, and let P be the matrix representing orthogonal projection onto the real cut space of X (as in Section 14.8). Show that the edge e is of cut-type if and only if the diagonal ee-entry of $\tau(X)P$ is odd.

15. Suppose e is an edge of X, and $e = c + f$, where c is a cut and f is a flow in X. Show that if e is of cut-type, then $|c|$ is odd and $|f|$ is even, and if e is of flow-type, then $|c|$ is even and $|f|$ is odd.

Notes

The material in the first four sections of this chapter is quite standard. Our results on lattices come mostly from Bacher, de la Harpe, and Nagnibeda [1] and the related work of Biggs [2]. Section 14.14 is based on [1].

Chip firing is a topic with a surprising range of ramifications. There is a considerable physics literature on the subject, in the guise of "abelian sandpiles." For graph theorists, the natural starting points are Björner and Lovász [4], Björner, Lovász, and Shor [5], and Lovász and Winkler [8]. The bound on the length of a terminating chip-firing game given as Theorem 14.10.2 is due to G. Tardos [9]. The eigenvalue bound, given as Corollary 14.10.5, is from [5].

Biggs [3] establishes a connection between chip-firing and the rank polynomial. We treat the rank polynomial at length in the next chapter, but unfortunately, we will not be able to discuss this work.

Our treatment of chip firing generally follows the work of Biggs. The very useful concept of a diffuse state comes from Jeffs and Seager [6] though. Our approach in Section 14.13 is based in part on ideas from Ahn-Louise Larsen's Master's thesis [7].

Bicycles and the principal tripartition will play a useful role in our work on knots and links in Chapter 17.

References

[1] R. Bacher, P. de la Harpe, and T. Nagnibeda, *The lattice of integral flows and the lattice of integral cuts on a finite graph*, Bull. Soc. Math. France, 125 (1997), 167–198.

[2] N. Biggs, *Algebraic potential theory on graphs*, Bull. London Math. Soc., 29 (1997), 641–682.

[3] N. L. Biggs, *Chip-firing and the critical group of a graph*, J. Algebraic Combin., 9 (1999), 25–45.

[4] A. Björner and L. Lovász, *Chip-firing games on directed graphs*, J. Algebraic Combin., 1 (1992), 305–328.

[5] A. BJÖRNER, L. LOVÁSZ, AND P. W. SHOR, *Chip-firing games on graphs*, European J. Combin., 12 (1991), 283–291.

[6] J. JEFFS AND S. SEAGER, *The chip firing game on n-cycles*, Graphs Combin., 11 (1995), 59–67.

[7] A. L. LARSEN, *Chip firing games on graphs*, Master's thesis, Technical University of Denmark, 1998.

[8] L. LOVÁSZ AND P. WINKLER, *Mixing of random walks and other diffusions on a graph*, in Surveys in combinatorics, 1995 (Stirling), Cambridge Univ. Press, Cambridge, 1995, 119–154.

[9] G. TARDOS, *Polynomial bound for a chip firing game on graphs*, SIAM J. Discrete Math., 1 (1988), 397–398.

15

The Rank Polynomial

One goal of this chapter is to introduce the rank polynomial of a signed graph, which we will use in the next chapter to construct the Jones polynomial of a knot. A second goal is to place this polynomial in a larger context. The rank polynomial is a classical object in graph theory, with a surprising range of ramifications. We develop its basic theory and provide an extensive description of its applications.

We treat the rank polynomial as a function on matroids, rather than graphs. This certainly offers greater generality at a quite modest cost, but the chief advantage of this viewpoint is that there is a natural theory of matroid duality which includes, as a special case, the duality theory of planar graphs. This duality underlies many properties of the rank polynomial.

15.1 Rank Functions

Let Ω be a finite set. A function rk on the subsets of Ω is a *rank function* if it is nonnegative, integer-valued, and satisfies the following conditions:

(R1) If A and B are subsets of Ω and $A \subseteq B$, then $\text{rk}(A) \leq \text{rk}(B)$.
(R2) For all subsets A and B of Ω,

$$\text{rk}(A \cap B) + \text{rk}(A \cup B) \leq \text{rk}(A) + \text{rk}(B).$$

(R3) If $A \subseteq \Omega$, then $\text{rk}(A) \leq |A|$.

Functions satisfying R1 are called *monotone*, and functions satisfying R2 are called *submodular*. These are the important properties; Section 15.15

shows that any integer-valued monotone submodular function can be modified to produce a rank function.

A rather uninteresting example of a rank function is the function $\mathrm{rk}(A) = |A|$, which clearly satisfies the three conditions. The following result gives a family of more interesting examples.

Lemma 15.1.1 *Let D be an $m \times n$ matrix with entries from some field $\mathbb{F}$, and let Ω index the columns of D. For any subset $A \subseteq \Omega$, view the columns corresponding to the elements of A as vectors in $\mathbb{F}^m$ and let $\mathrm{rk}(A) = \dim(\mathrm{span}(A))$. Then* rk *is a rank function.* □

There are two particularly important special cases of this result, arising from graphs and from codes.

Let X be a graph with an arbitrary orientation σ and let D be the incidence matrix of the oriented graph X^σ. Reversing the orientation on an edge only multiplies the corresponding column of D by -1, and hence does not alter the rank function. Therefore, the rank function is independent of the orientation, and so is determined only by the graph X.

A *linear code* is a subspace of a vector space $\mathbb{F}^n$. A *generator matrix* for a linear code C is a $k \times n$ matrix whose rows form a basis for C. Although a linear code may have many generator matrices, the rank function is independent of which matrix is chosen, and hence we refer to the rank function determined by a code C.

All of these examples have the property that the elements of Ω are vectors in a vector space and the rank of a subset is the dimension of its span. However, there are rank functions that are not of this form, and we present one example. Consider the configuration of points and lines shown in Figure 15.1. Let Ω be the set of points, and define a function as follows:

$$\mathrm{rk}(A) = \begin{cases} |A|, & \text{if } |A| \le 2; \\ 2, & \text{if } A \text{ is the 3 points of a line;} \\ 3, & \text{otherwise.} \end{cases}$$

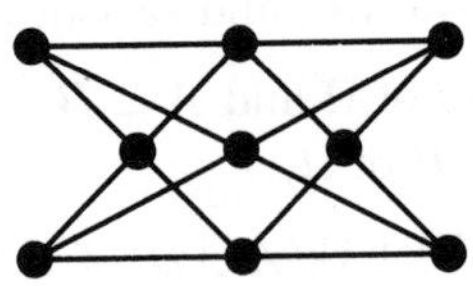

Figure 15.1. A configuration of points and lines

15.2 Matroids

A *matroid* is a set Ω together with a rank function rk on the subsets of Ω. The theory of matroids is an abstraction of the theory of dependence and independence in linear algebra, with the rank function playing the role of "dimension." Therefore, much of the language of matroid theory is analogous to that of linear algebra. A subset $A \subseteq \Omega$ is *independent* if $\mathrm{rk}(A) = |A|$, and *dependent* otherwise. A maximal independent set is called a *basis*. As well as the language, some of the fundamental properties of linear algebra also hold in the more abstract matroid setting.

Lemma 15.2.1 *If $A \subseteq \Omega$, then the rank of A is the size of any maximal (with respect to inclusion) independent set contained in A.*

Proof. If A is itself independent, then the result follows immediately. So let A' be any maximal independent set properly contained in A. Then for every element $x \in A \setminus A'$ we have $\mathrm{rk}(A' \cup \{x\}) = \mathrm{rk}(A)$. Now, consider any subset B such that $A' \subset B \subseteq A$, and let $x \in B \setminus A'$. Then by applying the submodularity property to $B \setminus \{x\}$ and $A' \cup \{x\}$ we get

$$\mathrm{rk}(A') + \mathrm{rk}(B) \leq \mathrm{rk}(B \setminus \{x\}) + \mathrm{rk}(A' \cup \{x\}),$$

and so $\mathrm{rk}(B) = \mathrm{rk}(B \setminus \{x\})$, which in turn implies that $\mathrm{rk}(B) = \mathrm{rk}(A')$. By taking $B = A$ this yields that $\mathrm{rk}(A) = \mathrm{rk}(A') = |A'|$. □

An important consequence of this result is that the collection of independent sets determines the rank function, and hence the matroid. Therefore, a matroid M can be specified just by listing its independent sets. In fact, since any subset of an independent set is independent, it is sufficient to list just the bases of M. Alternatively, a matroid can be specified by listing the minimal dependent sets of M; these are called the *circuits* of M.

Corollary 15.2.2 *If M is a matroid, then all bases of M have the same size* $\mathrm{rk}(\Omega)$. □

The common size of all the bases of M is called the rank of the matroid, and denoted by $\mathrm{rk}(M)$. This is again reminiscent of linear algebra, where all bases of a vector space have the same size.

Now, we consider the matroid determined by a graph in more detail. Let X be a graph on n vertices and D be the incidence matrix of an arbitrary orientation of X. The columns of D can be identified with the edges of X, and therefore the rank function determined by X yields a matroid on $\Omega = E(X)$. This matroid is called the *cycle matroid* of X and denoted by $M(X)$.

We identify a subset $A \subseteq E(X)$ with the subgraph of X with vertex set $V(X)$ and edge set A. If A has c components, then by Theorem 8.3.1,

$$\mathrm{rk}(A) = n - c.$$

If A does not contain any cycles, then $c = n - |A|$, and so A is independent. Conversely, if A does contain a cycle, then the corresponding columns of D are linearly dependent, and so A is a dependent set of $M(X)$. Therefore, the independent sets of $M(X)$ are precisely the sets of edges that contain no cycles. The bases of $M(X)$ are the maximal spanning forests of X—spanning trees if X is connected—and the circuits of $M(X)$ are the cycles of X.

A *loop* in a matroid is an element e such that $\text{rk}(\{e\}) = 0$. If a graph contains a loop, then the corresponding column of its incidence matrix D is defined to be zero. Therefore, this column forms a dependent set on its own, and so is a loop in the cycle matroid.

Distinct graphs may have the same cycle matroid. For example, any tree on n vertices has a cycle matroid with $n-1$ elements where every subset of Ω is independent. More generally, if X and Y are two graphs on disjoint vertex sets, then the disconnected graph $X \cup Y$ has the same cycle matroid as the graph obtained by identifying a single vertex of X with a vertex of Y. However, it is known that the cycle matroid determines the graph if it is 3-connected.

15.3 Duality

An important secret of elementary linear algebra is that the concepts "spanning set" and "independent set" are dual. Thus the fact that a minimal spanning set is a basis is dual to the assertion that a maximal independent set is a basis. This is a reflection of the result that if M is a matroid on a set Ω, then the complements of the bases of M are the bases of a second matroid on Ω.

We will denote the complement in Ω of the set A by $\overline{A}$. If f is a function on the subsets of Ω, define its *dual* to be the function $f^\perp$, given by

$$f^\perp(A) := |A| + f(\overline{A}) - f(\Omega).$$

A little thought shows that if $f(\emptyset) = 0$, then $(f^\perp)^\perp = f$, which provides some justification for the notation.

Theorem 15.3.1 *If* rk *is a rank function on* Ω, *then the function* $\text{rk}^\perp$ *given by*

$$\text{rk}^\perp(A) := |A| + \text{rk}(\overline{A}) - \text{rk}(\Omega)$$

is also a rank function on Ω.

Proof. If f is a function on the subsets of Ω, let $\overline{f}$ be the function (on subsets of Ω) defined by

$$\overline{f}(A) := f(\overline{A}).$$

If A and B are subsets of Ω and f is submodular, we have

$$f(\overline{A}) + f(\overline{B}) \geq f(\overline{A} \cap \overline{B}) + f(\overline{A} \cup \overline{B}) = f(\overline{A \cup B}) + f(\overline{A \cap B}),$$

which implies that $\overline{f}$ is submodular. It is immediate that the sum of two submodular functions is submodular. Since the size of a set is a submodular function, it follows that, if f is submodular, then so is $\overline{f} + |\cdot|$.

Next we show that if f is a rank function on Ω, then the function mapping a subset A to $\overline{f}(A) + |A|$ is monotone. If $A \subseteq B$, then

$$f(\overline{A}) = f(\overline{B} \cup (B \setminus A)) \leq f(\overline{B}) + f(B \setminus A) \leq f(\overline{B}) + |B \setminus A|.$$

Hence

$$\overline{f}(A) + |A| \leq f(\overline{B}) + |B \setminus A| + |A| = \overline{f}(B) + |B|.$$

Thus we have established that $\mathrm{rk}^{\perp}$ is monotone and submodular, and it remains only to show that $\mathrm{rk}^{\perp}$ is nonnegative and satisfies $\mathrm{rk}^{\perp}(A) \leq |A|$. Because rk is submodular and monotone,

$$\mathrm{rk}(\Omega) \leq \mathrm{rk}(\overline{A}) + \mathrm{rk}(A),$$

and so

$$0 \leq \mathrm{rk}(\Omega) - \mathrm{rk}(\overline{A}) \leq \mathrm{rk}(A) \leq |A|.$$

Therefore, if we rewrite $\mathrm{rk}^{\perp}$ as

$$\mathrm{rk}^{\perp}(A) = |A| - (\mathrm{rk}(\Omega) - \mathrm{rk}(\overline{A})),$$

we see that

$$0 \leq \mathrm{rk}^{\perp}(A) \leq |A|.$$

Therefore, $\mathrm{rk}^{\perp}$ is a rank function. □

If rk is the rank function of a matroid M, we call the matroid with rank function $\mathrm{rk}^{\perp}$ the *dual* of M and denote it by $M^{\perp}$. The rank of the dual matroid is $\mathrm{rk}^{\perp}(\Omega) = |\Omega| - \mathrm{rk}(\Omega)$.

Lemma 15.3.2 *The bases of $M^{\perp}$ are the complements of the bases of M.*

Proof. If A is a subset of Ω, then

$$\mathrm{rk}^{\perp}(\overline{A}) = |\overline{A}| + \mathrm{rk}(A) - \mathrm{rk}(\Omega),$$

and so if A is an independent set in M, then $\overline{A}$ is a spanning set in $M^{\perp}$. Conversely, if $\overline{A}$ is a spanning set in $M^{\perp}$, then A is independent in M. By duality, A is a spanning set in M if and only if $\overline{A}$ is independent in $M^{\perp}$. Therefore, A is a base of M if and only if $\overline{A}$ is a base of $M^{\perp}$. □

The dual of the cycle matroid of a graph X is called the *bond matroid* of X. For a subset $A \subseteq E(X)$, let $\bar{c}$ be the number of components of the spanning subgraph of X with edge set $\overline{A}$. Then

$$\mathrm{rk}^{\perp}(A) = |A| + (n - \bar{c}) - (n - 1) = |A| - \bar{c} + 1.$$

Thus A is independent in $M^{\perp}$ if and only if $\bar{c} = 1$ or, equivalently, if A does not contain a set of edges whose removal disconnects X. Therefore, the circuits of $M^{\perp}$ are the minimal edge cuts, or bonds, of X; hence justifying the name bond matroid. This also holds true for disconnected graphs, provided that we consider an edge-cut to be a set of edges whose removal increases the number of components of the graph.

If X is a graph embedded in the plane, then the cycle matroid of X is the bond matroid of the dual graph of X. A matroid is *graphic* if it can be expressed as the cycle matroid of a graph. The bond matroid of a graph X is known to be graphic if and only if X is planar.

An element e of a matroid M such that e is a loop in $M^{\perp}$ is called a *coloop* in M. In general, we see that e is a coloop if and only if

$$\operatorname{rk}(\Omega \setminus \{e\}) = \operatorname{rk}(\Omega) - 1.$$

In a graphic matroid, this implies that e is a coloop if and only if it is a cut-edge, or bridge.

15.4 Restriction and Contraction

Suppose that M is a matroid with rank function rk defined on a set Ω. If T is a subset of Ω, then the restriction of rk to subsets of T is a rank function on T. Hence we have a matroid on T, denoted by $M \restriction T$, and called the *restriction* of M to T. Sometimes this is also called the *deletion* of $\overline{T}$ from M. In particular, if $e \in \Omega$, we usually denote $M \restriction (\Omega \setminus e)$ by $M \setminus e$, and say that it is obtained by deleting e.

Now, let ρ denote the function on subsets of $\overline{T}$ given by

$$\rho(A) = \operatorname{rk}(A \cup T) - \operatorname{rk}(T).$$

Then ρ is a rank function, and hence determines a matroid on $\overline{T}$. This can be shown directly by showing that ρ satisfies the three conditions for a rank function, but we get more information if we proceed by considering $\rho^{\perp}$. Then

$$\begin{aligned}
\rho^{\perp}(A) &= |A| + \rho(\overline{T} \setminus A) - \rho(\overline{T}) \\
&= |A| + \operatorname{rk}((\overline{T} \setminus A) \cup T) - \operatorname{rk}(T) + \operatorname{rk}(T) - \operatorname{rk}(\Omega) \\
&= |A| + \operatorname{rk}(\Omega \setminus A) - \operatorname{rk}(\Omega) \\
&= \operatorname{rk}^{\perp}(A).
\end{aligned}$$

Consequently, $\rho^{\perp}$ is a rank function on $\overline{T}$, and therefore ρ is a rank function on $\overline{T}$. The corresponding matroid is called the *contraction* of T from M, and is denoted by M/T. Our argument has also shown that

$$(M/T)^{\perp} = M^{\perp} \restriction \overline{T}$$

and, by duality,

$$(M \upharpoonright T)^\perp = M^\perp / \overline{T}.$$

You are invited to verify that if e and f are distinct elements of M, then

$$(M \setminus e)/f = (M/f) \setminus e.$$

Our next result implies that for the cycle matroids of graphs, the terminology just introduced is consistent with the graph-theoretic terminology defined in Section 13.2.

Theorem 15.4.1 *Let X be a graph, and let $e \in E(X)$. Then $M(X) \setminus e = M(X \setminus e)$ and $M(X)/e = M(X/e)$.*

Proof. The result for deletion is a direct consequence of the definitions of deletion, so we need only consider the result for contraction. If e is a loop, then $X/e = X \setminus e$, and it is easy to see that $M(X)/e = M(X) \setminus e$, and so the result follows from the result for deletion. So suppose that e is not a loop. From the definition of ρ, a set $A \subseteq E(X) \setminus e$ is independent in $M(X)/e$ if and only if $A \cup e$ is independent in $M(X)$. It is not hard to see that $A \cup e$ contains no cycles of X if and only if A contains no cycles of X/e. Therefore, $M(X)/e$ and $M(X/e)$ have the same independent sets. □

If there is a subset T of Ω with $0 < |T| < |\Omega|$ such that for all subsets A of Ω

$$\mathrm{rk}(A) = \mathrm{rk}(A \cap T) + \mathrm{rk}(A \cap \overline{T}),$$

then we say that M is the *direct sum* of the matroids $M \upharpoonright T$ and $M \upharpoonright \overline{T}$. A matroid is *connected* if it is not a direct sum. If the graph X is not connected, then the cycle matroid $M(X)$ is a direct sum of the cycle matroids of each connected component. If X is connected but has a cut-vertex, then $M(X)$ is again not connected; it is the direct sum of the cycle matroids of the maximal cut-vertex-free subgraphs, or *blocks*, of X. If e is a loop or a coloop in a matroid M, then M is the direct sum of $M \upharpoonright \{e\}$ and $M \setminus e = M/e$.

15.5 Codes

Recall that a linear code of length n is simply defined to be a subspace of a vector space $\mathbb{F}^n$. Let C be a linear code with generator matrix G, and consider the rank function defined on the columns of G. The rank function is independent of the choice of generator matrix, and so we denote the corresponding matroid by $M(C)$. The columns of G correspond to the coordinate positions of vectors in $\mathbb{F}^n$, so we may assume that $M(C)$ is a matroid on the set $\Omega = \{1, \ldots, n\}$.

An element $i \in \Omega$ is a loop in $M(C)$ if and only if the ith column of G is zero. An element i is a coloop if the rank of the matrix obtained by deleting column i from G is less than the rank of G. If $M(C)$ has a coloop i, then by using row operations on G if necessary, we can assume that C has a generator matrix of the form

$$G = \begin{pmatrix} 0 & \cdots & 0 & 1 & 0 & \cdots & 0 \\ & & & 0 & & & \\ & G_1 & & \vdots & & G_2 & \\ & & & 0 & & & \end{pmatrix}. \tag{15.1}$$

From this we see that if i is a coloop, then C contains the ith standard basis vector e_i. It is not hard to verify the converse, and thus we conclude that i is a coloop if and only if $e_i \in C$.

If C is a code, then its *dual code*, $C^\perp$, is defined by

$$C^\perp := \{v \in \mathbb{F}^n : \langle u, v \rangle = 0 \text{ for all } u \text{ in } C\}.$$

If C is the row space of the generator matrix G, then $C^\perp$ is the null space of G.

Lemma 15.5.1 *If the linear code C has a generator matrix*

$$G = (\, I_k \quad M \,),$$

then the matrix

$$H = (\, -M^T \quad I_{n-k} \,)$$

is a generator matrix for $C^\perp$.

Proof. Clearly, $GH^T = 0$, so the columns of H^T are in the null space of G. Since G has rank k and H^T has rank $n-k$, the columns of H^T form a basis for the null space of G. □

The next result shows that matroid duality corresponds to duality in codes as well as graphs.

Lemma 15.5.2 *If C is a code of dimension k in $\mathbb{F}^n$, then $M(C^\perp) = M(C)^\perp$.*

Proof. Suppose that $A \subseteq \Omega$ is a base of $M(C)$. Then there is a generator matrix for C such that the matrix formed by the columns corresponding to A is I_k. Hence by Lemma 15.5.1, there is a generator matrix H for $C^\perp$ where the matrix formed by the columns corresponding to $\overline{A}$ is I_{n-k}. Therefore, $\overline{A}$ is a base of $M(C^\perp)$. By duality, if $\overline{A}$ is a base of $M(C^\perp)$, then A is a base of $M(C)$. Hence the bases of $M(C^\perp)$ are precisely the complements of the bases of $M(C)$, and the result follows. □

A coding theorist *punctures* a code C at coordinate i by deleting the ith entry of each codeword. The resulting code is denoted by $C \setminus i$. It is clear that puncturing a code corresponds to deleting an element, and that the matroid of the resulting code is $M(C)\setminus i$.

A code is *shortened* at coordinate i by taking the subcode formed by the vectors x such that $x_i = 0$ and then puncturing the resulting subcode at i. The resulting code is denoted by C/i, and the next result justifies this choice of terminology.

Theorem 15.5.3 *If C/i is the code obtained by shortening C at coordinate position i, then $M(C/i) = M(C)/i$.*

Proof. Let G be the generator matrix for C. If i is a loop, then the ith column of G is zero, and puncturing and shortening the code are the same thing, and so $M(C/i) = M(C)\setminus i = M(C)/i$. So suppose that i is not a loop. Using row operations if necessary, we can assume that the ith column of G contains a single nonzero entry in the first row and that the remaining $k-1$ rows of G form a generator matrix for C/i. Recall that if rk is the rank function for $M(C)$, then $\text{rk}(A \cup i) - 1$ is the rank function for $M(C/i)$. Then it is clear that A is independent in $M(C)/i$ if and only if $A \cup i$ is independent in $M(C)$ if and only if A is independent in $M(C/i)$. Hence $M(C/i) = M(C)/i$. □

A matroid is called $\mathbb{F}$-*linear* if it is determined by the columns of a matrix over $\mathbb{F}$, and *binary* if it is determined by the columns of a matrix over $GF(2)$. By Lemma 14.15.1 the cycle matroid of a graph is binary, and thus Lemma 15.5.1 implies that the bond matroid is also binary.

15.6 The Deletion–Contraction Algorithm

In this section we will consider several graph parameters that can be computed using a similar technique: the deletion–contraction algorithm.

We start with one that we have seen several times already. In Section 13.2 we gave an expression for the number of spanning trees $\tau(X)$ of a graph X:

$$\tau(X) = \begin{cases} \tau(X\setminus e), & \text{if } e \text{ is a loop;} \\ \tau(X/e), & \text{if } e \text{ is a cut-edge;} \\ \tau(X\setminus e) + \tau(X/e); & \text{otherwise.} \end{cases}$$

Since both $X\setminus e$ and X/e are smaller graphs than X, this yields a recursive algorithm for computing $\tau(X)$ by applying this equation until the only graphs in the expression are trees (which of course have one spanning tree). This algorithm is called the *deletion–contraction algorithm*.

The deletion–contraction algorithm is not a good algorithm for computing the number of spanning trees of a graph, because it can take exponential time, while the determinant formula given in Section 13.2 can be computed in polynomial time. However, there are other graph parameters for which the deletion–contraction algorithm is essentially the only available algorithm. Let $\kappa(X)$ denote the number of acyclic orientations of X. Then we can find an expression for $\kappa(X)$ that depends on whether an edge is a loop, coloop, or neither.

Theorem 15.6.1 *If X is a graph, then the number $\kappa(X)$ of acyclic orientations of X is given by*

$$\kappa(X) = \begin{cases} 0, & \text{if } X \text{ contains a loop;} \\ 2\kappa(X/e), & \text{if } e \text{ is a cut-edge;} \\ \kappa(X \setminus e) + \kappa(X/e); & \text{if } e \text{ is not a loop or a cut-edge.} \end{cases}$$

Proof. If X contains a loop, then this loop forms a directed cycle in any orientation. If e is a cut-edge in X, then any acyclic orientation of X can be formed by taking an acyclic orientation of X/e and orienting e in either direction. Suppose, then, that $e = uv$ is neither a loop nor a cut-edge. If σ is an acyclic orientation of X, then $\tau = \sigma\lceil(V(X)\setminus e)$ is an acyclic orientation of $X \setminus e$. We partition the orientations of $X \setminus e$ according to whether or not there is a path respecting the orientation joining the end-vertices of e (either from u to v, or v to u; clearly, there cannot be both). If there is no path joining the end-vertices of e, then τ is an acyclic orientation of X/e, and so there are $\kappa(X/e)$ such acyclic orientations, leaving $\kappa(X\setminus e) - \kappa(X/e)$ remaining acyclic orientations. If τ is in the former category, then it is the restriction of two acyclic orientations of X (because e may be oriented in either direction), and if it is in the latter category, then it is the restriction of one acyclic orientation of X (since e is oriented in the same direction as the directed path joining its end-vertices). Therefore,

$$\kappa(X) = 2\kappa(X/e) + (\kappa(X \setminus e) - \kappa(X/e)) = \kappa(X/e) + \kappa(X \setminus e),$$

as claimed. □

Consider now the number of acyclic orientations $\kappa(X, v)$ with a particular vertex v as the unique source. It is not at all obvious that this is independent of v. However, this is in fact true, and is a consequence of the following result.

Theorem 15.6.2 *If X is a graph, then the number $\kappa(X, v)$ of acyclic orientations of X where v is the unique source is given by*

$$\kappa(X, v) = \begin{cases} 0, & \text{if } X \text{ contains a loop;} \\ \kappa(X/e, v), & \text{if } e \text{ is a cut-edge;} \\ \kappa(X \setminus e, v) + \kappa(X/e, v); & \text{if } e \text{ is not a loop or a cut-edge.} \end{cases}$$

Proof. This is left as Exercise 8. □

15.7 Bicycles in Binary Codes

If C is a code over a finite field $\mathbb{F}$, then the intersection $C \cap C^{\perp}$ may be nontrivial. Coding theorists call it the *hull* of the code. If C is the binary code formed by the row space of the incidence matrix B of a graph X, then the hull of C is just the bicycle space of X. In this section we consider only binary codes, and therefore call $C \cap C^{\perp}$ the *bicycle space* of C, and its elements the bicycles of C (or of $C^{\perp}$).

We start by extending the tripartition of Section 14.16 to binary codes.

Theorem 15.7.1 *Assume that C is a binary code of length n and let e_i denote the ith standard basis vector of $GF(2)^n$. Then exactly one of the following holds:*

(a) *i lies in the support of a bicycle,*
(b) *i lies in the support of a codeword γ such that $\gamma + e_i \in C^{\perp}$, or*
(c) *i lies in the support of a word φ in $C^{\perp}$ such that $\varphi + e_i \in C$.*

Proof. If i does not lie in the support of a bicycle, then $\langle e_i, \beta \rangle = 0$ for each element β of $C \cap C^{\perp}$, and consequently

$$e_i \in (C \cap C^{\perp})^{\perp} = C^{\perp} + C.$$

Thus $e_i = \gamma + \varphi$, where $\gamma \in C$ and $\varphi \in C^{\perp}$.

Suppose we also have $e_i = \gamma' + \varphi'$, where $\gamma' \in C$ and $\varphi' \in C^{\perp}$. Then

$$\gamma + \gamma' = \varphi + \varphi' \in C \cap C^{\perp}.$$

Therefore, i is in the support of both or neither of γ and γ', and similarly for φ and φ'. Hence precisely one of the last two conditions holds. □

We say that an element i is of *bicycle-type*, *cut-type*, or *flow-type* according as (a), (b), or (c) respectively holds. This classification therefore gives us a tripartition of the elements of the matroid $M(C)$. It is straightforward to check that any loop is of flow-type, and any coloop is of cut-type.

Our next aim is to relate the dimension of the bicycle space of C with that of $C \setminus i$ and C/i.

Theorem 15.7.2 *Let C be a binary code with a bicycle space of dimension b. If i is an element of $M(C)$, then the following table gives the dimension of the bicycle space of $C \setminus i$ and C/i.*

Type of i	$C \setminus i$	C/i
Loop or coloop	b	b
Bicycle-type	$b-1$	$b-1$
Cut-type, not coloop	$b+1$	b
Flow-type, not loop	b	$b+1$

Proof. To verify this table, we need only consider the effect of deletion, since once we know what happens in this case, the effect of contraction is determined by duality.

First we consider loops and coloops. Deleting a loop clearly does not affect the dimension of the bicycle space. If i is a coloop, then no bicycle of C has i in its support. Any bicycle of C is a bicycle of $C\backslash i$ once the ith coordinate is deleted, and conversely, any bicycle of $C\backslash i$ yields a bicycle of C by inserting an ith coordinate with value 0.

For any codeword $\alpha \in C$, let $\hat{\alpha}$ denote the codeword in $C \backslash i$ obtained by deleting the ith coordinate of α. Provided that i is not a coloop, every codeword in $C\backslash i$ is the image under the map $\alpha \mapsto \hat{\alpha}$ of a unique codeword in C. We will analyze when $\hat{\alpha}$ is a bicycle, which depends both on α and the type of i.

If α is a bicycle in C, then $\hat{\alpha}$ is a bicycle in $C\backslash i$ if and only if $\alpha_i = 0$. If α is not a bicycle and $\alpha_i = 0$, then $\hat{\alpha}$ is not a bicycle. However, if α is not a bicycle but $\alpha_i = 1$, then $\hat{\alpha}$ is a bicycle if and only if $\alpha + e_i \in C^{\perp}$.

Therefore, the map $\alpha \mapsto \hat{\alpha}$ may map both bicycles and non-bicycles of C to bicycles and non-bicycles of $C\backslash i$. For each type of edge we must account for the bicycles lost and gained.

If i is of bicycle type, then only the bicycles of C with $\alpha_i = 0$ map to bicycles of $C\backslash i$, and therefore the dimension of the bicycle space drops by one.

If i is not of bicycle type, then every bicycle of C maps to a bicycle of $C\backslash i$, and so the only question is whether any bicycles are gained. If i is of flow-type, then there are no codewords α for which $\alpha + e_i \in C^{\perp}$, and so the dimension of the bicycle space of $C\backslash i$ is equal to the dimension of the bicycle space of C. However, if i is of cut-type, then there is a codeword α, with i in its support, such that $\alpha + e_i \in C^{\perp}$. Therefore, $\hat{\alpha}$ is a bicycle in $C\backslash i$. Now, if β is another codeword such that $\hat{\beta}$ is a bicycle in $C\backslash i$, then $\beta + e_i \in C^{\perp}$, and so $\alpha + \beta \in C^{\perp}$, and so is a bicycle. Therefore, every vector with i in its support that maps to a bicycle has the form $\gamma + \alpha$, where γ is a bicycle. Hence the dimension of the bicycle space of $C\backslash i$ is just one greater than the dimension of the bicycle space of C. □

We can use the information we have just obtained to show that the dimension of the bicycle space can be obtained by deletion and contraction. If C is a binary code, let $\ell(C)$ denote its length and let $b(C)$ be the dimension of its bicycle space. Define the parameter bike(C) by

$$\text{bike}(C) = (-1)^{\ell(C)}(-2)^{b(C)}.$$

Up to sign, this is simply the number of bicycles in C.

Lemma 15.7.3 *Let i be an element of the code C. Then*

$$\operatorname{bike}(C) = \begin{cases} (-1)\operatorname{bike}(C \setminus i), & \text{if } i \text{ is a loop;} \\ (-1)\operatorname{bike}(C/i), & \text{if } i \text{ is a coloop;} \\ \operatorname{bike}(C \setminus i) + \operatorname{bike}(C/i), & \text{otherwise.} \end{cases}$$

Proof. If i is a loop or a coloop, then $b(C \setminus i) = b(C/i) = b(C)$, whence the first two claims follow. So suppose that i is neither a loop nor coloop, and assume that $\ell(C) = n$ and $b(C) = d$. If i is of cut-type, then

$$\begin{aligned} \operatorname{bike}(C \setminus i) + \operatorname{bike}(C/i) &= (-1)^{n-1}(-2)^{d+1} + (-1)^{n-1}(-2)^{d} \\ &= (-1)^{n}(-2)^{d}. \end{aligned}$$

If i is of flow-type, then

$$\begin{aligned} \operatorname{bike}(C \setminus i) + \operatorname{bike}(C/i) &= (-1)^{n-1}(-2)^{d} + (-1)^{n-1}(-2)^{d+1} \\ &= (-1)^{n}(-2)^{d}. \end{aligned}$$

Finally, if i lies in the support of a bicycle, then

$$\begin{aligned} \operatorname{bike}(C \setminus i) + \operatorname{bike}(C/i) &= 2(-1)^{n-1}(-2)^{d-1} \\ &= (-1)^{n}(-2)^{d}. \end{aligned}$$

Thus in all cases, $\operatorname{bike}(C \setminus i) + \operatorname{bike}(C/i) = \operatorname{bike}(C)$. □

Although the dimension of the bicycle space can be calculated by the deletion–contraction algorithm, this leads to an exponential algorithm. In practice, the dimension of the bicycle space can be computed more easily using Gaussian elimination. Along with the number of spanning trees, this provides an example showing that evaluating the rank polynomial at a specific point can be easy, whereas finding the rank polynomial itself is NP-hard.

15.8 Two Graph Polynomials

Numerical parameters are not the only graphical parameters that can be calculated by the deletion–contraction algorithm. In this section we consider two graph polynomials: the chromatic polynomial and the reliability polynomial of a graph.

Given a graph X, let $P(X, t)$ be the number of proper colourings of the vertices of X with t colours. For some graphs we can compute this function easily. If $X = K_n$, then it is easy to see that

$$P(K_n, t) = t(t-1)\cdots(t-n+1),$$

because there are t choices of colour for the first vertex, $t-1$ for the second vertex, and so on. If T is a tree, then there are t choices of colour for the first vertex, and then $t-1$ choices for each of its neighbours, $t-1$ choices for each of their neighbours, and so on. Therefore,

$$P(T,t) = t(t-1)^{n-1}.$$

The function $P(X,t)$ is called the *chromatic polynomial* of the graph X. For complete graphs and trees we have just seen that the function actually is a polynomial in t. The next result shows that this is true for all graphs.

Theorem 15.8.1 *Let $P(X,t)$ denote the number of proper colourings of X with t colours. Then*

$$P(X,t) = \begin{cases} 0, & \text{if } X \text{ contains a loop;} \\ P(X\setminus e,t) - P(X/e,t), & \text{if } e \text{ is not a loop.} \end{cases}$$

Proof. If X is empty, then $P(X,t) = t^n$. If X contains a loop, then it has no proper colourings. Otherwise, suppose that u and v are the ends of the edge e. Any t-colouring of $X\setminus e$ where u and v have different colours is a proper t-colouring of X, while colourings where u and v are assigned the same colour are in one-to-one correspondence with colourings of X/e. Therefore,

$$P(X\setminus e,t) = P(X,t) + P(X/e,t),$$

and the result follows. □

This result shows that the chromatic polynomial of a graph can be computed by deletion and contraction, and therefore that it actually is a polynomial in t. Using the deletion–contraction algorithm on the Petersen graph yields the chromatic polynomial

$$t\,(t-1)\,(t-2)\left(t^7 - 12\,t^6 + 67\,t^5 - 230\,t^4 + 529\,t^3 - 814\,t^2 + 775\,t - 352\right).$$

Our next example is based on the idea of using a graph to represent a computer network, where the edges represent possibly unreliable links that may fail with some probability. Let X be a graph and suppose we independently delete each edge of X with fixed probability p where $0 \le p \le 1$. Let $C(X,p)$ denote the probability that no connected component of X is disconnected as a result—the probability that the network "survives." For example, if T is a tree on n vertices, then it remains connected if and only if every edge survives, so

$$C(T,p) = (1-p)^{n-1}.$$

The function $C(X,p)$ is called the *reliability polynomial* of X. As with the chromatic polynomial, we justify its name by showing that it actually is a polynomial.

Theorem 15.8.2 *Let $C(X,p)$ denote the probability that no component of X is disconnected when each edge of X is deleted independently with* *probability p. Then*

$$C(X,p) = \begin{cases} C(X \setminus e, p), & \text{if } e \text{ is a loop;} \\ (1-p)C(X/e,p), & \text{if } e \text{ is a cut-edge;} \\ pC(X \setminus e, p) + (1-p)C(X/e,p), & \text{otherwise.} \end{cases}$$

Proof. In each case we consider the two possibilities that e is deleted or not deleted and sum the conditional probabilities that no component of X is disconnected given the fate of e. For example, if e is a cut-edge, then the number of components of X remains the same if and only if e survives (probability $(1-p)$) and the deleted edges do not form a cut in the rest of the graph (probability $C(X/e,p)$). Arguments for the other cases are broadly similar. □

Using the deletion–contraction algorithm on the Petersen graph yields the reliability polynomial

$$\left(704\,p^6 + 696\,p^5 + 390\,p^4 + 155\,p^3 + 45\,p^2 + 9\,p + 1\right)(1-p)^9 .$$

Figure 15.2 shows a plot of the reliability polynomial for the Petersen graph over the range $0 \le p \le 1$.

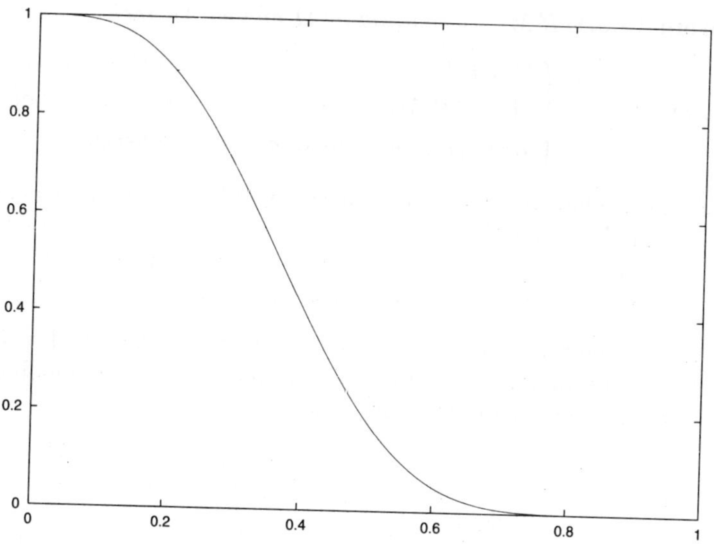

Figure 15.2. Plot of the reliability polynomial of the Petersen graph

15.9 Rank Polynomial

The similarity of the deletion–contraction formulas for the parameters and polynomials introduced in the previous sections strongly suggests that we should search for a common generalization. This generalization is the rank polynomial, which we introduce in this section.

The *rank polynomial* of a matroid M on Ω with rank function rk is the polynomial

$$R(M; x, y) = \sum_{A \subseteq \Omega} x^{\mathrm{rk}(\Omega)-\mathrm{rk}(A)} y^{\mathrm{rk}^{\perp}(\Omega)-\mathrm{rk}^{\perp}(\Omega \backslash A)}.$$

From this we see at once that

$$R(M^{\perp}; x, y) = R(M; y, x).$$

Since $\mathrm{rk}^{\perp}(\Omega \setminus A) = |\Omega \setminus A| + \mathrm{rk}(A) - \mathrm{rk}(\Omega)$ and $\mathrm{rk}^{\perp}(\Omega) = |\Omega| - \mathrm{rk}(\Omega)$, it also follows that

$$R(M; x, y) = \sum_{A \subseteq \Omega} x^{\mathrm{rk}(\Omega)-\mathrm{rk}(A)} y^{|A|-\mathrm{rk}(A)}.$$

This shows that the rank polynomial is essentially a generating function for the subsets of Ω, enumerated by size and rank.

We have the following important result.

Theorem 15.9.1 *If M is a matroid on Ω and $e \in \Omega$, then*

$$R(M; x, y) = \begin{cases} (1+y)R(M \setminus e; x, y), & \text{if } e \text{ is a loop;} \\ (1+x)R(M/e; x, y), & \text{if } e \text{ is a coloop;} \\ R(M \setminus e; x, y) + R(M/e; x, y), & \text{otherwise.} \end{cases}$$

Proof. If e is a loop or coloop, then this follows directly from the definition of the rank polynomial. If e is neither a loop nor a coloop, then the subsets of Ω that do not contain e contribute $R(M \backslash e; x, y)$ to the rank polynomial, while the subsets that do contain e contribute $R(M/e; x, y)$. □

Therefore, the rank polynomial of any matroid can be calculated using the deletion–contraction algorithm. The coefficients of the rank polynomial of the cycle matroid of the Petersen graph are

$$R(P; x, y) = \begin{array}{c} \\ x^0 \\ x^1 \\ x^2 \\ x^3 \\ x^4 \\ x^5 \\ x^6 \\ x^7 \\ x^8 \\ x^9 \end{array} \begin{array}{c} \begin{array}{ccccccc} y^0 & y^1 & y^2 & y^3 & y^4 & y^5 & y^6 \end{array} \\ \begin{pmatrix} 2000 & 2172 & 1230 & 445 & 105 & 15 & 1 \\ 4680 & 2765 & 816 & 135 & 10 & 0 & 0 \\ 5805 & 1725 & 240 & 15 & 0 & 0 & 0 \\ 4875 & 630 & 30 & 0 & 0 & 0 & 0 \\ 2991 & 130 & 0 & 0 & 0 & 0 & 0 \\ 1365 & 12 & 0 & 0 & 0 & 0 & 0 \\ 455 & 0 & 0 & 0 & 0 & 0 & 0 \\ 105 & 0 & 0 & 0 & 0 & 0 & 0 \\ 15 & 0 & 0 & 0 & 0 & 0 & 0 \\ 1 & 0 & 0 & 0 & 0 & 0 & 0 \end{pmatrix} \end{array}$$

Lemma 15.9.2 *Suppose that M is a matroid on Ω. Then*

(a) $R(M;1,1) = 2^{|\Omega|}$.

(b) $R(M;0,0)$ *is the number of bases of* M.

(c) $R(M;1,0)$ *is the number of independent sets in* M.

(d) $R(M;0,1)$ *is the number of spanning sets in* M. □

This result shows that various properties of a matroid can be found as evaluations of its rank polynomial at specific points. However, we can say something much stronger: The rank polynomial is the most general matroid parameter that can be computed by deletion–contraction. The key to this observation is the following converse of Theorem 15.9.1. Despite the power of this result, the proof is quite straightforward, and so we leave it as an exercise.

Theorem 15.9.3 *Let $\mathcal{F}$ be a function defined on matroids such that for the empty matroid $\emptyset$, we have $\mathcal{F}(\emptyset) = 1$, and for all other matroids*

$$\mathcal{F}(M) = \begin{cases} a(1+y)\mathcal{F}(M\setminus e), & \text{if } e \text{ is a loop;} \\ b(1+x)\mathcal{F}(M/e), & \text{if } e \text{ is a coloop;} \\ a\mathcal{F}(M\setminus e) + b\mathcal{F}(M/e), & \text{otherwise.} \end{cases}$$

If M is a matroid on Ω, then $\mathcal{F}(M) = a^{|\Omega|-\mathrm{rk}(\Omega)} b^{\mathrm{rk}(\Omega)} R(M;x,y)$. □

15.10 Evaluations of the Rank Polynomial

In this section we present several applications of Theorem 15.9.3, showing that all of the parameters that we have considered in the last few sections are evaluations of the rank polynomial.

Lemma 15.10.1 *The number of spanning trees in a connected graph X is*

$$\tau(X) = R(M(X);0,0).$$ □

Lemma 15.10.2 *The number of acyclic orientations of a graph X is*

$$\kappa(X) = R(M(X);1,-1).$$

Proof. Using Theorem 15.9.3 and the deletion–contraction expression given in Theorem 15.6.1, we see that

$$\begin{aligned} a(1+y) &= 0, \\ b(1+x) &= 2, \\ a &= 1, \\ b &= 1, \end{aligned}$$

and so $x = 1$ and $y = -1$. □

Theorem 15.10.3 *Let C be a binary code of length n, with a bicycle space of dimension d. Then*

$$(-1)^n(-2)^d = R(M(C); -2, -2).$$

Proof. This follows directly from Theorem 15.9.3 and Lemma 15.7.3. □

Theorem 15.10.4 *If X is a graph with n vertices, c components, and chromatic polynomial $P(X,t)$, then*

$$P(X,t) = (-1)^{n-c}t^c R(M(X); -t, -1).$$

Proof. We cannot work directly with $P(X,t)$ because it does not satisfy the conditions of Theorem 15.9.3. Instead, consider the function

$$\hat{P}(X,t) := t^{-c}P(X,t),$$

which takes the value 1 on the empty matroid, that is, on a graph with no edges. If e is a coloop, then some thought shows that $P(X \setminus e, t) = tP(X/e,t)$, and therefore putting $\hat{P}$ into Theorem 15.9.3, we get

$$\begin{aligned} a(1+y) &= 0, \\ b(1+x) &= (t-1), \\ a &= 1, \\ b &= -1, \end{aligned}$$

and therefore $y = -1$ and $x = -t$. Multiplying by t^c to recover $P(X,t)$ gives the stated result. □

We leave the proof of the final result as an exercise.

Theorem 15.10.5 *If X is a graph with n vertices, e edges, c components, and reliability polynomial $C(X,p)$, then*

$$C(X,p) = (1-p)^{n-c}p^{e-n+c}R\left(M(X); 0, \frac{1}{p} - 1\right). \quad \square$$

15.11 The Weight Enumerator of a Code

The *weight* of a codeword is the number of nonzero entries in it. If C is a code of length n over a finite field, then the number of codewords of any weight is finite. For a code C over a finite field, the *weight enumerator* of the code is the polynomial

$$W(C,t) := \sum_{i=0}^{n} n_i t^i,$$

where n_i is the number of codewords of weight i in C.

Theorem 15.11.1 *Let C be a linear code over the finite field $GF(q)$ with weight enumerator $W(C,t)$. Then*

$$W(C,t) = \begin{cases} W(C\setminus i,t), & \text{if } i \text{ is a loop in } M(C); \\ (1+(q-1)t)W(C/i,t), & \text{if } i \text{ is a coloop in } M(C); \\ tW(C\setminus i,t)+(1-t)W(C/i,t), & \text{otherwise.} \end{cases}$$

Proof. The result is clear if i is a loop. If i is a coloop, then consider a generator matrix G in the form of (15.1). Any codeword of C is either zero or nonzero in coordinate position i. The codewords that are zero in this position contribute $W(C/i,t)$ to the weight enumerator. All other codewords consist of a codeword in C/i plus a nonzero multiple of the first row of G. These codewords contribute $(q-1)tW(C/i,t)$ to the weight enumerator. Now, suppose that i is neither a loop nor a coloop, and so there is a one-to-one correspondence between codewords of $C\setminus i$ and C. If every codeword of C had nonzero ith coordinate, then the weight enumerator of C would be $tW(C\setminus i,t)$. However, every codeword does not have nonzero ith coordinate, and those with ith coordinate zero actually contribute $W(C/i,t)$ to the weight enumerator rather than $tW(C/i,t)$. Subtracting the overcount gives the stated result. □

Therefore, the weight enumerator of a code is an evaluation of the rank polynomial of the associated matroid.

Theorem 15.11.2 *Let C be a code of dimension k and length n over the finite field $GF(q)$ and let M be the matroid determined by C. Then*

$$W(C,t) = (1-t)^k t^{n-k} R\left(M; \frac{qt}{1-t}, \frac{1-t}{t}\right). \qquad \square$$

Given that $R(M;x,y) = R(M^{\perp};y,x)$, this result implies the following important theorem from coding theory.

Theorem 15.11.3 (MacWilliams) *Let C be a code of length n and dimension k over the finite field $GF(q)$. Then*

$$W(C^{\perp},t) = q^{-k}(1+(q-1)t)^n W\left(C, \frac{1-t}{1+(q-1)t}\right). \qquad \square$$

15.12 Colourings and Codes

Let D be the incidence matrix of an arbitrarily oriented graph X^{σ}. We can choose to view the row space of D as a code over any field, although this does not change the associated matroid.

So suppose that we take D to be a matrix over the finite field $GF(q)$. If X has n vertices, then any vector $x \in GF(q)^n$ can be viewed as a function on $V(X)$. The entries of $x^T D$ are all nonzero if and only if x takes distinct values on each pair of adjacent vertices, and so is a proper q-colouring of X. Therefore, if X has e edges, the row space of D contains vectors of weight e if and only if X has a q-colouring. (This shows that determining the maximum weight of a code is as difficult a problem as determining the chromatic number of a graph; hence it is NP-hard.)

By Theorem 15.10.4, the number of q-colourings of a graph is an evaluation of the rank polynomial, and hence we can find the number of codewords of weight e in the row space of D. Our next result extends this to all codes.

Lemma 15.12.1 *If M is the matroid of a linear code C of length n and dimension k over the field $GF(q)$, then the number of codewords of weight n in C is*

$$(-1)^k R(M; -q, -1).$$

Proof. Let $\sigma(C)$ denote the number of codewords with no zero entries in a code C. It is straightforward to see from Theorem 15.11.1 that

$$\sigma(C) = \begin{cases} 0, & \text{if } i \text{ is a loop in } M(C); \\ (q-1)\sigma(C/i), & \text{if } i \text{ is a coloop in } M(C); \\ \sigma(C\backslash i) - \sigma(C/i), & \text{otherwise.} \end{cases}$$

The result then follows immediately from Theorem 15.9.3. □

Suppose C is a code of dimension k and length n over $GF(q)$, with generator matrix G. Then we may view G as a matrix over $GF(q^r)$, in which case it generates a code C' with the same dimension k, and so

$$|C'| = q^{rk} = |C|^r.$$

Even though C and C' are different codes, $M(C) = M(C')$. Hence, if $M = M(C)$, then $|R(M; -q^r, -1)|$ equals the number of words of weight n in C'.

There is a second interpretation of $|R(M; -q^r, -1)|$. Suppose that D is the incidence matrix of an oriented graph as above. If X has c connected components, then the number of proper q^r-colourings of X is $(q^r)^c|R(M; -q^r, -1)|$, because $(q^r)^c$ vectors x in $GF(q^r)^n$ map onto each word $x^T D$ of weight e. Since $GF(q^r)$ and $GF(q)^r$ are isomorphic as vector spaces over $GF(q)$, there is a one-to-one correspondence between vectors $x \in GF(q^r)^n$ and $r \times n$ matrices A over $GF(q)$ such that $x^T D$ has no zero entries if and only if AD has no zero columns. Therefore, the number of such matrices is $q^{rc}|R(M; -q^r, -1)|$. Interpreting this in terms of an arbitrary code yields the following result.

Lemma 15.12.2 *Let C be a code of length n over a field q and let M be the matroid of C. Then the number of ordered r-tuples of codewords in C such that the union of the supports of the codewords in the r-tuple has size n is*

$$(-1)^k R(M; -q^r, -1).$$ □

Applying this to the incidence matrix of an oriented graph, we conclude that a graph X is q^r-colourable if and only if it is the union of r graphs, each q-colourable. This may appear surprising, but is very easy to prove directly, without any reference to matroids. (Algebraic methods often allow us access to results we cannot prove in any other way; still, there are occasions such as this where they provide elephantine proofs of elementary results.)

We have established a relation between colouring problems on graphs and a natural problem in coding theory. We now go a step further by describing a geometric view. Let G be a $k \times n$ matrix of rank k over $GF(q)$ and let M be the matroid on its columns. The columns of G may be viewed as a set Ω of points in projective space of dimension $k-1$. Each vector $x \in GF(q)^k$ determines a hyperplane

$$\{y \in GF(q)^k : x^T y = 0\},$$

which is disjoint from Ω if and only if no entry of $x^T G$ is zero. All nonzero scalar multiples of x determine the same hyperplane; hence Lemma 15.12.1 yields that the number of hyperplanes disjoint from Ω is

$$(q-1)^{-1}(-1)^k R(M; -q, -1).$$

(Thus we might say that the $GF(q)$-linear matroid M is q-colourable if and only if there is a hyperplane disjoint from Ω.) Further, there is an $r \times k$ matrix A over $GF(q)$ such that no column of AG is zero if and only if there are r hyperplanes whose intersection has dimension $k-r$ and is disjoint from Ω. Therefore, $|R(M; -q^r, -1)| \neq 0$ if and only if there is a space of dimension $k-r$ disjoint from Ω.

We note one interesting application of this. Let Ω be the set of all nonzero vectors with weight at most d in $GF(q)^k$. Then a subspace disjoint from Ω is a linear code with minimum distance at least $d+1$. If M is the matroid associated with Ω, then the smallest value of r such that $R(M; -q^r; -1) \neq 0$ is the minimum possible codimension of such a linear code. Determining this value of r is, of course, a central problem in coding theory.

15.13 Signed Matroids

Let $\Omega = \Omega(+) \cup \Omega(-)$ be a partition of Ω into two parts. In this situation we call Ω a *signed set*, and refer to the positive and negative elements of Ω. If M is a matroid on Ω, then its rank polynomial counts the subsets of Ω according to their size and rank. In this section we consider a generalization of the rank polynomial that counts subsets of Ω according to their size, rank, and number of positive elements. This polynomial will play an important role in our work on knots in the next chapter.

If $A \subseteq \Omega$, then define $A(+)$ to be $A \cap \Omega(+)$ and $A(-)$ to be $A \cap \Omega(-)$. As usual, $\overline{A}$ refers to $\Omega \setminus A$. Then if M is a matroid on Ω and α and β are two commuting variables, we define the rank polynomial of the signed matroid M by

$$R(M;\alpha,\beta,x,y) := \sum_{A\subseteq\Omega} \alpha^{|\overline{A}(+)|+|A(-)|}\beta^{|A(+)|+|\overline{A}(-)|}x^{\mathrm{rk}(\Omega)-\mathrm{rk}(A)}y^{\mathrm{rk}^{\perp}(\Omega)-\mathrm{rk}^{\perp}(\overline{A})}.$$

If e is a loop, then

$$R(e)=\begin{cases}(\alpha+\beta y), & e\in\Omega(+);\\ (\alpha y+\beta), & e\in\Omega(-).\end{cases}$$

If e is a coloop, then

$$R(e)=\begin{cases}(\alpha x+\beta), & e\in\Omega(+);\\ (\alpha+\beta x), & e\in\Omega(-).\end{cases}$$

Theorem 15.13.1 *Let M be a matroid on the signed set Ω and let e be an element of Ω. If e is neither a loop nor a coloop, then*

$$R(M)=\begin{cases}\alpha R(M\setminus e)+\beta R(M/e), & e\in\Omega(+);\\ \beta R(M\setminus e)+\alpha R(M/e), & e\in\Omega(-).\end{cases}$$

If e is a loop, then $R(M)=R(e)R(M\setminus e)$, and if e is a coloop, $R(M)=R(e)R(M/e)$. □

The signing of Ω provides a new parameter for each subset of Ω, but leaves the rank function unchanged. If Ω is a signed set, then we define the dual of M to be $M^{\perp}$, but swap the signs on Ω. With this understanding, we obtain the following:

Corollary 15.13.2 *If M is a matroid on a signed set and $M^{\perp}$ its dual, then*

$$R(M^{\perp};\alpha,\beta,x,y)=R(M;\alpha,\beta,y,x).$$ □

If M^{-} is the matroid obtained from M simply by swapping the signs on Ω but leaving the rank function unchanged, then

$$R(M^{-};\alpha,\beta,x,y)=R(M;\beta,\alpha,y,x).$$

If M is the cycle matroid of a graph and e is a coloop, then $M\backslash e = M/e$, but $X\setminus e\neq X/e$. If X is a signed graph with c components, define the modified polynomial $\widehat{R}(X)$ by

$$\widehat{R}(X;\alpha,\beta,x,y):=x^{c-1}R(M(X);\alpha,\beta,x,y).$$

If X is a connected planar graph, then $\widehat{R}(X;\alpha,\beta,x,y)=\widehat{R}(X^{*};\alpha,\beta,y,x)$, where as above, the sign on an edge is reversed on moving to the dual.

Theorem 15.13.3 *Let $\widehat{R}(X)$ be the modified rank polynomial of the signed graph X. Then*

$$\widehat{R}(X) = \begin{cases} \widehat{R}(e)\widehat{R}(X \setminus e), & \text{if } e \text{ is a loop;} \\ \alpha\widehat{R}(X \setminus e) + \beta\widehat{R}(X/e), & \text{if } e \text{ is positive and not a loop;} \\ \beta\widehat{R}(X \setminus e) + \alpha\widehat{R}(X/e), & \text{if } e \text{ is negative and not a loop.} \end{cases}$$

Proof. If e is a coloop in X, then

$$x^{-c}\widehat{R}(X \setminus e) = R(X \setminus e) = R(X/e) = x^{-c+1}\widehat{R}(X/e),$$

and so $\widehat{R}(X \setminus e) = x\widehat{R}(X/e)$. Since $R(X) = R(e)R(X/e)$, it follows that if e is a positive coloop, then $\widehat{R}(X) = (\alpha x + \beta)\widehat{R}(X/e)$, and so

$$\widehat{R}(X) = \alpha\widehat{R}(X \setminus e) + \beta\widehat{R}(X/e).$$

Similarly, if e is a negative coloop, then

$$\widehat{R}(X) = \beta\widehat{R}(X \setminus e) + \alpha\widehat{R}(X/e).$$

It is easy to verify that the last two recurrences hold true for positive and negative edges that are neither loops nor coloops. □

We conclude with two small examples, both based on the graph K_3. If all three edges are positive, then

$$R(K_3; \alpha, \beta, x, y) = 3\alpha\beta^2 + 3\alpha^2\beta x + \beta^3 y + \alpha^3 x^2,$$

while if two are positive and one negative, we get

$$R(K_3; \alpha, \beta, x, y) = \beta^3 + 2\alpha^2\beta + (2\alpha\beta^2 + \alpha^3)x + \alpha\beta^2 y + \alpha^2\beta x^2.$$

15.14 Rotors

Although the rank polynomial of the cycle matroid of a graph X contains a lot of information about X, it does not determine the graph. This can easily be demonstrated by selecting nonisomorphic graphs with isomorphic cycle matroids, which obviously must yield the same rank polynomial. However, it is also possible to find nonisomorphic pairs of 3-connected graphs with cycle matroids—necessarily not isomorphic—having the same rank polynomial. An example is provided in Figure 15.3, which shows the smallest cubic planar 3-connected graphs whose cycle matroids have the same rank polynomial. We describe a method of constructing such pairs of graphs.

For $n \geq 3$, a *rotor* of order n is a graph R together with a set N of n vertices that is an orbit of a cyclic subgroup of $\mathrm{Aut}(R)$ of order n. Therefore, there is some element $g \in \mathrm{Aut}(R)$ of order n and some vertex $x_0 \in N$ such that

$$N = \{x_0, \ldots, x_{n-1}\},$$

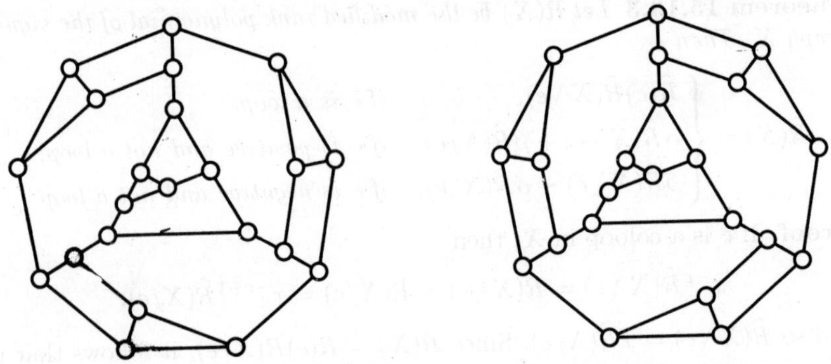

Figure 15.3. Two graphs with the same rank polynomial

where $x_{i+1} = x_i^g$. The first graph in Figure 15.4 is a rotor of order five with the five vertices of N highlighted.

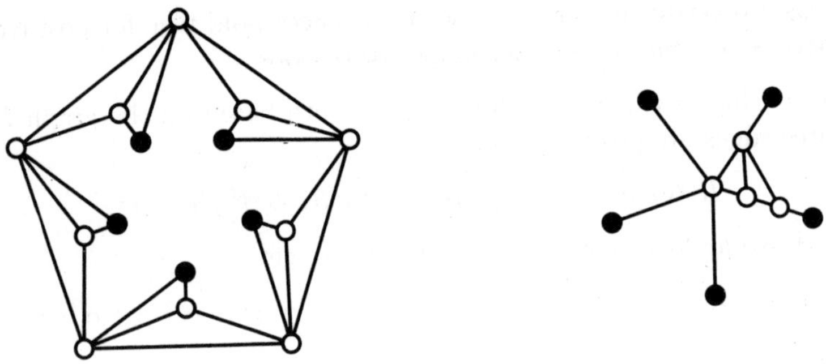

Figure 15.4. The rotor R and graph S

Let S be an arbitrary graph with vertex set disjoint from $V(R)$, together with a set $N' = \{y_0, \ldots, y_{n-1}\}$ of distinguished vertices. Then define two graphs X and X^φ, where X is obtained from $R \cup S$ by identifying vertices x_i and y_i for all i, and X^φ is obtained from $R \cup S$ by identifying vertices x_i and y_{n-i} for all i. If S is the second graph of Figure 15.4, then Figure 15.5 shows the graph X.

For obvious reasons, we say that X^φ is obtained from X by *flipping the rotor* R; this is shown in Figure 15.6.

Theorem 15.14.1 *If R is a rotor of order less than six, then the cycle matroids of X and X^φ have the same rank polynomial.*

Proof. We describe a bijection between subsets of edges of X and subsets of edges of X^φ that preserves both size and rank.

Let $A = A_R \cup A_S$ be a subset of $E(R) \cup E(S)$, where $A_R \subseteq E(R)$ and $A_S \subseteq E(S)$. The connected components of the subgraph of R with edge-set

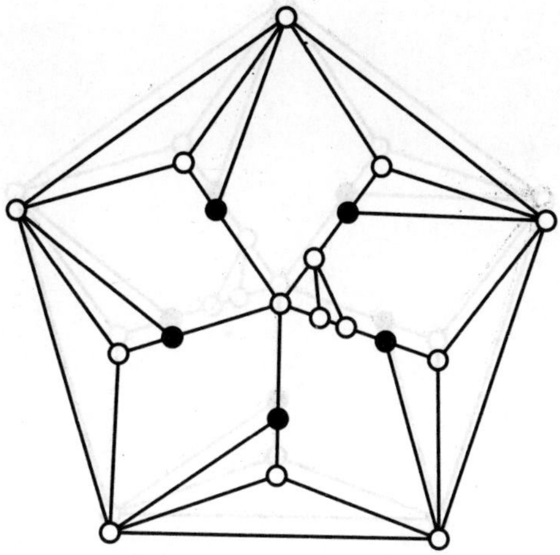

Figure 15.5. The graph X obtained from R and S

A_R induce a partition ρ on the vertices $\{x_0, \dots, x_{n-1}\}$, and the connected components of the subgraph of S with edge-set A_S induce a partition σ on the vertices $\{y_0, \dots, y_{n-1}\}$.

The number of connected components of A depends on the number of components that do not contain any vertices of N, and the number of cells in the join $\rho \vee \sigma$.

It is straightforward to check that any partition of a set of size five or less is invariant under at least one of the "reflections"

$$\tau(i) : y_{i+j} \mapsto y_{i-j}.$$

(If $n = 4$, then it may be necessary to take i to be nonintegral; in this case i will be an odd multiple of $\frac{1}{2}$, which causes no difficulty.) Therefore, there is at least one i such that σ is invariant under $\tau(i)$. Choose the smallest such i, set $h = g^{-2i}$, and define

$$A' = A_R^h \cup A_S.$$

We claim that the rank of A in the cycle matroid of X is equal to the rank of A' in the cycle matroid of X^φ. Establishing this claim will prove the theorem, because since the same $\tau(i)$ is chosen every time a particular A_S is used, the mapping from A to A' is a bijection between subsets of $E(X)$ and subsets of $E(X^\varphi)$.

Let ρ' be the partition determined by A_R^h on $\{x_0, \dots, x_{n-1}\}$. It is clear that A and A' have the same number of components that do not contain any vertices of N. Therefore, to show that they have the same rank, we need to show that $\rho' \vee \sigma$ and $\rho \vee \sigma$ have the same number of cells.

In X^φ the vertex x_{j-2i} is identified with

$$y_{n-(j-2i)} = y_{2i-j} = y_{i+i-j} = y_j^{\tau(i)},$$

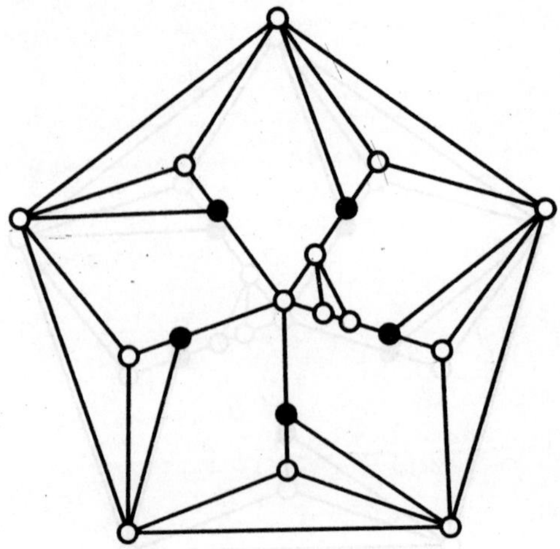

Figure 15.6. The graph X^{φ} obtained from X by flipping the rotor

and therefore expressed as a partition of $\{y_0, \dots, y_{n-1}\}$,

$$\rho' = \rho^{\tau(i)}$$

Since σ is invariant under $\tau(i)$, we have

$$\rho' \vee \sigma = \rho^{\tau(i)} \vee \sigma^{\tau(i)} = (\rho \vee \sigma)^{\tau(i)},$$

and so the two partitions have the same number of cells. Clearly, A and A' contain the same number of edges, and so the claim is proved. □

If R and S are chosen so that neither has an automorphism of order two, then usually X and X^{φ} will not be isomorphic.

We can define a *signed rotor*—a rotor with signed edges—in an analogous fashion to a rotor, with the understanding that an automorphism is a mapping that preserves both sign and incidence. The rank polynomial of a signed graph counts subsets of edges according to rank and the number of positive edges. The bijection given in Theorem 15.14.1 preserves the number of positive edges, and so can be used unchanged to prove the following result.

Theorem 15.14.2 *If X and Y are signed graphs such that Y is obtained from X by flipping a signed rotor of order less than six, then the rank polynomial of X is equal to the rank polynomial of Y.* □

15.15 Submodular Functions

Suppose that f is an integer-valued function on the subsets of Ω that is monotone and submodular. Although f may not be a rank function itself, we can nevertheless associate a matroid with f by devising a rank function based on f.

By adding a constant function to f, we can assume that it is nonnegative and that $f(\emptyset) = 0$. The final modification requires more work.

Theorem 15.15.1 *Let f be an integer-valued monotone submodular function on the subsets of Ω such that $f(\emptyset) = 0$. The function $\hat{f}$ on the subsets of Ω defined by*

$$\hat{f}(A) := \min\{f(A \setminus B) + |B| : B \subseteq A\}$$

is a rank function.

Proof. Clearly, $\hat{f}$ is nonnegative. We show that it is monotone, submodular and that $\hat{f}(A) \le |A|$. First, suppose that $A \subseteq A_1$. If $B \subseteq A_1$, then $A \setminus B \subseteq A_1 \setminus B$, and since f is monotone,

$$f(A_1 \setminus B) + |B| \ge f(A \setminus B) + |A \cap B|.$$

Therefore, $\hat{f}(A_1) \ge \hat{f}(A)$, which shows that $\hat{f}$ is monotone.

Now, let A and B be subsets of Ω. Suppose that $C \subseteq A$ and $D \subseteq B$ are the subsets that provide the values for $\hat{f}(A)$ and $\hat{f}(B)$, respectively. Let $W = (C \setminus B) \cup (D \setminus A) \cup (C \cap D)$ and $X = (C \cup D) \cap (A \cap B)$, and observe that $|C| + |D| = |W| + |X|$. Then

$$\begin{aligned}
\hat{f}(A) + \hat{f}(B) &= f(A \setminus C) + f(B \setminus D) + |C| + |D| \\
&\ge f((A \setminus C) \cup (B \setminus D)) + f((A \setminus C) \cap (B \setminus D)) + |C| + |D| \\
&= f((A \cup B) \setminus W) + f((A \cap B) \setminus X) + |W| + |X| \\
&\ge \hat{f}(A \cup B) + \hat{f}(A \cap B),
\end{aligned}$$

and so $\hat{f}$ is submodular.

Finally, by taking B equal to A in the definition of $\hat{f}$, we see that $\hat{f}(A) \le |A|$, and therefore $\hat{f}$ is a rank function. □

Therefore, any integer-valued monotone submodular function f such that $f(\emptyset) = 0$ determines a matroid M whose rank function is $\hat{f}$. We can describe the independent sets of M directly in terms of the original function f.

Corollary 15.15.2 *Let f be an integer-valued monotone submodular function on the subsets of Ω such that $f(\emptyset) = 0$, and let M be the matroid determined by the rank function $\hat{f}$. A subset A of Ω is independent in M if and only if $f(B) \ge |B|$ for all subsets B of A.* □

Proof. A subset A of Ω is independent in M if and only if $f(A \setminus B) + |B| \ge |A|$ for all subsets B of A, or, equivalently, if and only if $f(A \setminus B) \ge |A \setminus B|$ for all subsets B. □

We consider some examples of rank functions arising in this manner. It is reassuring to observe that if rk is a rank function already, then $\widehat{\mathrm{rk}} = \mathrm{rk}$.

Let X be a bipartite graph and let (Ω, Φ) be a bipartition of its vertex set. If $A \subseteq \Omega$, define $N(A)$ to be the set of vertices in Φ adjacent to a vertex in A. Define $f(A)$ to be $|N(A)|$. It is easy to see that f is submodular (and monotone and integer-valued and that $f(\emptyset) = 0$), so $\hat{f}$ is a rank function determining a matroid on Ω. By the corollary, a subset A is independent in this matroid if and only if $|N(B)| \geq |B|$ for all subsets $B \subseteq A$. The following result, known as Hall's theorem, shows that this happens if and only if there is a matching covering the vertices in A.

Lemma 15.15.3 *Let X be a bipartite graph with bipartition (Ω, Φ), and let $A \subseteq \Omega$. There is a matching in X that covers the vertices in A if and only if $|N(B)| \geq |B|$ for all subsets B of A.* □

(It should be clear that Hall's condition for the existence of a matching that covers A is necessary; the point of the result is that this condition is sufficient.)

The last example can be generalized. Suppose that X is a bipartite graph as above and that r is the rank function of a matroid M on Φ. If we define $f(A)$ to be $r(N(A))$, once again we find that f is an integer-valued monotone submodular function with $f(\emptyset) = 0$, and so $\hat{f}$ is the rank function of a matroid M' on Ω. We say that M' is the matroid *induced* from M by X. A subset $A \subseteq \Omega$ is independent in M' if and only if there is a matching that covers A and pairs the vertices of A with the vertices of an independent set of M'. (This claim is a modest, but nonetheless very significant, generalization of Hall's theorem due to Rado.)

If rk_1 and rk_2 are the respective rank functions of matroids M_1 and M_2 on the same set Ω, then their sum $\mathrm{rk}_1 + \mathrm{rk}_2$ is a monotone nonnegative integer-valued submodular function on Ω. Hence, by the theorem, it determines a rank function; the corresponding matroid is called the *union* of M_1 and M_2 and denoted by $M_1 \vee M_2$.

Lemma 15.15.4 *Let M be the union of the matroids M_1 and M_2 on Ω. A subset A of Ω is independent in M if and only if $A = A_1 \cup A_2$, where A_i is independent in M_i.*

Proof. Let rk_1 and rk_2 be the rank functions of M_1 and M_2 respectively. Suppose that A is the union of the sets A_1 and A_2, where A_i is independent in M_i. For any $B \subseteq A$, set $B_i := B \cap A_i$. Then

$$|B| \leq |B_1| + |B_2| = \mathrm{rk}_1(B_1) + \mathrm{rk}_2(B_2) \leq \mathrm{rk}_1(B) + \mathrm{rk}_2(B),$$

and by Corollary 15.15.2, A is independent in $M_1 \vee M_2$.

The converse takes more work. Let N denote the direct sum $M_1 \oplus M_2$, with ground set Φ consisting of two disjoint copies of Ω. Let Γ denote a third copy of Ω and let X be the bipartite graph with bipartition (Γ, Φ), where the ith element of Γ is joined to the ith elements in each of the two copies of Ω in Φ. This is not a very interesting graph, but the matroid induced from $M_1 \oplus M_2$ by X is $M_1 \vee M_2$. From this it follows that a subset A of Γ is independent in $M_1 \vee M_2$ if and only if there is a matching in X that pairs the vertices of A with an independent set in Φ, whence we infer that A is independent if and only if it is the union of two sets, one independent in M_1 and the other in M_2. □

We use this to prove a graph-theoretic result whose alternative proofs seem to require rather detailed arguments. If π is a partition of the vertex set of X, then let $e(X, \pi)$ denote the number of edges going between distinct cells of π.

Theorem 15.15.5 *A connected graph X has k edge-disjoint spanning trees if and only if for every partition π of the vertices,*

$$e(X, \pi) \geq k(|\pi| - 1).$$

Proof. For any partition π, it is easy to see that a spanning tree contributes at least $|\pi| - 1$ edges to $e(X, \pi)$, and so if X has k edge-disjoint spanning trees, the inequality holds.

For the converse, suppose that X has n vertices, and that rk is the rank function of the cycle matroid $M(X)$. Let $kM = M(X) \vee \cdots \vee M(X)$ denote the k-fold union of $M(X)$ with itself. By the previous result, X has k edge-disjoint spanning trees if and only if the rank of kM is $k(n - 1)$. By the definition of the rank function of kM, this is true if and only if for all $A \subseteq E(X)$,

$$k\,\mathrm{rk}(A) + |\overline{A}| \geq k(n - 1).$$

If A is a subset of $E(X)$, then let π be the partition whose cells are the connected components of the graph with vertex set $V(X)$ and edge set A. Then $\mathrm{rk}(A) = n - |\pi|$ and $e(X, \pi) = |\overline{A}|$, and so

$$k\,\mathrm{rk}(A) + |\overline{A}| \geq k(n - |\pi|) + k(|\pi| - 1) = k(n - 1),$$

and thus the result follows. □

Exercises

1. Suppose that A and B are independent sets in a matroid with $|A| < |B|$. Show that there is an element $x \in B \setminus A$ such that $A \cup \{x\}$ is independent.

2. Suppose that C and D are distinct circuits in a matroid on Ω, and that $x \in C \cap D$. Prove that there is a circuit in $C \cup D$ that does not contain x.

3. A subset A in a matroid M is a *flat* if $\mathrm{rk}(A \cup x) > \mathrm{rk}(A)$ for all $x \in \Omega \setminus A$. Show that the set of flats ordered by inclusion forms a

lattice (that is, a partially ordered set in which every set of elements has a least upper bound and a greatest lower bound).

4. A *hyperplane* is a maximal proper flat. Show that the rank of a hyperplane is $\mathrm{rk}(\Omega) - 1$, and that a subset is a hyperplane if and only if its complement is a circuit in $M^{\perp}$.

5. If M is a matroid defined on the columns of a matrix D, then show that the complements of the supports of the words in the row space of D are flats.

6. Show that a matroid is connected if and only if each pair of elements lies in a circuit.

7. If the matroid M is the direct sum of matroids M_1 and M_2, show that $R(M) = R(M_1)R(M_2)$.

8. Derive the expression given in Theorem 15.6.2 for the number of acyclic orientations of a graph with a given vertex as the unique source. Express this as an evaluation of the rank polynomial, and find the number of such orientations for the Petersen graph.

9. For any $e \in E(X)$, show that
$$|\mathrm{Hom}(X, Y)| = |\mathrm{Hom}(X \setminus e, Y)| - |\mathrm{Hom}(X/e, Y)|.$$

10. Use the results of this chapter to prove that a connected graph X has no bicycles if and only if it has an odd number of spanning trees.

11. Prove the converse to the deletion–contraction formula for the rank polynomial, as given in Theorem 15.9.3.

12. Let X be a graph with a fixed orientation σ, and let D be the incidence matrix of the oriented graph X^σ Then for any integer $q > 1$, a nowhere-zero q-flow on X is a vector $x \in (\mathbb{Z}_q)^{E(X)}$ such that $x_e \neq 0$ for any $e \in E(X)$ and
$$Dx \equiv 0 \pmod{q}.$$
Show that the number $F(X, q)$ of nowhere-zero q-flows satisfies the equation
$$F(X,q) = \begin{cases} (q-1)F(X \setminus e, q), & \text{if } e \text{ is a loop;} \\ 0, & \text{if } X \text{ contains a bridge;} \\ F(X/e, q) - F(X \setminus e, q), & \text{if } e \text{ is not a loop or a bridge.} \end{cases}$$
The function $F(X, q)$ is known as the *flow polynomial* of a graph.

13. Use the results of the previous exercise to express $F(X, q)$ as an evaluation of the rank polynomial.

14. An *eulerian orientation* of an even graph is an orientation such that the in-valency of every vertex is equal to its out-valency. Show that

if X is 4-regular, then there is a one-to-one correspondence between eulerian orientations of X and nowhere-zero 3-flows of X. Use the result of the previous exercise to obtain an expression for the number of eulerian orientations of a 4-regular graph.

15. Let X be a connected graph with edge set E, and assume we are given a total ordering of E. Let T be a spanning tree of X. An edge g of X is *internally active* relative to T if it is contained in T and is the least element in the cut of X determined by g and T. It is *externally active* if it is not contained in T and it is the least edge in the unique cycle in $T \cup g$. Let $t_{ij}(X)$ denote the number of spanning trees of X with exactly i internally active edges and j externally active edges. If edge e immediately precedes f in the order, show that swapping e and f does not change the numbers $t_{ij}(X)$. (Hence these numbers are independent of the ordering used.)

16. We continue with the notation of the previous exercise. If f is the last edge relative to the given ordering, and is not a loop or a bridge, show that

$$t_{ij}(X) = t_{ij}(X \setminus e) + t_{ij}(X/e).$$

(Note that a bridge lies in each spanning tree and is always internally active, while a loop does not lie in any spanning tree and is always externally active.)

17. Define the *Tutte polynomial* $T(X; x, y)$ of the graph X by

$$T(X; x, y) := \sum_{i,j} t_{ij} x^i y^j.$$

Use the previous exercise to determine the relation between the Tutte and rank polynomials.

18. If M is a matroid on the signed set Ω, then show that $R(M \restriction \Omega(+))$ and $R(M \restriction \Omega(-))$ can both be determined from $R(M)$.

19. If M_1 and M_2 are matroids on Ω with rank functions rk_1 and rk_2, respectively, then show that the rank in $M = M_1 \vee M_2$ of a subset $A \subseteq \Omega$ is given by

$$\max\{\mathrm{rk}_1(B) + \mathrm{rk}_2(A \setminus B) : B \subseteq A\}.$$

Notes

One of the most useful references for the material in this chapter is Biggs [1]. Two warnings are called for. If M is the cycle matroid of a graph, then our $\mathrm{rk}^{\perp}$ is not what Biggs calls the corank. Secondly, our rank polynomial is his modified rank polynomial (see p. 101 of [1]) and not his rank polynomial. Our usage follows Welsh [10]. Oxley's book [7] is an interesting and current reference for background on matroid theory.

Theorem 15.9.3 is encoded as Theorem 3.6 in Brylawski [4]. It is strongly foreshadowed by work of Tutte [8, 9].

The Tutte polynomial is the rank polynomial with a simple coordinate change, and so any results about one can immediately be phrased in terms of the other. Oxley and Brylawski [3] give an extensive survey of applications of the Tutte polynomial including many open problems. Welsh [11] considers several questions related to the complexity of evaluating the Tutte polynomial at specific points, or along specific curves.

The rank polynomial of a matroid on a signed set is based on the graph polynomial introduced by Murasugi in [5, 6]. Perhaps the ultimate generalization of this approach is given in [2].

We offer a warning that the solution of Exercise 15 requires a large number of cases to be considered.

We will use the rank polynomial of a signed matroid in the next chapter to obtain a knot invariant. It would be interesting to find further combinatorial applications for this polynomial.

References

[1] N. Biggs, *Algebraic Graph Theory*, Cambridge University Press, Cambridge, second edition, 1993.

[2] B. Bollobás and O. Riordan, *A Tutte polynomial for coloured graphs*, Combin. Probab. Comput., 8 (1999), 45–93.

[3] T. Brylawski and J. Oxley, *The Tutte polynomial and its applications*, in Matroid applications, Cambridge Univ. Press, Cambridge, 1992, 123–225.

[4] T. H. Brylawski, *A decomposition for combinatorial geometries*, Trans. Amer. Math. Soc., 171 (1972), 235–282.

[5] K. Murasugi, *On invariants of graphs with applications to knot theory*, Trans. Amer. Math. Soc., 314 (1989), 1–49.

[6] ——, *Classical numerical invariants in knot theory*, in Topics in knot theory (Erzurum, 1992), Kluwer Acad. Publ., Dordrecht, 1993, 157–194.

[7] J. G. Oxley, *Matroid Theory*, The Clarendon Press Oxford University Press, New York, 1992.

[8] W. T. Tutte, *A ring in graph theory*, Proc. Cambridge Philos. Soc., 43 (1947), 26–40.

[9] ——, *The dichromatic polynomial*, Congressus Numeratium, XV (1976), 605–635.

[10] D. J. A. Welsh, *Matroid Theory*, Academic Press, London, 1976.

[11] ——, *Complexity: Knots, Colourings and Counting*, Cambridge University Press, Cambridge, 1993.

16
Knots

A knot is a closed curve of finite length in $\mathbb{R}^3$ that does not intersect itself. Two knots are equivalent if one can be deformed into the other by moving it around without passing one strand through another. (We will define equivalence more formally in the next section). One of the fundamental problems in knot theory is to determine whether two knots are equivalent. If two knots are equivalent, then this can be demonstrated: For example, we could produce a video showing one knot being continuously deformed into the other. Unfortunately, if two knots are not equivalent, then it is not at all clear how to prove this. The main approach to this problem has been the development of knot invariants: values associated with knots such that equivalent knots have the same value. If such an invariant takes different values on two knots, then the two knots are definitely not equivalent.

Until very recently, most work on knots was carried out using topological tools, particularly fundamental groups and homology. One notable exception to this was some important early work of Alexander, which has a highly combinatorial flavour. More recently Vaughan Jones discovered the object now known as the Jones polynomial, a new knot invariant that has stimulated much research and has been used to answer a number of old questions about knots.

In this chapter we derive the Jones polynomial from the rank polynomial of a signed graph. Thus we have a graph-theoretical construction of an important invariant from knot theory.

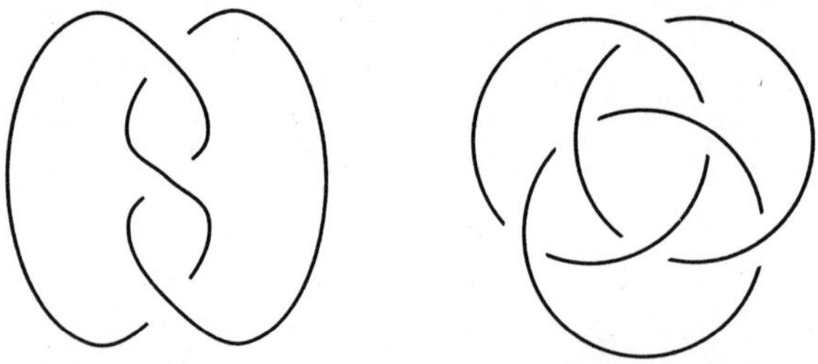

Figure 16.1. A knot diagram and a link diagram

16.1 Knots and Their Projections

A *knot* is a piecewise-linear closed curve in $\mathbb{R}^3$. A *link* is a collection of pairwise disjoint knots; the knots constituting a link are called its *components*. The assumption that a curve is piecewise linear ensures that it consists of a finite number of straight line segments. In practice, we will draw our links with such a large number of small straight line segments that they look like continuous curves, and there is no harm in viewing a knot as a smooth curve. (It is easier to work rigorously in the piecewise-linear category.)

Although knots and links live in $\mathbb{R}^3$, we usually represent them by *link diagrams*, which live in $\mathbb{R}^2$. Figure 16.1 provides examples. We can view a diagram as the shadow that a link would cast onto a wall if a light was shone through it, together with extra "under-and-over" information at the crossings indicating which strand is further from the light. Some care is needed, we insist:

(a) That at most two points on the link correspond to a given point in its diagram; neither of these points can be the end of a segment.

(b) That only finitely many points in the diagram correspond to more than one point on the link.

If a diagram has at least one crossing, then we may represent it by a 4-regular plane graph, together with information about over- and undercrossings at each vertex. For obvious reasons, we call this plane graph the *shadow* of the link diagram; an example is shown in Figure 16.2. A knot or link has infinitely many diagrams, and hence shadows associated with it.

A *homeomorphism* is a piecewise-linear bijection. Two links L_1 and L_2 are *equivalent* if there is an orientation-preserving homeomorphism ϕ from $\mathbb{R}^3 \cup \infty$ to itself that maps L_1 onto L_2. (Topologists prefer to view links as sitting in the unit sphere S^3 in $\mathbb{R}^4$, which is equivalent to $\mathbb{R}^3 \cup \infty$.) The map ϕ is necessarily an *isotopy*. This means that there is a family ϕ_t of piecewise linear homeomorphisms where $t \in [0,1]$ such that ϕ_0 is the identity, $\phi_1 = \phi$, and the map

$$(x,t) \mapsto (\phi_t(x), t)$$

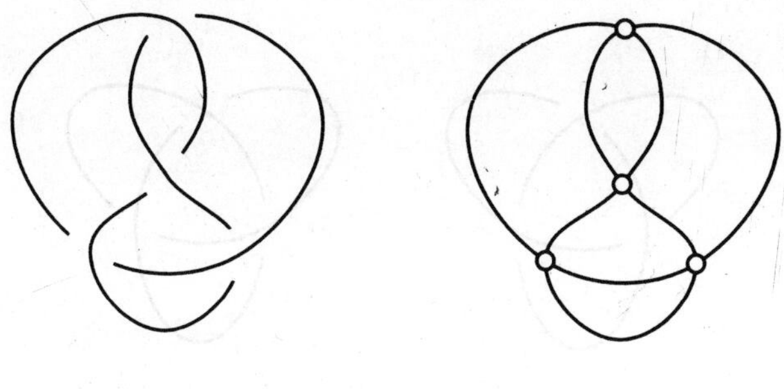

Figure 16.2. A knot diagram and its shadow

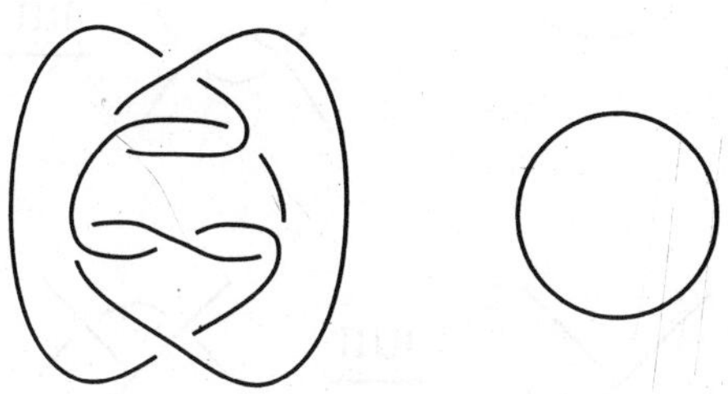

Figure 16.3. Two projections of the unknot

is a piecewise-linear homeomorphism from $S^3 \times [0,1]$ to itself. (This justifies our analogy with the video in the introduction above.)

A knot is the *unknot* if it is isotopic to a circle lying in a plane.

The fundamental problem of knot theory is to determine when two links are equivalent. As we are primarily interested in the way a link is tangled, rather than its location in $\mathbb{R}^3$, we will often informally refer to links as being the 'same' if they are equivalent. Given a diagram of a link L, it is straightforward to construct a link equivalent to L. The difficulty (and challenge) of knot theory arises from the fact that a link can have several diagrams that look quite different. It is far from obvious that the two diagrams of Figure 16.3 are both diagrams of the unknot. (Convince yourself that they are by using a piece of string!) Even recognizing whether a given knot is the unknot is an important problem.

Given a link L, we can form its *mirror image* L' by reflecting L in a plane through the origin. Although such a reflection is a bijective linear map, it is not orientation preserving. (A bijective linear map is orientation preserving if and only if its determinant is positive.) Therefore, it may happen that L and L' are inequivalent. A diagram of L' can be obtained by reflecting a diagram of L in a line in the plane or, equivalently, by swapping all the under-crossings and over-crossings. A link that is equivalent to its mirror

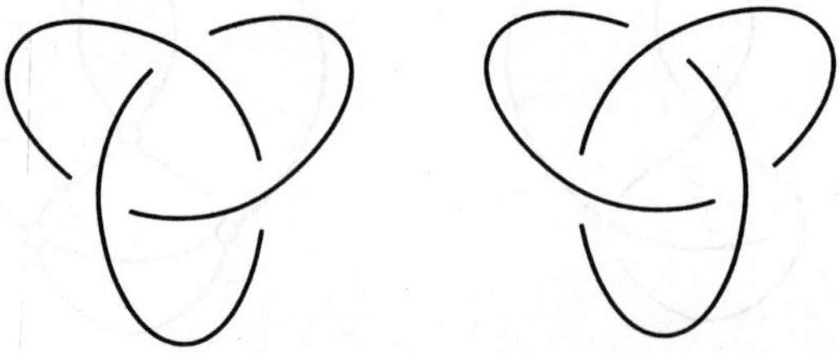

Figure 16.4. The right-handed and left-handed trefoil

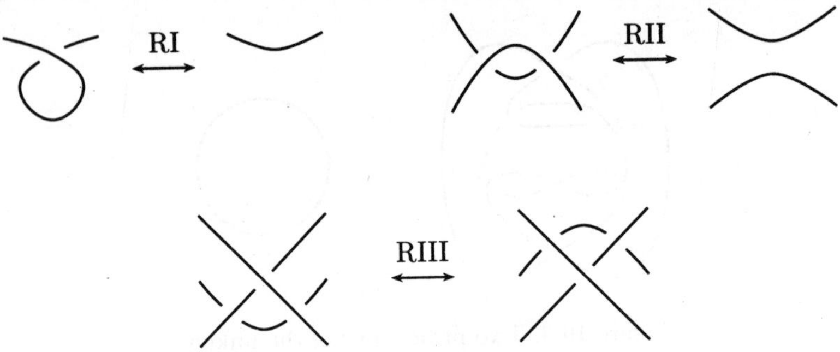

Figure 16.5. The three Reidemeister moves

image is called *achiral*. Figure 16.4 shows the trefoil and its mirror image. In Section 16.3 we will see how to prove that the trefoil is not achiral.

16.2 Reidemeister Moves

If two diagrams are related by an isotopy of the plane to itself, then they determine the same knot. In addition, there are three further operations that can be applied to a link diagram without changing the knot it represents, even though they do alter the number of crossings. These operations are known as the *Reidemeister moves* of types I, II, and III, and are shown in Figure 16.5. Each involves replacing a configuration of strands and crossings in a link diagram with a different configuration, while leaving the remainder of the link diagram unchanged.

It should be intuitively clear that diagrams related by a Reidemeister move represent the same link. Therefore, any two link diagrams related by a sequence of Reidemeister moves and planar isotopies are diagrams of the same link. Much more surprising is the fact that the converse is true.

Figure 16.6. An arc in a link projection

Theorem 16.2.1 *Two link diagrams determine the same link if and only if one can be obtained from the other by a sequence of Reidemeister moves and planar isotopies.* □

We omit the proof of this because it is entirely topological; nonetheless, it is neither long nor especially difficult. This result shows that equivalence of links can be determined entirely from consideration of link diagrams. We regard this as justification for our focus on link diagrams, which are 2-dimensional combinatorial objects, rather than links themselves, which are 3-dimensional topological objects. Therefore, to show that two links are equivalent it is sufficient to present a sequence of Reidemeister moves leading from a diagram of one link to a diagram of the other.

Demonstrating that two links are not equivalent requires showing that there is no sequence of Reidemeister moves relating the two link diagrams. Doing this directly, for example by systematically trying all possible sequences, is not even theoretically possible because there is no known limit to how many moves may be required. A more promising approach is to find properties of link diagrams that are not affected by Reidemeister moves. Such a property is then shared by every diagram of a link, and so determines a *link invariant*. If a link invariant takes different values on two link diagrams, then the diagrams represent different links. We consider one important example of a link invariant.

Define an *arc* of a link diagram to be a piece that is maximal subject to having no under-crossing; one is shown in Figure 16.6. At each crossing one arc starts and another arc ends, so the number of arcs equals the number of crossings. (The unknot, as usual, provides an anomaly, which, as usual, we ignore.)

A 3-colouring of a link diagram is an assignment of one of three colours to each arc such that the three arcs at each crossing are all the same

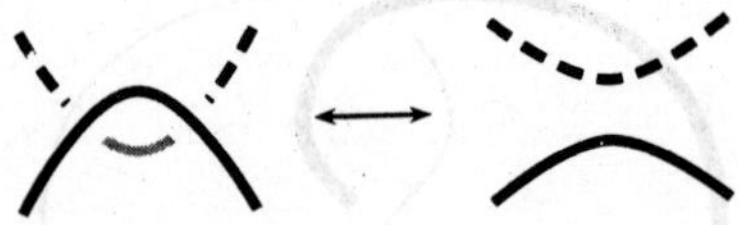

Figure 16.7. RII preserves a 3-colouring

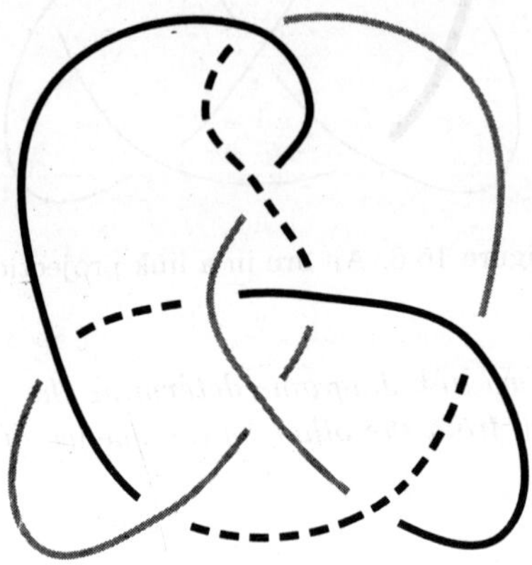

Figure 16.8. A 3-colouring of a knot diagram

colour or use all three colours. (See Figure 16.8.) Clearly, any link diagram has a 3-colouring where all the arcs receive the same colour, but not all link diagrams have proper (that is, nonconstant) 3-colourings. Moreover, if a link diagram has a proper 3-colouring, then so does any link diagram obtained after performing a single Reidemeister move. Figure 16.7 shows a pictorial proof of this for RII; we leave the proof for moves RI and RIII as an exercise.

Therefore, the property of having a proper 3-colouring is a link invariant. In particular, the unknot does not have a proper 3-colouring, and so the knot of Figure 16.8 is actually knotted. We can use 3-colourings to show that large classes of knots are knotted, but there are knots with no 3-colourings that are not equivalent to the unknot.

This idea can be generalized to colourings with n colours for any odd integer n. An *n-colouring* of a link diagram is a map γ from its arcs to the set $\{0, 1, \ldots, n-1\}$ such that if the arc x goes over the crossing where arcs y and z end, then

$$2\gamma(x) \equiv \gamma(y) + \gamma(z) \pmod{n}.$$

It is an interesting exercise to show that the number of proper n-colourings of a link diagram is invariant under Reidemeister moves.

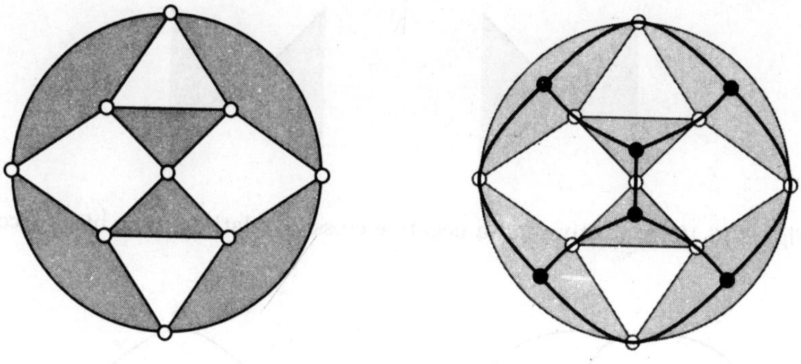

Figure 16.9. The graph on the black faces of a 4-valent plane graph

16.3 Signed Plane Graphs

The shadow of a knot diagram is a connected 4-regular plane graph, and hence its dual graph is bipartite. Thus we may 2-colour the faces of the shadow of a knot diagram with two colours, say black and white. Given such a colouring, we can define two graphs, one on the faces of each colour. Let B denote the graph with the black faces as its vertices, and the vertices of the shadow as its edges, with adjacency given as follows. Every vertex v of the shadow yields an edge of B joining the black faces that contain v. If v lies in only one black face, then the corresponding vertex of B has a loop, while if distinct black faces share more than one common vertex, then the corresponding vertices of B are joined by a multiple edge. Figure 16.9 shows a 4-regular plane graph, together with the graph on its black faces. We define the graph W on the white faces of the shadow analogously.

The two graphs B and W are called the *face graphs* of the shadow. Since the shadow of a knot diagram is connected, the graphs B and W are also connected, and it is not hard to see that $W = B^*$ and $B = W^*$, that is, the two face graphs are planar duals of each other. We will study the relation between the shadow of a knot diagram and its face graphs in some detail in the next chapter. For now, we content ourself with the observation that it is not hard to convince oneself (if not someone else) that a connected 4-regular plane graph is determined by either of its face graphs.

If L is a link diagram of a link with more than one component, then the shadow of L may not be connected. If X is a disconnected shadow, then we can still colour the faces of X with two colours. However, if we were to define the face graphs precisely as above, then one of them would be connected, and we would be unable to reconstruct X uniquely from it. To overcome this, we consider each component of X separately, and define the black face graph B of X to be the union of the black face graphs of its components, and analogously for W. In this way, any 4-regular plane graph is determined by either of its face graphs. If Y is a plane graph, then the *componentwise*

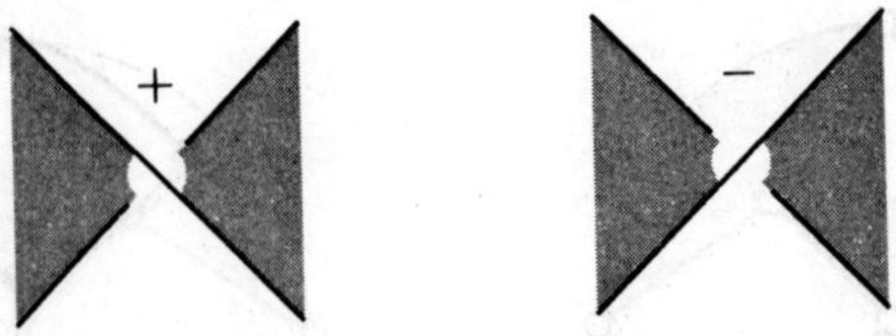

Figure 16.10. A positive and a negative crossing relative to the black faces

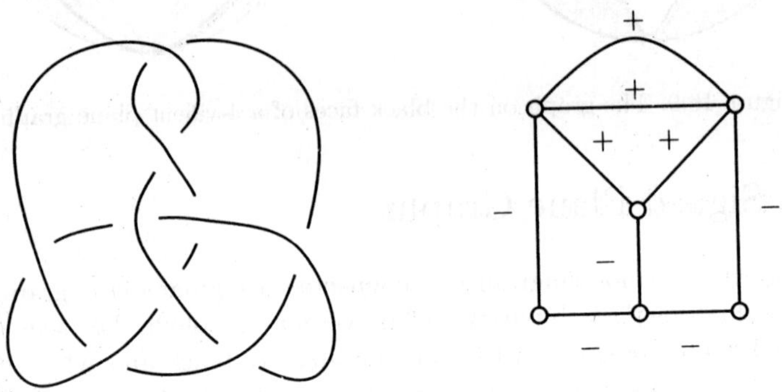

Figure 16.11. A knot diagram and one of its signed face graphs

planar dual Y^* of Y is the graph whose connected components are the planar duals of the components of Y considered separately. With this notion of planar duality, once again we have $W = B^*$ and $B = W^*$.

Now, let L be a link diagram and let X be the shadow of L, with a face graph Y. Then L is determined by X, together with the under-and-over information at the crossings. Therefore, since X is determined by its face graph Y (or Y^*), it is natural to consider a way of representing the link diagram by adding the under-and-over information to Y. Given a crossing in a link diagram, consider rotating the over-crossing anticlockwise until it lies parallel to the under-crossing. In doing this, the over-crossing moves over two faces of the same colour, and we define the crossing to be *positive* relative to the faces of this colour, and *negative* relative to the faces of the other colour. For example, Figure 16.10 shows crossings that are positive and negative with respect to the black faces.

Each crossing in the link diagram L is a vertex of X and an edge in each of the face graphs of X. If the crossing is positive relative to the black faces, then the corresponding edge is declared to be positive in B and negative in W. Therefore, in this fashion, a link diagram determines a dual pair of signed plane graphs, where the dual is the componentwise planar dual and signs are swapped on moving to the dual. The link diagram can be recovered uniquely from either of these signed plane graphs.

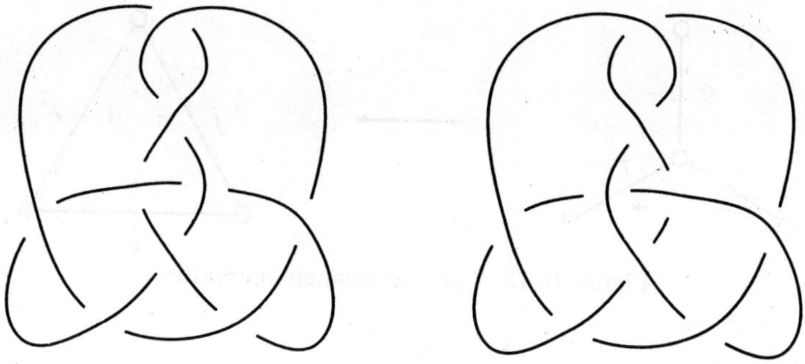

Figure 16.12. An alternating knot and a nonalternating knot

A knot is *alternating* if it has a knot diagram such that each of the signed face graphs has edges of only one sign. If it seems that this is the opposite of alternating, consider the knots in Figure 16.12 and observe that in the first knot, following the strand results in using over-crossings and under-crossings alternately.

16.4 Reidemeister moves on graphs

Since a link diagram can be represented as a dual pair of signed plane graphs, the Reidemeister moves on link diagrams can be described purely as operations on signed plane graphs.

Let L be a link diagram and Y and Y^* the two signed face graphs determined by L. Each Reidemeister move corresponds to an operation and its inverse, since the configuration on either side of the arrow can be replaced by the other.

A Reidemeister move of type I corresponds to deleting a positive loop from a vertex of Y or adding a positive loop to a vertex of Y. In the dual graph Y^*, this corresponds to contracting a negative coloop incident with a vertex of valency one, or adding a negative coloop attached to a new vertex of valency one.

We say that two edges are *parallel* if they are not loops, but share the same end-vertices. Then RII corresponds to deleting a pair of parallel edges of opposite sign, or adding a pair of parallel edges of opposite sign between two vertices of Y. In the dual graph, this operation is more complicated to describe, and involves either contracting or inserting a path of length two. A path uvw of length two can be contracted to a single vertex if uv and vw have opposite sign and v is a vertex of valency two. After contraction, v is adjacent to the neighbours of both u and w. The inverse operation involves replacing a vertex x with a path uvw of length two, where uv and vw have

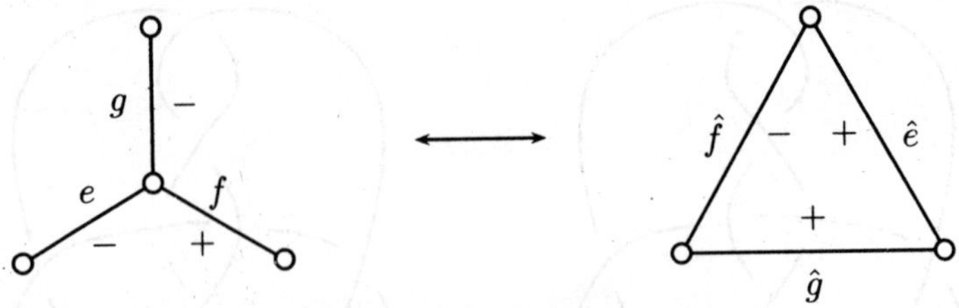

Figure 16.13. The star-triangle operation

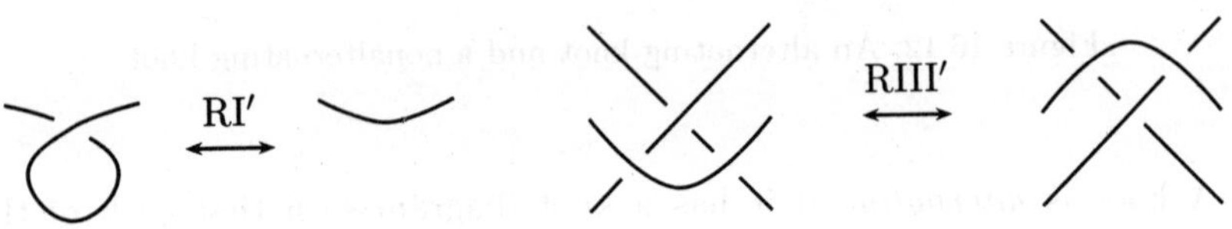

Figure 16.14. The mirror images of RI and RIII

opposite sign, v has valency two, and the edges that were incident to x are made incident to either u or w, in such a way that a plane graph results.

The operation RIII corresponds to replacing a star with two negative edges with a triangle with two positive edges, or replacing a triangle with star. In replacing a star with a triangle, the centre vertex and three edges e, f, and g are removed, and the triangle $\hat{e}$, $\hat{f}$, and $\hat{g}$ inserted. The edges $\hat{e}$, $\hat{f}$, and $\hat{g}$ have opposite sign to e, f, and g respectively, and join the two vertices of the triangle not incident with e, f, and g, respectively. (We offer Figure 16.13 to aid in decoding this description.) In the dual graph, this corresponds to replacing a triangle with two positive edges with a star with two negative edges.

Reidemeister moves RI and RIII have mirror images RI′ and RIII′ as shown in Figure 16.14. In graphical terms, RI′ corresponds to deleting or adding a negative loop, or dually, contracting or adding a positive end-edge. The move RIII′ is the star–triangle operation where the star has two positive edges, and the triangle two negative edges. It is a worthwhile exercise to show that RI′ and RIII′ are consequences of the Reidemeister moves. Once you have convinced yourself of this, you can freely use RI′ and RIII′ as additional "moves" in trying to show Reidemeister equivalence of two knots or signed plane graphs.

The next result shows that, at least in theory, it does not matter which face graph you choose to work with.

Theorem 16.4.1 *There is a sequence of Reidemeister moves leading from any signed plane graph Y to its componentwise dual Y^*.* □

16.5 Reidemeister Invariants

In the remainder of this chapter we develop a link invariant based on the rank polynomial of a signed graph. In this section we show that a suitable evaluation of the modified rank polynomial of the signed face graph associated with a link diagram is invariant under the Reidemeister moves RII and RIII. Although not invariant under RI, it is not hard to see the effect of moves of type RI, and Section 16.7 describes how to account for these.

Lemma 16.5.1 *Let Y be a signed graph and let e and f be parallel edges of opposite sign, not loops. If $\alpha\beta = 1$ and $y = -(\alpha^2 + \alpha^{-2})$, then $\widehat{R}(Y) = \widehat{R}(Y \setminus \{e, f\})$.*

Proof. Without loss of generality, suppose that e is positive. Then by Theorem 15.13.3,

$$\widehat{R}(Y) = \alpha\widehat{R}(Y \setminus e) + \beta\widehat{R}(Y/e).$$

Since f has the opposite sign to e and is an edge in $Y \setminus e$ and a loop in Y/e, we have

$$\widehat{R}(Y) = \alpha(\beta\widehat{R}(Y \setminus \{e, f\}) + \alpha\widehat{R}((Y \setminus e)/f) + \beta(\alpha y + \beta)\widehat{R}((Y/e) \setminus f).$$

Since $(Y \setminus e)/f = (Y/e) \setminus f$, the lemma follows, provided that $\alpha\beta = 1$ and

$$\alpha^2 + \beta^2 + \alpha\beta y = 0,$$

which follow from the given conditions. □

A similar argument also yields the following result:

Lemma 16.5.2 *Let Y be a signed graph and let e and f be edges of opposite sign having a common vertex of valency two. If $\alpha\beta = 1$ and $x = -(\alpha^2 + \alpha^{-2})$, then $\widehat{R}(Y) = \widehat{R}(Y/\{e, f\})$.* □

Therefore, we have shown that if $\alpha\beta = 1$ and $x = y = -(\alpha^2 + \alpha^{-2})$, then the modified rank polynomial of a signed plane graph is invariant under Reidemeister moves of type II. It is extremely surprising that these conditions are sufficient for $\widehat{R}$ to be invariant under type-III moves as well.

Lemma 16.5.3 *If Y is a signed graph, $\alpha\beta = 1$, and $x = y = -(\alpha^2+\alpha^{-2})$, then $\widehat{R}(Y; \alpha, \beta, x, y)$ is invariant under Reidemeister moves of type III.*

Proof. Let Y be a graph containing a star with two negative legs and a positive leg, and Y' the graph obtained by performing the star–triangle move on Y as shown in Figure 16.13. Let Z be the graph we get from $Y \setminus e$ by contracting f and g. Alternatively, it is the graph obtained from Y' by contracting $\hat{e}$ and deleting $\hat{f}$ and $\hat{g}$.

By applying Theorem 15.13.3 to the edge e in Y we find that

$$\widehat{R}(Y) = \beta\widehat{R}(Y \setminus e) + \alpha\widehat{R}(Y/e).$$

Applying Lemma 16.5.2 to the edges f and g in $Y\backslash e$, we see that $\widehat{R}(Y\backslash e) = \widehat{R}(Z)$, and therefore

$$\widehat{R}(Y) = \beta\widehat{R}(Z) + \alpha\widehat{R}(Y/e).$$

By applying Theorem 15.13.3 to the edge $\hat{e}$ in Y' we find that

$$\widehat{R}(Y') = \alpha\widehat{R}(Y'\backslash\hat{e}) + \beta\widehat{R}(Y'/\hat{e}).$$

In $Y'/\hat{e}$, the edges $\hat{f}$ and $\hat{g}$ are parallel with opposite sign. Hence Lemma 16.5.1 yields that $\widehat{R}(Y'/\hat{e}) = \widehat{R}(Z)$. Since $Y'\backslash\hat{e} = Y/e$, we conclude that $\widehat{R}(Y) = \widehat{R}(Y')$. □

Let L be a link diagram with signed face graph Y. Then the *Kauffman bracket* of L is defined to be

$$[L] := \widehat{R}(Y; \alpha, \alpha^{-1}, -(\alpha^2 + \alpha^{-2}), -(\alpha^2 + \alpha^{-2})).$$

An operation on link diagrams that is a combination of planar isotopy and Reidemeister moves of types II and III is known as a *regular isotopy*. Therefore, the results above show that the Kauffman bracket of a link diagram is invariant under regular isotopies.

Although a link diagram has two signed face graphs, the Kauffman bracket is well-defined, because using either Y or Y^* gives the same result. If Y is connected, this follows immediately because

$$R(Y^*; \alpha, \beta, x, y) = R(Y; \alpha, \beta, y, x),$$

and Y^* is connected. If Y is not connected and has components Y_1 and Y_2, then

$$R(Y) = R(Y_1)R(Y_2),$$

and so we get

$$\widehat{R}(Y) = x\widehat{R}(Y_1)\widehat{R}(Y_2).$$

If Y^* is the componentwise dual of Y, and $x = y$, then $\widehat{R}(Y^*) = \widehat{R}(Y)$.

Now, consider the knot diagrams for the right-handed trefoil and its mirror image, the left-handed trefoil, shown in Figure 16.4. The diagram of the right-handed trefoil has K_3 with all edges positive as a face graph, and the diagram of the left-handed trefoil K_3 with all edges negative. Using the expression given at the end of Section 15.13 yields that the Kauffman brackets of these two diagrams are

$$\alpha^7 - \alpha^3 - \alpha^{-5}$$

and

$$\alpha^{-7} - \alpha^{-3} - \alpha^5,$$

respectively.

Figure 16.15. A crossing in standard position

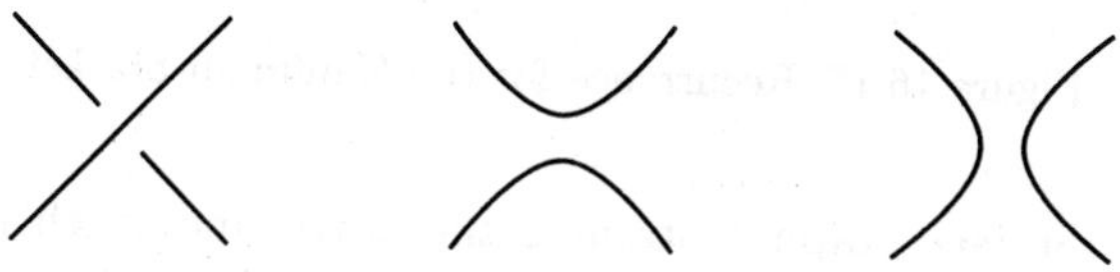

Figure 16.16. Link diagrams L, $L^=$, and $L^{\|}$, respectively

Even though the Kauffman bracket of a link diagram is not a full Reidemeister invariant, it can still be useful. Reidemeister moves of type I correspond to adding or deleting a positive loop in one of the face graphs. By Theorem 15.13.3, if e is a loop in the signed graph Y, then

$$\widehat{R}(Y) = \widehat{R}(e)\widehat{R}(Y \setminus e).$$

If e is a positive loop, $\alpha\beta = 1$, and $x = y = -(\alpha^2 + \alpha^{-2})$, then

$$\widehat{R}(e) = \alpha + \beta y = -\alpha^{-3}.$$

We conclude that if two signed graphs Y and Z are related by a sequence of Reidemeister moves, then $\widehat{R}(Y)/\widehat{R}(Z)$ is ± 1 times a power of α^3. Since the Kauffman brackets of the link diagrams of the trefoil and its mirror image are not related in this fashion, we deduce that there is no sequence of Reidemeister moves that takes the right-handed trefoil to the left-handed trefoil. Therefore, the trefoil knot is not achiral.

16.6 The Kauffman Bracket

In this section we provide another approach to the Kauffman bracket as developed in the previous section. We show that the deletion–contraction recurrence for the rank polynomial can be described directly in terms of the link diagram.

Let L be a link diagram with signed face graphs Y and Y^*. Given a crossing, the link diagram can be rotated (a planar isotopy) so that the crossing has the form shown in Figure 16.15.

Let $L^=$ and $L^{\|}$ denote the link diagrams obtained by replacing the crossing with two noncrossing strands, as shown in Figure 16.16.

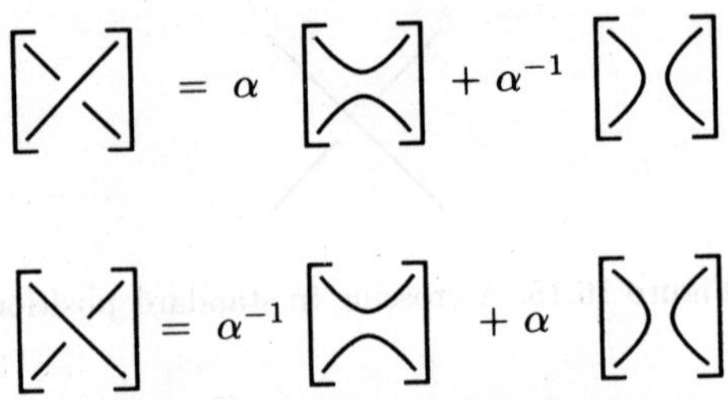

Figure 16.17. Recurrence for the Kauffman bracket

Let Y be the face graph containing the north and south faces and Y^* the face graph containing the east and west faces. If e is the edge of Y and Y^* corresponding to the crossing, then e is positive in Y and negative in Y^*. The link diagram $L^=$ has $Y \setminus e$ and Y^*/e as its two face graphs, and the link diagram $L^{\|}$ has Y/e and $Y^* \setminus e$ as its face graphs. If e is not a loop in Y, then by Theorem 15.13.3, we get

$$[L] = \alpha\widehat{R}(Y \setminus e) + \beta\widehat{R}(Y/e) = \alpha[L^=] + \alpha^{-1}[L^{\|}].$$

If e is a loop in Y, then it is not a loop in Y^*, and so by Theorem 15.13.3, we get

$$[L] = \alpha\widehat{R}(Y^*/e) + \alpha^{-1}\widehat{R}(Y^* \setminus e) = \alpha[L^=] + \alpha^{-1}[L^{\|}].$$

Therefore, in all cases we have

$$[L] = \alpha[L^=] + \alpha^{-1}[L^{\|}]. \tag{16.1}$$

This expression is usually presented in picturesque fashion, as shown in Figure 16.17. The second relation is equivalent to the first (rotate the page through a quarter turn) and given only for convenience.

Since both $L^=$ and $L^{\|}$ have fewer crossings than L, it follows that repeated application of (16.1) yields an expression where every term is the Kauffman bracket of a link diagram with no crossings, that is, a diagram whose components are all isotopic to circles. If L is such a diagram with d components, then

$$[L] = (-\alpha^2 - \alpha^{-2})^{d-1}.$$

Figure 16.18 shows one step of this process on the knot diagram for the right-handed trefoil. In this case, the second recurrence from Figure 16.17 applies, and so $[L] = \alpha^{-1}[L^=] + \alpha[L^{\|}]$.

16.7 The Jones Polynomial

In this section we show how to convert the Kauffman bracket of a link diagram L into a link invariant.

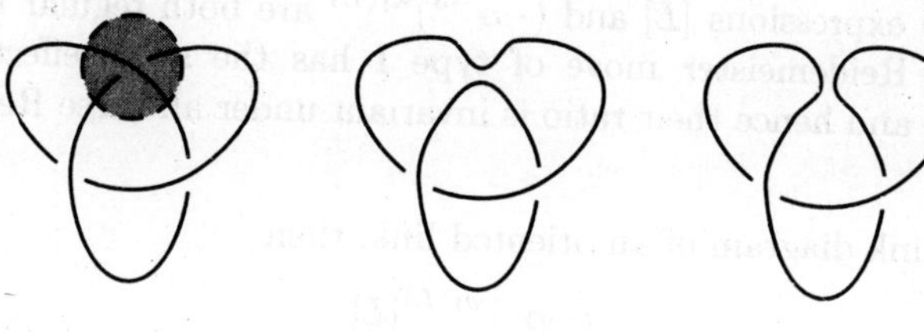

Figure 16.18. Link diagrams L, $L^{=}$, and $L^{\|}$

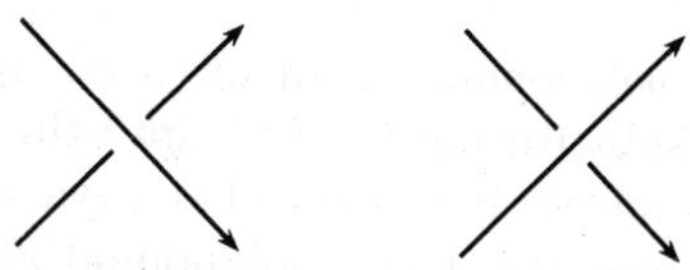

Figure 16.19. Left-hand, and right-hand, crossings

We start by introducing another parameter that is invariant under Reidemeister moves of type II and III, but not under type-I moves. We are obliged to work with *oriented links*—links in which a direction has been assigned to each component—but will show that the particular choice of orientation does not affect the result when the link is a knot. In a link diagram of an oriented link, define a crossing to be *left-handed* if when we move along the under-crossing branch in the direction of the orientation, the over-crossing runs from left to right. If a crossing is not left handed it is *right-handed*. The *writhe* of a link diagram L is the number of left-handed crossings, less the number of right-handed crossings; we denote it by $\mathrm{wr}(L)$.

The diagrams of the trefoil and its mirror image shown in Figure 16.4 have writhes 3 and -3, respectively.

It is not difficult to verify that the writhe of a link diagram is invariant under regular isotopy, but we leave the proof of this as an exercise. If the orientation of every component of a link is reversed, then the writhe of its link diagram does not change. Hence the writhe of a knot diagram is independent of the orientation.

A Reidemeister move of type I that is equivalent to deleting a positive loop decreases the writhe by one, because the number of left-handed crossings goes down by one. Adding a positive loop increases the writhe by one.

Theorem 16.7.1 *Let L be a link diagram of an oriented link. Then*

$$(-\alpha^3)^{\mathrm{wr}(L)}[L]$$

is invariant under all Reidemeister moves, and hence is an invariant of the oriented link.

Proof. The expressions $[L]$ and $(-\alpha^{-3})^{\mathrm{wr}(L)}$ are both regular isotopy invariants. A Reidemeister move of type I has the same effect on both expressions, and hence their ratio is invariant under all three Reidemeister moves. □

It L is a link diagram of an oriented link, then

$$(-\alpha^3)^{\mathrm{wr}(L)}[L]$$

is an integral linear combination of the indeterminates

$$\{\ldots, \alpha^{-8}, \alpha^{-4}, \alpha^0, \alpha^4, \alpha^8, \ldots\}.$$

We define the *Jones polynomial* of an oriented link_to be the polynomial in t obtained by substituting $t = \alpha^{1/4}$ into this expression. Although this is standard terminology, it is somewhat cavalier with the use of the term "polynomial". Since the Jones polynomial is an expression in the indeterminates

$$\{\ldots, t^{-2}, t^{-1}, 1, t, t^2, \ldots\},$$

it is a *Laurent polynomial* rather than a polynomial. Nevertheless, the standard terminology causes no difficulty, and we continue to use it.

As an example, consider the knot diagram L for the right-handed trefoil as shown in Figure 16.4. We have seen that $[L] = \alpha^7 - \alpha^3 - \alpha^{-5}$ and that $\mathrm{wr}(L) = 3$. Therefore, the Jones polynomial of the right-handed trefoil is

$$(-\alpha^3)^3(\alpha^7 - \alpha^3 - \alpha^{-5}) = -t^4 + t^3 + t.$$

The Jones polynomial of the left-handed trefoil is

$$-t^{-4} + t^{-3} + t^{-1},$$

and in general the Jones polynomial of the mirror image of a knot is obtained by substituting t^{-1} for t into the Jones polynomial for the knot. Therefore, any knot with a Jones polynomial that is not invariant under this substitution is not achiral.

Although the Jones polynomial is a powerful knot invariant, it is not a complete invariant, because there are pairs of distinct knots that it cannot distinguish. We will see examples of this in the next section.

16.8 Connectivity

A link is *split* if it has a link diagram whose shadow is not connected. Deciding whether a particular shadow is connected is easy, but it appears to be a hard problem to determine whether a link is split or not. Thus, it is difficult to determine if a signed planar graph is Reidemeister equivalent to a disconnected graph. (It is also difficult to determine whether a signed planar graph is Reidemeister equivalent to K_1. This is the problem of deciding

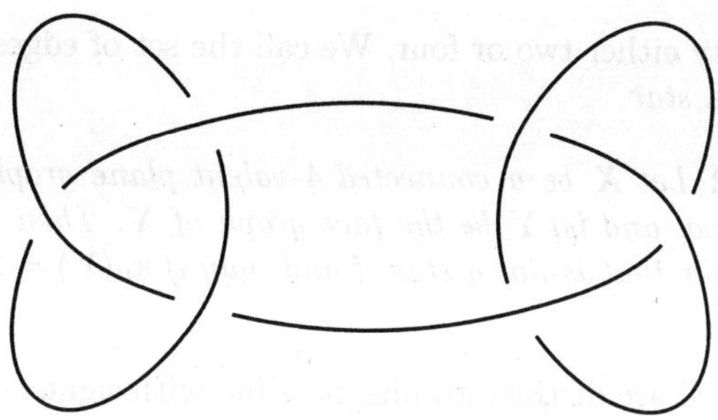

Figure 16.20. The sum of two knots

whether a knot is the unknot.) In this section we are going to study the relationship between the connectivity properties of 4-valent plane graphs and the links they represent.

First we introduce an operation on knots that allows us to build "composite" knots from smaller ones. Suppose that K_1 and K_2 are two knot diagrams, which we shall assume are oriented. Then their *sum* $K = K_1+K_2$ is the oriented knot diagram obtained by cutting one strand of K_1 and K_2, and joining them up to form one component in such a way that the orientations match up. It is necessary to use orientations to ensure that the sum is well-defined, because if K_2' is obtained by reversing the orientation on K_2, then it may occur that K_1+K_2 is not equivalent to K_1+K_2'. However, for our current purposes, the orientation plays no role, and can safely be ignored. Figure 16.20 shows one of the knot diagrams that results when two trefoils are added.

Despite its name, the sum of two knots behaves more like a product, and topologists define a knot to be *prime* if it does not have a knot diagram that is the sum of two nontrivial knot diagrams. Most published tables of knots list only prime knots. The shadow of the sum of two nontrivial knot diagrams has an edge cutset of size two, and therefore has edge connectivity two. Conversely, if the shadow of a knot diagram has edge connectivity two, then it is the sum of two smaller knot diagrams.

Recall that $\kappa_0(X)$ denotes the vertex connectivity of X and that $\kappa_1(X)$ denotes the edge connectivity of X. We leave the proof of the following result as an exercise.

Lemma 16.8.1 *Let X be a connected 4-valent plane graph, with face graph Y. Then $\kappa_1(X) = 2$ if and only if $\kappa_0(Y) = 1$.* □

It follows from this result that the Kauffman bracket of the sum of two knots is the product of the Kauffman brackets of the components.

The edge connectivity of any eulerian graph is even, and since the shadow of a knot diagram is 4-valent, the set of edges incident with any one vertex is an edge cutset of size four. Therefore, the shadow of a knot diagram has edge connectivity either two or four. We call the set of edges incident with a single vertex a *star*.

Lemma 16.8.2 *Let X be a connected 4-valent plane graph with no edge cutset of size two, and let Y be the face graph of X. Then X has an edge cutset of size four that is not a star if and only if $\kappa_0(Y) = 2$.* □

Suppose Z and Z' are distinct graphs, possibly with signed edges, and that $\{u_1, u_2\}$ and $\{v_1, v_2\}$ are pairs of distinct vertices in Z and Z', respectively. Let Y_1 be the graph obtained by identifying u_1 with v_1 and u_2 with v_2, and let Y_2 be the graph obtained by identifying u_1 with v_2 and u_2 with v_1. Then both Y_1 and Y_2 have a vertex cutset of size two, and any graph with a vertex cutset of size two can be constructed in this fashion. In this situation we say that Y_1 and Y_2 are related by a *Whitney flip*. If there is an automorphism of Z that swaps u_1 and u_2, then Y_1 and Y_2 are isomorphic, and similarly for Z'. In general, though, Y_1 and Y_2 are not isomorphic. We provide one example in Figure 16.21, where the shaded vertices form the pair $\{u_1, u_2\} = \{v_1, v_2\}$.

Figure 16.21. The Whitney flip

Theorem 16.8.3 *Let Y_1 and Y_2 be signed graphs that are related by a Whitney flip. Then their rank polynomials are equal.*

Proof. The graphs Y_1 and Y_2 have the same edge set, and it is clear that a set $S \subseteq E(Y_1)$ is independent in $M(Y_1)$ if and only if it is independent in $M(Y_2)$. Therefore, the two graphs have the same cycle matroid. □

Two knot diagrams whose face graphs are related by a Whitney flip are said to be *mutants* of one another. Mutant knots have the same Kauffman bracket, and hence the same Jones polynomial. Figure 16.22 shows a famous mutant pair of knots.

If we view a Whitney flip as flipping a "rotor of order two," then flipping rotors of order $n > 2$ can be viewed as a generalization of knot mutation. Rotor-flipping permits the construction of many pairs of knots with the same Jones polynomial. However, we note that in general it is not easy to determine whether the resulting knots are actually inequivalent.

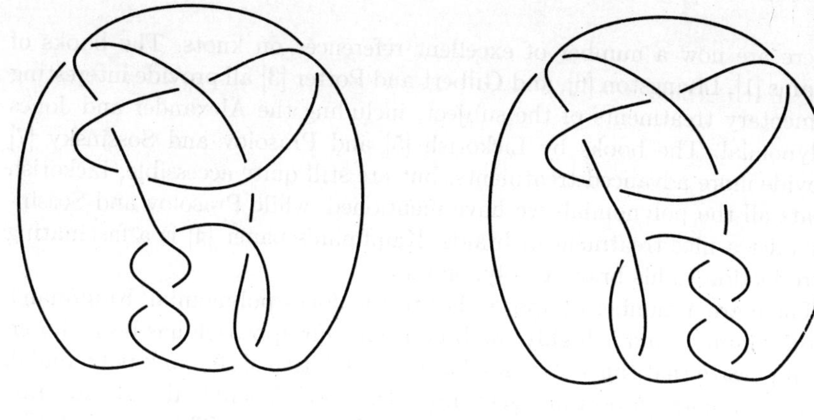

Figure 16.22. Two knots with the same Jones polynomial

Exercises

1. Show that the number of n-colourings of a link diagram is invariant under the Reidemeister moves.

2. Do the knots of Figure 16.22 have 3-colourings?

3. We label the arcs of an oriented knot diagram with elements of a group G by assigning elements of G to the arcs, subject to the following condition. Suppose that at a given crossing the arc ending at the crossing is labelled h and the arc starting at it is labelled k. Then, if the crossing is left-handed, the label g of the overpass must be chosen so that $k = g^{-1}hg$; if the crossing is right-handed, then we require $h = g^{-1}kg$. Thus all elements used in a proper labelling must be conjugate in G. Show that the property of having a labelling using a given conjugacy class of G is invariant under Reidemeister moves.

4. Show that the moves RI′ and RIII′ are consequences of planar isotopy and the Reidemeister moves RI, RII, and RIII. Why is there no move RII′?

5. Prove Theorem 16.4.1.

6. Prove Lemma 16.5.2.

7. Show that the writhe of a link is invariant under regular isotopy.

8. Show that the Jones polynomial, when expressed in terms of α, involves only powers of α^4.

9. Show that the Jones polynomial of a knot evaluated at $t = 1$ has absolute value 1.

Notes

There are now a number of excellent references on knots. The books of Adams [1], Livingston [6], and Gilbert and Porter [3] all provide interesting elementary treatments of the subject, including the Alexander and Jones polynomial. The books by Lickorish [5] and Prasolov and Sossinsky [7] provide more advanced treatments, but are still quite accessible. Lickorish treats all the polynomials we have mentioned, while Prasolov and Sossinsky offer a nice treatment of braids. Kauffman's paper [4] is a fascinating introduction to his bracket polynomial.

There are a number of approaches to the Jones polynomial. Kauffman's is the simplest, and is highly combinatorial. Our approach has been chosen to emphasize that this polynomial is a close relative of the rank polynomial, which is a central object in algebraic graph theory. There is also an algebraic approach, based on representations of the braid group. This is, essentially, Jones's original approach, and has proved the most fruitful.

There are other useful knot polynomials of combinatorial interest that we have not discussed. In particular, the Alexander polynomial is quite accessible and would fit in well with what we have done. Alexander's original paper on his polynomial is very readable, and has a highly combinatorial flavour. Then there is the HOMFLY polynomial, a polynomial in two variables that is more or less the least common multiple of the Alexander and Jones polynomials. The most convenient reference for these polynomials is Lickorish [5].

The term "mutant" was coined by Conway, and the two knots of Figure 16.22 are called the Conway knot and the Kinoshita–Terasaka knot. Anstee, Przytycki, and Rolfsen [2] consider some generalizations of mutation that are based essentially on rotor-flipping. Their aim was to see whether the unknot, which has a Jones polynomial equal to 1, could be mutated into a knot. This would provide an answer to the unsolved question of whether the Jones polynomial can be used to determine knottedness. A resolution of this question, at least in the affirmative, would be a major advance in knot theory.

References

[1] C. C. Adams, *The Knot Book*, W. H. Freeman and Company, New York, 1994.

[2] R. P. Anstee, J. H. Przytycki, and D. Rolfsen, *Knot polynomials and generalized mutation*, Topology Appl., 32 (1989), 237–249.

[3] N. D. Gilbert and T. Porter, *Knots and Surfaces*, The Clarendon Press Oxford University Press, New York, 1994.

[4] L. H. Kauffman, *New invariants in the theory of knots*, Amer. Math. Monthly, 95 (1988), 195–242.

[5] W. B. R. LICKORISH, *An Introduction to Knot Theory*, Springer-Verlag, New York, 1997.

[6] C. LIVINGSTON, *Knot Theory*, Mathematical Association of America, Washington, DC, 1993.

[7] V. V. PRASOLOV AND A. B. SOSSINSKY, *Knots, Links, Braids and 3-manifolds*, American Mathematical Society, Providence, RI, 1997.

17
Knots and Eulerian Cycles

This chapter provides an introduction to some of the graph theory associated with knots and links. The connection arises from the description of the shadow of a link diagram as a 4-valent plane graph. The link diagram is determined by a particular eulerian tour in this graph, and consequently many operations on link diagrams translate to operations on eulerian tours in plane graphs. The study of eulerian tours in 4-valent plane graphs leads naturally to the study of a number of interesting combinatorial objects, such as double occurrence words, chord diagrams, circle graphs, and maps. Questions that are motivated by the theory of knots and links can often be clarified or solved by being reformulated as a question in one of these different contexts.

17.1 Eulerian Partitions and Tours

We defined walks in graphs earlier, but then we assumed that our graphs had no loops and no multiple edges. We now need to consider walks on the shadow of a link diagram, and so we must replace our earlier definition by a more refined one: A *walk* in a graph X is an alternating sequence of vertices and edges that starts and finishes with a vertex, with the property that consecutive vertices are the end-vertices of the edge between them. A walk is *closed* if its first and last elements are equal, and *eulerian* if it uses each edge at most once.

We will be concerned with closed eulerian walks, but first we consider two operations on this set. A *rotation* of a closed eulerian walk is the closed eulerian walk obtained by cyclically shifting the sequence of vertices and edges. A *reversal* of a closed eulerian walk is obtained by reversing the sequence of vertices and edges. For our purposes, only the cyclic ordering of vertices and edges determined by a closed eulerian walk is important, rather than the starting vertex. Therefore, we wish to regard all rotations of a closed eulerian walk as being the same. Usually, the direction of a closed eulerian walk is also not important, and so we define an *eulerian cycle* to be an equivalence class of closed eulerian walks under rotation and reversal. We normally treat an eulerian cycle as a specific closed eulerian walk, but with the understanding that any other member of the equivalence class could equally well be used.

Note that the subgraph spanned by the set of vertices and edges of an eulerian cycle need not be a cycle in the usual sense, but will be an eulerian subgraph of X.

An *eulerian partition* of X is a collection of eulerian cycles such that every edge of X occurs in exactly one of them. An *eulerian tour* of X is an eulerian cycle of X that uses every edge of X or, equivalently, an eulerian partition with only one eulerian cycle. If X is a 4-valent plane graph, then an eulerian cycle is *straight* if it always leaves a vertex by the edge opposite the edge it entered by.

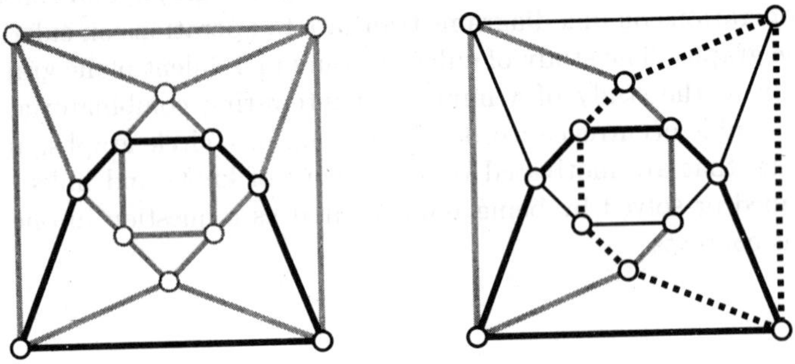

Figure 17.1. A straight eulerian cycle and a straight eulerian partition

Given a link, choose a starting point and imagine following the strand (in a fixed, but arbitrarily chosen, direction) until we return to the starting position. Visualizing this same process on the link diagram, it is easy to see that it corresponds to tracing out a straight eulerian cycle in the shadow of the link diagram. Therefore, each component of a link determines a straight eulerian cycle, and the link itself determines an eulerian partition into straight eulerian cycles. We call this the *straight eulerian partition*. If the link has only one component, then it determines a straight eulerian tour.

Suppose now that we are given an eulerian partition of a 4-valent graph. This eulerian partition induces a partition on the four edges incident to any given vertex into two pairs of consecutive edges. Conversely, an eulerian partition can be specified completely by giving the induced partition at each vertex. (So a 4-valent graph with n vertices has exactly 3^n eulerian partitions.) We say that two eulerian partitions *differ* at x if they do not determine the same partition at x; if they do not differ at x, we say that they *agree* there. An eulerian partition that differs at each vertex from the straight eulerian partition of a 4-valent plane graph is said to be *bent*.

Lemma 17.1.1 *Let $\mathcal{T}$ be an eulerian tour of the 4-valent graph X. If $u \in V(X)$, then there is unique eulerian tour $\mathcal{T}'$ in X that differs from $\mathcal{T}$ at u and agrees with $\mathcal{T}$ at all other vertices of X.*

Proof. There are three eulerian partitions that agree with $\mathcal{T}$ at the vertices in $V(X) \setminus u$. We show that one of these is an eulerian tour and that the other has two eulerian cycles.

Suppose that $u \in V(X)$ and that a, b, c, and d are the edges on u and that $\mathcal{T}$ contains the subsequence (a, u, b, S, c, u, d), where S is an alternating sequence of incident edges and vertices. Thus $\mathcal{T}$ partitions the edges on u into the pairs $\{a, b\}$ and $\{c, d\}$. Let S' be the reverse of the eulerian walk S. If we replace the subsequence (b, S, c) of $\mathcal{T}$ by its reversal (c, S', b), the result is a new eulerian tour that partitions the edges on u into the pairs $\{a, c\}$ and $\{b, d\}$.

There is a unique eulerian partition that agrees with $\mathcal{T}$ at each vertex other than u, and partitions the edges at u into the pairs $\{a, d\}$ and $\{b, c\}$. But this eulerian partition has two components, one of which is the eulerian cycle (u, b, S, c, u). □

If $\mathcal{T}'$ is the unique eulerian tour that differs from $\mathcal{T}$ only at u, then we say that $\mathcal{T}'$ is obtained from $\mathcal{T}$ by *flipping* at u.

Lemma 17.1.2 *Let $\mathcal{S}$ and $\mathcal{T}$ be two eulerian tours in the 4-valent graph X. Then there is a sequence of vertices such that $\mathcal{T}$ can be obtained from $\mathcal{S}$ by flipping at each vertex of the sequence in turn.*

Proof. Suppose $\mathcal{S}$ and $\mathcal{T}$ are two eulerian tours that do not agree at a vertex u. Let $\mathcal{S}'$ and $\mathcal{T}'$ denote the tours obtained from $\mathcal{S}$ and $\mathcal{T}$, respectively, by flipping at u. Since there are only three partitions of the edges on u into two pairs, and since $\mathcal{S}$ and $\mathcal{T}$ do not agree at u, one of the three pairs of tours

$$\{\mathcal{S}', \mathcal{T}\}, \quad \{\mathcal{S}, \mathcal{T}'\}, \quad \{\mathcal{S}', \mathcal{T}'\}$$

must agree at u, and therefore differ at one fewer vertex than $\{\mathcal{S}, \mathcal{T}\}$. A straightforward induction on the number of vertices at which two tours differ yields the result. □

Let $\mathcal{E}(X)$ denote the graph on the eulerian tours of X, where two eulerian tours are adjacent if and only if they are obtained from each other by flipping at a vertex. The previous result shows that $\mathcal{E}(X)$ is connected.

17.2 The Medial Graph

The shadow of a link diagram is a 4-valent plane graph, which gives rise to a dual pair of face graphs. In Section 16.3 we asserted that this graph is determined by either of its face graphs, and now we formalize the procedure that takes us from a face graph back to the 4-valent plane graph.

Given a connected plane graph Y, its *medial graph* $\mathcal{M}(Y)$ is defined as follows. The vertices of $\mathcal{M}(Y)$ are the edges of Y. Each face $F = e_1, \dots, e_r$ of length r in Y determines r edges

$$\{e_i e_{i+1} : 1 \le i \le r-1\} \cup \{e_r e_1\}$$

of $\mathcal{M}(Y)$. In this definition, a loop e that bounds a face is viewed as a face of length one, and so determines one edge of $\mathcal{M}(Y)$, which is a loop on e. If Y has an edge e adjacent to a vertex of valency one, then the face containing that edge is viewed as having two consecutive occurrences of e, and so once again there is a loop on e. Figure 17.2 gives an example of a plane graph and its medial graph.

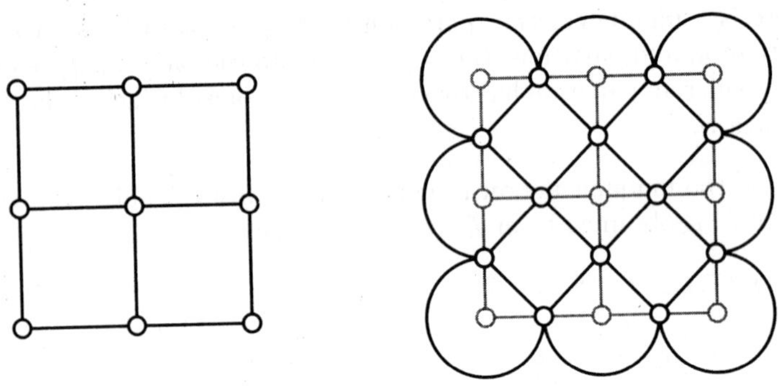

Figure 17.2. A plane graph and its medial graph

There are two important consequences of this definition. Firstly, it is clear that $\mathcal{M}(Y)$ is a 4-valent graph. Secondly, if two edges are consecutive in some face of Y, then they are consecutive in some face of Y^*, and hence $\mathcal{M}(Y) = \mathcal{M}(Y^*)$. Finally, we note that Y and Y^* are the face graphs of $\mathcal{M}(Y)$.

There is a more topological way to approach this rather awkward definition of a medial graph. Consider placing a small disk of radius ϵ_1 over each vertex, and a thin strip of width ϵ_2 over each edge, where ϵ_2 is much smaller than ϵ_1. The union of the disks and the strips forms a region of the plane. At the midpoint of each thin strip, pinch the two sides of the strip together to a single point. Then the boundary of this region is a collection of curves meeting only at the points where the strips were pinched together. The medial graph is the graph whose vertices are these points, and whose edges are these curves. Figure 17.3 should help convince you that these two definitions are equivalent. Although this is straightforward, care is needed with edges adjacent to a vertex of valency one and loops that bound a face.

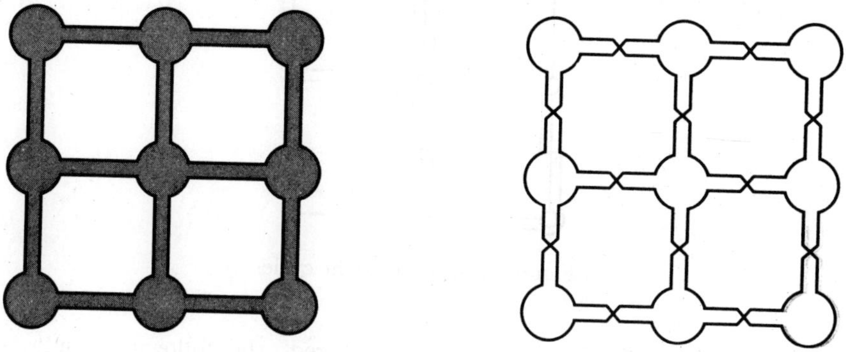

Figure 17.3. Disks, strips, and the medial graph

We now consider a special kind of walk in a plane graph Y. As above, view each vertex of Y as a small disk, and each edge as a thin strip. Since each edge is a strip, it has two distinct sides, and we can visualize travelling along the side of an edge. Select a starting point on the graph where the side of a strip meets the boundary of a disk. We call such a vertex–edge–side triple a *flag*. From there, walk along the side of the edge crossing to the opposite side of the edge when you reach the point on the edge halfway between its endpoints. On reaching the neighbouring vertex, walk around the boundary of the disk representing the vertex, leaving the vertex along the side of the edge lying in the same face as the side of the edge you have just arrived on. Extend the walk by using the same rules for negotiating edges and vertices. A *left right walk* is the alternating sequence of vertices and edges encountered during such a walk, together with the starting flag. (This definition can be given more formally—but less intuitively—purely in terms of flags; we develop some of the necessary machinery in Section 17.10.)

The underlying sequence of vertices and edges in a left–right walk is a walk in the usual sense, but distinct left–right walks may have the same underlying walk if they start at flags on opposite sides of the same edge. A flag in the plane graph Y determines a flag in its dual graph Y^*, and with the usual identification between the edges of Y and its dual Y^*, a left–right walk in Y is also a left–right walk in Y^*.

A *closed left–right walk* is a left–right walk that starts and ends at the same flag. A *left–right cycle* is an equivalence class of closed left–right walks under rotation and reversal; an example is shown in Figure 17.4. Thus in a left–right cycle, the cyclic order of the vertices and edges is important and which sides of the edges are used is important, but the direction and starting vertex are not.

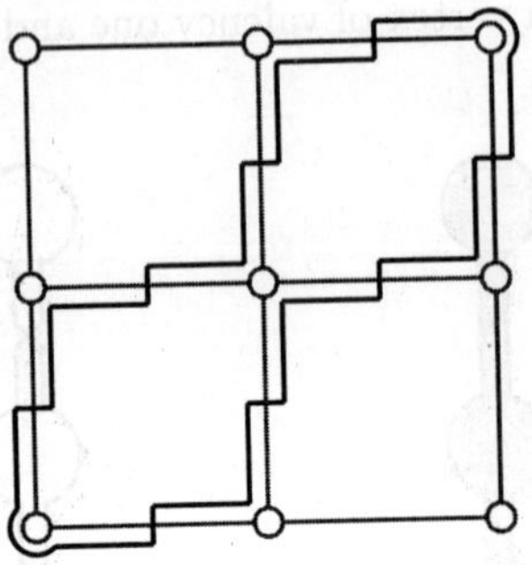

Figure 17.4. A left–right cycle

After these definitions have been mastered, the following result is immediate.

Lemma 17.2.1 *Let Y be a connected plane graph and let X be its medial graph. Then there is a bijection between straight eulerian cycles in X and left–right cycles in Y.* □

17.3 Link Components and Bicycles

If X is the shadow of a link diagram, then the number of components of the link is the number of eulerian cycles in the straight eulerian partition of X. In this section we relate the number of components of a link to the dimension of the bicycle space of the face graph of X.

Let Y be a plane graph, and let P be a left–right cycle in Y. If P uses an edge, then it uses it at most twice. The set of edges that are used exactly once is called the *core* of P.

Lemma 17.3.1 *The core of a left–right cycle in a plane graph Y is a bicycle of Y.*

Proof. Let P be a left–right cycle and let Q be its core. Let u be a vertex of Y, and consider the edges on u. The total number of occurrences of these edges in P is even, and so the total number of these edges in Q is also even. Therefore, Q determines an even subgraph of Y. Since P is also a left–right cycle in Y^*, it follows that Q is also an even subgraph of Y^*. Therefore, Q is a bicycle of Y. □

Every edge of Y either occurs once in each of two left–right cycles or occurs twice in one left–right cycle. Therefore, each edge of Y lies either in two cores or no cores at all.

Lemma 17.3.2 *Let Y be a connected plane graph and let $\mathcal{P}$ be the set of all the left–right cycles of Y. Each edge of Y occurs in an even number of members of $\mathcal{P}$, but no proper subset of $\mathcal{P}$ covers every edge an even number of times.*

Proof. The first claim follows directly from the comments preceding this result. For the second claim, suppose that $\mathcal{P}'$ is a proper subset of $\mathcal{P}$ containing every edge an even number of times. Then there is at least one edge that does not occur at all in the walks in $\mathcal{P}'$. Since Y is connected, we can find a face of Y containing two consecutive edges e and f, where e does not occur in the walks in $\mathcal{P}'$ but f occurs twice. This, however, is impossible, because one of the left–right cycles through f must use the neighbouring edge e. □

An immediate consequence of this result is that if any left–right cycle has an empty core, then it contains every edge an even number of times, and so is the only left–right cycle in Y.

Lemma 17.3.3 *Let Y be a connected plane graph with exactly c left–right cycles. Then the subspace of $GF(2)^{E(Y)}$ spanned by the characteristic vectors of the cores has dimension $c-1$.*

Proof. Let $\mathcal{P} = \{P_1, \ldots, P_c\}$ be the set of left–right cycles in Y, and let $\mathcal{Q} = \{Q_1, \ldots, Q_c\}$ be the corresponding cores. Identify a core with its characteristic vector, and suppose that there is some linear combination of the cores equal to the zero vector. Let $\mathcal{Q}'$ be the set of cores with nonzero coefficients in this linear combination, and let $\mathcal{P}'$ be the corresponding left–right cycles. Then $\mathcal{Q}'$ covers every edge an even number of times, and so $\mathcal{P}'$ covers every edge an even number of times. By the previous result $\mathcal{P}'$ is either empty or $\mathcal{P}' = \mathcal{P}$. Therefore, there is a unique nontrivial linear combination of the cores equal to the zero vector, and so the subspace they span has dimension $c-1$. □

We now consider a partition of the edges of a plane graph Y into three types. An edge is called a *crossing* edge if some left–right cycle of Y uses it only once. Any edge e that is not a crossing edge is used twice by some left–right cycle P. If P uses it twice in the same direction, we say that e is a *parallel* edge, and otherwise it is a *skew* edge. Any coloop in Y is a skew edge, any loop is a parallel edge, and any edge is parallel in Y if and only if it is skew in Y^*. We let $c(Y)$ denote the number of left–right cycles in a graph Y.

Lemma 17.3.4 *If Y is a plane graph, and e is an edge of Y, then*

(a) *If e is a loop or coloop, then $c(Y\setminus e) = c(Y/e) = c(Y)$.*
(b) *If e is a parallel edge, then $c(Y\setminus e) = c(Y) = c(Y/e) - 1$.*
(c) *If e is a skew edge, then $c(Y/e) = c(Y) = c(Y\setminus e) - 1$.*
(d) *If e is a crossing edge, then $c(Y\setminus e) = c(Y/e) = c(Y) - 1$.*

Proof. We leave (a) as an exercise.

If $e = uv$ is a parallel edge, then we can assume that the left–right cycle containing e has the form $P = (u, e, v, S, u, e, v, T)$. The graph $Y\setminus e$ contains all the left–right cycles of Y except for P, and in addition the left–right cycle (u, S', v, T) where S' is the reverse of S. The graph Y/e has also lost P, but it has gained two new left–right cycles, namely (x, S) and (x, T), where x is the vertex that resulted from merging u and v.

If e is a skew edge, then it is parallel in Y^*, and the result follows directly from (b).

Finally, if e is a crossing edge, then the two left–right cycles through e are merged into a single walk in both Y/e and $Y\setminus e$. Therefore, both graphs have one fewer left–right cycle than Y. □

Theorem 17.3.5 *If Y is a connected plane graph with exactly c left–right cycles, then the bicycle space has dimension $c - 1$ and is spanned by the cores of Y.*

Proof. We prove this by induction on the number of left–right cycles in Y. First we consider the case where Y has a single left–right cycle. Assume by way of contradiction that Y has a nonempty bicycle B. Then B is both an even subgraph and an edge cutset whose shores we denote by L and R. Since B is even, it divides the plane into regions that can be coloured black and white so that every edge in B has a black side and a white side, and every other edge of Y has two sides of the same colour. Consider a left–right cycle starting from a vertex in L on the black side of an edge in B. After it uses this edge, it is on the white side of the edge at a vertex in R. Every time the walk returns to L it uses an edge in B, and therefore returns to the black side of that edge. Therefore, the edges in B are used at most once by this walk, contradicting the assumption that Y has only one left–right cycle.

Now, let Y be a plane graph with $c > 1$ left–right cycles. The dimension of its bicycle space is at least $c-1$ by Lemma 17.3.3. If e is a crossing edge of Y, then $Y\setminus e$ has $c-1$ left–right cycles and by the inductive hypothesis, a bicycle space of dimension $c - 2$. Since deleting an edge cannot reduce the dimension of the bicycle space by more than one, the bicycle space of Y has dimension at most $c - 1$, and the result follows. □

Corollary 17.3.6 *If a link has c components, then the face graphs of any link diagram L have a bicycle space of dimension $c - 1$. In particular, the face graphs of a knot diagram are pedestrian.* □

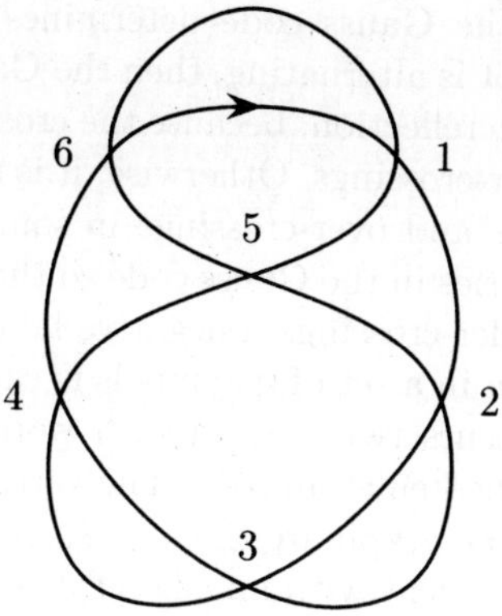

Figure 17.5. Crossing eulerian tour of a knot shadow

In addition, the results of this section imply that the partition of the edges of a plane graph into crossing edges, parallel edges, and skew edges is actually the principal tripartition of Section 14.16 in disguise. The crossing edges are the bicycle edges, the parallel edges are the flow-type edges, and the skew edges are the cut-type edges. We finish with one application of this observation to knot diagrams.

Theorem 17.3.7 *Let Y be the signed face graph of a knot diagram K. Let ℓ be the number of negative cut-type edges plus the number of positive flow-type edges of Y, and let r be the number of positive cut-type edges plus the number of negative flow-type edges of Y. Then the writhe of K is equal to $\ell - r$.* □

17.4 Gauss Codes

We have seen that the face graphs of a link diagram provide a way to represent the link, and also yield useful information about the link itself. However, as representations of link diagrams, face graphs are not particularly simple. In dealing with knots there is a useful alternative representation that goes back to the work of Gauss.

Consider a knot diagram where the crossings are arbitrarily labelled. The shadow of the knot diagram has a unique straight eulerian tour, which can be recorded simply by writing down the crossing points in the order in which they are visited by the tour. The resulting string of symbols is called the *Gauss code* of the shadow, and unsurprisingly, it is determined only up to rotation and reversal. The shadow of the knot diagram shown in Figure 17.5 has Gauss code 1 2 3 4 5 1 6 5 2 3 4 6.

We emphasize that the Gauss code determines only the shadow of a knot diagram. If the knot is alternating, then the Gauss code is sufficient to determine the knot up to reflection, because the crossings alternate between under-crossings and over-crossings. Otherwise, it is necessary to distinguish between under-crossings and over-crossings in some fashion. One method would be to sign the entries in the Gauss code so that over-crossings receive a positive label and under-crossings a negative label.

A word ω of length $2n$ in a set of n symbols is called a *double occurrence word* if each symbol occurs twice. Each such word determines a 4-valent graph with a distinguished eulerian tour. The vertices of the graph are the symbols. The edges of the graph are the subsequences of length two in the word, and a vertex is incident with an edge if it is contained in it. (So our graph is undirected even though each edge is associated with an ordered pair of symbols.) It is immediate that the word itself describes an eulerian tour. Conversely, an eulerian tour in a 4-valent graph determines a double occurrence word on the vertices of the graph.

Our ultimate aim is to characterize the double occurrence words that arise as the straight eulerian tour of the shadow of a knot diagram. For the moment we note two useful properties of such words.

Lemma 17.4.1 *If ω is the Gauss code of the shadow of a knot diagram, then*

(a) *Each symbol occurs twice in ω.*

(b) *The two occurrences of each symbol are separated by an even number of other symbols.*

Proof. The Gauss code gives the unique straight eulerian tour of the knot shadow, and since an eulerian tour of a 4-valent graph visits each vertex twice, the first part of the claim follows immediately.

The two occurrences of the symbol v partition the Gauss code into two sections. The edges of the shadow of the knot diagram determined by one of these sections form a closed curve C in the plane starting and ending at the crossing labelled v. The edges determined by the other section form another closed curve starting and ending at v, and the first and last edges of this curve are both inside or both outside C. Therefore, by the Jordan curve theorem these two curves intersect in an even number of points other than v. □

A double occurrence word is called *even* if the two occurrences of each symbol are separated by an even number of other symbols. If X is a 4-valent graph embedded in any surface, then it has a unique straight eulerian partition. If this has only one component, then it is a double occurrence word. Our next result indicates when a double occurrence word obtained in this manner is even.

An embedding of a graph in a surface is called a 2-*cell embedding* if every face is homeomorphic to an open disk; for the remainder of this chapter we will assume that all embeddings are 2-cell embeddings. If X is a graph embedded in a surface, then we *orient* a cycle of X by giving a cyclic ordering of its vertices. If the surface is the sphere, for example, we can orient each face of X clockwise about a nonzero vector pointing outwards from the centre of the face. By orienting a cycle we also give an orientation to each edge in it. The orientations of two cycles are *compatible* if no edge in their union occurs twice with the same orientation. We say that an embedding of X in a surface is *orientable* if there is an orientation of the faces of X such that every pair of faces is compatibly oriented. Finally, a surface is orientable if there is an orientable embedding of some graph in it. (It can be shown that if one graph has an orientable embedding in a surface, then all embeddings are orientable. Thus orientability is a global property of a surface.)

Theorem 17.4.2 *Let X be a 4-valent graph embedded in a surface and suppose ω is the double occurrence word corresponding to a straight eulerian tour of X. If the surface is orientable and X^* is bipartite, then ω is even.*

Proof. Assume that the surface is orientable. Since X^* is bipartite, we may partition the faces of X into two classes, so that two faces in the same class do not have an edge in common. Choose one of these classes and call the faces in it white; call the other faces black. The straight eulerian tour given by ω has the property that as we go along it, the faces on the left alternate in colour. Since the embedding is orientable, we can orient the white faces. This gives an orientation of the edges of X so that at each vertex there are two edges pointing in and two pointing out. This orientation has the further property that any two consecutive edges in ω point in opposite directions, which implies that ω is even. □

This result shows that merely being even is not sufficient to guarantee that a double occurrence word is the Gauss code of the shadow of a knot diagram. We provide a complete characterization of Gauss codes in Section 17.7.

17.5 Chords and Circles

We will use a pictorial representation of a double occurrence word, known as a *chord diagram*. The chord diagram of ω is obtained by arranging the $2n$ symbols of the word on the circumference of a circle and then joining the two occurrences of each symbol by a chord of the circle. Figure 17.6 shows the chord diagram corresponding to the Gauss code given above.

We may also view a chord diagram as a graph in its own right. It is a cubic graph with a distinguished perfect matching (the chords); the remaining edges form a hamilton cycle called the *rim* of the chord diagram. If we write

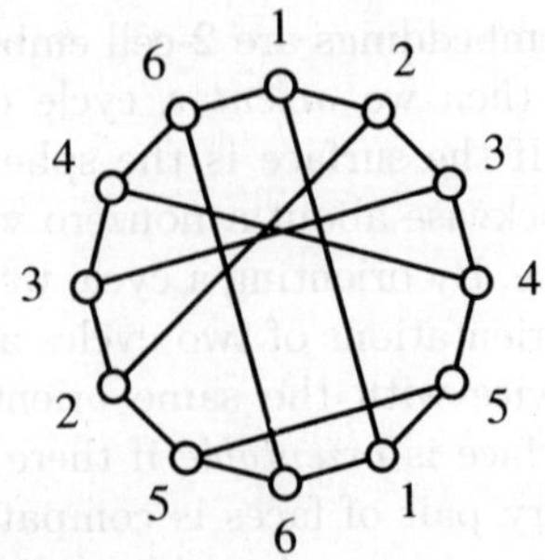

Figure 17.6. Chord diagram of a double occurrence word

down the sequence of vertices in the rim in order, we recover the double occurrence word.

If we contract each chord in a chord graph, we get a 4-valent graph, and the image of the rim in this is the eulerian tour from which the chord graph was constructed. Also, we can represent a signed Gauss code by orienting each chord in its chord diagram. For example, a chord could always be oriented from the over-crossing to the under-crossing.

A *circle graph* is a graph whose vertices are the chords of a circle, where two vertices are adjacent if the corresponding chords intersect. The chord diagram of any double occurrence word therefore immediately yields a circle graph, as shown in Figure 17.7. Two double occurrence words related by rotation and/or reversal determine the same circle graph. See Exercise 8 for an example indicating that the converse of this statement is not true.

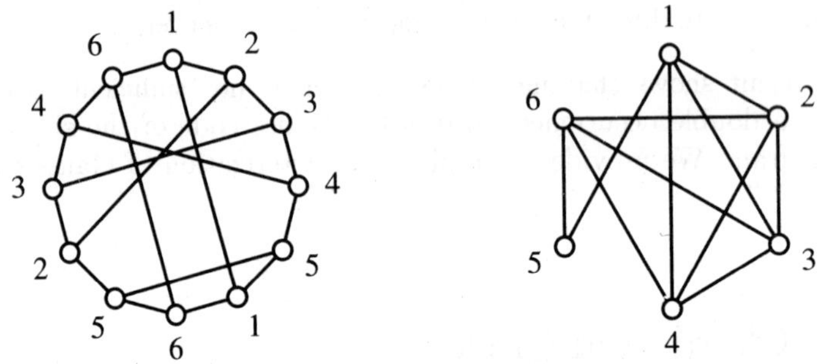

Figure 17.7. Chord diagram and associated circle graph

Lemma 17.5.1 *A chord diagram is a planar graph if and only if its circle graph is bipartite.*

Proof. Let Z be a chord diagram with a bipartite circle graph. We can map the rim of Z onto a circle in the plane. The chords in one colour class of the circle graph can be embedded inside the circle without intersecting, and the chords in the other colour class similarly embedded outside the circle.

Conversely, suppose we have an embedding of a chord diagram in the plane. The rim of the chord diagram is a continuous closed curve in the plane, which divides the plane into two disjoint parts: inside and outside. Thus there are two classes of chords, those embedded on the inside and those on the outside, and no two chords in the same class are adjacent in the circle graph. Hence the circle graph of Z is bipartite. □

17.6 Flipping Words

If ω is a double occurrence word, then we *flip* it at the symbol v by reversing the subsequence between the two occurrences of v. You might object that there are two such sequences, since double occurrence words are cyclically ordered. But since we do not distinguish between ω and its reversal, both choices yield the same result. If ω corresponds to an eulerian tour in a graph X, then the word we get by flipping at v corresponds to the eulerian tour we get by flipping at the vertex v in the sense of Section 17.1. Thus it is the unique eulerian tour that differs from the original eulerian tour at v, and agrees with it everywhere else.

We can also interpret a flip in terms of the associated chord diagram and circle graph. In the chord diagram, flipping at the chord joining the two occurrences of v can be viewed as cutting out the section of the rim between the two occurrences of v, flipping it over (with any other chords still attached), and then replacing it in the rim (see Figure 17.8). The rim of the flipped chord diagram is the flipped double occurrence word.

In the circle graph, flipping at the vertex v has the effect of replacing all the edges in the neighbourhood of v with nonedges, and all the nonedges in the neighbourhood of v with edges. Recall that this operation is known as local complementation at v (see Figure 17.9).

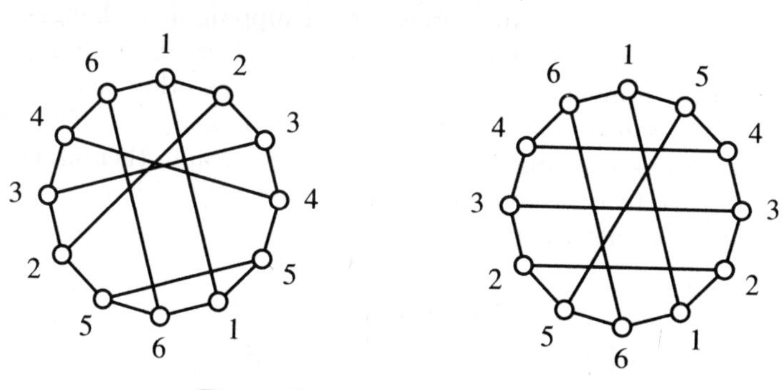

Figure 17.8. Fiipping on chord 1

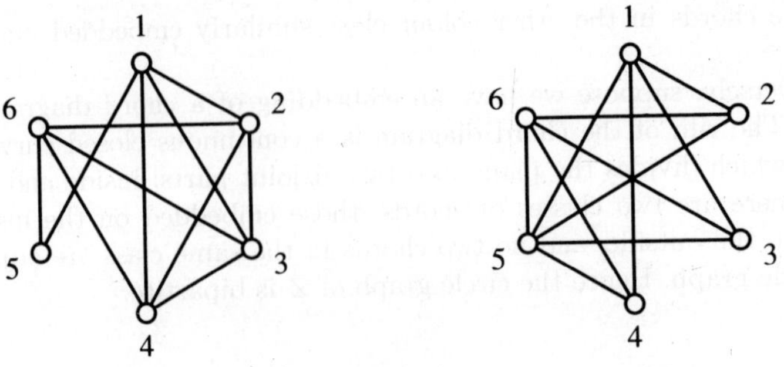

Figure 17.9. Local complementation at vertex 1

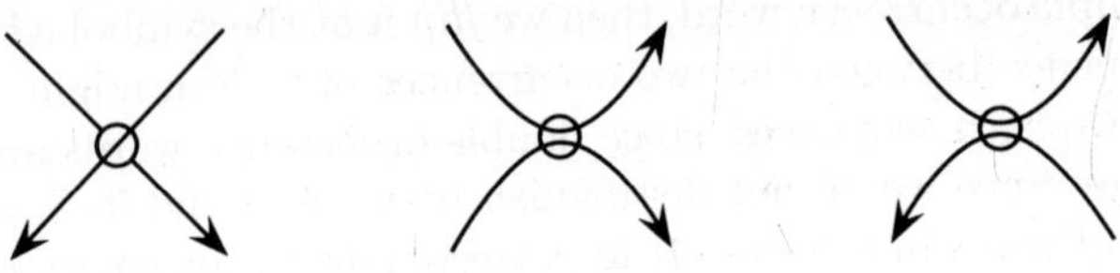

Figure 17.10. Overpass, one-way, and two-way vertices

The vertices in an arbitrary eulerian tour of a 4-valent plane graph may be classified into three types: *overpass*, *one-way*, and *two-way*. A vertex v is an overpass if the eulerian tour enters and leaves v through opposite edges. A vertex is one-way if the two portions of the eulerian tour are oriented in the same direction as they pass through v, and two-way if they are oriented in opposite directions as they pass through v. Figure 17.10 should help in interpreting this.

Flipping an eulerian tour at a vertex alters the types of the vertices in a predictable manner. Suppose that ω is the double occurrence word describing an eulerian tour of a 4-valent plane graph. We consider the effect of flipping at v, first on the type of v, and then on the types of the remaining vertices. If v is an overpass, then flipping at v changes it to a two-way vertex, and vice versa. If v is a one-way vertex, then its type is not changed by flipping at v. The type of any vertex other than v changes only if it is adjacent to v in the circle graph determined by ω. If w is a one-way or two-way vertex adjacent to v in the circle graph, then flipping at v changes its type to two-way or one-way, respectively. If it is an overpass, then its type does not change.

17.7 Characterizing Gauss Codes

We now have the necessary tools to characterize which double occurrence words are Gauss codes. Any double occurrence word describes an eulerian

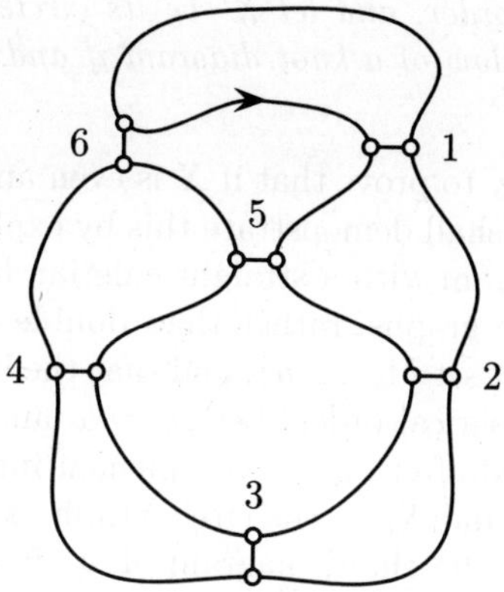

Figure 17.11. Plane chord diagram of a bent eulerian tour

tour in some 4-valent graph (possibly with multiple edges and loops). If the 4-valent graph is not the shadow of a knot diagram or the eulerian tour is not the straight eulerian tour, then the double occurrence word is not a Gauss code. Lemma 17.4.1 shows that a Gauss code is necessarily even, but this condition is not sufficient; any attempt to reconstruct a shadow with Gauss code 1 2 3 4 5 3 4 1 2 5 will not succeed. We describe another property of Gauss codes, which will enable us to provide a characterization.

If we flip on each symbol of a Gauss code in some order, the double occurrence word that results corresponds to an eulerian tour with no overpasses, and so it is a bent tour. Suppose that we draw this bent tour on the shadow of the knot diagram, split each vertex into two vertices a short distance apart, and join each pair with a short edge. The resulting graph is clearly a plane graph. The bent tour does not cross itself, and the newly introduced edges are so short that no problems arise. Moreover, this plane graph is actually the chord diagram of the bent eulerian tour—just embedded in an unusual manner.

For example, if we start with the word 1 2 3 4 5 1 6 5 2 3 4 6, associated with the knot of Figure 17.5, then flipping at the symbols 1 through 6 in order results in the double occurrence word 1 5 2 3 4 5 6 4 3 2 1 6. This is a bent tour of the shadow pictured in Figure 17.5, and its chord diagram is shown in Figure 17.11.

Recall that a chord diagram is planar if and only if its associated circle graph is bipartite. We conclude that if ω is a Gauss code and ω' is the double occurrence word obtained from ω by flipping once on each symbol, then the circle graph of ω is even and the circle graph of ω' is bipartite. The surprise is that this easily checked condition is sufficient to characterize Gauss codes.

Theorem 17.7.1 *Let ω be a double occurrence word with circle graph X. Let ω' be the double occurrence word obtained by flipping once at each* *symbol of ω in some order, and let X' be its circle graph. Then ω is the Gauss code of the shadow of a knot diagram if and only if X is even and X' is bipartite.*

Proof. It remains only to prove that if X is even and X' is bipartite, then ω is a Gauss code. We shall demonstrate this by explicitly constructing the shadow of a knot diagram with a straight eulerian tour given by ω.

We work with circle graphs, rather than double occurrence words. Assume that X has vertex set $\{1, \ldots, n\}$, and that the local complementations are carried out in numerical order. Let $\omega_0 = \omega$ and ω_i be the double occurrence word obtained after locally complementing at vertices $\{1, \ldots, i\}$. Similarly, let $X_0 = X$ and X_i be the circle graph associated with ω_i. Since $X' = X_n$ is bipartite, the chord diagram of ω_n has a planar embedding, and by shrinking the chords we can find a plane 4-valent graph with a bent eulerian tour C_n given by ω_n. By considering what happens to the rim of a chord diagram as each chord is shrunk to a vertex, it is easy to see that every vertex is two-way in C_n.

Now, we aim to construct a series of eulerian tours C_n, C_{n-1}, ..., C_0, where C_i has double occurrence word ω_i, so that the final eulerian tour is straight. The first condition is easy to satisfy: Starting with C_n we merely need to flip on the vertices n to 1 in reverse numerical order to get the eulerian tours C_{n-1} to C_0. The only remaining question is whether this necessarily creates a straight eulerian tour. Since each flip can create at most one overpass, C_0 is a straight eulerian tour if and only if every one of the n flips creates one overpass each. This occurs if and only if the vertex i is two-way in C_i, for all i.

Since every vertex in C_n is two-way, the type of i in C_i depends on whether its type has been changed an even or odd number of times as a consequence of the flips at vertices $\{n, \ldots, i+1\}$. Now, X is even, and an easy induction argument shows that for all i, the subgraph of X_i induced by $\{i+1, \ldots, n\}$ is also even. Therefore, i is adjacent to an even number of vertices of $\{i+1, \ldots, n\}$ in X_i.

Now, we claim that for all $j \geq i$, either i is two-way in C_j and is adjacent to an even number of vertices from $\{j+1, \ldots, n\}$ in X_j or i is one-way in C_j and is adjacent to an odd number of vertices from $\{j+1, \ldots, n\}$ in X_j. This is true when $j = n$, and since i changes type when j is flipped if and only if $i \sim j$ in X_j, it is true for $j = n-1$, ..., $j = i$. Since we know that i is adjacent to an even number of vertices from $\{i+1, \ldots, n\}$ in X_i, we conclude that i is two-way in C_i, and the result follows. □

17.8 Bent Tours and Spanning Trees

In this section we consider some more properties of eulerian tours of 4-valent plane graphs. In particular, we determine a relationship between the bent eulerian tours of a 4-valent plane graph and the spanning trees of its face graphs.

Suppose that X is a 4-valent plane graph and that its faces are coloured black and white. There are three partitions of the four edges at a vertex $v \in V(X)$ into two pairs of edges. We will call the partition at a vertex *straight* if the two cells are pairs of opposite edges, *black* if the two edges in each cell lie in different black faces, and *white* if the two edges in each cell lie in different white faces (see Figure 17.12).

Figure 17.12. Straight, black, and white partitions at a vertex

A bent eulerian partition of X induces a partition at each vertex that is either black or white. Conversely, by specifying at each vertex whether the black or white partition is to be used, we determine a bent eulerian partition. (Thus if X has n vertices, then it has 2^n bent eulerian partitions.)

Lemma 17.8.1 *Let X be a 4-valent plane graph with face graph Y. Then there is a bijection between the bent eulerian tours of X and the spanning trees of Y.*

Proof. Suppose that Y is the graph on the white faces, and that Y^* is the graph on the black faces. Let T be a spanning tree of Y, and then $\overline{T} = E(Y) \setminus T$ is a spanning tree of Y^*. We can identify $E(Y)$ and $E(Y^*)$ with $V(X)$, and so $T \cup \overline{T}$ is a partition of $V(X)$. Now, define an eulerian partition of X by taking the white partition at every vertex in T and the black partition at every vertex in $\overline{T}$. (That is, take the white partition for every edge in the white spanning tree, and the black partition for every edge in the black spanning tree.) From our earlier remarks this is a bent eulerian partition, so it remains to show that it has just one component.

Consider an adjacency relation on the faces of X, with two white faces being adjacent if they meet at a vertex where the induced partition is white, and two black faces being adjacent if they meet at a vertex where the induced partition is black. It follows immediately that this adjacency relation is described by T and $\overline{T}$, and so has two components, being all the white faces and all the black faces. An eulerian cycle of X determines a closed curve C in the plane, and so has an inside and an outside. The adjacency relation cannot connect a face inside C to a face outside C. If the bent eulerian partition has more than one component, then there are either two white or two black faces on opposite sides of an eulerian cycle, which is a contradiction.

Conversely, given a bent eulerian tour, partition the vertices into two sets T and $\overline{T}$, according to whether the bent eulerian partition induces the white partition or the black partition, respectively, at that vertex. Then T is a spanning tree of Y, and $\overline{T}$ is a spanning tree of Y^*. □

Figure 17.13 shows a 4-valent plane graph, together with a spanning tree on the white faces, and Figure 17.14 shows the corresponding bent eulerian tour.

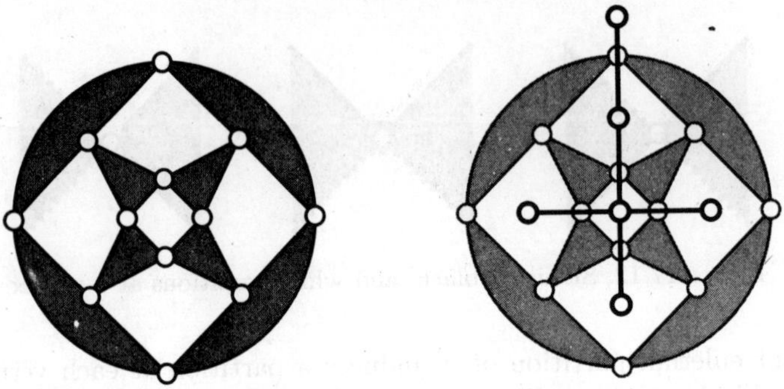

Figure 17.13. A 4-valent plane graph and spanning tree on the white faces

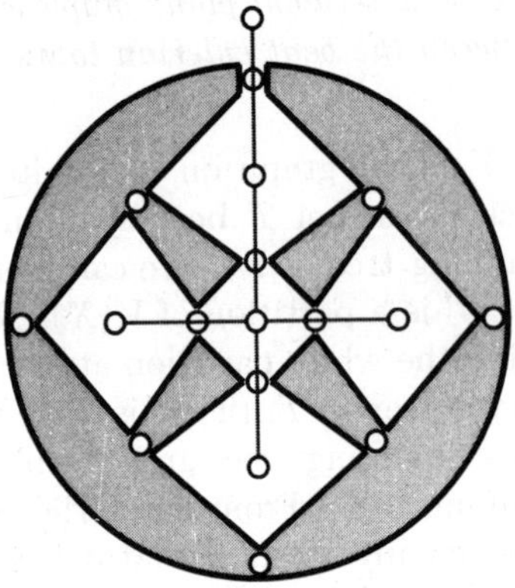

Figure 17.14. Spanning tree and associated bent eulerian tour

In Section 17.7 we saw that flipping once on each vertex of the straight eulerian tour of X yields a bent eulerian tour. The next result shows that every bent eulerian tour of X arises in this fashion.

Lemma 17.8.2 *Suppose that X is a 4-valent plane graph with a straight eulerian tour $\mathcal{S}$. For every bent eulerian tour $\mathcal{T}$ of X, there is a sequence σ containing each vertex of X once such that $\mathcal{T}$ is obtained from $\mathcal{S}$ by flipping on the elements of σ in turn.*

Proof. See Exercise 7. □

Now, suppose that Y is a plane graph, and that X is its medial graph. Any spanning tree of Y determines a bent eulerian tour of X, which determines a planar chord diagram, which in turn has a bipartite circle graph. Therefore, any plane graph Y determines a collection of bipartite circle graphs.

The adjacency matrix of any bipartite graph can be written in the form

$$A = \begin{pmatrix} 0 & M^T \\ M & 0 \end{pmatrix}.$$

However, when A is the adjacency matrix of a bipartite circle graph arising from a spanning tree of a plane graph, then it can be described in terms of the spanning tree T. We declare that the rows of M are indexed by the edges of Y in T and the columns of M are indexed by the edges of Y not in T. Every edge f of Y that is not in T determines a unique cycle in $T \cup \{f\}$. We claim that if $e \in T$ and $f \notin T$, then the ef-entry of M is 1 if and only if e lies on the unique cycle in $T \cup \{f\}$.

This implies that the rows of the matrix

$$\begin{pmatrix} I & M^T \end{pmatrix}$$

are a basis for the flow space of Y (over $GF(2)$). We described this basis earlier in Section 14.2; from our work there it follows that the rows of

$$\begin{pmatrix} M & I \end{pmatrix}$$

are a basis for the cut space. It follows that the kernel of $A + I$ is the intersection of the cut space and flow space of Y. In other words, it is the bicycle space of Y. Therefore, the rank of $A + I$ depends only on the dimension of the bicycle space of Y, and in particular $A + I$ is invertible over $GF(2)$ if and only if Y is pedestrian.

17.9 Bent Partitions and the Rank Polynomial

In the last section we saw that the number of bent eulerian tours in a 4-valent plane graph is equal to the number of spanning trees of one of its face graphs. The number of spanning trees is an evaluation of the rank polynomial, and a bent eulerian tour is a bent eulerian partition with one component. Our next task is to show that the rank polynomial actually counts the bent eulerian partitions with any number of components.

Lemma 17.9.1 *Let X be a 4-valent plane graph with face graphs Y and Y^*, and let T be a bent eulerian partition of X with c components. Then T determines a set of edges S in Y and the complementary set of edges $\overline{S}$ in Y^*. If c_w is the number of components in the subgraph of Y with edge* *set S, and c_b the number of components in the subgraph of Y^* with edge set $\overline{S}$, then*

$$c = c_b + c_w - 1.$$

Proof. The eulerian cycles of the bent eulerian partition $\mathcal{T}$ form a collection of disjoint closed cycles in the plane, and hence divide the plane into $c+1$ regions. Every face of X lies completely within one of these regions, and each region contains faces of only one colour. Two white faces of X lie in the same region if and only if there is a path in Y connecting them using only edges of S, and two black faces lie in the same region if and only if there is a path joining them in Y^* using only edges of $\overline{S}$. Therefore, the partition into regions is the same as the partition into components of Y and Y^*. Hence $c+1 = c_b + c_w$, and the result follows. □

Corollary 17.9.2 *Let X be a 4-valent plane graph with face graph Y. Let $R(Y; x, y)$ be the rank polynomial of Y. Then the number of bent eulerian partitions of X with c components is the coefficient of x^{c-1} in $R(Y; x, x)$.*

Proof. Let E denote the edge set of Y, and identify it with the edge set of Y^*. By the definition of the rank polynomial we see that

$$R(Y; x, x) = \sum_{S \subseteq E} x^{\text{rk}(E)-\text{rk}(S)} x^{\text{rk}^\perp(E)-\text{rk}^\perp(E \setminus S)}.$$

From the results of Section 15.2, the expression $\text{rk}(E) - \text{rk}(S)$ is equal to the number of components of the subgraph of Y with edge set S, and $\text{rk}^\perp(E) - \text{rk}^\perp(E \setminus S)$ is the number of components of the subgraph of Y^* with edge set $\overline{S}$. □

17.10 Maps

In previous sections we have been content to use an intuitive topologically based notion of the embedding of a graph in a surface. We are going to present a purely combinatorial approach, which describes maps by ordered triples of suitable permutations.

Before giving the formal definition, we present an extended example showing how our usual concept of a graph embedded in a surface leads to three permutations. We will use the Cartesian product $K_3 \,\square\, K_3$ embedded in the torus as shown in Figure 17.15.

Define a *flag* to be an ordered triple consisting of a vertex, an edge, and a face such that the vertex is contained in the edge, which in turn is contained in the face. Let Φ denote the set of flags in our embedding of $K_3 \,\square\, K_3$. If $(v, e, f) \in \Phi$, then there are unique flags (v', e, f), (v, e', f), and (v, e, f') that each differ from (v, e, f) in only one element. We define three

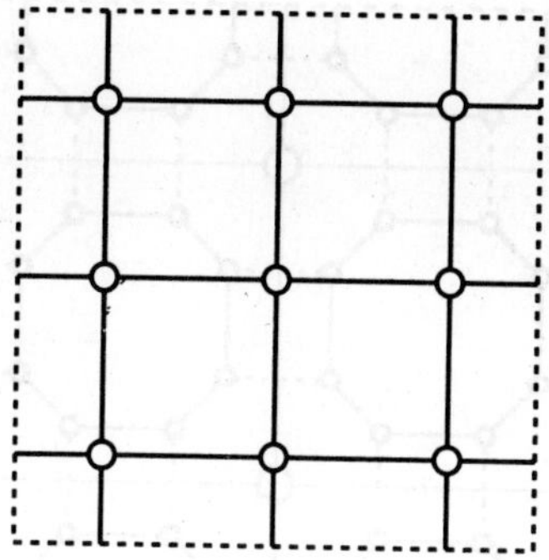

Figure 17.15. $K_3 \,\square\, K_3$ embedded in the torus

involutions τ_0, τ_1, and τ_2 on Φ as follows:

$$\begin{aligned}\tau_0 &: (v, e, f) \leftrightarrow (v', e, f),\\ \tau_1 &: (v, e, f) \leftrightarrow (v, e', f),\\ \tau_2 &: (v, e, f) \leftrightarrow (v, e, f').\end{aligned}$$

We can visualize these permutations on a drawing of the graph by representing each flag (v, e, f) as a small vertex near v, towards the end of e and inside f. If two flags are exchanged by τ_0, τ_1, or τ_2, then join the corresponding vertices with an edge of colour 0, 1, or 2, respectively. Figure 17.16 shows such a representation for our example.

Now, we take a more formal stance, and give a combinatorial definition of a map purely in terms of the action of three permutations. Let Φ be an arbitrary set whose elements we call flags, and let (τ_0, τ_1, τ_2) be an ordered triple of permutations of Φ. We say that $M = (\tau_0, \tau_1, \tau_2)$ is a *map* if:

I. τ_0, τ_1, and τ_2 are fixed-point-free involutions,

II. $\tau_0\tau_2 = \tau_2\tau_0$, and $\tau_0\tau_2$ is fixed-point free,

III. $\langle \tau_0, \tau_1, \tau_2 \rangle$ is transitive.

The *vertices*, *edges*, and *faces* of the map M are defined to be the respective orbits on Φ of the subgroups

$$\langle \tau_1, \tau_2 \rangle, \quad \langle \tau_0, \tau_2 \rangle, \quad \langle \tau_0, \tau_1 \rangle.$$

A vertex is incident with an edge or a face if the corresponding orbits have a flag in common; similarly, an edge and face are incident if they have a flag in common.

Given a map M, we can also define an edge-coloured cubic graph Z with vertex set Φ, where two vertices are joined by an edge of colour 0, 1 or 2 if they are exchanged by τ_0, τ_1, or τ_2, respectively. The edges of a given colour form a perfect matching in Z, and so the union of the edges of two colours is a 2-regular subgraph whose components are even cycles. If the two colours

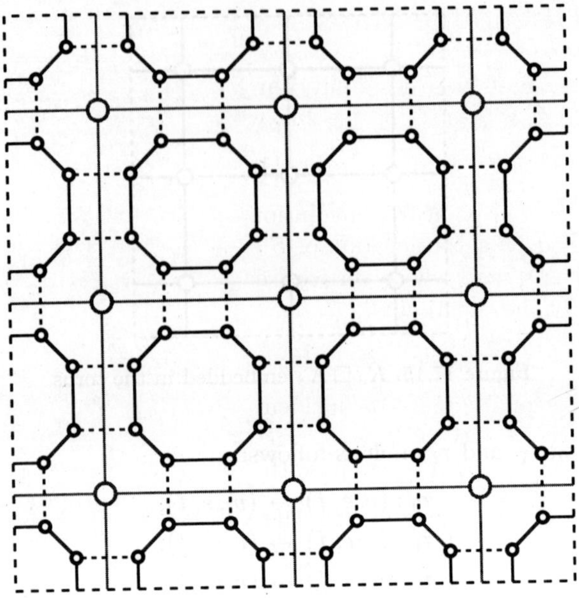

Figure 17.16. The three permutations as an edge-coloured graph

concerned are i and j, then these components are called ij-cycles. We can interpret the axioms above in terms of Z. For example, Axiom II above is equivalent to the requirement that each 02-cycle should have length four, and Axiom III is equivalent to requiring that Z be connected. It would be quite reasonable to use these conditions as the definition of a map. We define M to be *orientable* if Z is bipartite.

The *dual* of the map $M = (\tau_0, \tau_1, \tau_2)$ is the map specified by the triple (τ_2, τ_1, τ_0); it is easy to see that this coincides with our intuitive notion of the dual of a graph embedded in a surface.

The *medial graph* of M has the edges of M as its vertices and has the orbits of τ_1 as its edges. An orbit of τ_1 contains two flags, and so joins the edges of M that contain those flags. (Each edge is an orbit of $\langle\tau_0, \tau_2\rangle$ and consists of four flags.)

If $M = (\tau_0, \tau_1, \tau_2)$ is a map, then

$$(\tau_0\tau_2, \tau_1, \tau_2)$$

is also a map, called the *Petrie map* of M. The Petrie map of M has the same vertices and edges as M, but its faces are the orbits of the subgroup $\langle\tau_0\tau_2, \tau_1\rangle$, which are the left–right cycles of M. Since the left–right cycles are the faces of the Petrie map, it follows that they cover each edge of the map exactly twice, but no proper subset covers each edge an even number of times. Thus we have a generalization of Lemma 17.3.2.

17.11 Orientable Maps

For orientable maps, we can simplify our machinery: Only two permutations are needed. Let S be a set of size $2m$, with elements

$$\{1, 2, \ldots, m\} \cup \{1', 2', \ldots, m'\}.$$

Let θ be the fixed-point-free involution of S that exchanges i and i' for all i, and let σ be a permutation of S such that $\langle \sigma, \theta \rangle$ is transitive. Then $M = (\sigma, \theta)$ is a map whose vertices, edges, and faces are the respective orbits on S of the subgroups

$$\langle \sigma \rangle, \quad \langle \theta \rangle, \quad \langle \sigma\theta \rangle.$$

Two elements of the map are incident if they have an element of S in common.

We consider an example. Suppose that σ and θ are given by

$$\sigma = (1, 2, 3, 4)(1', 3', 4', 2'),$$
$$\theta = (1, 1')(2, 2')(3, 3')(4, 4').$$

Since the group $\langle \sigma, \theta \rangle$ is transitive, these permutations determine an orientable map M, whose faces are the orbits of

$$\sigma\theta = (1, 2')(1', 3, 4', 2, 3', 4).$$

If n, e, and f are the numbers of vertices, edges, and faces of M, then we see that $n = 2$, $e = 4$, and $f = 2$. Putting these values into the left-hand side of Euler's formula, we see that

$$n - e + f = 0,$$

and so this is not a map on the plane. However, Figure 17.17 shows this map on the torus. In general, the value $n - e + f$ determines the surface in which an orientable map can be embedded.

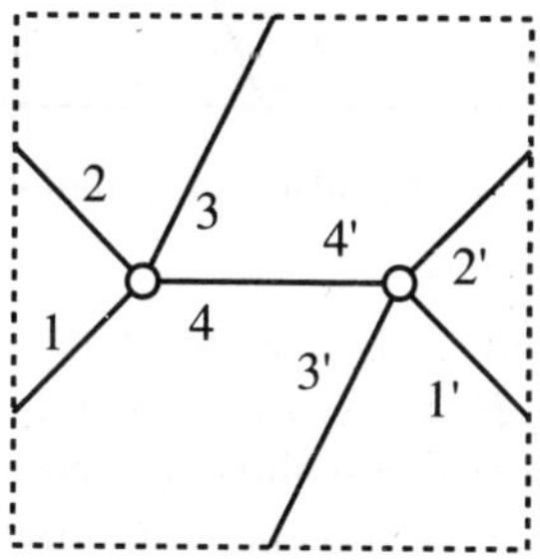

Figure 17.17. A map on the torus

It is not too difficult to translate between the two definitions of an orientable map. Let $M = (\tau_0, \tau_1, \tau_2)$ be an orientable map according to the definition given using three permutations. Then the edge 3-coloured cubic graph associated with M is bipartite, and we can define S to be one of the two colour classes. If we take

$$\sigma := \tau_1\tau_2, \qquad \theta := \tau_0\tau_2,$$

then it is not hard to see that σ and θ satisfy the requirements for a map under the second definition. Moreover, the vertices, edges, and faces of this map are the intersections of the vertices, edges, and faces of M with S. It is a worthwhile exercise to show that any map according to this second definition is a map according to the first definition.

There is also a natural way to represent an orientable map in terms of a cubic graph. If M is an orientable map, then we can form a cubic graph Z by *truncating* the vertices of the map (in the same sense as in Section 6.14). Then the vertices of Z can be identified with the elements of S. The distinguished perfect matching of Z consisting of the original edges of the map is the permutation θ. The remaining edges of Z form a collection of disjoint cycles, and if these are oriented consistently (all clockwise or all anticlockwise), then they form the cycles of σ. The embedding of the map obviously leads directly to an embedding of Z in the same surface. Figure 17.18 shows this for the map of Figure 17.17.

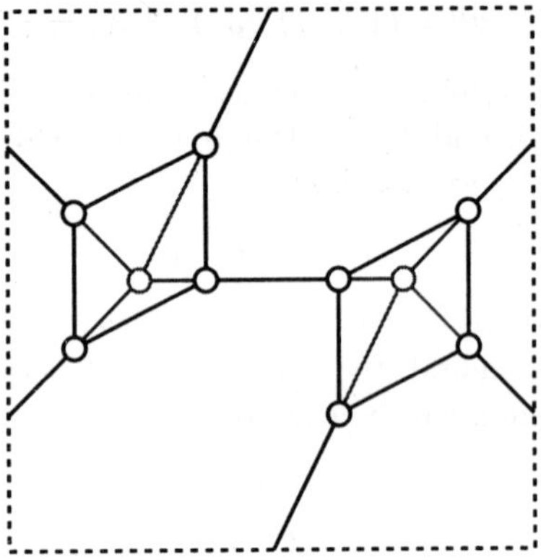

Figure 17.18. Truncation of a map on the torus

Conversely, suppose that Z is a connected cubic graph with a distinguished perfect matching F. Label the two ends of the ith edge of F with i and i', and set θ to be the involution exchanging all such pairs. The edges not in this matching form a disjoint union of cycles, and if each of these is oriented, then they determine a permutation σ of $V(Z)$. Since Z is connected, the group generated by these permutations is transitive, and so (σ, θ) is an orientable map. The faces of the map associated with Z can easily be identified in Z itself. Define an *alternating cycle* to be an oriented cycle in Z that alternately uses edges from $Z \setminus F$ in the direction in which they are oriented, and edges from the perfect matching F. There is a bijection between the faces of the map (σ, θ) and the alternating cycles of Z.

Note that we can form a cubic graph by truncating a map on any surface, but the map itself can be reconstructed from the graph, the perfect matching, and the orientation of the cycles only if the surface is orientable.

17.12 Seifert Circles

In this section we give a connection between an important invariant of an oriented link and the faces of an associated map.

Before we start this, however, we need to extend our terminology to account for the orientation of the link. Define an *oriented eulerian cycle* to be an equivalence class of closed eulerian walks under rotation only. Thus we are now viewing an eulerian cycle and its reversal as different. If X^ρ is an oriented graph, then an *oriented eulerian partition* of X^ρ is a collection of oriented eulerian cycles such that each edge of X occurs—correctly oriented—in exactly one of the oriented eulerian cycles.

Lemma 17.12.1 *If X is a 4-valent plane graph and ρ is an orientation such that every vertex is the head of two edges and the tail of two edges, then X^ρ has a unique straight oriented eulerian partition $\mathcal{S}$ and a unique bent oriented eulerian partition $\mathcal{S}^*$.* □

The conditions of this lemma are satisfied if X is the shadow of the link diagram of an oriented link and ρ is the orientation of X inherited from the orientation of the link. Since $\mathcal{S}^*$ is bent, it forms a collection of noncrossing closed curves in the plane, which are known as the *Seifert circles* of the link diagram. The minimum number of Seifert circles in any diagram of an oriented link is an important invariant in knot theory.

A knot can be oriented in only two ways, and both orientations yield the same collection of Seifert circles up to reversal. In particular, the number of Seifert circles is an invariant of the knot diagram. Figure 17.19 shows the shadow of a knot diagram and its Seifert circles.

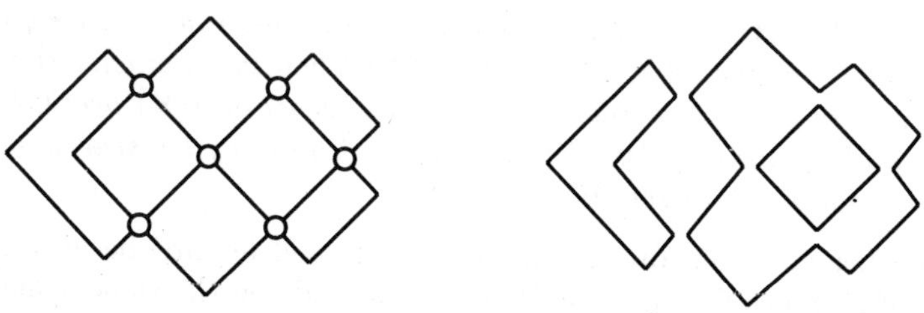

Figure 17.19. The shadow of a knot diagram and its Seifert circles

The oriented eulerian cycles of S use each vertex of X twice. If we arbitrarily label one occurrence of each vertex x with x', then S determines a permutation σ of the set

$$S = \{1, 2, \ldots, n\} \cup \{1', 2', \ldots, n'\}.$$

If we take θ to be the usual involution exchanging i and i', then since X is connected, $M = (\sigma, \theta)$ is an orientable map. The faces of M are the Seifert circles of the oriented link diagram. Similarly, S^* determines an orientable map; this map is the dual of M.

17.13 Seifert Circles and Rank

We can say more about Seifert circles if we restrict our attention to knot diagrams. In particular, we will show that the number of Seifert circles of a knot diagram is determined by the binary rank of the associated circle graph.

Suppose that X is the shadow of a knot diagram, and let Z be the chord diagram whose rim is the straight eulerian tour of X, and where for convenience we assume that the rim is oriented clockwise. For each vertex x, arbitrarily relabel one of its two occurrences with x'. Then Z is a cubic graph where the chords form the distinguished perfect matching, and the rim is a single oriented cycle. The Seifert circles of X are the faces of the associated orientable map, and so are determined by the alternating cycles of Z.

Each alternating cycle of Z alternately uses rim edges in a clockwise direction and chords. The collection of all alternating cycles of Z uses each rim edge once in the clockwise direction, and each chord twice, once in each direction. If C is an alternating cycle of Z, then define its *core* to be the set of chords that it uses once only.

Our remaining results are based on the following simple property of an alternating cycle. Say that an alternating cycle *crosses* the chord $\alpha\alpha'$ if successive rim edges are on opposite sides of $\alpha\alpha'$. It is straightforward to verify that an alternating cycle can cross $\alpha\alpha'$ only if the chord separating the successive rim edges is a chord that crosses $\alpha\alpha'$.

Lemma 17.13.1 *Let Z be a chord diagram with associated circle graph Y, and let $A(Y)$ be the adjacency matrix of Y. Then the characteristic vector of the core of any alternating cycle is in the kernel of $A(Y)$ over $GF(2)$. If Z has s alternating cycles, then the cores of these cycles span a subspace of dimension $s-1$ in* $\ker(A(Y))$.

Proof. Let C be an alternating cycle of Z, and let $\alpha\alpha'$ be an arbitrary chord of Z. Since C is a cycle, it starts and ends on the same side of $\alpha\alpha'$, and so crosses $\alpha\alpha'$ an even number of times in total. A chord not in the core of C is used exactly twice, and so the number of chords in the core of

C that cross $\alpha\alpha'$ is even. Hence the core of C contains an even number of neighbours of any vertex α, and so its characteristic vector is in $\ker(A(Y))$.

Let $C_1, \ldots, C_s$ be the collection of alternating cycles of Z, and let $c_1, \ldots, c_s$ be the characteristic vectors of the cores of $C_1, \ldots, C_s$, respectively. Clearly,

$$c_1 + c_2 + \cdots + c_s = 0,$$

and so we must show that no other linear combination of the vectors is equal to zero. So suppose instead for a contradiction that there is some set of indices $I \subset \{1, \ldots, s\}$ such that

$$\sum_{i \in I} c_i = 0.$$

Then the corresponding set of alternating cycles $\mathcal{C}_I = \{C_i : i \in I\}$ contains every chord an even number of times. As this collection of alternating cycles does not contain every chord twice, there is a pair of chords $\alpha\alpha'$ and $\beta\beta'$ with α and β successive vertices of the rim of Z, and such that $\alpha\alpha'$ occurs twice in $\mathcal{C}_I$ and $\beta\beta'$ does not occur at all. This is clearly impossible, because $\beta\beta'$ must be the chord used immediately after $\alpha'\alpha$ in an alternating cycle, and so we have the desired contradiction. □

Lemma 17.13.2 *Let Z be a chord diagram with associated circle graph Y. If Z has only one alternating cycle, then $A(Y)$ is invertible over $GF(2)$.*

Proof. If C is an alternating cycle, then form a word ω' by listing the vertex at the end of each "chord" step. Since C is the only alternating cycle of Z, it uses every chord twice, and so ω' is the rim of another chord diagram Z' on the same set of vertices as Z. If Y' is the circle graph of Z', then we claim that over $GF(2)$,

$$A(Y)A(Y') = I,$$

which clearly suffices to show that $A(Y)$ is invertible.

Let $N_Y(\alpha)$ denote the set of vertices adjacent to α in Y; we identify this set of vertices with the corresponding set of chords of Z. Then the (α, β)-entry of $A(Y)A(Y')$ is determined by the parity of the set $N_Y(\alpha) \cap N_{Y'}(\beta)$. We aim to show that this is odd if $\alpha = \beta$ and even if $\alpha \neq \beta$. Let C' be the portion of C that leaves β along a rim edge and ends at β' having used the chord $\beta\beta'$. This determines the subword of ω' between β and β'. Therefore, the neighbours of β in Y' are the chords that are used exactly once in C', other than β itself. Every time C' uses a chord from $N_Y(\alpha)$ it crosses from one side of $\alpha\alpha'$ to the other.

First suppose that $\alpha = \beta$. Then C' crosses $\alpha\alpha'$ an odd number of times (because the first rim edge used is the one immediately clockwise of α, while the last rim edge used is the one immediately anticlockwise of α), and so $|N_Y(\alpha) \cap N_{Y'}(\alpha)|$ is odd.

Now, suppose that $\alpha \neq \beta$. The number of times that C' crosses $\alpha\alpha'$ depends on whether β and β' are on the same side of $\alpha\alpha'$. If β and β' are on the same side, then C' crosses $\alpha\alpha'$ an even number of times, and so $|N_Y(\alpha) \cap N_{Y'}(\beta)|$ is even. If β and β' are on opposite sides of $\alpha\alpha'$, then C' crosses $\alpha\alpha'$ an odd number of times. However, one of the chords used by C' is $\beta\beta'$ itself, which is not in $N_{Y'}(\beta)$, and so again $|N_Y(\alpha) \cap N_{Y'}(\beta)|$ is even. □

To illustrate this result consider the chord diagram with rim

$$\omega = 1\ 3\ 4\ 2\ 1'\ 2'\ 3'\ 4'.$$

This has a single alternating cycle

$$C = 1\ 3\ 3'\ 4'\ 4\ 2\ 2'\ 3'\ 3\ 4\ 4'\ 1\ 1'\ 2'\ 2\ 1',$$

and so

$$\omega' = 1\ 3'\ 4\ 2'\ 3\ 4'\ 1'\ 2.$$

Exercise 10 asks you to verify that the adjacency matrices of the circle graphs associated with ω and ω' are indeed the inverses of each other.

Theorem 17.13.3 *Let Z be a chord diagram with n chords, and with associated circle graph Y. If Z has s alternating cycles, then* $\mathrm{rk}_2(A(Y)) = n + 1 - s$.

Proof. We will prove this by induction; our previous result shows that it is true when $s = 1$.

So suppose that Z has $s \geq 2$ alternating cycles $C_1, \ldots, C_s$. Select a chord $\alpha\alpha'$ that is in two distinct cores, say C_{s-1} and C_s, and consider the chord diagram $Z \setminus \alpha\alpha'$ obtained by deleting α and α' from the rim of Z. The alternating cycles of this chord diagram are $C_1, \ldots, C_{s-2}$ and an alternating cycle obtained by merging C_{s-1} and C_s, whose core has characteristic vector $c_{s-1} + c_s$. Therefore, $Z \setminus \alpha\alpha'$ has $s - 1$ alternating cycles, and by induction

$$\mathrm{rk}_2(A(Y \setminus \alpha)) = (n-1) + 1 - (s-1) = n + 1 - s.$$

Since $A(Y \setminus \alpha)$ is a principal submatrix of $A(Y)$, we see that $\mathrm{rk}_2(A(Y))$ is at least $n + 1 - s$. By Lemma 17.13.1, $\mathrm{rk}_2(A(Y)$ is at most $n + 1 - s$, and so the result follows. □

Corollary 17.13.4 *Let Y be the circle graph associated with the diagram of a knot. If the diagram has n crossings and s Seifert circles, then*

$$s = 1 + n - \mathrm{rk}_2(A(Y)).$$ □

Exercises

1. The goal of this exercise is to present Tutte's proof of Smith's result: Each edge in a cubic graph lies in an even number of Hamilton cycles. Let X be a cubic graph and let $\mathcal{S}$ be the set of subgraphs formed by the disjoint union of two perfect matchings. Each 3-edge colouring of X gives rise to three elements of $\mathcal{S}$; we call them a balanced triple. Let us use the same symbol for a subset of $E(X)$ and its characteristic vector over $GF(2)$. If S_1, S_2, and S_3 form a balanced triple, then
$$S_1 + S_2 + S_3 = 0. \tag{17.1}$$
If $S \in \mathcal{S}$, let $c(S)$ denote the number of components of S. We see that S can be expressed as the disjoint union of two perfect matchings in exactly $2^{(c(S)-1)}$ ways. Since the set of edges not in S is a perfect matching, this means that S gives rise to $2^{(c(S)-1)}$ distinct balanced triples. Given all this, use (17.1) to show that
$$\sum H_i = 0,$$
where the sum is over all Hamilton cycles in X.

2. Show that a cubic graph with a Hamilton cycle has at least three Hamilton cycles.

3. Show that a planar graph is hamiltonian if and only if the vertex set of its dual can be partitioned into two sets, each of which induces a tree.

4. Let X be a cubic plane graph on n vertices. Show that there is a bijection between the perfect matchings of the medial graph of X and the orientations of X such that each vertex has odd out-valency. Hence deduce that the number of perfect matchings in the medial graph of X is $2^{n/2+1}$ if $n \equiv 0 \bmod 4$, and 0 otherwise.

5. Show that the embedding of K_6 in the projective plane is not orientable.

6. Let X be a 4-valent graph embedded in a surface and suppose ω is a double occurrence word corresponding to a straight eulerian tour in X. Prove that any two of the following assertions imply the third:
   (a) The surface is orientable.
   (b) X^* is bipartite.
   (c) ω is even.

7. Let X be a 4-valent plane graph with a straight eulerian tour S. If Y and Y^* are the face graphs of X, show that every vertex of X determines an edge of either Y or Y^*, according to whether flipping at that vertex would result in the white or the black partition at that

vertex. Use this to prove the assertion of Lemma 17.8.2 that every bent eulerian tour can be obtained from $\mathcal{S}$ by flipping at each vertex in some order.

8. Consider the two mutant knots of Figure 16.22. Show that the associated double occurrence words are not related by rotation, reversal, or relabelling, but that they yield isomorphic circle graphs.

9. Let Z be the truncation of an orientable map, with associated perfect matching F, and with the cycles of $Z \setminus F$ oriented clockwise. If we contract the edges in F, the graph Z/F that results is 4-valent and has an oriented eulerian partition, $\mathcal{S}$ say. Describe how to recover the original map from Z/F and the partition. If $\mathcal{S}^*$ is the partition dual to $\mathcal{S}$, show that the map determined by Z/F and $\mathcal{S}^*$ is the dual of the original map. Show that $\mathcal{S}^*$ is straight.

10. Show that the circle graphs that arise from the double occurrence words

$$1\,3\,4\,2\,1\,2\,3\,4$$

and

$$1\,3\,4\,2\,3\,4\,1\,2$$

have adjacency matrices that are inverses over $GF(2)$.

11. Let X be the 4-valent plane graph shown in Figure 17.19. Draw the chord diagram Z corresponding to the straight eulerian partition of X, and find the associated circle graph Y. Show that Z has three alternating cycles and that the core of each is in $\ker(A(Y))$.

Notes

There are several definitions of medial graph in the literature, some of which do not account for the subtleties raised by face-bounding loops or vertices of valency one. In evaluating any proposed definition, a useful rule of thumb is to try it on the "tadpole"—a two-vertex graph equal to K_2 with a loop on one vertex—which encapsulates the difficulties.

Section 17.3 is based on Shank's paper [8]. The main result—the relation between the bicycle space and the cores—is ascribed there to J. D. Horton.

A number of papers have presented characterizations of Gauss codes. The fundamental idea of flipping words is due to Dehn, but was extended by Rosenstiehl and Read [7], and then given in a manner essentially equivalent to ours by de Fraysseix and Ossona de Mendez [3].

Although lists of knots could be given by Gauss codes, it is not usual to do so. The Gauss code is essentially a description of the rim of a chord diagram, from which the chords are implicit. However, it is quite possible to describe the chords instead, leaving the rim implicit. This is done by labelling the symbols on the rim 1 through $2n$ in order as we move clockwise around the rim. Then the chords give a perfect matching on the set $\{1, \ldots, 2n\}$. If the chord diagram comes from a Gauss code, then this matching pairs each odd integer with an even integer; hence we can describe it by listing the even integers paired with 1, 3, 5, etc. The latter is known to topologists as the *Dowker code* and appears to be their preferred numerical representation.

The correspondence between bent eulerian tours and spanning trees in Section 17.8 is due to de Fraysseix [2]. Our work in Section 17.9 largely follows Las Vergnas [5]. The relationship between the alternating cycles of a chord diagram and the binary rank of a circle graph is due to Bouchet [1].

For treatments of maps along the lines we followed, see the work of Lins [6] and Vince [9]. Graph theorists tend to prefer using two permutations (and much baroque terminology) to describe maps, rather than three involutions.

Exercise 4 is a slight extension of an observation from Section VI B of Kuperberg [4].

References

[1] A. Bouchet, *Unimodularity and circle graphs*, Discrete Math., 66 (1987), 203–208.

[2] H. de Fraysseix, *Local complementation and interlacement graphs*, Discrete Math., 33 (1981), 29–35.

[3] H. de Fraysseix and P. Ossona de Mendez, *On a characterization of Gauss codes*, Discrete Comput. Geom., 22 (1999), 287–295.

[4] G. Kuperberg, *An exploration of the permanent-determinant method*, Electron. J. Combin., 5 (1998), Research Paper 46, 34 pp. (electronic).

[5] M. Las Vergnas, *Eulerian circuits of 4-valent graphs imbedded in surfaces*, in Algebraic methods in graph theory, Vol. I, II (Szeged, 1978), North-Holland, Amsterdam, 1981, 451–477.

[6] S. Lins, *Graph-encoded maps*, J. Combin. Theory Ser. B, 32 (1982), 171–181.

[7] R. C. Read and P. Rosenstiehl, *On the Gauss crossing problem*, in Combinatorics (Proc. Fifth Hungarian Colloq., Keszthely, 1976), Vol. II, North-Holland, Amsterdam, 1978, 843–876.

[8] H. Shank, *The theory of left-right paths*, Lecture Notes in Math., 452 (1975), 42–54.

[9] A. Vince, *Combinatorial maps*, J. Combin. Theory Ser. B, 34 (1983), 1–21.

Glossary of Symbols

The entries in this index are divided into two lists. Entries such as $A(X)$ and $\chi(X)$ that have fixed letters as part of their representation occur in the first list, in alphabetic order (phonetically for Greek characters). Entries such as $X^\bullet$ that involve only variables and mathematical symbols occur in the second list, grouped somewhat arbitrarily.

$\chi^*(X)$	fractional chromatic number of X, 136
$\chi^\circ(X)$	circular chromatic number of X, 157
$C(v,r)$	cylic interval graph, 145
$\mathcal{C}_n(X)$	n-colouring graph of X, 155
$\chi(X)$	chromatic number of X, 7
D_n	root system, 266
$\Delta(X)$	diagonal matrix of valencies, 166
$\dim U$	dimension of U, 231
$d(v)$	valency of vertex v, 321
$d^+(v)$	out-valency of vertex v, 321
∂A	set of edges with one end in A, 38
$\det A$	determinant of A, 187
$d(x,y)$	distance from x to y, 5
$d_X(x,y)$	distance from x to y in X, 5
$E(X)$	edge set of X, 1
E_6	root system, 272
E_7	root system, 272
E_8	root system, 272
E_θ	principal idempotent, 185
e_i	ith standard basis vector, 180
$\text{ev}(A)$	eigenvalues of A, 185
$F(X,q)$	flow polynomial of X, 370
$\text{fix}(g)$	fixed points of permutation g, 22
$GL(3,q)$	general linear group, 81
$\text{Hom}(X,Y)$	set of homomorphisms from X to Y, 107
J	all-ones matrix, 95
$J(v,k,i)$	generalized Johnson graph, 9
K_n	complete graph on n vertices, 2
$K_{1,n}$	star graph, 10
$K_{m,n}$	complete bipartite graph, 12

$\kappa(X)$	number of acyclic orientations of X, 350
$\kappa_0(X)$	vertex connectivity of X, 39
$\kappa_1(X)$	edge connectivity of X, 37
$\ker A$	kernel of A, 177
$K_{v:r}$	Kneser graph, 135
$L(X)$	line graph of X, 10
$\lambda_i(Q(X))$	ith smallest eigenvalue of $Q(X)$, 280
$M(C)$	matroid defined by code C, 347
$M(X)$	cycle matroid of X, 343
M/T	matroid obtained by contracting T, 346
$M \setminus e$	matroid obtained by deleting e, 346
$\mathcal{M}(Y)$	medial graph of Y, 398
m_θ	multiplicity of eigenvalue θ, 220
$N(x)$	neighbours of x, 5
$n^+(A)$	number of positive eigenvalues of A, 205
$n^-(A)$	number of negative eigenvalues of A, 205
$OA(k,n)$	orthogonal array, 224
$\omega(X)$	size of largest clique of X, 3
$\omega^*(X)$	fractional clique number of X, 137
$P(X,t)$	chromatic polynomial of X, 353
$PG(2,q)$	classical projective plane, 80
$PG(3,q)$	3-dimensional projective space, 83
P_n	path on n vertices, 10
$\Phi(X)$	conductance of X, 292
$\phi(A,x)$	characteristic polynomial of A, 164
$\phi(X,x)$	characteristic polynomial of X, 164
$Q(X)$	Laplacian of X, 279
Q_k	k-dimensional cube, 33
$R(M;x,y)$	rank polynomial of M, 356
rk	rank function, 341

$\text{rk}\, B$	rank of B, 166
$\text{rk}_2(X)$	binary rank of X, 181
$\rho(A)$	spectral radius of A, 177
$S(X)$	Seidel matrix of X, 250
$S(X)$	subdivision graph of X, 45
$\text{supp}(v)$	support of v, 176
$\sigma_u(X)$	local complement of X at u, 182
$\text{Sw}(X)$	switching graph of X, 255
$\text{Sym}(V)$	symmetric group, 4
$\text{Sp}(2r)$	symplectic graph, 184
$\tau(X)$	number of spanning trees of X, 282
$\theta_i(A)$	ith largest eigenvalue of A, 193
$\theta_{\max}(X)$	maximum eigenvalue of X, 174
$\theta_{\min}(X)$	minimum eigenvalue of X, 174
$\text{tr}\, A$	trace of A, 165
$V(X)$	vertex set of X, 1
$W(C,t)$	weight enumerator of C, 358
$\text{wr}(L)$	writhe of L, 387
$X(G,C)$	Cayley graph for G, 34
$X(\mathbb{Z}_n,C)$	circulant graph, 8
$X(\mathcal{I})$	incidence graph of $\mathcal{I}$, 78
$H < G$	H is a proper subgroup of G, 28
$H \leq G$	H is a subgroup of G, 28
$x \sim y$	x is adjacent to y, 1
$[L]$	Kauffman bracket of L, 384
$\overline{X}$	complement of X, 5
$g \circ f$	composition of homomorphisms, 103
$\mathcal{I}^*$	dual incidence structure, 78
$\mathcal{L}^*$	dual lattice, 316

$C^{\perp}$	dual code, 348
$u^{\perp}$	subspace of vectors orthogonal to u, 83
X^*	dual graph, 14
$\|x\|$	Euclidean length of x, 285
F^X	the map graph, 108
$\langle x \rangle$	subspace spanned by x, 266
$X \cong Y$	X is isomorphic to Y, 2
(n, k, a, c)	parameters of a strongly regular graph, 218
$t\text{-}(v, k, \lambda_t)$	parameters of a t-design, 94
$A \otimes B$	Kronecker product, 206
$X * Y$	strong product, 155
$X[Y]$	lexicographic product, 17
$X \square Y$	Cartesian product, 154
$X \times Y$	product of X and Y, 106
$Q[u]$	matrix Q with row and column u deleted, 282
A/π	quotient matrix, 203
X/π	quotient graph, 104
X/π	quotient graph, 196
$\mathbb{R}^{V(X)}$	space of functions from $V(X) \to \mathbb{R}$, 171
$f \restriction Y$	restriction of f to Y, 7
G_S	stabilizer of S in G, 20
G_x	stabilizer of x in G, 20
$\mathbb{R}^E$	space of real-valued functions on E, 307
v^g	image of v under mapping g, 4
x^G	orbit of x under G, 20
Y^g	image of Y under mapping g, 5
Ω^T	transpose of orbital Ω, 26
X/e	graph obtained from X by contracting e, 281
$X^\bullet$	core of X, 105
$X \setminus e$	graph obtained from X by deleting e, 281

$X \to Y$	there is a homomorphism from X to Y, 103
X^σ	oriented graph with orientation σ, 167
X^σ	graph obtained from X by switching on σ, 255
$X_i(u)$	vertices at distance i from u in X, 67

Index

Graduate Texts in Mathematics

(continued from page ii)

66 WATERHOUSE. Introduction to Affine Group Schemes.
67 SERRE. Local Fields.
68 WEIDMANN. Linear Operators in Hilbert Spaces.
69 LANG. Cyclotomic Fields II.
70 MASSEY. Singular Homology Theory.
71 FARKAS/KRA. Riemann Surfaces. 2nd ed.
72 STILLWELL. Classical Topology and Combinatorial Group Theory. 2nd ed.
73 HUNGERFORD. Algebra.
74 DAVENPORT. Multiplicative Number Theory. 3rd ed.
75 HOCHSCHILD. Basic Theory of Algebraic Groups and Lie Algebras.
76 IITAKA. Algebraic Geometry.
77 HECKE. Lectures on the Theory of Algebraic Numbers.
78 BURRIS/SANKAPPANAVAR. A Course in Universal Algebra.
79 WALTERS. An Introduction to Ergodic Theory.
80 ROBINSON. A Course in the Theory of Groups. 2nd ed.
81 FORSTER. Lectures on Riemann Surfaces.
82 BOTT/TU. Differential Forms in Algebraic Topology.
83 WASHINGTON. Introduction to Cyclotomic Fields. 2nd ed.
84 IRELAND/ROSEN. A Classical Introduction to Modern Number Theory. 2nd ed.
85 EDWARDS. Fourier Series. Vol. II. 2nd ed.
86 VAN LINT. Introduction to Coding Theory. 2nd ed.
87 BROWN. Cohomology of Groups.
88 PIERCE. Associative Algebras.
89 LANG. Introduction to Algebraic and Abelian Functions. 2nd ed.
90 BRØNDSTED. An Introduction to Convex Polytopes.
91 BEARDON. On the Geometry of Discrete Groups.
92 DIESTEL. Sequences and Series in Banach Spaces.
93 DUBROVIN/FOMENKO/NOVIKOV. Modern Geometry—Methods and Applications. Part I. 2nd ed.
94 WARNER. Foundations of Differentiable Manifolds and Lie Groups.
95 SHIRYAEV. Probability. 2nd ed.
96 CONWAY. A Course in Functional Analysis. 2nd ed.
97 KOBLITZ. Introduction to Elliptic Curves and Modular Forms. 2nd ed.
98 BRÖCKER/TOM DIECK. Representations of Compact Lie Groups.
99 GROVE/BENSON Finite Reflection Groups. 2nd ed.
100 BERG/CHRISTENSEN/RESSEL. Harmonic Analysis on Semigroups: Theory of Positive Definite and Related Functions.
101 EDWARDS. Galois Theory.
102 VARADARAJAN. Lie Groups, Lie Algebras and Their Representations.
103 LANG. Complex Analysis. 3rd ed.
104 DUBROVIN/FOMENKO/NOVIKOV. Modern Geometry—Methods and Applications. Part II.
105 LANG. $SL_2(\mathbf{R})$.
106 SILVERMAN. The Arithmetic of Elliptic Curves.
107 OLVER. Applications of Lie Groups to Differential Equations. 2nd ed.
108 RANGE. Holomorphic Functions and Integral Representations in Several Complex Variables.
109 LEHTO. Univalent Functions and Teichmüller Spaces.
110 LANG. Algebraic Number Theory.
111 HUSEMÖLLER. Elliptic Curves.
112 LANG. Elliptic Functions.
113 KARATZAS/SHREVE. Brownian Motion and Stochastic Calculus. 2nd ed.
114 KOBLITZ. A Course in Number Theory and Cryptography. 2nd ed.
115 BERGER/GOSTIAUX. Differential Geometry: Manifolds, Curves, and Surfaces.
116 KELLEY/SRINIVASAN. Measure and Integral. Vol. I.
117 SERRE. Algebraic Groups and Class Fields.
118 PEDERSEN. Analysis Now.
119 ROTMAN. An Introduction to Algebraic Topology.
120 ZIEMER. Weakly Differentiable Functions: Sobolev Spaces and Functions of Bounded Variation.
121 LANG. Cyclotomic Fields I and II. Combined 2nd ed.
122 REMMERT. Theory of Complex Functions. *Readings in Mathematics*
123 EBBINGHAUS/HERMES et al. Numbers. *Readings in Mathematics*
124 DUBROVIN/FOMENKO/NOVIKOV. Modern Geometry—Methods and Applications. Part III.
125 BERENSTEIN/GAY. Complex Variables: An Introduction.
126 BOREL. Linear Algebraic Groups. 2nd ed.
127 MASSEY. A Basic Course in Algebraic Topology.
128 RAUCH. Partial Differential Equations.
129 FULTON/HARRIS. Representation Theory: A First Course. *Readings in Mathematics*
130 DODSON/POSTON. Tensor Geometry.

131 LAM. A First Course in Noncommutative Rings.
132 BEARDON. Iteration of Rational Functions.
133 HARRIS. Algebraic Geometry: A First Course.
134 ROMAN. Coding and Information Theory.
135 ROMAN. Advanced Linear Algebra.
136 ADKINS/WEINTRAUB. Algebra: An Approach via Module Theory.
137 AXLER/BOURDON/RAMEY. Harmonic Function Theory. 2nd ed.
138 COHEN. A Course in Computational Algebraic Number Theory.
139 BREDON. Topology and Geometry.
140 AUBIN. Optima and Equilibria. An Introduction to Nonlinear Analysis.
141 BECKER/WEISPFENNING/KREDEL. Gröbner Bases. A Computational Approach to Commutative Algebra.
142 LANG. Real and Functional Analysis. 3rd ed.
143 DOOB. Measure Theory.
144 DENNIS/FARB. Noncommutative Algebra.
145 VICK. Homology Theory. An Introduction to Algebraic Topology. 2nd ed.
146 BRIDGES. Computability: A Mathematical Sketchbook.
147 ROSENBERG. Algebraic *K*-Theory and Its Applications.
148 ROTMAN. An Introduction to the Theory of Groups. 4th ed.
149 RATCLIFFE. Foundations of Hyperbolic Manifolds.
150 EISENBUD. Commutative Algebra with a View Toward Algebraic Geometry.
151 SILVERMAN. Advanced Topics in the Arithmetic of Elliptic Curves.
152 ZIEGLER. Lectures on Polytopes.
153 FULTON. Algebraic Topology: A First Course.
154 BROWN/PEARCY. An Introduction to Analysis.
155 KASSEL. Quantum Groups.
156 KECHRIS. Classical Descriptive Set Theory.
157 MALLIAVIN. Integration and Probability.
158 ROMAN. Field Theory.
159 CONWAY. Functions of One Complex Variable II.
160 LANG. Differential and Riemannian Manifolds.
161 BORWEIN/ERDÉLYI. Polynomials and Polynomial Inequalities.
162 ALPERIN/BELL. Groups and Representations.
163 DIXON/MORTIMER. Permutation Groups.
164 NATHANSON. Additive Number Theory: The Classical Bases.
165 NATHANSON. Additive Number Theory: Inverse Problems and the Geometry of Sumsets.
166 SHARPE. Differential Geometry: Cartan's Generalization of Klein's Erlangen Program.
167 MORANDI. Field and Galois Theory.
168 EWALD. Combinatorial Convexity and Algebraic Geometry.
169 BHATIA. Matrix Analysis.
170 BREDON. Sheaf Theory. 2nd ed.
171 PETERSEN. Riemannian Geometry.
172 REMMERT. Classical Topics in Complex Function Theory.
173 DIESTEL. Graph Theory. 2nd ed.
174 BRIDGES. Foundations of Real and Abstract Analysis.
175 LICKORISH. An Introduction to Knot Theory.
176 LEE. Riemannian Manifolds.
177 NEWMAN. Analytic Number Theory.
178 CLARKE/LEDYAEV/STERN/WOLENSKI. Nonsmooth Analysis and Control Theory.
179 DOUGLAS. Banach Algebra Techniques in Operator Theory. 2nd ed.
180 SRIVASTAVA. A Course on Borel Sets.
181 KRESS. Numerical Analysis.
182 WALTER. Ordinary Differential Equations.
183 MEGGINSON. An Introduction to Banach Space Theory.
184 BOLLOBAS. Modern Graph Theory.
185 COX/LITTLE/O'SHEA. Using Algebraic Geometry.
186 RAMAKRISHNAN/VALENZA. Fourier Analysis on Number Fields.
187 HARRIS/MORRISON. Moduli of Curves.
188 GOLDBLATT. Lectures on the Hyperreals: An Introduction to Nonstandard Analysis.
189 LAM. Lectures on Modules and Rings.
190 ESMONDE/MURTY. Problems in Algebraic Number Theory.
191 LANG. Fundamentals of Differential Geometry.
192 HIRSCH/LACOMBE. Elements of Functional Analysis.
193 COHEN. Advanced Topics in Computational Number Theory.
194 ENGEL/NAGEL. One-Parameter Semigroups for Linear Evolution Equations.
195 NATHANSON. Elementary Methods in Number Theory.
196 OSBORNE. Basic Homological Algebra.

197 EISENBUD/HARRIS. The Geometry of Schemes.
198 ROBERT. A Course in p-adic Analysis.
199 HEDENMALM/KORENBLUM/ZHU. Theory of Bergman Spaces.
200 BAO/CHERN/SHEN. An Introduction to Riemann–Finsler Geometry.
201 HINDRY/SILVERMAN. Diophantine Geometry: An Introduction.
202 LEE. Introduction to Topological Manifolds.
203 SAGAN. The Symmetric Group: Representations, Combinatorial Algorithms, and Symmetric Function. 2nd ed.
204 ESCOFIER. Galois Theory.
205 FÉLIX/HALPERIN/THOMAS. Rational Homotopy Theory.
206 MURTY. Problems in Analytic Number Theory.
Readings in Mathematics
207 GODSIL/ROYLE. Algebraic Graph Theory.